THE PIG
as a Laboratory Animal

THE PIG
as a Laboratory Animal

L.E. MOUNT and D.L. INGRAM

Department of Applied Biology,
Agricultural Research Council
Institute of Animal Physiology,
Babraham, Cambridge, England

1971

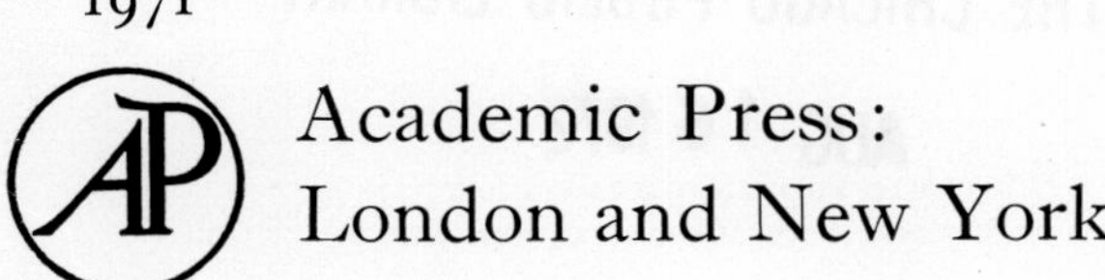

Academic Press:
London and New York

ACADEMIC PRESS INC. (LONDON) LTD

Berkeley Square House
Berkeley Square,
London, W1X 6BA

U.S. Edition published by
ACADEMIC PRESS INC.
111 Fifth Avenue,
New York, New York 10003

Library of Congress Catalog Card Number: 70-141722
ISBN: 0-12-509050-1

PRINTED IN GREAT BRITAIN
BY W & J MACKAY & CO LTD, CHATHAM, KENT

Preface

The aim of this book is to provide a practical guide to those wishing to work with pigs as laboratory animals. It is expected that many workers will have biological or medical backgrounds and will be familiar with pigs only in general terms. In this volume we have tried to provide essential information on the animal and on problems of supply, growth and handling of pigs of various ages; sufficient references are given to take the subject further where necessary. Those with veterinary qualifications will already be familiar with the pig, but they may not previously have considered this animal under laboratory rather than farm conditions.

We should like to express our gratitude to Professor M. de Burgh Daly (Department of Physiology, Medical College of St Bartholomew's Hospital, London) for his comments and criticism during the final stages of the preparation of this book. We should also like to thank some members of this Institute, Drs B. A. Baldwin, R. M. Binns, R. B. Heap, J. L. Linzell, Margaret W. Stanier and D. B. Stephens, for their useful suggestions and corrections during the course of the work, and the Library Staff and the Photographic Section for their valuable assistance. Any errors and omissions which remain are our responsibility. We are grateful to the Director and the Agricultural Research Council for their permission to undertake this work and to those who have given permission for the reproduction of figures and other material in the text.

A.R.C. Institute of Animal Physiology,
Babraham, Cambridge
February, 1971

L. E. Mount
D. L. Ingram

Contents

* A.R.C. Institute of Animal Physiology, Babraham, Cambridge.

CHAPTER 1

Introduction

There is a wide range of breeds of pig from which to choose for experimental work. In practical terms, however, the principal choice lies between using normal farm pigs or one of the breeds of so-called miniature pigs which have become popular for scientific work in recent years. Investigations into animal production and the growth of pigs for meat are more appropriately carried out on farm pigs. These animals are large in size (about 90 kg at 6 months), sometimes difficult to manage, and not of a particularly co-operative temperament. When investigations are not specifically directed towards animal production, therefore, it is sometimes easier to use miniature pigs (see Chapter 11). These animals have characteristics in common with full size pigs, but handling is much easier and food costs are less. At 6 months of age they weigh 30–40 kg.

In Britain the breeds of farm pig which are by far the most commonly available are the Large White and the Landrace. An indication of this can be seen in the following numbers of boar licences in England and Wales, by breed, for the year April 1965 to March 1966 (Pig Industry Development Authority):

Large White	10,419
Landrace	5832
Welsh	1766
Wessex Saddleback	352
All others	353

Selection over many generations has resulted in the modern pig being different from its wild ancestor both in body conformation and in temperament. Some aspects of the biology of the pig are discussed elsewhere (Mount, 1968).

The pig has been widely investigated in its capacity as a meat-producing farm animal, particularly in respect of its nutritional requirements. More recently it is also being selected with increasing frequency as the animal of choice for wider aspects of medical and physiological research. The use of the animal to provide information relative to human physiology prompts comparison of the two species in several other contexts (see Sillar and Meyler, 1961 and Pope, 1962).

Some of the similarities between man and pig are, however, more apparent than real. The relative lack of hair has contributed to the assumption that the skins of pig and man are similar, but whereas the dermis of the pig is relatively poor in blood vessels and contains only apocrine glands, man's skin is richly vascularized and contains many eccrine glands. Another distinction between man and pig lies in the pig's inability to sweat in response to heat, whereas man sweats abundantly in hot environments, and can lose more water in this way than any other mammal. Pigs also have a higher growth rate and a more acute sense of smell than humans. In spite of these differences, the similarities between man and the pig in respect of size and function are great enough to allow useful comparisons to be made. The animal is currently employed for studies on cardiovascular physiology and disease, renal function, effects of radiation, the gastrointestinal tract, transplantation of organs and the occurrence and treatment of dental caries. The use of the animal in immunological work has also been increased by the perfection of methods for the production of germ-free pigs.

In Britain, the term "pig" is commonly used to cover the species in general and younger animals in particular. Older breeding animals are termed boar, sow and gilt, the last name applying to the female up to the end of her first pregnancy. The castrated male pig is sometimes called a barrow. The term "swine" is not employed so commonly in this country, although in the U.S.A. it is the general term for the animal. According to the Shorter Oxford English Dictionary, "swine" is "an animal of the genus *Sus* or family Suidae . . . now only literary or dialectal, or generic term in zoology". "Pig" is given as "the young of swine: by extension, a swine of any age; any of various species of Suidae". Thus the two terms tend towards the same meaning; "pig" is used in this book.

Classification

The classification of the pig family is given by Morris (1965), and the numbers of pigs and their distribution throughout the world by the FAO Production Yearbook for 1964. The total pig population of the United Kingdom fluctuated between seven and eight million during the period 1963–1966. This number is overshadowed, however, by other farm livestock: about twelve million cattle, thirty million sheep, and nearly one hundred and twenty million poultry.

PHYLOGENETIC RELATIONSHIPS

The pig belongs to the order of mammals known as the *Artiodactyla*, or even-toed Ungulates. This order includes almost all the animals which have been husbanded by man such as cattle, sheep, goats, deer, buffalo, bison and camels, as well as giraffes and hippopotamuses. Members of the order are characterized by the fact that they walk on the tips of their digits (Digitigrade) and that the axis of the limb passes between the third and fourth digit; the resemblance to the *Perissodactyla* or odd-toed ungulates (horses) results not from a close relationship but from a similar course of evolution.

Within the order the pig represents a relatively small section of animals with a simple stomach, while the bulk of the members belong to the sub-order Ruminantia which have developed a complex digestive system (the Tylopoda, e.g. camels and llama, form an intermediate group).

The living members of the order Artiodactyla are divided by modern authorities (Rothschild, 1965; Morris, 1965) into three sub-orders, one of which, the Suiformes, is again divided into two infra orders, the Suina and the Ancodonta (Hippopotamuses). The Suina contains the superfamily Suoidae all the members of which are loosely called pigs, and superficially resemble the familiar domestic animal. There are, however, two families, the Tayassuidae or peccaries, which are natives of Central and South America, and the family Suidae, which are the true or old world pigs. The Suidae contain five living genera, of which the genus *Sus* contains several species including *S. scrofa*, the domestic pig. For more detailed information the reader is referred to Morris (1965), but a summary of the classification is given below.

The pig has been associated with man from the very earliest times, its bones having been found at the sites of Stone Age settlements. Darwin (1882) recorded that the pig has been domesticated in China for nearly 5000 years. With such a long history of domestication it is not surprising that very little is known about its origins.

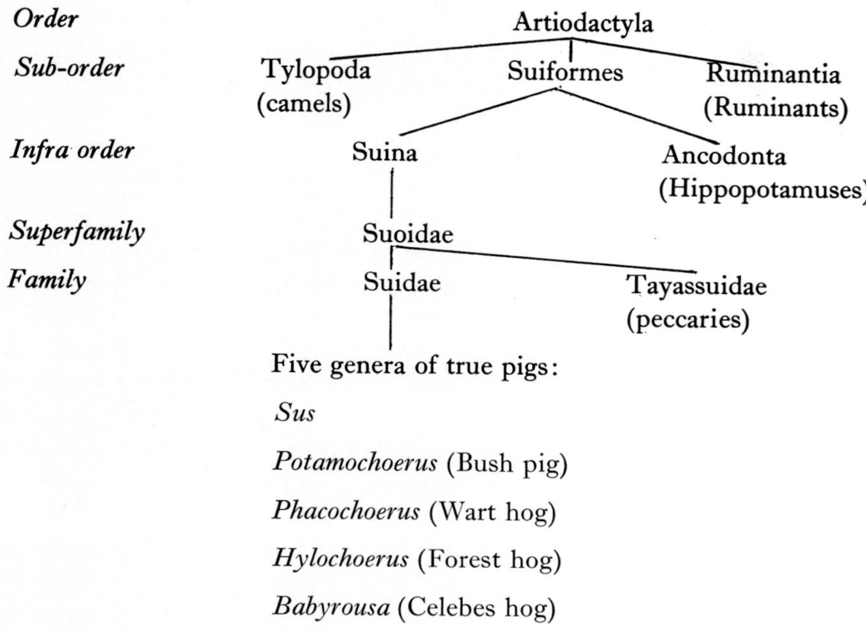

In the older literature it was stated that the European pig was derived chiefly from the Chinese *Sus vittatus* or *S. indicus* and the Indian *S. cristatus*. These species names along with *Sus wadituaneus* of Italy have now, however, been abandoned since it is considered that they all represent local variations of the same species, *Sus scrofa*. The wild pigs of some South Sea islands, New Zealand, Australia and the southern parts of the United States have all been derived from domestic pigs which originally escaped from early explorers in the sixteenth century. In the same way before the times of strict land enclosure in the old world it seems likely that numbers of domestic pigs would also have escaped to breed with any wild pigs. Thus the domestic pig, the wild pigs of China, Europe, parts of India, Australasia and the New World are all variations of the species *Sus scrofa*. Other members of the genus are given by Morris (1965) as

Sus salvanius (Pygmy hog)

Sus verrucosus (Javan pig)

Sus barbatus (Bornean pig)

Breeds

The term "breed" is not well defined, and in the earlier literature it may mean little more than a herd developed on a local estate. In more recent times breed societies have laid down certain standards and criteria which have to be met before a herd can be registered. The situation is thus similar to that of dogs where there are a large number of recognized breeds, and an even larger number of cross-breeds.

Over a period of a hundred years or so there has been a general tendency towards the production of a much leaner animal (Sidney, 1871; Spencer, 1921). Old publications illustrate pigs which are extremely rotund, and, although one cannot be certain, it would seem unlikely that this was due entirely to feeding. An exception appears to be the Old Irish Greyhound (illustrated by Darwin, 1882) which had long thin legs from which its name was derived. In Britain almost every county where pigs were kept seemed to produce its own breed, but in modern times these have been considerably reduced in number. From the point of view of the research worker, however, it may be of importance to note that these breeds have left a choice of animals of different skin colours (white, black, red and mixed) and also breeds with markedly different coats, and to a lesser extent different growth rates.

WHITE BREEDS

In the British Isles the most popular breeds are the Large White, (known as the Yorkshire in some countries), and the Landrace; both these breeds grow to a large size and have been produced chiefly for the bacon market. Although they are both white, the pigs are easily distinguished from each other by the fact that in the Large White the ears are held erect while in the Landrace they fall over the eyes. The Welsh pig is not unlike the Landrace in appearance, although rather shorter in the body; it has been bred mainly for hardiness, enabling it to be raised in the rigorous conditions of isolated Welsh valleys. The Middle White carries the ears erect and is distinguished by its short turned-up (dished) snout. It has been bred for the pork market and matures more rapidly than the bacon pig. Possibly as a result of the snout shape it is less inclined to root than some breeds, and is therefore more easily kept out of doors without excessive damage to the ground. The American breeds of pig (Zeller, 1964; Briggs, 1969) are similar, the Large White being known more usually as the

Yorkshire, while there is also an American Landrace. In addition the Chester White is an animal similar to the Large White with semi-erect ears.

BLACK BREEDS

The Large Black, like the Landrace, carries the ears down over the eyes. In farm practice it is more usually used as the dam to a cross litter (using a Large White, Welsh or Landrace boar). The litters of such crosses have hybrid vigour and produce high-grade carcasses, a fact which is in part related to the qualities of the Large Black as a mother. In addition the breed is said to be quiet and easy to handle. Another breed which is black apart from white legs is the Berkshire which has been developed chiefly for pork, and which, like the Middle White, also matures early. The Poland China in the U.S.A. has very similar colouring with the addition of a white snout. This breed has been produced chiefly from the Berkshire and partly from Russian stock.

RED PIGS

The colour of these breeds is a reddish brown, and in Britain the Tamworth is the best known. It matures late and is therefore used for bacon; its ears are carried erect, and it is hardy enough to live out of doors with a minimum of housing. In the U.S.A. the Duroc resembles the Tamworth, and in sheer weight of numbers is probably the most popular breed in the world.

MIXED COAT COLOUR

For some experiments, e.g. involving skin grafts, a pig of mixed skin colour may have advantages. The Saddleback is an animal which is mainly black with a white band round the body and forelimbs; in some individuals traces of white appear on the nose and hind limbs. It is a hardy breed which is usually kept out of doors. The American Hampshire appears to have been derived from the Wessex and has very similar markings. Another mixed coat colour animal is the Gloucester Old Spot, a white pig with black spots over the body. Commercially it is used as a dual purpose animal and is suitable for keeping out of doors. In the U.S.A. the Spotted or Spotted Poland China is also a white animal with black spots, and in so-called perfect specimens there should be 50% black and 50% white.

COAT TYPE

Most breeds of pigs have a relatively sparse hair coat, but the Mangalitza from Hungary has a thick curly coat which causes it to resemble a sheep at first glance. The breed is chiefly a lard pig having a thick layer of back fat, and is not kept commercially outside central Europe. The Lincolnshire Curly Coat is a British breed which also has a thick and long coat, but it is not so readily available as other breeds.

MINIATURE PIGS

The above outline is by no means a full account of all the breeds but should serve to indicate the varieties which are available in different parts of the world. As in any species, the variations which are to be seen in living animals represent only a small part of what might be developed by selective breeding. One such selection has been carried out entirely for the experimentalist and has resulted in various types of miniature pig, which are described in Chapter 11 in connection with their use in research. The name "miniature pig", however, has been applied to several strains of animals and the size of the fully grown adult varies; moreover, since these pigs have been developed only recently, the degree of genetic homogeneity is still being improved. Some care is, therefore, necessary in the selection of breeding stock if the small size is to be maintained. The animals' smaller size is an obvious advantage for laboratory work, particularly when sexually mature animals are needed. Against this must be weighed the fact that miniature pigs are purely experimental animals with very little, if any, commercial value. The pigs must be specially bred and any animals surplus to requirement can be sold only to other research workers.

Genetics

The animal breeder's interest in the selection of pigs for commercial use has been directed towards the heritability of litter size (Boylan *et al.*, 1961; Cox, 1966) and body conformation. Studies similar to those which have been made on mice with reference to qualitative traits have, however, been very limited. Hetzer (1945) has reviewed some information available about coat colour, but the only aspect of qualitative genetics which has received intensive study is that related to blood antigens, which are considered in Chapter 7.

Cytogenetics

The cytology of chromosomes in pigs has received some attention and the work has been reviewed by McFeely and Hare (1966). Descriptions are available from many sources (Bryden, 1933; Crew and Kollner, 1939; Makino, 1944; Spalding and Berry, 1956; Gimenez-Martin *et al.*, 1962; McConnell *et al.*, 1963) and it is generally agreed that the diploid number is 38, although McFee *et al.* (1966) found some variation in wild pigs. Illustrations of the karyotype of the Hampshire, Poland China, Berkshire and Large White are available in McConnell *et al.* (1963) and McFeely and Hare (1966), and the latter authors also indicate that workers in their laboratory have made progress towards the identification of the sex chromosomes.

CHAPTER 2

The Supply of Pigs for Experiment

The supply of pigs for the laboratory can be arranged basically in three ways: (i) from a pig production unit which is maintained primarily as a source of pigs for the laboratory; (ii) the required animals can be bought from farms producing pigs on a commercial scale; (iii) a few animals can be kept and bred purely for laboratory use. These alternatives will be discussed in turn.

The Larger Pig Unit

This offers the laboratory worker the greatest scope and involves him in the least effort in obtaining animals, but is possible only where the facilities for such a unit can be built. At the Agricultural Research Council Institute of Animal Physiology at Babraham, Cambridge, a farm is attached to the Institute; a herd of breeding sows and boars is maintained and the unit currently produces some 700 pigs each year. Up to 80% of these animals are taken for laboratory investigation; half of these are later returned to the herd since no surgery or drug treatment will have been involved, only measurements such as that of metabolic rate which do not require such interference with the animal. Pigs which are surplus to experimental requirements are sent to market in the usual way. The great advantages of this system are that the total number of animals can be kept in excess of laboratory requirements, and changes in the anticipated needs of the laboratory can usually be accommodated. Moreover, with a large enough number of sows, matings can be arranged to ensure a steady supply of piglets throughout the year. This type of unit, involving a large number of pigs, needs experienced pig stockmen to run it; the attention of members of the research staff as part-time pig keepers is not adequate, and in any case they are not likely to be able to cope with the situation.

A most important point, however, needs to be emphasized. In this system it must be understood by all concerned that the only reason for the existence of a pig breeding unit is to supply animals to the laboratory. All decisions about the breed of animal to be kept, the maintenance of a herd free of enzootic pneumonia (see Chapter 12), and the production of cross-bred animals, must be made in the light of requirements for the laboratory. For example, the maintenance of a small number of two other breeds in addition to the main herd may from the farmer's point of view be a nuisance, although it is valuable for the research workers. The farm staff may also understandably take little pride in any older animals which are kept for studies in gerontology. Potential problems such as these require explanation and discussion. On the other hand, decisions about the type of pig house, type of feed and date of weaning, where they do not interfere with laboratory investigations, are best left, after some consultation, in the hands of the pig stockman. Modern pig farming is very well informed, highly competitive, and pays close attention to detail. A good stockman has developed against this background, and his reasonable requirements for raising animals should be met as far as possible in order that he can willingly maintain an efficient service to the laboratory.

Consideration will now be given to what is needed to maintain a pig unit on the lines given above. If pregnant sows are kept in a field, this should ideally be large enough to allow the ground to be rested for one year in every two since the pig's habit of rooting rapidly turns pasture into what looks like a ploughed field, especially when the ground is wet; another reason for resting the ground is to avoid re-infestation with parasites. This means that allowing for the rotation of fields, one acre should be allowed for each five or six sows. An alternative is to keep all sows in yards with wooden or brick sleeping quarters, a system which has now been adopted at Babraham. This arrangement reduces the amount of land which is needed, but it increases the capital cost of the unit. If a field is used, it can be divided into sections by means of an electric fence; wooden huts should be provided for sleeping, and a piped, insulated water supply taken to the field. A set of boar pens with outside yards is also necessary. These pens are best contructed of brick, and the yards of concrete, but it is obviously possible to use wooden buildings. The fully grown boar is a strong animal which can lift a five-barred gate on the tip of its snout, and a strong enclosure thus has obvious advantages.

The farrowing house in the Babraham unit accommodates eighteen sows and litters, and there are also a few farrowing pens outside. This arrangement is more elaborate than absolutely necessary, since wooden farrowing huts could be used throughout. The permanent farrowing house, which is insulated and heated in winter, has obvious advantages, particularly where the research interest involves new-born animals. The enthusiasm of most investigators will be greatly taxed by repeatedly waiting up into the small hours on winter nights in order to obtain an unsuckled pig if the sow is confined in a relative cramped hut outside.

The gilt can be mated successfully at about 8 months of age; the gestation period is nearly 4 months, the litter of pigs numbering up to twelve, and the sow can be expected to produce two litters per year. A conservative estimate is to allow for fourteen pigs being raised each year for each sow kept; this number allows for losses due to infertile matings, early mortality and disease. Reproduction in the pig is discussed further in Chapter 4.

The young pigs are usually weaned at 5–8 weeks of age, and are then raised and sold for pork at 4–5 months old, or for bacon at 6 months. During this period they need to be housed in some type of fattening house. At Babraham a separate house with a covered run is used as intermediate housing between weaning and the main fattening house. This arrangement has advantages inasmuch as most users of pigs for experiments tend to take animals aged 2–3 months from this house, and those which are left over are mostly destined to be sold for meat.

The main "fattening" or "growing" house is of a commercial type. The area is divided into twelve pens, each holding eleven pigs. Food is dispensed automatically from a hopper at pre-set times, and the dung is removed by a mechanical device running beneath the dunging passage. This is provided with a slatted floor so that the animal's excreta fall through to the channel below. Pigs rapidly learn to use the passage, and to reserve the pen for eating and sleeping. It is most important that adequate arrangements are made for the disposal of waste; this subject is discussed in Chapter 3.

About 20% of the breeding herd is replaced each year, and the animals which are required for this purpose are drawn from stock. The surplus of pigs is sold.

With such a system, the pig stockman and his assistant need to live near the unit, and would in any event expect accommodation to be provided as part of the job. The costs of this system include those for dry sow accommodation, farrowing and fattening housing. Predictions about the way a unit would work financially are difficult to make, but it is possible to cover feeding costs by returns from the sale of pigs. The recovery of the capital outlay, however, is unlikely to be achieved when the herd is being run primarily for experimental purposes.

Animals Bought as Required

The second method of obtaining animals is to make arrangements with local farmers. The advantages of this arrangement are that no large capital outlay is needed and there are no difficulties over farm administration. Disadvantages are that there is no control over the supply of pigs either from the point of view of their genetic constitution, or, in the case of new-born animals, their date of birth. Under such a system, the animals will need to be transported to the laboratory. There is therefore an increased risk of introducing disease into the laboratory, both from the transporting vehicle and from a diseased animal.

In many laboratories, however, it is likely that space considerations will leave no alternative but to buy pigs. The best plan in this event is to make arrangements with a large pig farm where the standards of husbandry are high. This choice will ensure a reliable supply of a reasonable number of pigs at given ages. A well-run pig unit will also have records of animals, and all animals will be identified by numbers. Smaller farms may not offer these advantages.

A Small Number of Breeding Animals

Buying in animals may often be combined with the third possibility of breeding just a few animals for laboratory uses only. When very young animals are needed, or the work involves the lactating sow or the sow–litter relationship, the animals need to be kept near the laboratory. Given a concrete yard with access to the roadway, the sows can be kept one in each hut, each with its own outside run. Additional huts of the same pattern, all of which can be obtained from farm suppliers, can be used to house the litter after weaning. The pig does, however, grow at a remarkable rate, and the pen of six weanlings becomes far less manageable within a few weeks and needs frequent cleaning out; in this connection, provision must be made for the disposal of dung.

A better method involves some system of pens in a permanent building where each pen has its own outside run. Routine cleaning in such houses is much easier than in a wooden hut. If some infection is suspected, brick and plaster buildings can be scrubbed with disinfectant, dried and made ready for use again after a few days. A wooden building, on the other hand, is very difficult to disinfect, and may need to be left empty for some weeks, or to be re-treated with creosote. With such a system involving only a few sows it may be preferable to rely on artificial insemination rather than to keep a boar.

Simple arithmetic will reveal that only a few animals can be supplied in this way, and that litters will be born at infrequent intervals. If the requirement is for fifty pigs per year this will mean on average six or seven litters of eight weaners. Since a sow can be expected to farrow twice a year, five sows would be enough to allow for some failures and small litters. At best this will give an average of eighty weaner animals per year in ten litters or one litter every five weeks; three sows, on the other hand, could be expected to produce only forty-eight animals at intervals of one litter per two months, with no allowance for failures.

CONCLUSION

If the requirement is for pigs of a given age every week throughout the year, a herd of a minimum of thirty sows or more is needed depending on the numbers required. Such a requirement is best handled under the first system described above, where the services of a full time pig stockman are used, and where a surplus of pigs is produced. If, on the other hand, the experimental use of the pig is to be on a smaller scale or more sporadic in nature, then one of the other systems is satisfactory.

CHAPTER 3

Housing and Handling

The types of equipment and housing required for pigs depends on the ages and numbers of the animals to be used. In respect of size and behaviour, the pig passes through three recognizable stages during the course of its development. During the first month after birth, the piglet grows rapidly from the new-born weight of rather more than 1 kg to the region of 8 to 10 kg; over this range of body weight, the animal, although noisy in protest, can be picked up easily and handled without difficulty. This is not so during the subsequent period, however, when the animal grows to a full size bacon pig of 90 kg at about 6 months of age. Weaning often takes place at 8 weeks of age, with a body weight approaching 20 kg, but it can be effected quite successfully much earlier (see Chapter 5). At body weights exceeding 25–30 kg the animal can be unmanageable for experimental procedures unless crated or otherwise confined.

The third stage is that of the mature breeding animal, sow or boar; here the considerable size, around 150–250 kg, makes handling difficult without special equipment; in addition, the mature pig is not tractable, and can be vicious. In this stage of the pig's life, careful training of the animal and expert handling are essential if successful experimental results are to be obtained. The mature boar may attack a person for no apparent reason, and on this account should be approached with caution. The nursing sow can become unusually aggressive when her offspring are disturbed; if it is necessary to handle the piglets while they are still living with the sow, there should be a sow-proof creep which a man can enter. It is sound practice to seek advice from the stockman when dealing with mature breeding pigs.

General Effects of Environment

The new-born pig is susceptible to chilling on account of its small size and low thermal insulation (Mount, 1959, 1964). The larger pig is susceptible to heat since it is unable to thermoregulate effectively by sweating (Ingram, 1967); until the animal is of the order of 50 kg, however, it is still relatively susceptible to cold. The approximate critical temperature for the new-born pig is 35°C. The critical temperature is defined as the lower end of the zone of thermal neutrality, in which metabolic rate is at a minimum, and it falls progressively as the animal grows larger, until in the mature pig it has a value approaching 0°C. The critical temperature depends on the level of nutrition, being lower in a well fed animal than in a fasted animal.

Resistance to a cold environment is increased when pigs associate in a group, as they huddle, bringing the individual pig's micro-environment (as distinct from the general physical environment) closer to thermal neutrality. The behaviour of the group so modifies the need for a metabolic response to cold that a clear-cut critical temperature disappears, and there is instead a zone in which the animal's metabolism is nearly independent of environmental temperature (for further details of effects of environment, see Mount, 1968).

Housing

The changes in the animal-environment relation which take place during growth require that the housing of pigs should take into account the differing needs of older and younger pigs, single pigs and groups. If a single new-born animal is to be kept for a period of time, and artificially fed, it should be protected from draughts and the temperature of its enclosure should not fall below 30°C for the first 4 or 5 days. A temperature higher than about 35°C, may, however, cause distress. For single pigs of 10 kg body weight about 25°C is desirable, while for older pigs the temperature is lowered progressively until it can be in the region of 20°C for groups of animals each exceeding about 30 kg in weight. In practice, therefore, a temperature range from 20 to 30°C covers most requirements. These values pre-suppose insulated floors and draught-free conditions. Pigs will certainly live and grow at lower temperatures, although they will not grow as rapidly, they will be more prone to disease, and their apparent well-being will not be so evident.

In farm practice, insulated floors take the form of a cement screed, up to 4–5 cm thick, laid on hollow tiles or other material giving an air space under the screed. The insulation is then mainly due to these air spaces, since trapped air has a high thermal insulation. A hard-wearing but insulated floor, which may be washed and scrubbed, is thus provided. This is very suitable for growing pigs, but for the new-born pig a cement screed represents

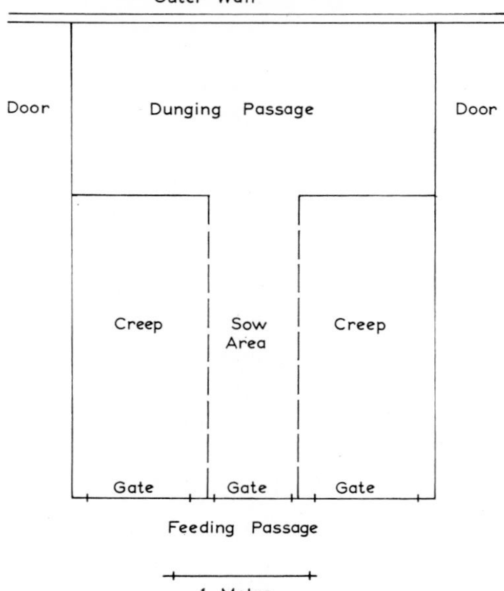

Fig. 1 The arrangement of a farrowing house pen: (a) the plan; each pen has three gates opening on to the central passage, which is unusual in farm practice but useful for experimental purposes since the piglets can be approached separately from the sow when they are shut off in either creep area;

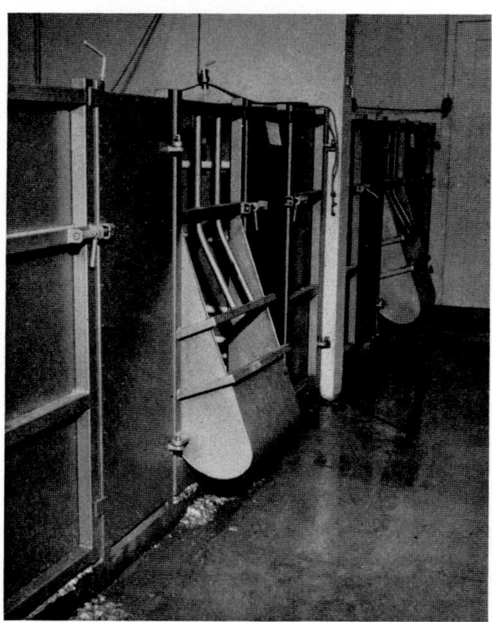

(b) a front oblique view of the closed pen, showing the sow's feed hopper mounted in the centre gate;

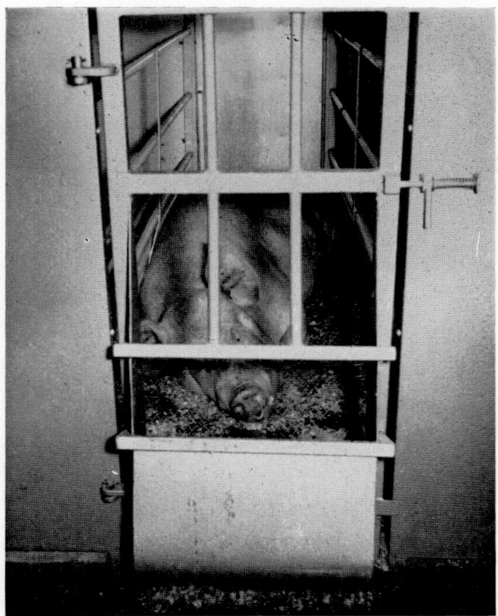

(c) a frontal view of the pen, showing the sow lying between creep divisions; these divisions are removable;

(d) piglets on wood shavings in the creep, showing on the left a hinged bar section which allows about 25 cm headroom for the piglets to move between sow and creep, and which can be lowered to the floor to confine the piglets to the creep.

a considerable heat capacity so that the underlying insulation cannot fill an effective role in preventing heat loss. In this case the use of bedding is strongly advised. From measurements of heat flow, 2–3 cm thickness of straw on concrete is equivalent to raising the floor temperature by 12°C (Mount, 1967) and wood wool is even better.

HOUSING OF PIGLETS

It is normally necessary to arrange accommodation for the sow and piglets together, although these have quite different environmental requirements. When the sow and litter share the same pen in a farrowing house or hut, as is normally the case, an infra-red heating lamp, often of 250 watts rating, is commonly used over the "creep" area (Fig. 1). In farm practice, this method is used to provide, in the same pen, temperatures to suit both

Fig. 2 A piglet living in a converted rabbit cage. Food and water are provided in open troughs, which are secured to the front of the cage by wing nuts; the troughs are cleaned out and refilled twice daily. In the photograph the pig is taking feed from the trough on the right. The dimensions of the cage shown are 57 × 41 × 46 cm high.

the sow and the litter; under-floor heating in the creep is sometimes used as an alternative. The same sort of lamp can be used with advantage for the single new-born pig, or whole litter, kept separately from the sow. Great care must be taken to ensure that the heat load is not too great on pigs which are unable to retreat to cooler quarters.

Piglets which have been weaned early from the sow can be housed individually in suitably adapted small cages, roughly the size of a rabbit cage, (Fig. 2), and groups of pigs can be kept in larger cages or boxes, equipped with troughs for food and water (Tribe, 1957). The animals do well if the floor is of galvanized iron wire in a mesh with spaces not exceeding 1 cm; this allows urine and much of the faeces to pass through, but the mesh is small enough to form a firm base for the animals' feet. Heating is provided by a 250 watt infra-red lamp over the cage, with air temperature about 20–25°C, leaving space for the pigs to retreat from the lamp. If radiant heat is not provided, the air temperature should be higher, around 30°C for new-born pigs weaned very early in life, decreasing to about 27°C for 3 kg pigs. The avoidance of draughts of air and the prevention of exposure to wetting are most important in reducing heat loss in these animals, which are so readily susceptible to chilling.

HOUSING OF OLDER PIGS

In farm practice growing pigs are kept in groups in pens in a fattening house (Fig. 3), or in a hut with either an enclosed or an outdoor run (Fig. 4). The aim is to provide a sleeping and eating area and a dunging area; urination and defaecation normally take place in the latter, so that the sleeping area is kept clean. In farm practice "slatted floors" are often provided, particularly in the dunging area, so that waste matter drops through longitudinal slots in the floor to a collecting area underneath (see Sainsbury (1963) for details). Such floors ensure that the pigs remain cleaner than would otherwise be the case.

The "dry" sow, that is the non-lactating animal, is preferably kept in either a hut with an outdoor run (see Fig. 4), a paddock with wooden arks or other simple shelter, or in an enclosed yard with adequate exercise area. The boar can be housed in a separate enclosure, or in the sow yard; with the latter arrangement it is easier to detect oestrus in the sows, and so to ensure fertile matings (see Chapter 4). Maintaining the sow and boar in these ways, however, presupposes the existence of a pig herd managed for the production of pigs for experiments (see Chapter 2).

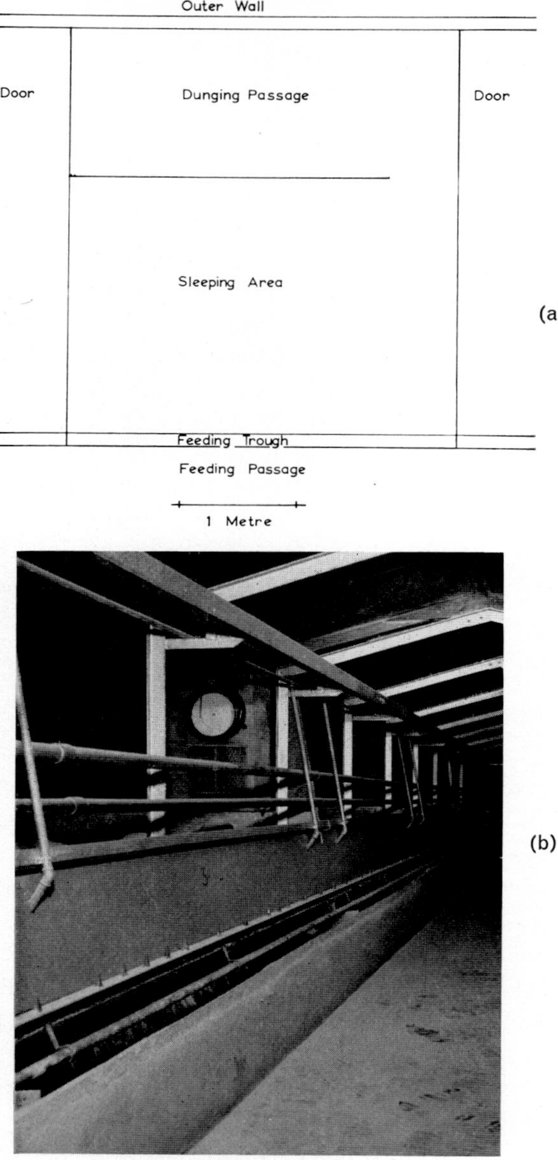

Fig. 3 A house for holding growing pigs from weaning to bacon weight (90 kg); (a) a plan of a section of the house; (b) a view of the fronts of the pens from the feeding passage, showing the feeding trough and, on the extreme left, the opening between the pen and the dunging passage. In both fattening and farrowing houses, the doors in the dunging passage between adjacent pens can be opened, so that the passage can then be cleaned by access from either end of the house.

Fig. 4 A wooden hut and outdoor run for housing pigs. If the run is on a concrete base, this is washed down as required. If, however, the hut and run are kept on grass, the whole assembly must be moved from time to time.

EFFLUENT DISPOSAL

It is most important that adequate arrangements are made for the disposal of waste effluent from pig houses. On a farm, the effluent is often spread on the land, as convenient. If a concentrated system of pig production is carried on within the confines of a small area, however, the disposal of effluent can become a major embarrassment if it is not specially allowed for.

As a rough guide, each pig produces about 2 litres of liquid and 0·5 kg of solid effluent each day. Direct discharge into the main sewer is unacceptable to local authorities if a large number of pigs is involved. The possibilities for disposal vary between different localities, and consequently if there is a problem it should be discussed with local agricultural and sewage officers.

If there is any possibility that radioactive material is present in either the urine or faeces, disposal may not take place through the usual channels, but will instead be subject to the appropriate regulations governing such substances.

HEATING

The radiant heaters referred to for piglets are not in general appropriate for the larger animal, although they have been used with success. For either the individual larger pig, or groups of pigs, it is preferable to provide a suitable air temperature by convection heaters, which can be arranged in conjunction with ventilation (see below). An insulated floor is necessary, but if this should not be available a slightly raised wooden sleeping platform should be provided. The pigs will usually quickly learn to urinate and defaecate on the concrete base floor and reserve the wooden area for lying. Wood soaks up water, and can become foul, although hardwood slats have been employed successfully over several years in the large calorimeter used at Babraham for the measurement of heat losses from groups of pigs (Mount *et al.*, 1967). Straw is very useful as bedding, and can transform a cold bare room into a warm area. The animals make a nest in the straw, so that the temperatures in their immediate surroundings are high, owing to the thermal insulation so provided. Straw must be changed, but this work is minimized if a dunging area is provided. A slight fall in the floor, of 1:80–1:40, allows urine to drain off.

Under-floor heating is sometimes used with pigs. This requires construction especially for this purpose, but the expense is justified if the research programme with pigs is to be extensive, and provided that such heating does not conflict with the conduct of experiments.

VENTILATION

This must be arranged so that in an enclosed compartment, whether it is a small box or a whole house, excessive water vapour is removed, while at the same time air movement at pig level is kept at a minimum. In respect of ventilation and air movement there are three requirements: (i) overall ventilation, i.e. change of air, must be adequate to remove water vapour, and, in an air-tight chamber, sufficient to provide oxygen and to keep the carbon dioxide concentration at a low level; (ii) the air in the house or chamber must be kept mixed, so that unventilated pockets do not develop; (iii) air movement in the vicinity of the pigs themselves should be kept to a minimum, below 10 cm/sec (20 ft/min) if possible.

At first sight these requirements conflict one with another, but in practice the combination of factors (i), (ii) and (iii) is achieved by baffles on air inlets, and by arranging the main air stream to move peripherally round the chamber. This keeps the air mixed in the middle of the chamber, but low air velocities occur in the vicinity of the animals, away from the main peripherally directed stream.

Ventilation requirements vary with the number of pigs housed and the state of the ventilating air. Cold air drawn in from outside in the winter usually has a lower absolute humidity than the air intake in summer; this is because the quantity of water vapour which can be carried by air rises steeply with its temperature. When the cold air is heated on entry, its relative humidity falls to a low level. In winter lower ventilation rates are required than in summer; in the summer, the air intake is both warmer and has a higher absolute humidity, so that its drying power, *at house temperature*, is less, volume for volume, than in the winter. The removal of water vapour from the house therefore calls for higher ventilation rates in summer than in winter, even though the internal house temperature is not greatly changed. In addition to its effect on the humidity, the level of ventilation can also have a considerable effect on house temperature. An obvious example of this is where a high ventilation rate in winter cools the house by exceeding the capability of heaters in the air inlets and by removing too rapidly the considerable quantities of heat produced by numbers of pigs housed intensively.

Practical aspects of housing pigs under farm conditions, which are needed when experimental work requires large numbers of animals, are discussed by Hicks (1965) and by Sainsbury (1963, 1967). Improvised accommodation, however, is quite satisfactory for small numbers of animals provided that it fulfils their requirements. As an example of what can be done in practice, holding sheds have been made at Babraham by the conversion of pre-fabricated concrete garages, insulated with expanded polystyrene, lined with hardboard and given an insulated concrete floor. Heating is by thermostatically controlled fan heaters, and ventilation by variable speed extractor fans. Such sheds have been in use for several years, holding pigs in mobile cages (see Fig. 5).

If the accommodation is to be improvised without the use of cages, it is important to consider the type of wall used to enclose the pigs. The wall should be readily cleaned, of firm surface and firmly secured. In practice, a brick wall rendered with plaster, or a wooden wall clad with galvanized iron sheeting, are satisfactory; an alternative is a strong welded wire mesh.

(a)

(b)

FIGURE 5

(c)

Fig. 5 Three mobile cages used for the housing and transport of experimental pigs. One cage is equipped with a water bowl and food hopper. The cages used at Babraham vary from 100 to 180 cm in length, 55 to 85 cm in width, and from 65 to 105 cm in internal height; the larger cages are used for holding pigs for longer periods, since then sufficient space is allowed for the animal to be unrestrained and to be able to turn round with ease.

Handling and Restraint

The restraint of pigs can be conveniently combined with the collection of urine and faeces and the taking of blood samples, in suitably designed metabolism crates. Such a unit was described by Hansard *et al.* (1951b); care was taken in the design of this equipment to prevent contamination by radioactive isotopes when these were in use. Another unit, made largely of wood, is described by Mayo (1961). The separate collection of total urine and faeces from male pigs is easier than from the female, since the opening of the urethra in the male is some distance cranial to the anus, making separate harness attachments feasible. Morgan and Davey (1965) designed an adjustable wooden cage with a stainless steel screen floor, so that urine can be collected below the screen while faeces are caught in a polyethylene bag secured to the pig by an adjustable harness. They give details of the harness and its mode of attachment to the pig. Crates for the restraint of large pigs are described by Pugh and Penny (1966) and by Baker and Andresen (1964). These allow for the animal to be secured and then turned on its back, into the supine position, while it is held in the framework of the crate. This is a very useful position for the collection of blood samples (see Chapter 9) and other procedures. Some of the cages used at Babraham for holding and transporting pigs of different sizes are shown in Fig. 5; a pig can also be trained to stand in a harness (Chapter 9, Fig. 2) in a framework of the sort illustrated in Chapter 10, Fig. 3. Pekas (1968) has described a metabolism apparatus made of steel and plastic; it is adjustable for pigs between 23 and 100 kg body weight, and it provides convenient access to any part of the animal. An instrumentation harness, primarily designed for use with miniature pigs, has been described for pigs weighing 35 to 90 kg (Karagianes, 1968).

It is an advantage to have the animal in its stand in a partially soundproofed room or chamber, and to have recording and other equipment and personnel outside, with connecting leads and an observation window. Habituation to such apparatus takes several days, but patient handling and repeated short exposures lead to the animal's acceptance of the routine. The experimenters should not be dissuaded by the pig's noisy protest on its first exposures to an unusual situation; the animal can often be quietened by feeding, and by scratching. It is imperative that the pig should be dealt with in a calm, unexcited manner, without raising the voice or moving suddenly; if necessary, the experimenter should wait for the animal to move voluntarily in the required direction. If the pig is required to enter a cage or enclosure, it is an advantage to have the far end open so that the pig can see through and move forward. Both front and back can then be closed together. A little food thrown into the cage encourages the animal to enter.

A number of references to detailed descriptions of metabolism crates and cages by various authors are given by Robinson and Coey (1957).

With pigs larger than about 40 kg, simple manual restraint by an assistant becomes impractical when, for example, it becomes necessary to take blood samples or to make intravenous injections (see Chapter 9). A slip-noose round the upper jaw is useful, since when the line is tied to a point somewhat above the pig's head the animal pulls back, and tends to remain relatively stationary (Carle and Dewhirst, 1942). It may then be held against a wall, or it can be housed in a narrow cage or crate with access openings at convenient points. A pig may also be held against a wall, or in a corner, with a hurdle.

The application of radiography to pigs, together with methods of restraint during the exposures, are discussed by Jarmoluk and Fredeen (1965). They have developed methods for handling both piglets and animals of 90 kg.

WEIGHING

The piglet can be weighed in a sack hung from a spring balance, or in a box resting on such a balance. The spring balance is particularly useful since it comes to rest so rapidly. When a pig is put into a sack or box, it tends to move more or less continuously, so that the needle of the attached balance oscillates violently. An accurate reading may be obtained, however, even if the animal is motionless for only a second or so, since the moving parts of the balance have so little momentum.

The larger pig is most conveniently weighed in a cage which is either wheeled onto the platform of a weigh-bridge, or, as in usual pig-farming practice, in a cage suspended from a spring balance (Fig. 6). This equipment

Fig. 6 A mobile weighing machine for pigs.

is readily available from farm supply stores.

MARKING

Pigs may be marked for identification by the ear punches used by pig farmers. Alternative methods for shorter term identification in the laboratory include marking the back with felt pens or other freely flowing innocuous material. Marking crayons which are designed for the purpose can be bought.

TRANSPORT

Pigs can be moved locally in cages on wheels (Fig. 5). If such trolleys have wire mesh or slotted floors, it is important to provide an underslung tray to catch urine and faeces, and a lid to prevent the animal jumping out. It is also important that the route to be taken by trolleys should include slopes rather than steps; the negotiation of steps with a heavily laden pig trolley is difficult.

In confined spaces, or where equipment is installed, the movement of the animals from place to place should be permitted only when they are confined in a suitable vehicle. An excited pig can do a lot of damage, simply on account of random activity, if it is free in a laboratory. The method of moving pigs from place to place by allowing them to walk and to be driven should be reserved for the farm.

Transport over greater distances, in motor vehicles, may be expected to disturb the animals. They lose weight, and may exhibit diarrhoea and behavioural disturbances. Such transport should be arranged so that a few days can elapse before the animals are used in experiments, particularly in work where any habituation or training is to be involved.

CHAPTER 4

Reproduction

General accounts of the reproductive physiology of the pig are given by Asdell (1964) and Dukes (1955), while more detailed information can be found in individual chapters in Parkes (1956) and Cole and Cupps (1959, 1969).

The Female

The nulliparous female, which is termed a gilt, reaches sexual maturity at about the seventh month of age, there being very little difference between different breeds (Phillips and Zeller, 1943; Burger, 1952; Self *et al.*, 1955). According to Squiers *et al.* (1952) and Foote *et al.* (1956), inbreeding can lead to delayed puberty, but this is probably related more to a general lowering of the vitality of the stock rather than to some specific factor. Undernutrition as the result of a restricted food intake also delays sexual maturity (Burger, 1952), while very severe underfeeding will delay puberty indefinitely (McCance, 1960). Even at 12 months of age, however, such grossly underfed pigs will nevertheless grow and reach puberty when they they are fed *ad libitum*.

A further factor which has been reported to influence the onset of maturity is the time of year at which the gilt is born. Both Robertson *et al.* (1951) and Wiggins *et al.* (1950) find that animals born in late spring reach maturity faster than those animals born during the winter. This effect may, however, be related to nutrition since even with the same food intake animals exposed to a warm environment will be relatively better fed since less food is required to maintain body temperature than is the case in the cold.

OESTROUS CYCLE

Once the gilt is sexually mature, oestrous cycles are repeated regularly throughout the year in the absence of pregnancy, there being no period of anoestrus. The mean length of the cycle is generally given as 21 days, and even though there is some variation between breeds, the differences from the mean do not on average exceed 1 day (Burger, 1952). Within a breed the standard deviation of the cycle length is usually 1 or 2 days (Boda, 1959; Asdell, 1964). The occurrence of "silent" heat, i.e. oestrous cycles with very little sign of oestrus, has been reported by Burger (1952), and may well account for the occasional reports of oestrous cycles of 40 days. Age of the animal makes very little difference to the length of the cycle (McKenzie and Miller, 1930).

OESTRUS

Externally, oestrus is marked by an increase in the size of the vulva and behaviourally by the fact that the sow will tolerate mounting by other pigs, either male or female. As a practical test the stockman applies pressure with the hand to the sow's back; the animal in oestrus stands still and adopts a lordotic posture. In pre-pubertal gilts swelling of the vulva may begin some weeks before oestrus, but in the sow the enlargement is confined to the 3 or 4 days before true oestrus.

According to McKenzie and Miller (1930), the duration of true oestrus, i.e. the period during which the female will accept the male, is 40 to 60 hours, but the duration of the first oestrus after weaning or after parturition may be prolonged to 65 hours. Other workers give slightly different figures for the duration of oestrus in different breeds, but there is agreement on the longer duration after parturition and weaning. The duration of eostrus is not affected by mating, whether fertile or not. The onset of oestrus occurs equally during the day or night and ovulation follows about 36 hours later. The onset of oestrus can be controlled by injection of luteinizing hormone (Radford, 1965) or by feeding compound 33828, also known as methallibure (Imperial Chemical Industries, Ltd.). In this method oestrus is first suppressed by feeding the compound and then induced on withdrawal. Using this technique several sows can be brought into oestrus at the same time (Dziuk, 1960; Dziuk and Baker, 1962; Polge, 1965; Dziuk and Polge, 1965; Groves, 1966; Barker, 1967).

The mating behaviour of pigs is described in Chapter 10.

POST PARTUM AND LACTATION OESTRUS

After parturition, the sow usually comes into oestrus in 2 days, but matings are not often fertile. During lactation, on the other hand, oestrus rarely occurs, although if it does mating is usually fertile. More recently it has proved possible to induce oestrus during lactation by injections of 1500 i.u. pregnant mare serum gonadotrophin on the 23rd day post partum. By this method, oestrus can be induced in about 75% of the cases even when the treatment has been repeated in four successive lactations (Crighton, 1970). In the usual way, once a litter has been weaned, the sow again comes into oestrus in 1–3 weeks, but although the time is variable it does not appear to be related to the loss of weight during lactation, nor does the age of the litter at weaning appear to influence the time before the next oestrus period.

HISTOLOGY OF THE OVARY AND THE NUMBER OF OVA SHED

The development of the ovary from birth to 3·5 months has been investigated by Casida (1935) who states that chords of eggs can still be seen up to the 4th week of age. Follicles with more than one layer of granulosa cells do not appear until the 7th week, and the first antrum-bearing follicles are seen at 11 weeks, becoming numerous by 15 weeks of age.

The basic changes in the histology of the ovary and reproductive tract of adults have been fully described by Corner (1921).

The follicles which are destined to ovulate begin to develop rapidly 2 or 3 days before oestrus, the theca interna hypertrophies and the cumulus oophorus begins to dissolve. Externally at this time the follicle is shell pink in colour, and the appearance of a transparent region indicates that it is about to rupture, an event which is spontaneous and not dependent on copulation. At the time of ovulation the first polar body has already been separated off and the second spindle has been formed. Development of the corpus luteum begins immediately after ovulation, and by the 7th day it is fully differentiated. If pregnancy does not occur, the corpus luteum begins to regress at about the 15th day, by which time it is 11 mm in diameter and has caused a noticeable increase in the weight of the ovary. As the corpus luteum ages it changes in colour from dark red to pale purple by day 15; the total length of the life of the corpus luteum in the absence of pregnancy is probably about 40 days.

Ovulation occurs over a period of 6–7 hours, its time of onset being related to the duration of external oestrus. Thus in the Large Black breed with oestrus lasting 68 hours the optimum time for mating with respect to litter size is about 12 hours later than in the Large White with an eostrus of 48 hours duration (see p. 34). There is also some evidence suggesting that eggs which are not fertilized soon after ovulation, although still viable, may be subject to early intra-uterine death (Burger, 1952).

The number of ova shed at each oestrus increases with age and has been estimated by Lasley (1957) as between 10 and 35, the left ovary contributing rather more than the right (Warwick, 1926; Robertson *et al.*, 1951). Figures published by Burger (1952) and Squiers *et al.* (1952) suggest that gilts of 7–8 months shed 11–14 ova, while in 15–17 month old animals the number rises to 15–17 ova. Breed differences are important, as is nutrition; Zimmerman *et al.* (1957) showed that more ova are shed from animals on a high plane of nutrition, and Self *et al.* (1955) have demonstrated that animals deficient in vitamin B_{12} shed fewer eggs.

HISTOLOGY OF THE REPRODUCTIVE TRACT

The uterus (Chapter 6, Fig. 13) is bicornuate with a single cervix, but the common cavity is relatively larger than in the bitch or cow (Kelly and Eckstein, 1969). The histology of the uterus has been described by Corner (1921) and McKenzie (1926). The changes in the histology of the vagina during the oestrous cycle have been reviewed by Asdell (1964) and described by McKenzie (1926) and Wilson (1926), while information about the vaginal smear has been published by Altmann (1939).

PREGNANCY

The duration of pregnancy varies with breed (Cox, 1964a). Perry (1956) gives a value of 114 days for the Large White, with a range of 104–126 days. The actual time of conception is, however, often not known if the sow is served more than once, and hence the range given is probably greater than the actual range. The same mean length of pregnancy has also been recorded by Carmichael and Rice (1920), and by Braude *et al.* (1955), although the ranges were not the same. There is little tendency for a more constant gestation period in any one pig than there is between different pigs, and the length of pregnancy does not vary consistently with the age of the sow or the ordinal number of the litter (Perry, 1956). The relation between litter size and length of gestation has been studied by Cox (1964b).

The placenta is of the epithelio-chorial type (Kelly and Eckstein, 1969).

SIZE OF LITTER

As with most variables, litter size varies with the breed of pigs. Within a breed the number of pigs born in a litter increases at first with the ordinal number of the litter and then slowly declines (Ellinger, 1921; Lush and Mollin, 1942; Perry, 1956), the decline in litter size in older animals being related at least in part to intra-uterine mortality. The average litter size at birth is often given as 12, but litters of between 1 and 24 are possible with a more general range of 8–16; amongst these, however, one or two still-born animals can be expected.

SEX RATIO

Although more males appear to be conceived, they have a higher mortality rate both before and after birth, so that by the time a litter is weaned the mean sex ratio is near to 1.

SURVIVAL OF YOUNG

Survival of young depends much on management of the herd, but as would be expected large litters tend to have more casualties than small ones. A litter size of 12 at birth will, on average, be reduced to 8 at weaning.

Many factors influence the incidence of death amongst litters, such as type of farrowing pen and creep, heating and draughts, handling of animals, the individual sow which may be clumsy and lie on the young and the number of teats available on the mother.

LENGTH OF REPRODUCTIVE LIFE

Females may become sterile at any age of life, and 3–5% never become pregnant. Probably a similar percentage become sterile after each pregnancy, but herd records are not ideal sources of information since females which fail to become pregnant are soon discarded from the herd. The maximum duration of reproductive life is difficult to determine since sows are usually disposed of long before the end of their life span. Sows which have been kept in this Institute for much longer than is usual practice have produced litters at 10 years of age. Under husbandry conditions, however, few animals are kept after 4 years.

The Male

The sexual development of the boar has been described by Hauser *et al.*, (1952), Niwa and Hizuho (1954), and by Asdell (1964), while more detailed information on the biochemistry of semen is given by Mann (1959). Development of the epididymis occurs rapidly between 4 and 7 months.

The first spermatozoa appear in the testis at 4 months, and at 6 months about 85% of the seminiferous tubules contain sperm. Although ejaculation occurs at 6–7 months, the ejaculate contains many immature sperm, and it is not until 11–12 months of age that a considerable improvement in both quantity and quality of the sperm occurs.

The most obvious characteristic of the boar's ejaculate is its very large volume (200–500 ml), and the correspondingly low sperm density. This ejaculate is produced over a period of up to half an hour, and is roughly divisible into three fractions. The first contains fluid from the glands, the second contains the sperm, and the last fraction carries a gel-like substance which probably serves to plug the cervix.

The viability of the sperm as measured by motility depends, among other things, on the frequency of ejaculation. Sperm collected from boars at intervals more frequent than 3 days exhibit reduced motility, although the animals are not rendered infertile.

Once copulation has taken place, the first sperm arrive at the site of fertilization about 2 hours later and are assisted on their way by antiperistaltic movements in the female tract.

Since ovulation is believed to occur about 36 hours after the onset of oestrus, mating at 24 hours after the beginning of heat should be fertile. The ova are, however, shed over a period of some hours, and ovulation may be delayed; for these reasons a second mating on the following day is believed to produce larger litters. On average, mating over rather less than 1·5 oestrus periods per pregnancy can be expected.

ARTIFICIAL INSEMINATION AND COLLECTION OF SPERM

The collection of semen for use in artificial insemination is carried out routinely at AI centres. It is possible that semen collection may be required for experimental purposes and a general description is therefore included here. Expert help, however, would not be difficult to obtain at a local AI centre.

The boar is first introduced to a dummy sow, which is a barrel-shaped object with a platform on either side to receive the boar's front feet. Some pretraining may be needed, and this usually consists of allowing the boar to serve sows at a given place and then introducing the dummy at the same place, or if a trained boar is already available the sight of it serving the dummy may stimulate the inexperienced animal to mount. The boar should not be allowed to mount the dummy if no collection of semen is made, and for this reason the dummy must be kept out of sight and reach of the boar when it is not in use.

Once the boar has mounted, an artificial vaginal is pushed over the penis. The equipment used consists of a rubber tube, about 30 cm long, with its end connected to a 500 ml flask. Various types of artificial vagina have been described, some with double walls which hold warm water during collection, while in another type the vagina has to be warmed before collection starts. One type has two rubber bulbs connected to the vagina which when manipulated serve to raise the internal pressure and to transmit pulsations to the penis. Another simpler type of artificial vagina is pushed over the penis and is pulsated by hand.

At the beginning of collection the boar thrusts the penis forward as in natural mating, but once the flow of semen has started the boar remains quiet. If desired the 3 fractions of the ejaculate can be collected separately.

Full descriptions of the methods used have been published by Wallace (1949), Glover (1955), Polge (1956) and Aamdal and Högset (1957). Artificial insemination in pigs has been the subject of numerous reviews, papers and notes including those by Derivaux (1959), Rowson (1956, 1959), Smidt (1962), Rothe (1963) and Eibl and Wettke (1964).

INSEMINATION OF THE SOW

The most difficult task involved here, especially for those not trained in pig keeping, is to detect the onset of oestrus in the sow, and more so in the gilt. The swelling of the vulva and the animal's willingness to be mounted have already been described, and a study of the oestrous cycle with reference to the best time for mating has been made by Ito *et al.* (1959), Smidt (1963) and Schilling and Röstel (1964). Although there appears to be some relation between body temperature and ovulation this does not seem to be definite enough to be of practical help (Sanders *et al.*, 1964). Methods of controlling the time of the onset of oestrus are available as mentioned earlier, and these may be helpful when artificial insemination is used.

The semen are deposited within the cervical canal with the aid of a catheter some 45 cm in length. The catheter, of which various types are available, is inserted in the vagina with the top of the catheter pushing up towards the backbone in order to avoid the possibility of entering the bladder. The semen may then be introduced under slight pressure from a syringe, or other device, or allowed to run in under gravity. It is recommended that the semen should be warmed to at least 20°C before insemination.

CHAPTER 5

Growth and Nutrition

The major constituent of pig feeds in Britain is barley. This cereal can be fed in relatively large amounts to the young animal with resulting rapid growth and the avoidance of excessive fat deposition, a factor of great importance in pig farming since the relative lean and fat content of the carcass determine its market value. Many corn merchants throughout the country provide satisfactory pig diets in ground or pelleted form, and the experimenter who wishes to keep a relatively small number of pigs is well advised to obtain the animals' feed from such merchants.

The type of feed fall into three main categories: feed for sucking pigs, as a supplement while they are still on the sow (this is often known as "creep feed"); feed for the growing animal and feed for the mature animal (sow and boar). Pelleted feeds offer a considerable advantage in handling and cleanliness, with less dust than with meal. The feeds are supplied in bags; provision should be made for storage space.

Extensive information on the nutrient requirements of pigs at all stages of growth is contained in a recent Agricultural Research Council (1967) publication. Reference should be made to this and to Fishwick's book (Hicks, 1965) for detailed guidance to the pig's nutritional requirements; the feeding of pigs under American conditions is dealt with by the National Academy of Sciences (1968). As an approximate indication, however, the quantity of ready-mixed feed from a corn merchant which is needed by the growing pig ranges from about 5% of the body weight daily at 20 kg body weight to 4% at 90 kg. The *ad libitum* intake amounts to rather more than 5% of body weight per day, and is higher in the smaller animal than in the larger. The composition of the feed used at Babraham is given in Table I.

If pigs are allowed to feed *ad libitum* for long periods of time they become too fat. For this reason in farming practice the food intake of growing pigs is usually restricted to a level which allows rapid growth without undue fat deposition, so that the resulting carcass is relatively lean. Similarly, as the

Table I

The composition of the feed used at Babraham

Ingredient Composition of Meal (% dry weight)		
Barley meal	45	
Wheat meal	10	
Wheatings	24	
Flaked maize	4	
Soya bean meal	4	
Fish meal	4	
Meat meal	4	
Yeast	1	
Grass meal	2·5	
Vitamins, minerals	1·25	
Spice	0·25	
Mean Chemical Analysis of Meal		
Moisture 13·3%		
Composition (%) on a dry matter basis		Range
Crude protein	19·5	17·4–21·0
Ether extract	2·8	1·8– 3·1
Crude fibre	5·5	4·1– 8·1
N.F.E.	66·0	63·7–68·3
Ash	6·3	5·4– 7·0
Silicon free ash	5·7	
Silica	0·6	
Calcium	1·02	
Phosphorous	0·91	
Sodium chloride	0·53	
Magnesium	0·19	
Potassium	0·79	
Sodium	0·22	
Copper (ppm)	15·2	
Zinc (ppm)	104	
Amino Acid Composition		
dry matter of meal (%)		Requirements as a % of dry matter[a]
Lysine	0·91	0·90–0·95
Methionine and Cystine	0·65	0·60–0·70
Tryptophan	0·18	0·15–0·20
Threonine	0·75	0·50–0·60
Isoleucine	0·90	0·75

[a] From Agricultural Research Council (1967).

animal becomes larger, a ceiling intake is imposed, which ranges from 2 to 3 kg of feed per animal per day. Since in the majority of research applications the presence of excessive fat is undesirable, the imposition of a similar dietary regime is advisable for experimental animals. Early-weaned pigs are initially fed *ad libitum* in usual practice.

The level of feed intake controls the growth rate; this can be carried to the extreme typified by maintaining pigs at a body weight of about 6 kg for periods of up to 1 year, by allowing them only a small intake of feed (McCance, 1960).

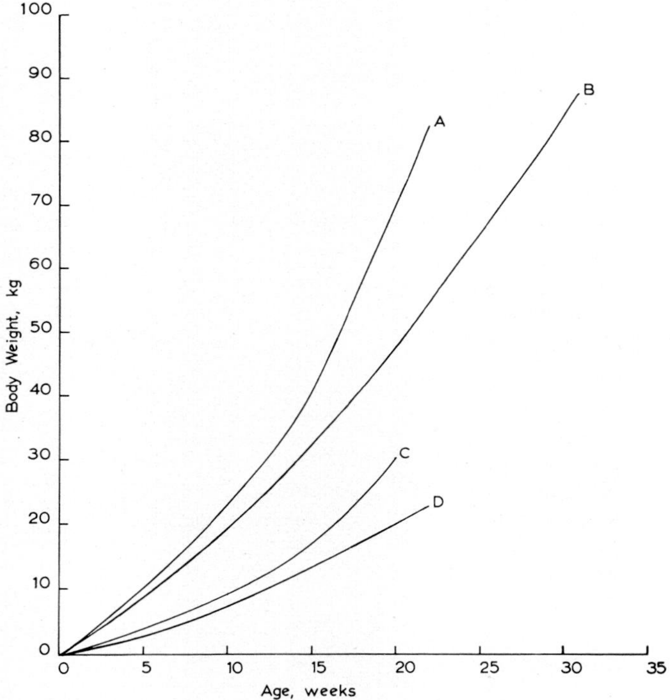

Fig. 1 The relation between age and body weight in four groups of pigs. The curves are averaged and take no account of the check in growth rate which commonly occurs at weaning. A, Large White pigs fed twice per day that amount which was eaten in 20 min on each feeding occasion; this quantity of feed was equal to 5% of body weight per day at 20 kg body weight, falling to 3·5% at 90 kg (Agricultural Research Council, 1967). B, Large White pigs at Babraham, fed at a restricted level of feed intake equal to approximately 4% of body weight per day up to a daily maximum of 2·3 kg per animal. C, miniature pigs (Rempel and Dettmers, 1966). D, miniature pigs (Haring, *et al.*, 1966).

The relation between age and body weight on the more usual levels of feed intake is given diagrammatically in Fig. 1.

New-born Pigs

It is important that the new-born pig should have access to the sow during the first hours after birth, so that it receives the high-protein colostrum secreted by the mammary gland at this time, before the onset of typical milk secretion. The importance of colostrum to the new-born pig is that passage of the large antibody protein molecules does not occur from the sow to the foetus *in utero*; this transferance takes place in the rabbit, guinea pig and man, but not in ruminants, horses and pigs (Hemmings and Brambell, 1961). The new-born pig acquires its antibodies in the colostrum from the sow, and the general health and survival of the pigs depend on this. The antibodies pass straight through the gut wall, and in addition to their immunological effects they have important physiological effects as well. Thus in one case the ingestion of maternal colostrum led to the absorption of globulins which, in the pig, increased the plasma volume by 30%, and at the same time the concentration of globulins in the plasma rose from 0·93 to 3·58 g/100 ml (McCance and Widdowson, 1959).

The neonatal period in which pigs can absorb globulins from colostrum is limited (Asplund *et al.*, 1962). Gamma-globulins have been detected electrophoretically in the sera of piglets fed colostrum 21–27 hours after birth, but not in the sera of pigs fed colostrum after that time. This suggests that piglets must receive colostrum during the first extra-uterine day if they are to absorb globulins. The composition of colostrum and milk from the sow is given in Table II.

Table II

The mean composition of colostrum (1st day milk) and milk later in the sow's lactation (from Bowland, 1966)

	Stage of lactation		
	1st day	3rd day	mean: 1–8 weeks
Total solids (%)	22·8	24·3	20·1
Fat (%)	6·6	12·0	6·8
Protein (%)	12·8	8·6	7·3
Lactose (%)	3·2	4·3	5·1
Ash (%)	0·73	0·89	0·99

Gut closure (that is, when the absorption of intact protein ceases) is influenced not only by age but also by a number of other factors (Ullrey *et al.*, 1966). Anti-trypsin activity in the colostrum decreases as lactation progresses, and this together with a rise in proteinase activity in the new-born pig's gut tends towards increasing hydrolysis of protein. The between litter variation in the uptake of antibody protein has been attributed to maternal effects, associated with the variable secretion of agents into the colostrum (Perry and Watson, 1967). Starving prolongs time to closure.

Pierce and Smith (1967) found that the efficiency with which bovine immune lactoglobulin was absorbed increased as the quantity fed was increased up to 2 g, after which the efficiency remained constant for higher doses. Thus about 1% was absorbed when 0·5 g was given; 5% when 1 g was given, and 10% when 2 g was given; this may be related to the saturation of proteolytic enzyme activity in the gut by the larger doses, so that a larger proportion is absorbed intact. Pierce and Smith found only very low levels of lactoglobulin in the serum of pigs before suckling, of the order of 1–4 mg/100 ml.

WEANING

Under farm conditions, pigs are weaned at 6–8 weeks of age, and sometimes earlier. Very early weaning, down to 2 days of age, is occasionally practised, and it is possible to rear piglets by artificial feeding from birth, although in that case it will be necessary to administer gamma-globulin for normal protection from disease. The alternative to this is to raise piglets in isolation from birth. The pigs are removed aseptically from the sow's uterus, by techniques described by Landy and Ledbetter (1966). Such animals, which can be reared "disease-free", are termed gnotobiotic pigs; the word "gnotobiotic" means "known life", as applied to contamination of the animals with micro-organisms. If the animals contain no organisms which can be cultivated, they are termed "germ-free"; if they are kept free from a wide range of pathogenic organisms, they are termed "specific pathogen free" (SPF). SPF pigs have been used to populate farms with pigs free from disease; techniques for their production have been described by Betts *et al.* (1960).

Suitable feed for early weaning can be obtained from several of the larger feed manufacturers, particularly for weaning piglets at about 10 days of age. The investigation of such diets revealed that food containing 56·6% sucrose caused diarrhoea and high mortality when given to pigs 1–10 days old; these undesirable effects disappeared when the sucrose was replaced by glucose

(Becker *et al.*, 1954). Sucrose, however, proved to be perfectly satisfactory in the diet of older pigs. Later it was found that the enzyme sucrase (invertase) is absent from the gut of the new-born pig (Walker, 1959; Dahlqvist, 1961). By 17 days of age the capacity to hydrolyse sucrose is developed (Kidder *et al.*, 1963); before that age, sucrose acts as a purgative.

Pregnancy and Lactation

A pregnant sow achieves a given weight gain on less feed than a non-pregnant sow of the same weight; the greater weight gain by the pregnant animal is largely the result of extra fat deposition (Heap and Lodge, 1967). If a sow receives a higher level of food intake during pregnancy, her intake during lactation is lower than it would be if she had received less food in pregnancy. Thus if a sow receives 3 kg feed per day during pregnancy she may consume no more than 5 kg per day during lactation. If, however, she eats only 2 kg per day during pregnancy, she may require 7 kg or so during lactation. The efficiency of feed utilization during the reproductive cycle may be influenced by the quantitative pattern of feed intake during pregnancy and lactation (Lodge, 1969; Elsley *et al.*, 1969).

Sows may therefore be fed a relatively large amount during pregnancy, when their requirements are less, and so accumulate reserves to meet the

Table III

The estimated protein requirements of lactating sows (*from Agricultural Research Council, 1967*)

Week of lactation	Average daily milk yield[a] (kg)	Average protein content of milk (%)	Output of milk protein (g/day)	Daily crude protein requirement (g/day)[b]
1	5·2	5·5	286	858
2	7·0	5·0	350	1050
3	7·3	5·3	387	1161
4	7·3	5·4	394	1182
5	7·3	5·7	416	1248
6	7·0	5·9	413	1239
7	5·5	6·2	341	1023
8	4·8	6·6	317	951
Average	6·4	5·7	363	1089

[a] For a sow suckling 8 pigs.

[b] Assuming a gross conversion efficiency of 33% for dietary crude protein into milk protein.

demand of lactation, or they may be fed according to their current requirements in order to reduce fluctuations in bodyweight.

Table III gives the estimated protein requirements for lactating sows and their milk yields. The sow suckling 8 pigs appears to need an average of about 1 kg of crude protein per day.

Water Requirements

There are considerable variations in water intake from one pig to another when free access to water is allowed. Water is given either mixed with the feed, the so-called "wet feed" system, or is allowed *ad libitum* from an automatic water bowl. For experimental animals water should always be allowed *ad libitum*, unless the experiment demands otherwise. For farming purposes, water is sometimes restricted, like feed intake, in order to obtain carcasses of better conformation.

As a guide to the animal's requirements at usual temperatures, the growing animal consumes a weight of water of the order of 10% of its body weight per day, and more than this under hot conditions. The larger animal requires proportionately rather less than the smaller pig; the non-pregnant sow requires 5 kg or so per day, the pregnant sow about 5–8 kg, and the lactating sow 15–20 kg. The large requirement of the lactating sow should be noted; this animal should always have an unrestricted water supply. Apart from the lactating sows, the voluntary water intake under usual conditions lies between 2 and 3 parts by weight of water for each part of feed.

Food Troughs and Water Bowls

Food and water can be offered to young pigs in cages in metal troughs of the type used for poultry. In order to prevent overturning it is useful to fix two bars under each trough, at right angles to its length.

For larger animals, and groups of pigs, galvanized iron troughs can be fastened in the cage or pen. In a pen, the trough is usually made of glazed earthenware, cemented in position, which makes cleaning a simple operation.

There are a number of different types of proprietary water bowl, obtainable from agricultural supply stores. These operate on a constant level system with a float valve, or with a demand valve which is pressed by the pig; they may be mounted in pens and in cages. Another type of water delivery equipment is a nozzle from which water flows when the pig sucks;

this type is more suitable for the dunging area of a pen. In order to prevent splashing of water from bowls, and fouling of the bowls, it is useful to provide a restricted opening and a hinged cover, which can be operated by the pig (see p. 108).

The behaviour of the pig in relation to feeding and drinking is discussed in Chapter 10.

CHAPTER 6

Anatomy

It is the intention here to indicate the main features of the anatomy of the pig, without pursuing any particular section in great detail. When it is necessary to carry out experiments which call for knowledge of certain anatomical details, what is needed is the most detailed account available; this allows the experimenter to embark on his own observations, which of themselves often add to anatomical knowledge of the part concerned. To deal with the subject at this level would be to attempt to duplicate the excellent treatises which already exist, a task which is not the aim of this book. It is intended rather to give the broad outline of the animal's anatomy, so that the experimenter may recognize landmarks, perhaps already familiar to him in other animals, in their setting in the pig.

The detailed works which are available include the "Ellenberger-Baum Handbuch der Vergleichenden Anatomie der Haustiere" (Berlin: Springer-Verlag, 1943), and the "Anatomie Régionale des Animaux Domestiques, III, Porc", by E. Bourdelle (Paris: Baillière, 1920). Grossman's (1953) revision of Sisson's "Anatomy of the Domestic Animals", published by Saunders, contains a systematic account of the pig's anatomy, and a number of excellent diagrams and photographs, some of which are reproduced in this chapter. An account of the pig's anatomy is given in Dunne's (1970) textbook on diseases of swine, and there are several books available on the anatomy of the foetal pig, such as the pictorial anatomy by Gilbert (1963).

There are many other papers which deal with aspects of the pig's anatomy; most of these are not readily available in this country, but they are listed in Index Veterinarius.

Skeleton and Topography

The skeleton of the pig in relation to the animal's profile is shown in Fig. 1. The vertebrae comprise 7 cervical, 14 or 15 thoracic, 6 or 7 lumbar,

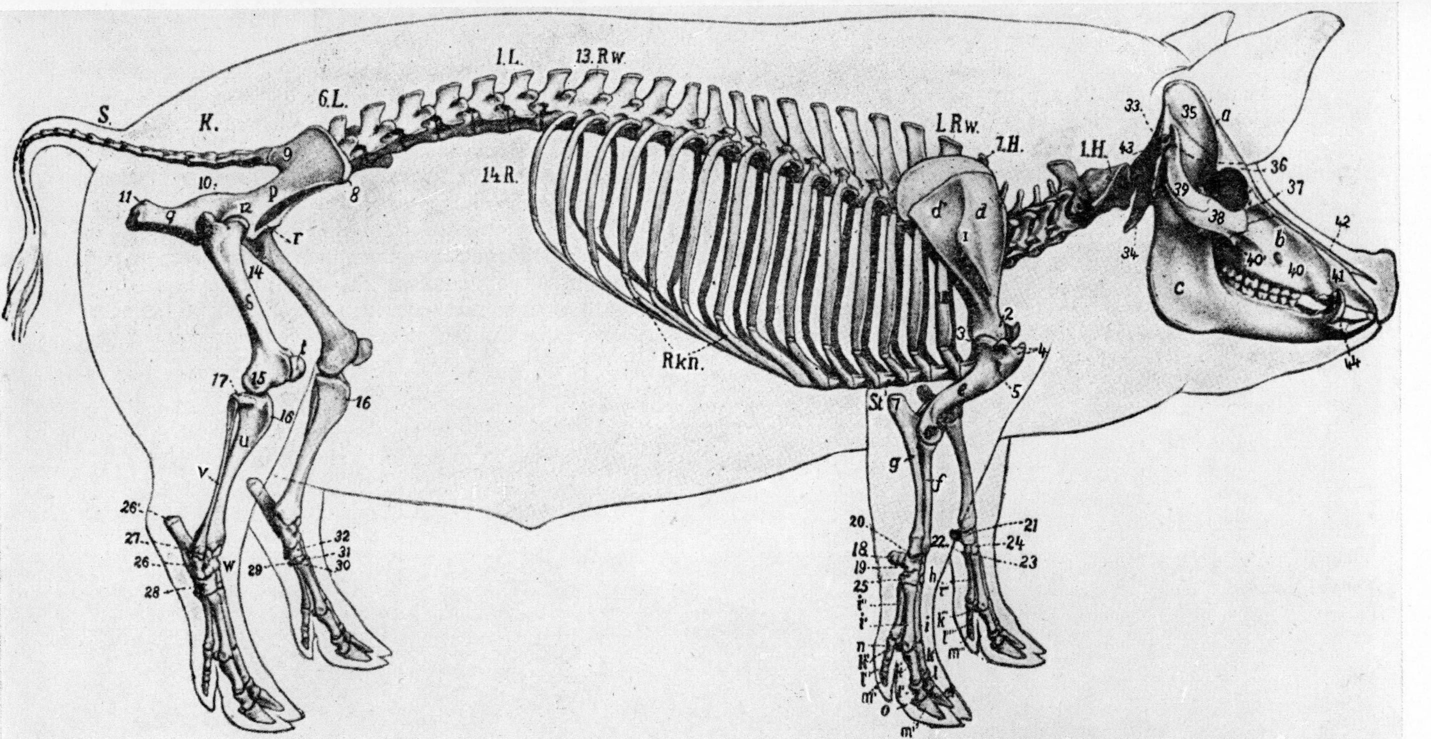

Fig. 1 Lateral view of the skeleton of the pig. a, Cranium; b, maxilla; c, mandible; 1H–7H., cervical vertebrae; 1 R.w., first thoracic vertebra; 13 R.w., thirteenth thoracic vertebra (next to last); 1L., first lumbar vertebra; 6L., sixth lumbar vertebra (next to last usually); K., sacrum; S., coccygeal vertebrae; 1R., first rib; 14R., last rib; R.kn., costal cartilages; St., sternum; d, supraspinous fossa; d′, infraspinous fossa; 1, spine of scapula; 2, neck of scapula; e, humerus; 3, head of humerus; 4, tuberosities of humerus; 5, deltoid tuberosity; 6, lateral epicondyle of humerus; f, radius; g, ulna; 7, olecranon; h, carpus; 18–25, carpal bones; i–i′′′′, metacarpus; k–k′′′′, proximal phalanges; l–l′′′′, middle phalanges; m–m′′′′, distal phalanges; n, o, sesamoids; p. ilium; 8, tuber coxae; 9, tuber sacrale; 10, superior ischiatic spine; q, ischium; 11, tuber ischii; r, pubis; 12, acetabulum; s, femur; 13, trochanter major; 14, trochanter minor; 15, lateral epicondyle; t, patella; u, tibia; 16, crest of tibia; 17, lateral condyle of tibia; v, fibula; w, tarsus; 26–31, tarsal bones; 26′, tuber calcis. After Ellenberger, in Leisering's Atlas (from "The Anatomy of Domestic Animals" by Sisson and Grossman, 1953, by permission of W. B. Saunders Company).

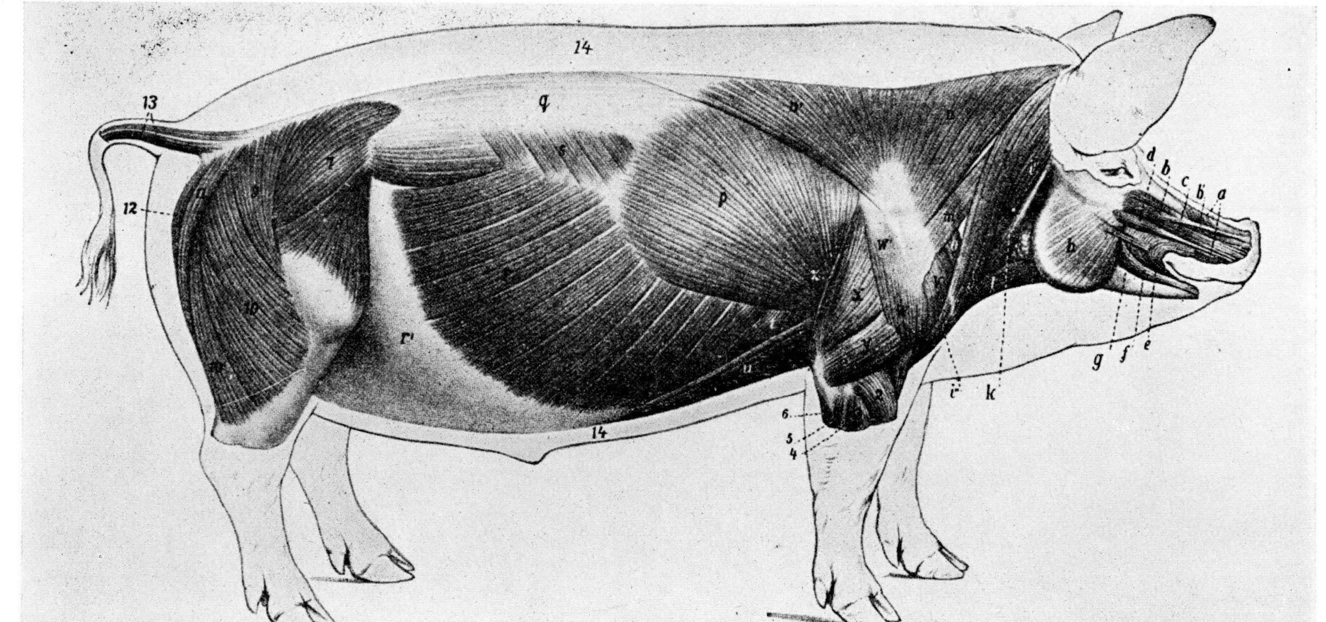

Fig. 2 The superficial muscles of the pig, after removal of M. cutaneus. a, Levator nasolabialis; b, levator labii superioris proprius; b′ fleshy slip of b which comes from premaxilla; c, dilator naris lateralis; d, depressor rostri; e, orbicularis oris; f, depressor labii inferioris; g, zygomaticus; h, masseter; i i′, i″, brachiocephalicus (cleido-occipitalis, cleido-mastoideus, pars clavicularis deltoidei); k, sterno-cephalicus; l, sterno-hyoideus; m, omo-transversarius; n, n′, trapezius; o, anterior deep pectoral; p, latissimus dorsi; q, lumbo-dorsal fascia; r, obliquus abdominis externus; r′, aponeurosis of r; s, serratus dorsalis; t, serratus ventralis; u, posterior deep pectoral; v, supraspinatus; w, w′, deltoideus; x, long head of triceps; y, lateral head of triceps; z, tensor fasciae antibrachii; 1, brachialis; 2, extensor carpi radialis; 3, extensor digiti quarti; 4, extensor digiti quarti; 4, extensor digiti quinti; 5, extensor carpi ulnaris; 6, ulnar head of deep flexor; 7, gluteus medius; 8, tensor fasciae latae; 9, 10, 10′, biceps femoris; 11, semitendinosus; 12, semimembranosus; 13, caudal muscles; 14, panniculus adiposus in section. After Ellenberger, in Leisering's Atlas (from "The Anatomy of Domestic Animals", by Sisson and Grossman, 1953, by permission of W. B. Saunders Company).

4 sacral and 20–23 coccygeal. The cervical region of the vertebral column is straight, and the thoracic and lumbar sections form a continuous single curve which is convex dorsally. The thoracic cage is made up by 14 or 15 pairs of ribs. The superficial muscles of the pig are illustrated in Fig. 2.

VISCERA

The viscera in relation to the skeleton and body outline are shown in Fig. 3, with views from both the left- and right-hand sides, male and female respectively. The topography of the viscera is discussed by Barone (1952).

SKIN

The skin of the pig has been referred to in contrast with the skin of man in Chapter 1; differences and similarities between the two species in respect of the integument have been discussed by Montagna and Yun (1964), Montagna (1966) and Weinstein (1966). Marcarian and Calhoun (1966) have described the microscopic anatomy of the integument of mature pigs, and Fowler and Calhoun (1964) and Smith and Calhoun (1964) have reported on the skin of the foetal and new-born pig respectively. It is characteristic of the pig that the animal can develop considerable quantities of subcutaneous fat depending on the plane of nutrition; the layers of back-fat in the pig have been described by Moody and Zobrisky (1966).

HEAD AND NECK

A semi-diagrammatic representation of a saggital section through the head is given in Fig. 4. An outstanding feature of the skull is the height of the nuchal surface and the breadth of the nuchal crest, which form attachment for the animal's powerful dorsal muscles. The lateral view of the skull given in Fig. 5 shows the relative massiveness of this attachment area and the consequently sharply cut away appearance of the caudal aspect of the skull, a feature which is not apparent in the whole animal owing to the presence of the musculature. Figure 5 also indicates the relatively large size of the mandible; anatomical and functional aspects of the jaw musculature of the pig have been discussed by Heinze (1961).

(a)

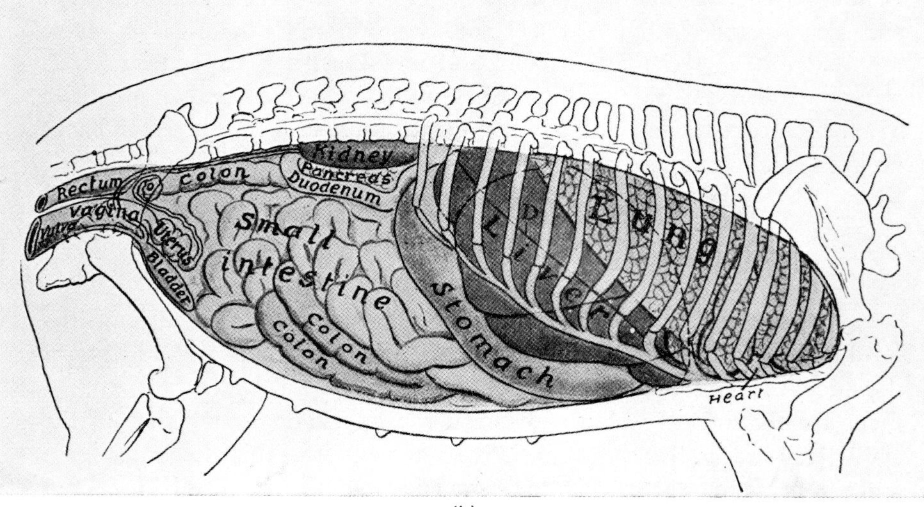

(b)

Fig. 3 The projection of the viscera of the pig on the body wall. (a) left side; (b) right side. D, costal line of diaphragm; O, ovary; U, ureter; V.S., seminal vesicle; B.g., bulbo-urethral gland; P, penis. The pancreas and duodenum are not in contact with the flank, but are covered laterally by small intestine (from "The Anatomy of Domestic Animals" by Sisson and Grossman, 1953, by permission of W. B. Saunders Company).

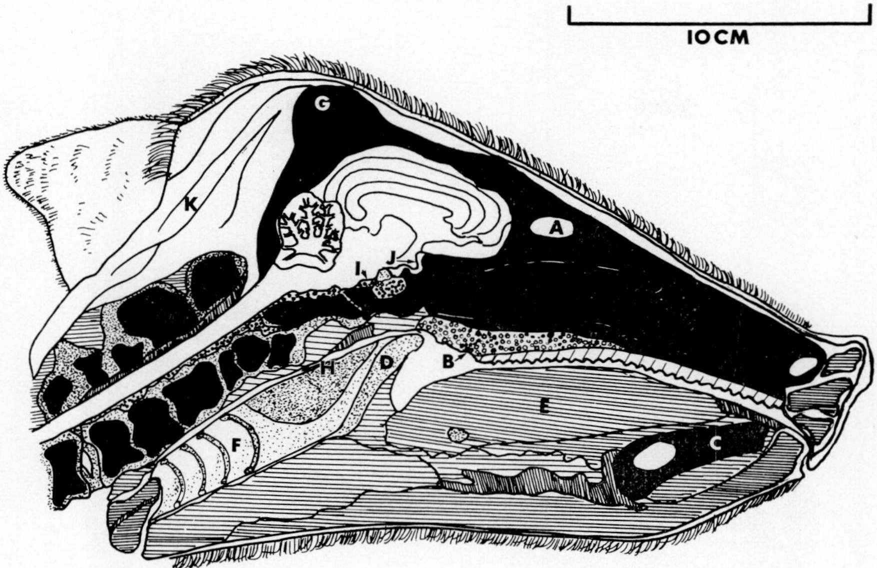

Fig. 4 Sagittal section through the head of a 3-month-old pig, partly diagrammatic. A, sinus; B, nasal septum; C, lower jaw; D, epiglottis; E, tongue; F, trachea; G, nuchal crest; H, oesophagus; I, pituitary; J, optic chiasma; K, nuchal ligament.

The dentition is indicated in Fig. 5. Ripke (1964) provides further discussion on the teeth of the pig. The dental formula for the pig is

$$2\left(\mathrm{I}\,\frac{3}{3}\,\mathrm{C}\,\frac{1}{1}\,\mathrm{P}\,\frac{4}{4}\,\mathrm{M}\,\frac{3}{3}\right) = 44$$

and the formula for the temporary teeth is

$$2\left(\mathrm{Di}\,\frac{3}{3}\,\mathrm{Dc}\,\frac{1}{1}\,\mathrm{Dp}\,\frac{4}{4}\right) = 32 \text{ (Sisson and Grossman, 1953).}$$

The tongue is long and narrow. At its caudal end, the entrance to the larynx over the epiglottis is almost at right angles to the line of the mouth cavity and the oesophagus. This feature, which causes difficulty in carrying out intubation for anaesthesia (see Chapter 8), is shown in detail in Fig. 6.

ENDOCRINE GLANDS

The position of the pituitary gland is indicated in Fig. 4; the anatomy

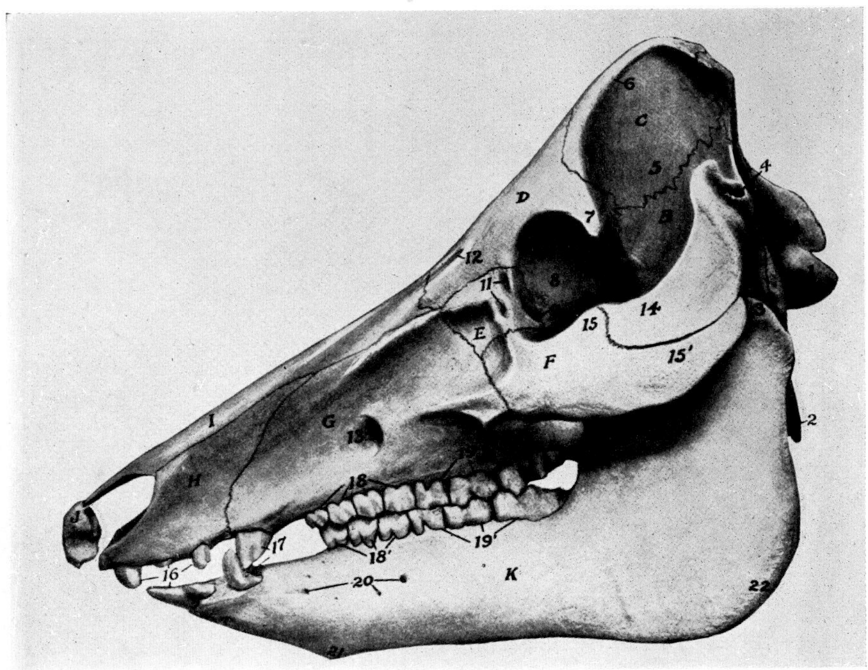

Fig. 5 Lateral view of the skull of the pig. A, Occipital bone; B, squamous temporal bone; C, parietal bone; D, frontal bone; E, lacrimal bone; F, malar bone; G, maxilla; H, premaxilla; I, nasal bone; J, os rostri; K, mandible; 1, occipital condyle; 2, paramastoid process; 3, condyle of mandible; 4, meatus acusticus externus; 5, temporal fossa; 6, parietal crest; 7, supraorbital process; 8, orbital part of frontal bone; 9, fossa for origin of ventral oblique muscle of eyeball; 10, orbital opening of supraorbital canal; 11, lacrimal foramina; 12, supraorbital foramen and groove; 13, infraorbital foramen; 14, zygomatic process of temporal bone; 15, temporal, and 15′, zygomatic process of malar bone; 16, incisor teeth; 17, canine teeth; 18, 18′, premolars; 19, 19′, molars; 20, mental foramina; 21, mental prominence; 22, angle of mandible (from "The Anatomy of Domestic Animals" by Sisson and Grossman, 1953, by permission of W. B. Saunders Company).

of the sella turcica has been described by Zintzsch and Flechsig (1966) and surgical approaches to hypophysectomy are referred to in Chapter 9.

The anatomical relations of the thyroid gland have been given by Romack *et al.* (1964). The gland is bilobed, with a large isthmus. Its position varies, and although it usually lies on the trachea some distance from the larynx, it may also be found in contact with the caudal boundary of the larynx; in some animals, thyroid tissue may extend caudally from the main

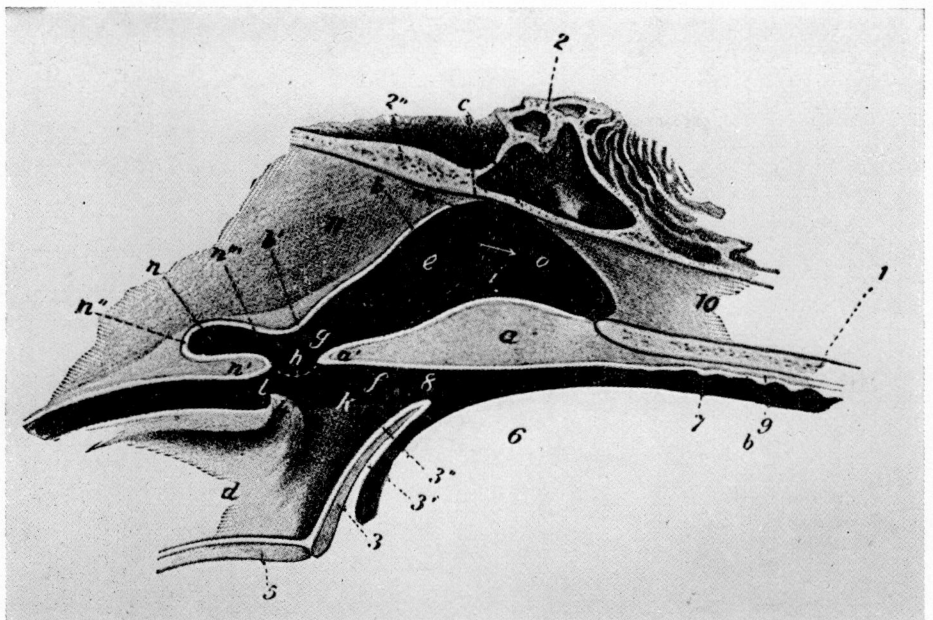

Fig. 6 A partly schematic saggital section through the pharyngeal region of the pig. 1, Palatine bone; 2, sphenoid bone; 2′, sphenoidal sinus; 2″, occipital bone; 3, epiglottis; 4, arytenoid cartilage; 5, thyroid cartilage; 6, root of tongue; 7, mouth cavity; 8, isthmus faucium; 9, hard palate; 10, septum nasi; 11, ventral muscles of head; a, soft palate; a′, free edge of a; b, dorsal wall of pharynx; c, fornix of pharynx; d, cavity of larynx; e, g, naso-pharynx; f, oro-pharynx; h, posterior pillar of soft palate; i, dotted line indicating lateral boundary between nasal cavity and pharynx; k, aditus laryngis; 1, aditus oesophagi; m, Eustachian orifice; n, pharyngeal diverticulum; o, posterior naris. After Ellenberger, in Leisering's Atlas (from "The Anatomy of Domestic Animals" by Sisson and Grossman, 1953, by permission of W. B. Saunders Company).

body of the gland. The thyroid artery lies at the right caudal border of the gland, and sometimes a second thyroid artery is found on the left caudal border. There are usually two thyroid veins.

THORAX

The thorax in the pig is rounded; the right lung has four lobes, although two of these are separated only by the cardiac notch (Fig. 7, and see Talanti, 1959).

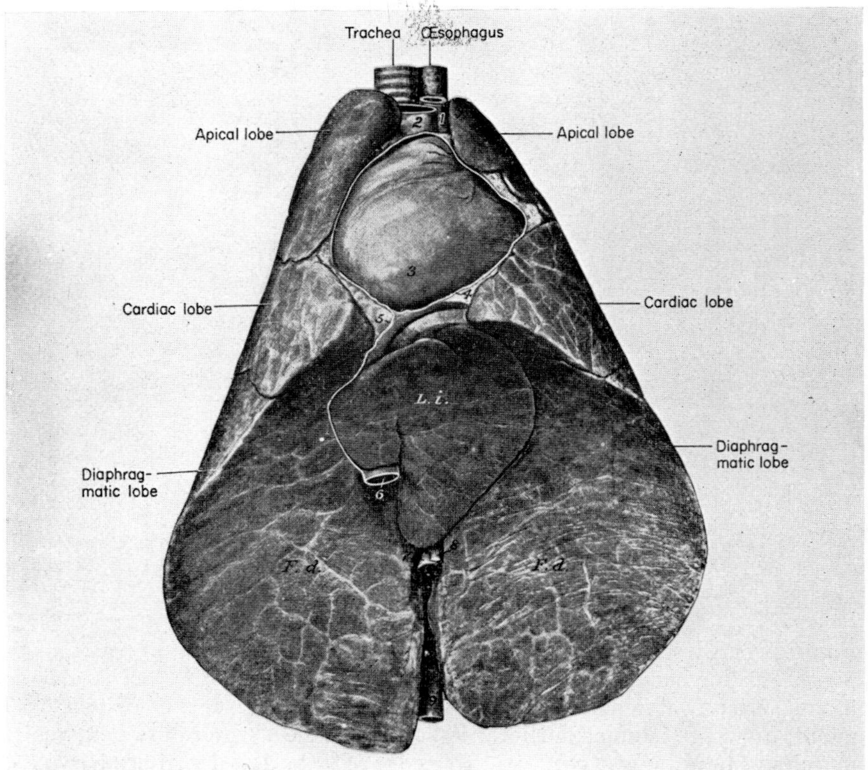

Fig. 7 A ventral view of the lungs and heart of the pig. L.i., Intermediate lobe of right lung; F.d., diaphragmatic surface of lungs; 1, left brachial artery; 2, brachiocephalic artery; 3, apex of heart; 4, pericardium (cut edge); 5, plica venae cavae; 6, posterior vena cava; 7, oesophagus; 8, ventral oesophageal nerve trunk; 9, aorta (from "The Anatomy of Domestic Animals" by Sisson and Grossman, 1953, by permission of W. B. Saunders Company).

The heart and the origins of the great vessels are shown in Fig. 8. Some of the blood-vessels of the pig have been subject to investigation, particularly the coronary arteries (Christensen and Campeti, 1959; Reiner *et al.*, 1961; Berg, 1964a,b, 1965). The aorta (Laing *et al.*, 1963), and accessory pulmonary arteries (Kaman and Červeńy, 1968) and the muscular architecture of the heart (Thomas, 1957, 1959) have also been studied.

Right and left lateral views of the heart, cardiac nerves and related ganglia are given in Fig. 9. Sympathetic cardiac nerves arise from the

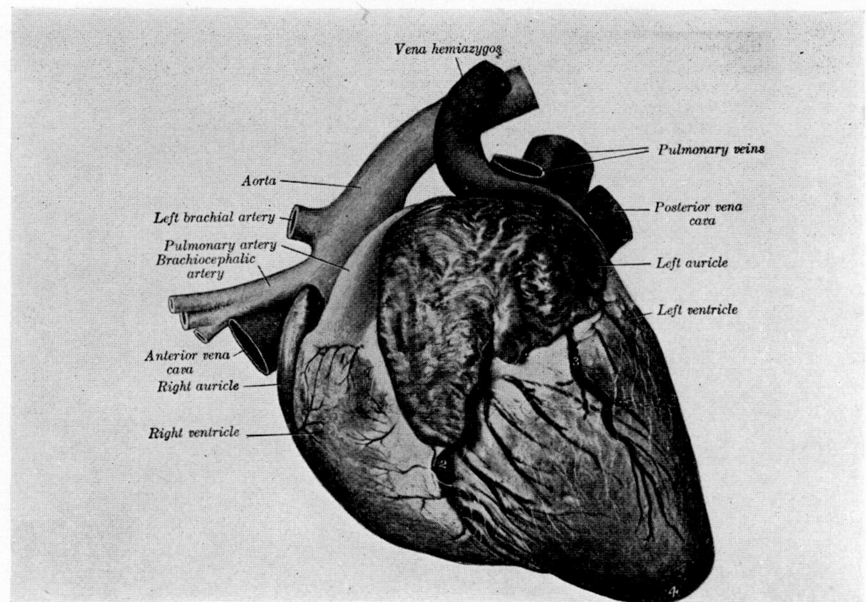

Fig. 8 Left view of the heart of the pig. 1, conus arteriosus; 2, great cardiac vein; 3, cardiac vein; 4, apex (from "The Anatomy of Domestic Animals" by Sisson and Grossman, 1953, by permission of W. B. Saunders Company).

thoracic, cervico-thoracic vertebral and middle cervical ganglia (McKibben and Getty, 1969).

LYMPH

The pattern of lymph flow from subcutaneous areas of the neck, thoracic and pelvic regions is indicated in Fig. 10. Lymph nodes and vessels and the drainage of lymph from the cranial end of the body has been investigated by Saar and Getty (1964), who found that the efferents from the dorsal superficial cervical lymph nodes (the main receivers for superficial afferent lymph vessels in this area) terminate in the brachiocephalic vein. Studies on the micro-anatomy of lymph nodes in the pig have been reported (Hunt, 1968).

ABDOMEN

The oesophagus is short and nearly straight; the stomach is large, with a capacity of 6–8 litres, and it has a diverticulum in the cardiac region (Fig. 11). The liver is large, and divided into four principal lobes; the greater part

1, ramus communicans
2, sympathetic trunk
3a, first thoracic ganglion
3b, second thoracic ganglion
3c, third thoracic ganglion
3d, fourth thoracic ganglion
3e, fifth thoracic ganglion
3f, sixth thoracic ganglion
3g, seventh thoracic ganglion
3h, eighth thoracic ganglion
4, cervicothoracic ganglion
5, ansa subclavia (caudal limb)
5′, ansa suclavia (cranial limb)
6, cardiac ganglion of the left intervascular triangle
7, intermediate ganglion
8, vertebral ganglion
9, middle cervical ganglion
11, vagus nerve
12, right recurrent laryngeal nerve
12′, left recurrent laryngeal nerve
13, thoracic cardiac nerve
14, cranial cervicothoracic cardiac nerve
14′′, caudoventral cervicothoracic cardiac nerve
16, vertebral nerve
17, intermediate cardiac nerve
18′, caudal vertebral cardiac nerve
19, middle cervical cardiac nerve
21, cranial vagal cardiac nerve
21′, caudal vagal cardiac nerve
22, recurrent cardiac nerve

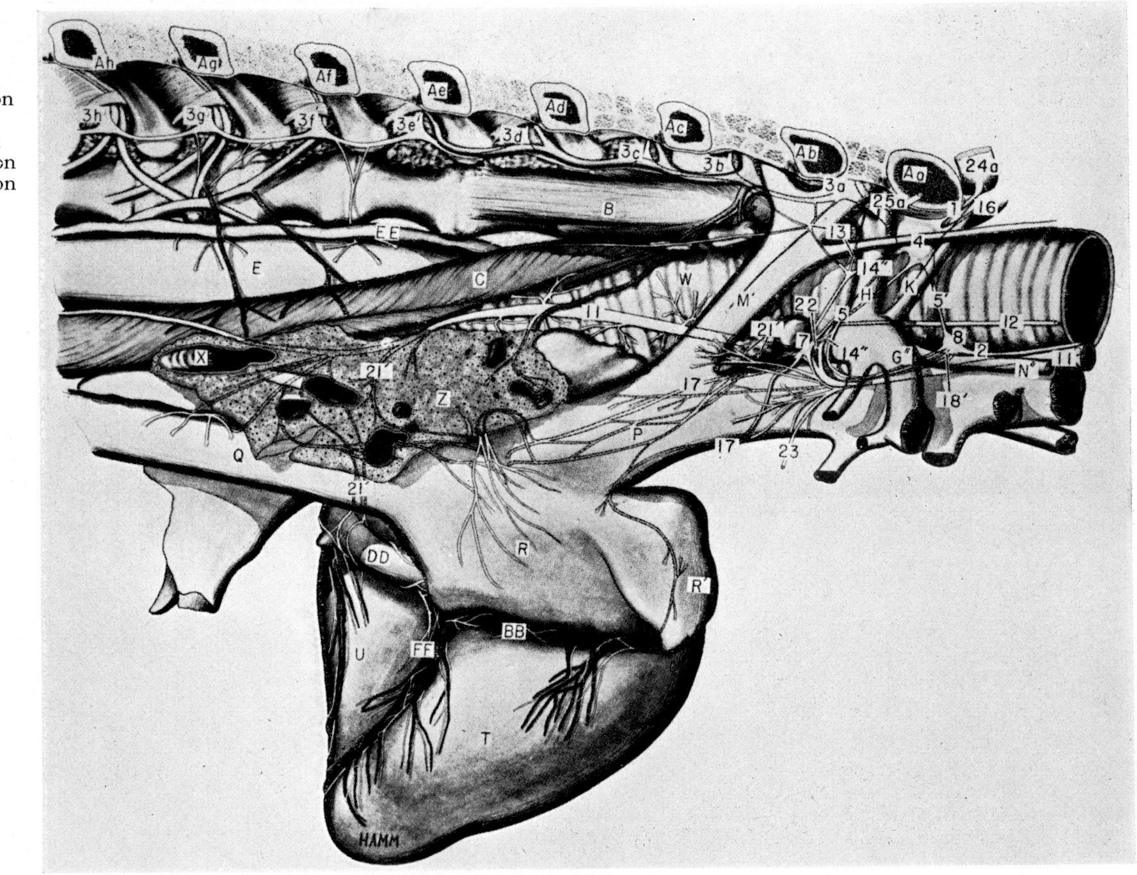

(a)

24a, eighth cervical spinal nerve
Aa, first rib
Ab, second rib
Ac, third rib
Ad, fourth rib
Ae, fifth rib
Af, sixth rib
Ag, seventh rib
Ah, eighth rib
B, longus colli muscle
C, esophagus
D, intercostal artery
D′, intercostal vein
E, aorta
F, brachiocephalic artery
G′, left subclavian artery
G″, right subclavian artery
H, costocervical artery
K, vertebral artery
L, deep cervical artery
M′, costocervico-vertebral vein
N′, left common carotid artery
N″, right common carotid artery
O′, left vena azygos (hemiazygos vein)
P, cranial vena cava
Q, caudal vena cava
R, right atrium
R′, right auricle
S′, left auricle
T, right ventricle
U, left ventricle
W, trachea
X, bronchus
Y, pulmonary artery
Z, lung
BB, right coronary artery
BB′, descending branch of left coronary artery
CC, great cardiac vein
DD, coronary sinus
EE, thoracic duct
FF, middle cardiac vein

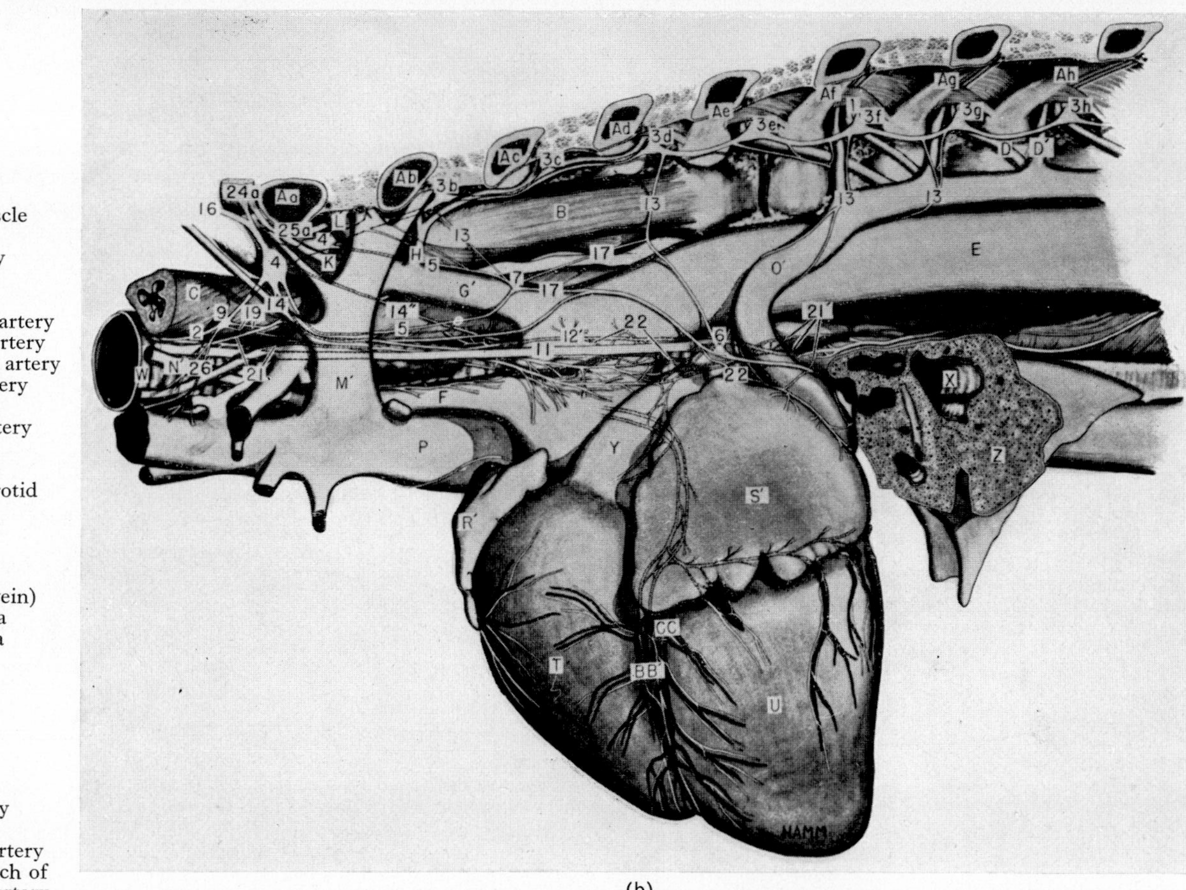

(b)

Fig. 9 Right (a) and left (b) views of dissections to show the heart and associated nerves and ganglia in the pig (by permission of the American Veterinary Medical Association from McKibben and Getty, 1969, *Am. J. vet. Res.* **30**, 779).

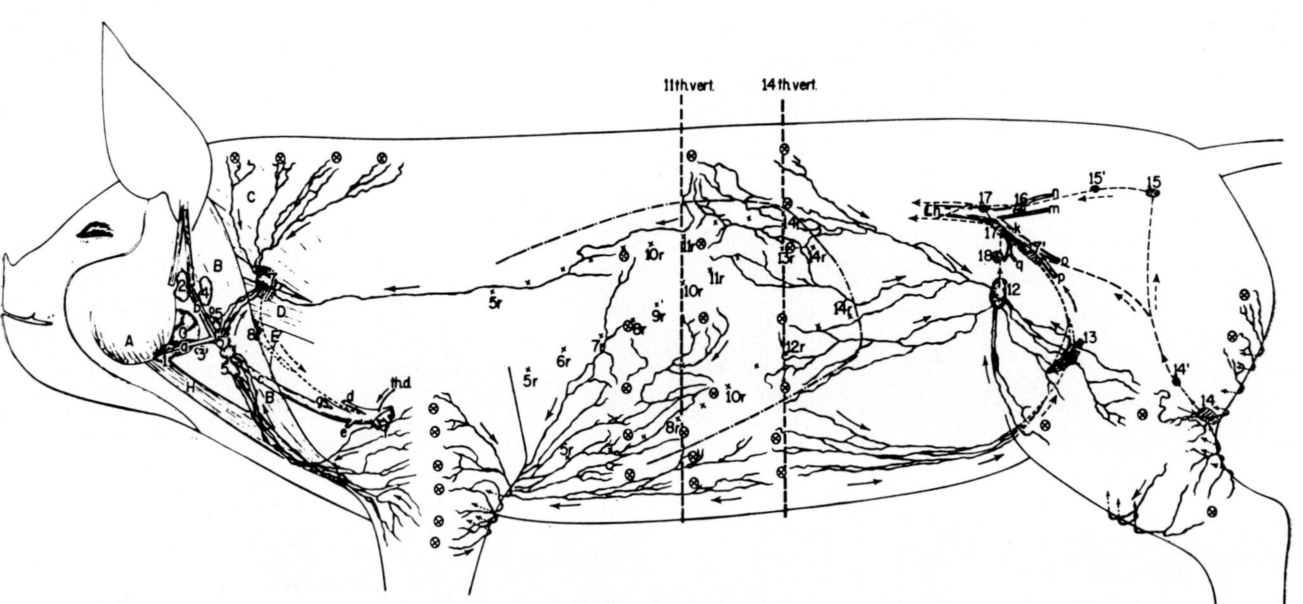

Fig. 10 The directions of lymph flow from subcutaneous areas in the pig. The rib cage is indicated in outline, and crosses indicate the positions of the ribs (5r to 14r). Two vertical lines pass through the 11th and 14th spinal processes of the thoracic vertebrae (by permission of the American Veterinary Medical Association from Saar and Getty, 1964, *Am. J. vet. Res.* **25**, 618).

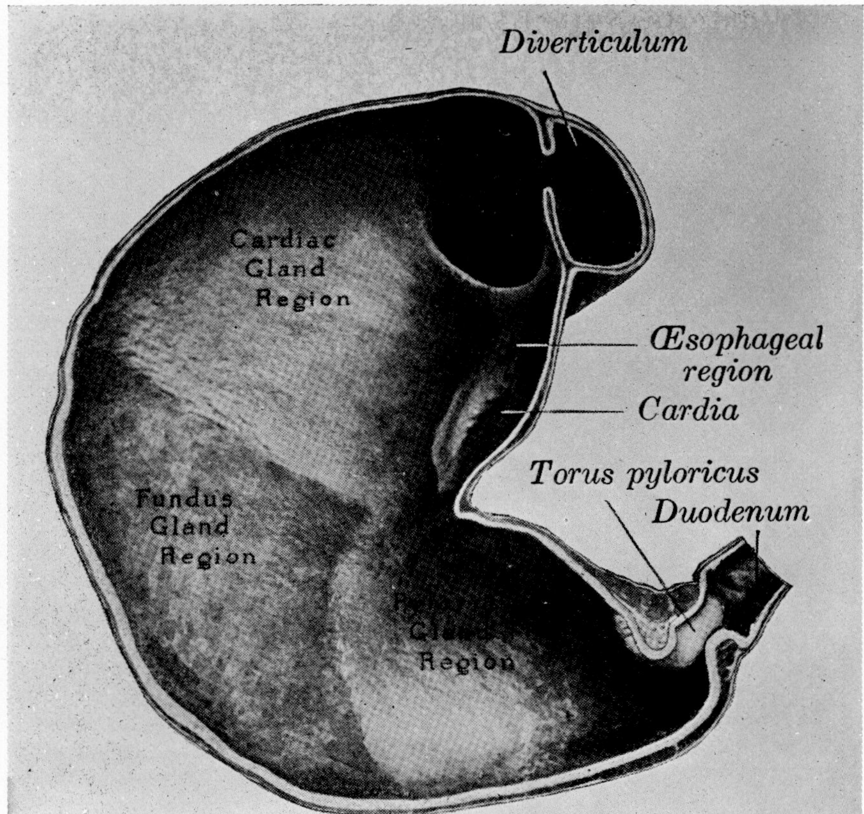

Fig. 11 Section through the stomach of the pig (from "The Anatomy of Domestic Animals" by Sisson and Grossman, 1953, by permission of W. B. Saunders Company).

is to the right of the median plane. The branching of the blood vessels in the liver of the pig has been studied by Kaman (1966).

The general relations of the abdominal viscera are indicated in the two diagrams of Fig. 3. The colon is spiral in form, a feature which is made more explicit in the diagram of Fig. 12 in which the coils of the colon have been pulled apart. The colon becomes progressively narrower in the latter part of its course and forms a returning spiral enmeshed with the initial part.

The uterus is shown in more detail in Fig. 13, as a dorsal view with partial dissection. The ovaries are concealed in the ovarian bursa; mature ovarian follicles may have a diameter of 7–8 mm, and corpora lutea may be

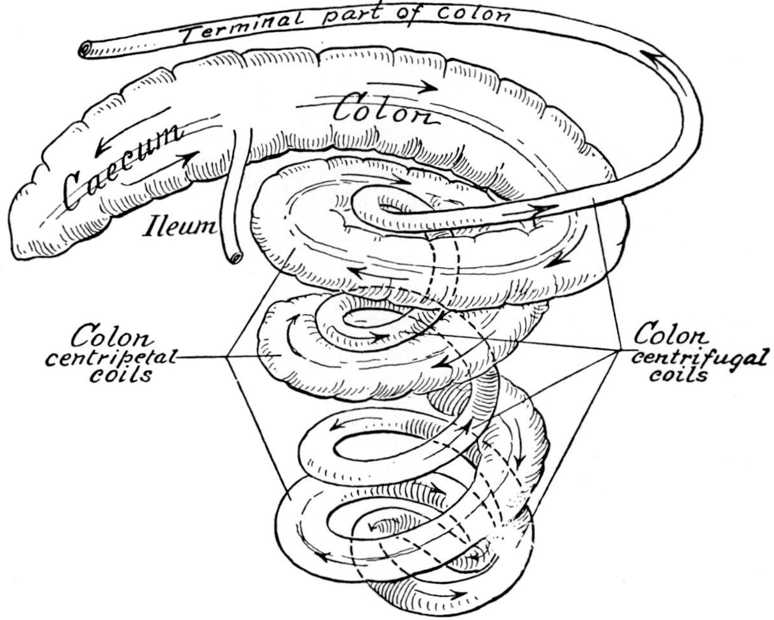

Fig. 12 Diagram of the caecum and colon of the pig; the coils of the colon have been pulled apart (from "The Anatomy of Domestic Animals" by Sisson and Grossman, 1953, by permission of W. B. Saunders Company).

found 12 mm in diameter (Sisson and Grossman, 1953). The uterus is bicornuate, with a short body (about 5 cm) and horns which may be 1–1·5 metres in length. The vasculature (Grahame and Morris, 1957; Yamashita, 1961; Oxenreider *et al.*, 1965) and morphology (Hadek and Getty, 1959a, b; Palmer *et al.*, 1964) of the internal genitalia have been studied.

LIMBS

The musculature of the forelimb and hindlimb of the pig is illustrated in Figs 14 and 15. The blood supply to the appendages has been examined in detail by Ghoshal and Getty (1968), and their schematic presentations are given in Figs 16 and 17.

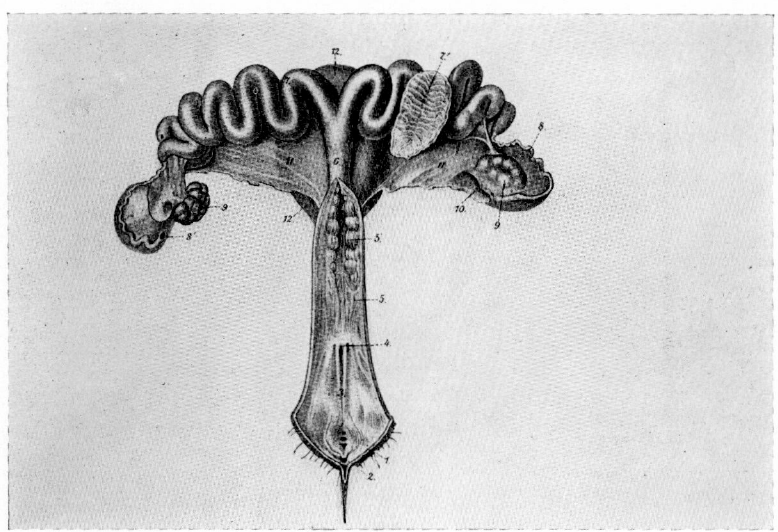

Fig. 13 Dorsal view of the genital organs of the sow; the vulva, vagina and cervix have been slit open. 1, labium vulvae, 2, glans clitoridis; 3, vulva; 4, external urethral orifice; 5, vagina; 5′, cervix uteri; 6, corpus uteri; 7, cornua uteri, one of which is opened at 7′ to show folds of mucous membrane; 8, uterine tube; 8′, abdominal opening of tube; 9,9, ovaries; 10, ovarian bursa; 11, 11, broad ligaments of uterus; 12, urinary bladder. From Leisering's Atlas (from "The Anatomy of Domestic Animals" by Sisson and Grossman, 1953, by permission of W. B. Saunders Company).

NERVOUS SYSTEM

The brain in an adult pig weighs about 125 g, and the spinal cord about 42 g. Some outstanding features are that the cerebellum is very wide and short, and the olfactory bulbs very large (Sisson and Grossman, 1953). Dorsal and lateral views of the brain are given in Figs 18 and 19.

The cytoarchitecture of the brain stem has been examined (Breazile, 1967), also myelinization of the nerve fibres in the spinal cord (Ziolo, 1965). A functional study has been made of the motor cortex in the domestic pig (Breazile *et al.*, 1966), and other studies of the sensory cortex (Adrian, 1943a, b; Woolsey and Fairman, 1946).

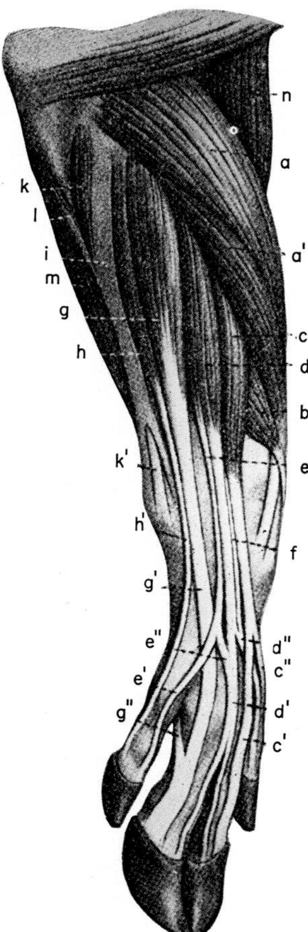

Fig. 14 Dorsolateral view of muscles of the forelimb of the pig. a, a′, extensor carpi radialis; b, extensor carpi obliquus (abductor pollicis longus); c, d, e, common digital extensor; c′, c″, tendons of insertion of c; d′, d″, tendons of d; e′, e″, tendons of e; f, tendon of extensor digiti secundi; g′, g″, extensor digiti quarti; h, extensor digiti quinti; h′, tendon of h; i, tendinous and k, fleshy, part of ulnaris lateralis; k′, tendon of k; l, ulnar head of deep digital flexor; m, superficial digital flexor; n, brachialis. After Ellenberger, in Leisering's Atlas (from "The Anatomy of Domestic Animals" by Sisson and Grossman, 1953, by permission of W. B. Saunders Company).

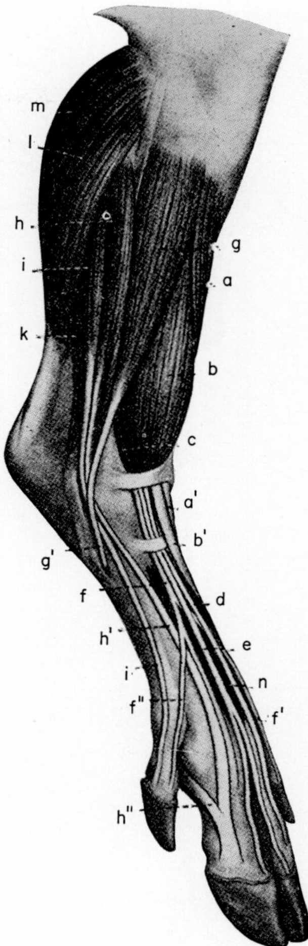

Fig. 15 Dorsolateral view of muscles of the hind limb of the pig. a, tibialis anterior; a′, tendon of preceding; b, peroneus tertius; b′, tendon of b; c, long digital extensor; d, e, f, f′, f″, tendons of c; g, peroneus longus; g′, tendon of g; h, extensor digiti quarti; h′, tendon of h, which receives h″ from the interosseus medius; i, extensor digiti quinti; k, deep digital flexor; l, soleus; m, gastrocnemius; n, extensor digitalis brevis. After Ellenberger, in Leisering's Atlas (from "The Anatomy of Domestic Animals" by Sisson and Grossman, 1953, by permission of W. B. Saunders Company).

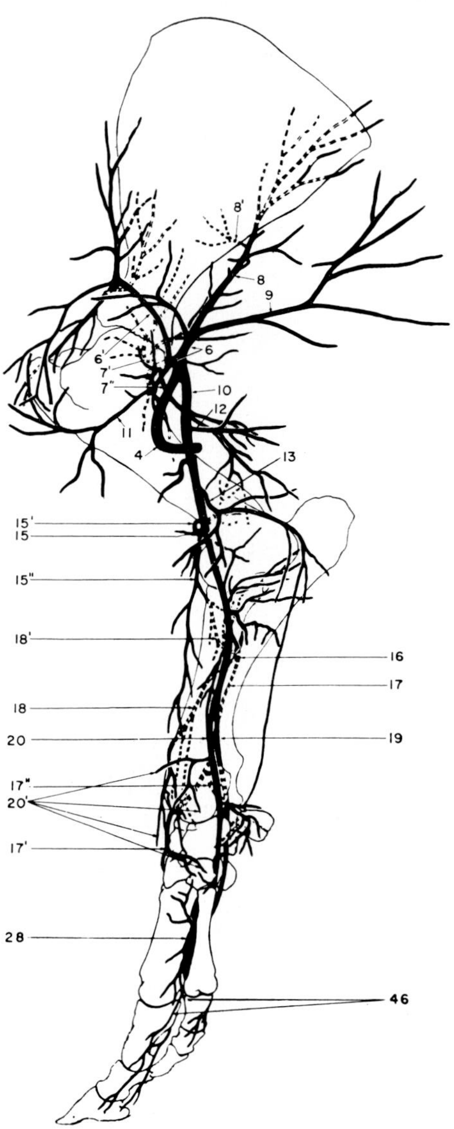

Fig. 16 Medial view of the arterial blood supply to the thoracic limb of the pig. 4, A. axillaris; 6, Truncus subscapularis; 6′, A. thoracoacromialis; 7′, Ramus proximalis of A. circumflexa humeri caudalis; 7″, Ramus distalis (A. collateralis radialis proximalis) of A.; 8, A. subscapularis; 8′, A. circumflexa scapulae; 9, A. thoracodorsalis; 10, A. brachialis; 11, A. circumflexa humeri cranialis; 12, A. profunda brachii; 13, A. collateralis ulnaris; 15, A. collateralis radialis distalis; 15′, Ramus proximalis; 15″, Ramus distalis; 16, A. interossea communis; 17, A. interossea caudalis; 17′, Ramus dorsalis; 17″, Ramus palmaris; 18, A. interossea cranialis; 18′, A. interossea recurrens; 19, A. mediana; 20, A. radialis; 20′, Rami-carpei dorsalia; 28, A. digitalis dorsalis communis II; 46, Aa, digitales II et III dorsales propriae (from Ghoshal and Getty, 1968, by permission of Iowa State Journal of Science, Vol. 43, pp. 125–152, "The arterial blood supply to the appendages of the domestic pig (Sus scrofa domesticus)").

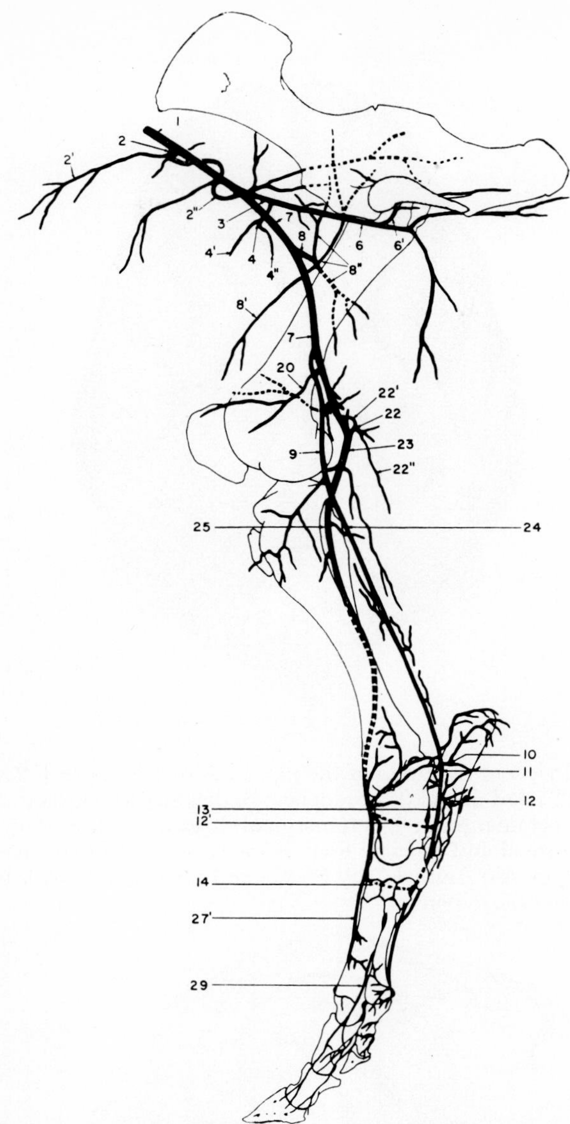

Fig. 17 Medial view of the arterial blood supply to the pelvic limb of the pig. 1, A. iliaca externa; 2, A. circumflexa ilium profunda; 2′, Ramus ascendens; 2″, Ramus descendens; 3, A. profunda femoris; 4, Truncus pudendo-epigastricus; 4′, A. epigastrica caudalis; 4″, A. pudenda externa; 6, A. circumflexa femoris medialis; 6′, Ramus obturatorius; 7, A. femoralis; 8, Truncus communis; 8′, femoris cranialis; 8″, A. circumflexa femoris lateralis; 9, A. saphena; 10, A. tarsea lateralis; 11, A. tarsea medialis; 12, A. plantaris lateralis; 12′, Ramus perforans proximalis accessorius; 13, A. plantaris medialis; 14, Ramus perforans proximalis; 20, A. genus descendens; 22, A. femoris caudalis; 22′, Ramus ascendens; 22″, Ramus descendens; 23, A. poplitea; 24, A. tibialis caudalis; 25, A. tibialis cranialis; 27′, A. metatarsea dorsalis III; 29, A. digitalis dorsalis pedis communis II (from Ghoshal and Getty, 1968, by permission of Iowa State Journal of Science, Vol. 43, pp. 125–152, "The arterial blood supply to the appendages of the domestic pig (Sus scrofa domesticus)").

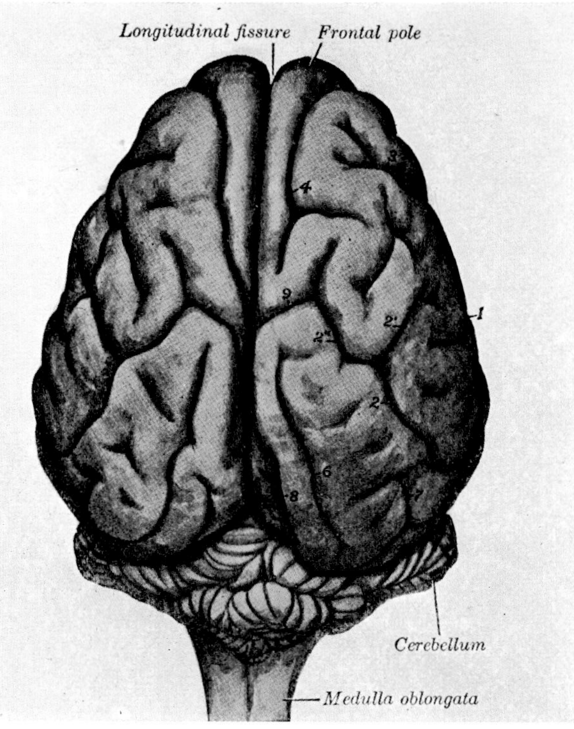

Fig. 18 Dorsal view of the brain of the pig. Fissures: 1, Lateral, 2, suprasylvian, with anterior (2′) and dorsal (2′) branches; 3, diagonal; 4, coronal; 5, presylvian; 6, marginal; 7, ectomarginal; 8, entomarginal; 9, cruciate. Two different arrangements of the coronal and cruciate fissures are seen on the two sides (from "The Anatomy of Domestic Animals" by Sisson and Grossman, 1953, by permission of W. B. Saunders Company).

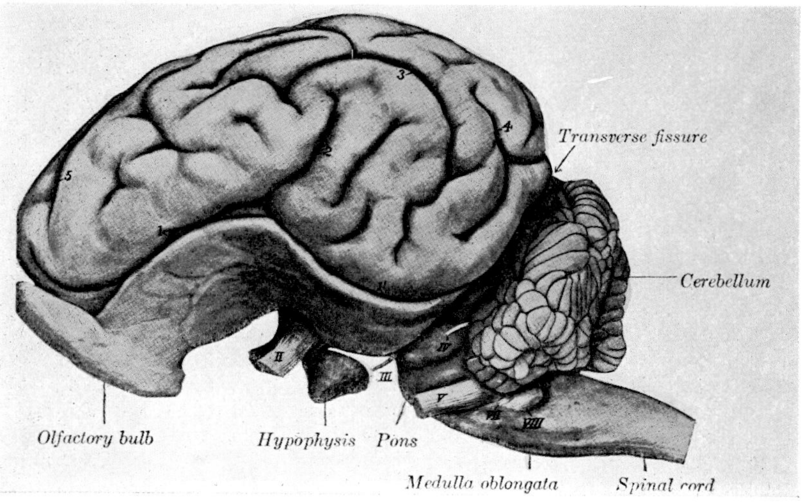

Fig. 19 Left view of the brain of the pig. Fissures: 1, 1′, sulcus rhinalis; 2, lateral; 3, suprasylvian; 4, ectomarginal; 5, coronal. Stumps of cranial nerves indicated by Roman numerals (from "The Anatomy of Domestic Animals" by Sisson and Grossman, 1953, by permission of W. B. Saunders Company).

CHAPTER 7

Physiology

In this section are outlined some aspects of the physiology of the pig which are of significance in the use of the animal for experimental purposes. There is consequently no attempt at a comprehensive treatment; the aim has been to give normal values for the pig where possible, and to indicate in which respects the pig differs from other animals. Other accounts of the pig's physiology are available (Dunne, 1970; Mount, 1968). Techniques for the cannulation of vessels and the collection of samples of blood and lymph are described in Chapter 9.

Blood Pressure

The literature relating to the measurement of blood pressure in the pig has recently been reviewed by Hörnicke (1966a, b), who has derived a formula for the prediction of systolic blood pressure in conscious animals, based on data from Ježková (1960), Engelhardt *et al.* (1961), and Engelhardt (1963). This formula predicts systolic blood pressure for pigs weighing between 1 and 400 kg, although direct information is lacking for animals weighing between 8 and 40 kg: Systolic blood pressure (mm Hg) $= 49 + [46 \times (\log \text{body weight in kg})]$.

The values predicted by this formula tend to be lower than those originally published by Dukes and Schwarte (1931), and Engelhardt (1966) suggests that this may be because the animals used by Dukes and Schwarte were excited, or because the technique they used affected the carotid sinus (Adams, 1958). Schmidt (1968) has investigated the response of the blood pressure to stimulation of the aortic nerve.

Pulse pressure is estimated by Engelhardt (1966) at 30–45 mm Hg on the basis of several observations including his own and those of Ježková (1960), Booth *et al.* (1960) and Evans *et al.* (1963). Nevertheless, Engelhardt believed that the results needed further confirmation, as some investigators had made only a few observations, and on account of the technical difficulties involved in measuring diastolic blood pressure. Calculation from the expression given for systolic blood pressure gives a value close to 120 mm Hg for the 40 kg pig; with a pulse pressure of 40 mm Hg the diastolic pressure is consequently about 80 mm Hg.

According to Wachtel *et al.* (1963), the pressure in the right ventricle is relatively high in the conscious pig as compared with man, cat and dog, but lower than for ruminants. However, in young pigs under barbiturate anaesthesia this does not appear to be so (Evans *et al.*, 1963). The need to take great care in recording the blood pressure of conscious pigs is emphasized by experiments in which responses to fear, disturbance and food supply have been investigated (Engelhardt *et al.*, 1961; Wachtel, 1963; Wachtel *et al.*, 1963; Smith *et al.*, 1964).

Pulse Rate

The rate at which the heart beats in the resting conscious animal depends in part, like blood pressure, on the degree of excitement. Engelhardt (1966) has collected results for unanaesthetized pigs from several sources and has arrived at a formula relating body weight to heart rate per min:

$$\text{Heart rate} = 216 \times (\text{body weight, kg})^{-0\cdot178}$$

The relation can be expressed as a straight line if both heart rate and weight are plotted on logarithmic scales (Fig. 1). The rate tends to decline from about 200/min at birth to less than 100/min at 300 kg, although there

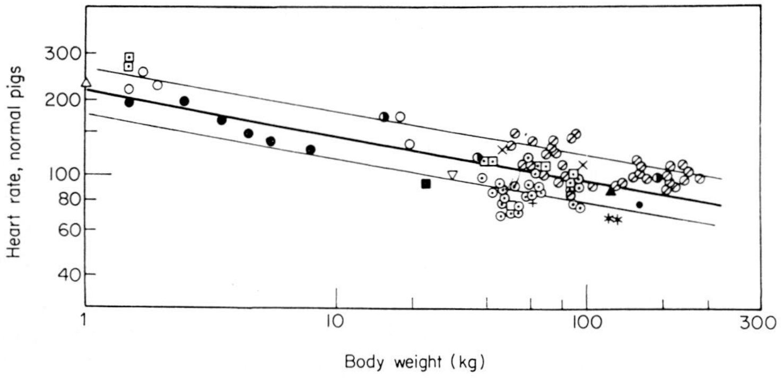

Fig. 1 The heart rate of resting pigs in relation to body weight. Both coordinates are logarithmic. The regression line has been calculated from data taken from various authors (by permission from Engelhardt, 1966).

is considerable variation. Part of the variation is related to disturbance of the animal; experience shows that considerable care is needed in order not to excite the pig (Ingram, 1964a). Even when the animal is lying down and apparently relaxed, examination with a stethoscope induces an initial increase in pulse rate which declines again within a minute. When taking records by electrocardiogram, electrodes which clip on to the animal are best avoided, and the animal must be well trained for the procedure. It should be housed in a separate room with the recording equipment outside. Under these conditions, the effects of noise or disturbance (e.g. opening the door of the animal room) are distinguishable when recordings are taken continuously, and allowance can therefore be made.

Two other factors which affect pulse rate, apart from excitement and age, are starvation (Smith *et al.*, 1964) and ambient temperature (Robinson and Lee, 1941; Heitman and Hughes, 1949; Ingram, 1964a). Starvation results in a decrease in heart rate, but the effects of ambient temperature depend on the conditions used. Short term exposure to heat seems to increase pulse rate (Ingram, 1964a), but after a long exposure heart rate declines (Heitman and Hughes, 1949).

Engelhardt (1966) has paid attention to factors affecting the relative length of systole and diastole. He believes that the low diastolic/systolic quotient (0·7–0·8) which is sometimes quoted for pigs is related chiefly to excitement with the result that heart rates are above normal. As in other animals, when heart rate is increased it is the duration of diastole which is reduced.

Cardiac Output

Determinations of cardiac output in pigs have been made chiefly on anaesthetized animals. The figures published by Widdowson and McCance (1955), Attinger and Cahill (1960), Wachtel (1963), Stowe and Good (1961) and Holt *et al.* (1962) cover animals weighing 7–240 kg, and are shown in tabular form in Table I. When expressed on a body weight basis, cardiac output decreases with weight increase. As Engelhardt points out, the figures published by Attinger and Cahill (1960) are much smaller than those of other workers, possibly on account of the anaesthetic used or because Attinger and Cahill used the dye dilution method while the others, with one exception, used the Fick principle.

Figures of cardiac output for conscious animals are available only for the miniature pig and have been published by Stone and Sawyer (1966). Their

Table I

Cardiac output in the pig (*by permission from Engelhardt, 1966*)

Body weight	Cardiac output (ml/kg.min)	No. of animals	Method	Reference
Domestic pig				
7·2	167	6	Fick[a]	Widdowson and McCance, 1955
28 ± 3·1	68·3 ± 19	19	Dye dilution[a]	Attinger and Cahill, 1960
20 to 24	144 ± 23	6	Fick[b]	Wachtel, 1963
32 ± 10·6	146 ± 53	7	Fick[c]	Stowe and Good, 1961
43 to 53	133 ± 8·7	10	Fick[b]	Wachtel, 1963
91 to 96	72 ± 5·1	5	Fick[b]	Wachtel, 1963
125 to 240	30 to 35[d]	3	Electrical conductivity[c]	Holt *et al.*, 1962
European wild pig				
41 to 58	82 ± 18	4	Fick[b]	Wachtel, 1963
75 to 90	66 ± 4·0	4	Fick[b]	Wachtel, 1963

Narcosis: [a]Pentobarbital; [b]Thiobarbiturate; [c]Chloralhydrate.
[d]Calculated from their graphs.

value of 5·34 litre/min for a 38 kg animal is at the upper end of the range for anaesthetized animals when expressed in ml/kg.min.

Blood Gases

Values for the percentage saturation with oxygen and the CO_2 content of blood at various levels of PO_2 and PCO_2 have been obtained by Bartels and Harms (1959) for the pig and other species. The oxygen dissociation curves which they calculated from their results are presented in Fig. 2.

Information about the relation between pH, CO_2 content and PCO_2 of blood can be found in the tables published by Bartels and Harms (1959). In general, however, the relation is similar to that for man and dog.

Blood Volume

Blood volume per unit of body weight falls, as the pig grows, from 9·5 to 5·6 ml/100 g according to Bush *et al.* (1955a), but in another breed the

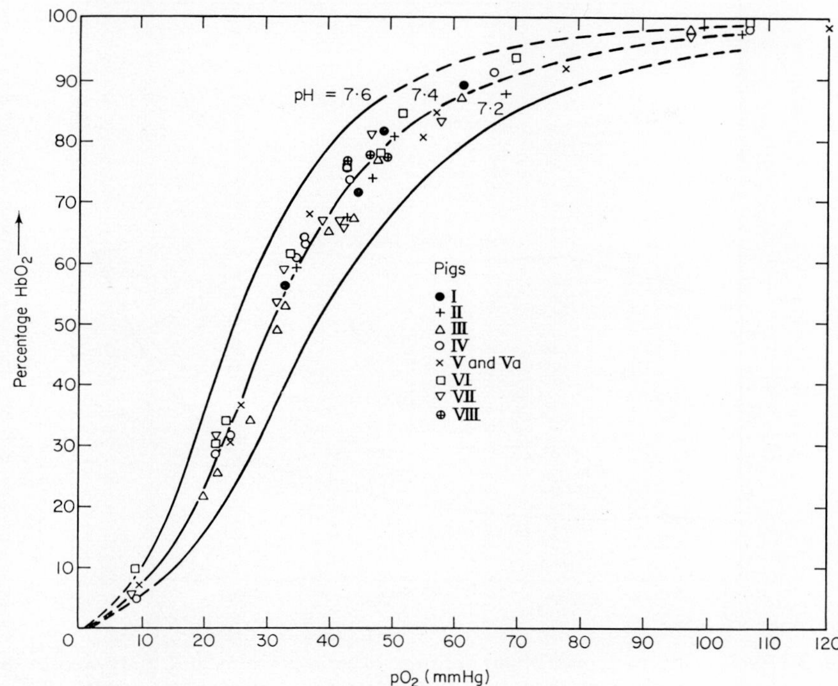

Fig. 2 Oxygen dissociation curves for the pig (by permission from Bartels, H. und H. Harms: Sauerstoffdissoziationskurven des Blutes von Säugetieren. *In* Pflügers Archiv, Bd. 268, S. 334–365. Berlin-Göttingen-Heidelberg: Springer 1959).

decrease was 7·4 to 4·6 ml/100g (Hansard *et al.*, 1951a). In the unweaned pig, Ramirez *et al.* (1962) have measured variations during the first few hours after birth when there is a slight rise (8·6 to 9·6 m./100 g), and then over the following 8 days when the volume is fairly steady at 9·6 to 10·1 ml/100 g. In another study by Ramirez *et al.* (1963), blood volumes were measured over the first 5 weeks of life (Fig. 3).

The relation of blood volume to bodyweight can be found from the following modifications of Engelhardt's (1966) formulae:

$$\text{Blood volume (litres)} = 9{\cdot}5\ (\text{body weight, kg})^{0{\cdot}932}$$

for animals up to 25 kg, and

$$\text{Blood volume (litres)} = 17{\cdot}9\ (\text{body weight, kg})^{0{\cdot}73}$$

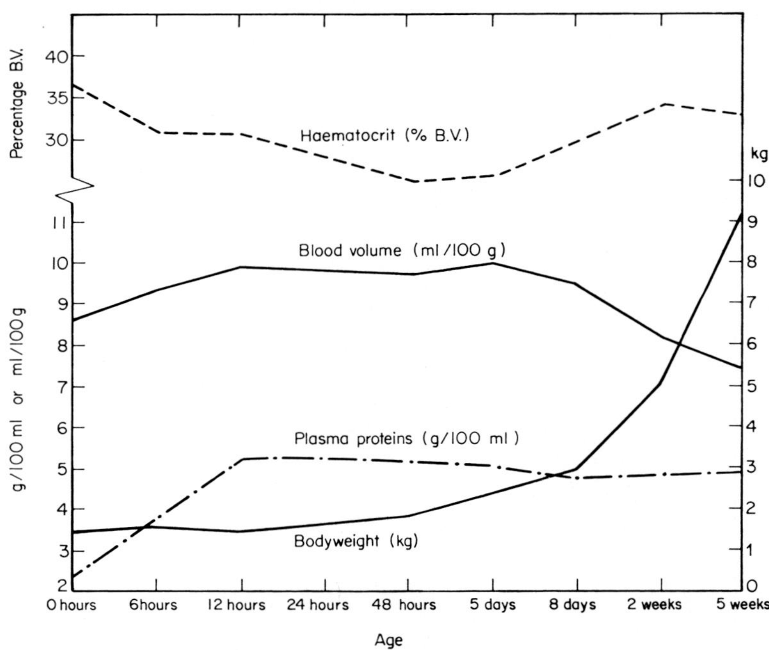

Fig. 3 Levels of haematocrit, blood volume, plasma proteins and body weight in the pig during the first 5 weeks following birth (from Ramirez, Miller, Ullrey and Hoefer, 1963, by permission of Journal of Animal Science).

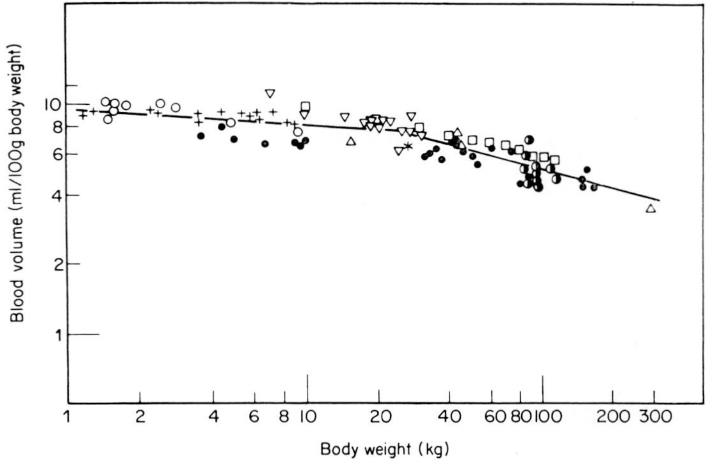

Fig. 4 The blood volume in swine in relation to body weight. Both co-ordinates are logarithmic (by permission from Engelhardt, 1966).

for animals weighing 25–300 kg. The results published by many authors (Hansard *et al.*, 1951a; Bush *et al.*, 1955a; Hansard, 1956; Doornenbal *et al.*, 1962; Kornegay *et al.*, 1963; Talbot and Swenson, 1963a) are summarized in Fig. 4 which is plotted on logarithmic co-ordinates.

New-born pigs are liable to develop an iron-deficiency anaemia unless they are treated; such treatment, or lack of it, is likely to influence the results of experiments, especially those involving measurement of blood volume or haematocrit.

Haematology

General accounts of the haematology of the pig have been published by Scarborough (1930), Miller *et al.* (1961a, b) and Schalm (1965). The variation in the blood pictures with age are summarized in Figs 5 and 6. Changes in the haematology of the perinatal pig have been discussed by Brooks and Davis (1969).

ERYTHROCYTES AND HAEMOGLOBIN

The erythrocyte number decreases rapidly along with a decrease in haemoglobin concentration after birth, and tends to fall again later at weaning.

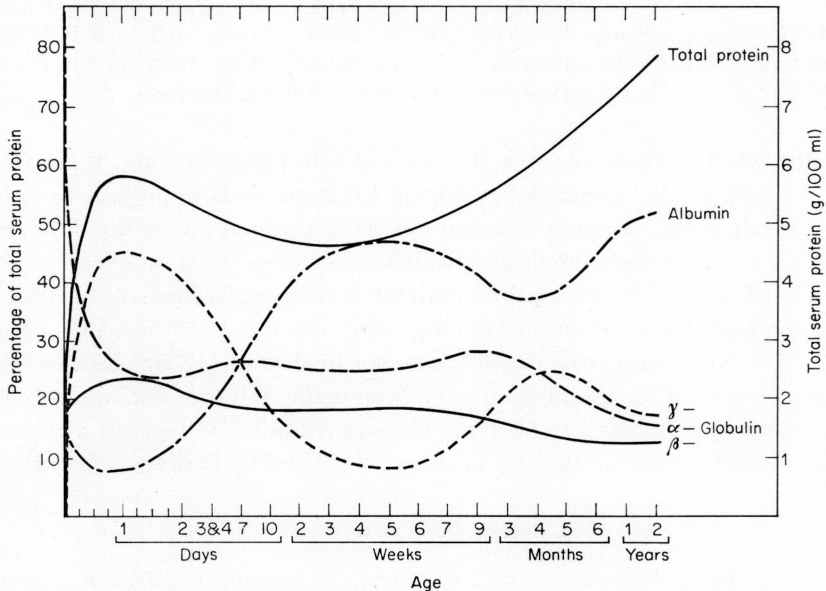

Fig. 5 Variations in serum proteins in the pig in relation to age (from Miller *et al.*, 1961a, by permission of Journal of Animal Science).

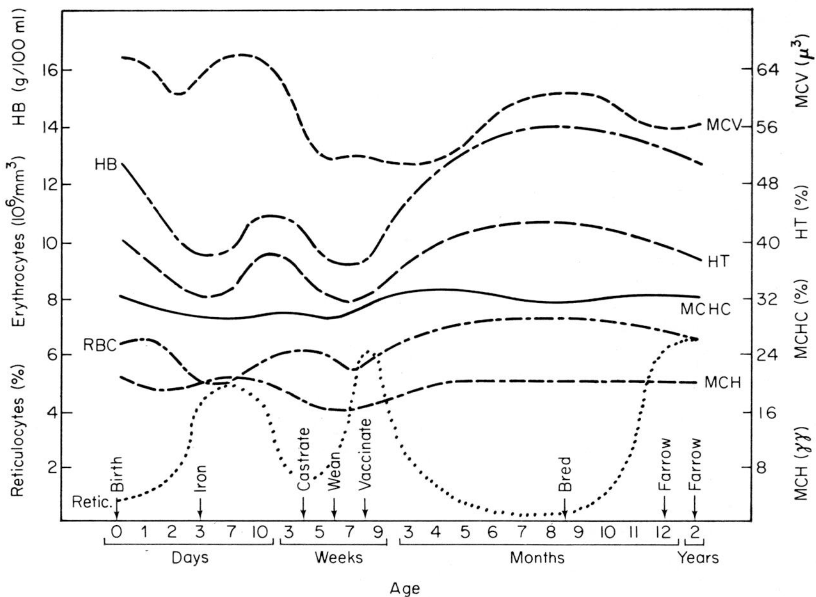

Fig. 6 Variations in the cellular elements of the blood of the pig in relation to age. MCV=mean corpuscular volume; MCH=mean corpuscular haemoglobin; MCHC=mean corpuscular haemoglobin concentration; RBC=red blood corpuscle; RETIC=reticulocytes; HT=haematocrit; HB=haemoglobin (from Miller *et al.*, 1961b, by permission of Journal of Animal Science).

In the adult a slight anaemia develops in late pregnancy and there is a fall in corpuscular haemoglobin during lactation. It is important to note that much of the variation reported may be accounted for by the different degrees to which the animals are disturbed in taking blood samples.

The life span of the pig erythrocyte has been studied by Bush *et al.* (1955b) and by Talbot and Swenson (1963b), using glycine-2-C^{14} and Cr^{51}. The half-life value appears to be 25–60 days, but Bush *et al.* (1955b) believe that after corrections for the method have been made, the true potential life of the erythrocyte is nearer to 86 days. The susceptibility of the erythrocyte to haemolysis has been studied by Hudson (1955) and by Perk *et al.* (1964).

SEDIMENTATION RATE

Values for sedimentation rate are given by Schalm (1965) for pigs of various ages during pregnancy and parturition. The rates are very variable but tend to be higher than for other species.

WHITE CELLS

The total leucocyte count is in the region of 20,000 per mm^3, but the figures which have been published indicate a wide variation. For detailed information see Fraser (1938), Venn (1944), Luke (1953), Gardiner *et al.* (1953), Dunne (1963), Schalm (1965) and Lie (1968). Differential counts indicate a high value for neutrophils shortly after birth, but the number then falls while the percentage of lymphocytes increases. In the adult, according to Scarborough (1930), the value for lymphocytes is 52%, while Luke (1953) gives 66%. Scarborough's value for polymorphs is 41% and Luke's 32%. Eosinophils make up 1·6–4·5% (see also Craft and Moe, 1932; Gardiner *et al.*, 1953).

BLOOD GROUPS

There is a considerable literature dealing with blood groups and serum types in animals, and reviews and recent findings are published by the European Society for Animal Blood Group Research which devotes a section to pigs. The subject is developing rapidly, and more blood factors and the genes which control them are described each year.

In the pig, fourteen blood group systems have been described which are labelled A to O, and well over thirty blood-typing reagents have been prepared. For practical purposes, the systems can be divided into two sorts: the A system, in which naturally occurring antibodies have been demonstrated, and other systems, B to O, in which the naturally occurring antibodies are very weak.

The A system

For the purposes of blood transfusion in man it is necessary to distinguish three primary blood factors which occur in any combination giving rise to groups A, B, AB and O. In the pig, the corresponding factors are limited to A and O (Andresen, 1962, 1963; Hŏjny and Hăla, 1964), just as in cattle there is a J–O system and in sheep an R–O system (Sprague, 1958). The two genes in the pig allelomorphic system are known as A^a and a^o, and since A is dominant only two phenotypes A and O would be expected. A^aA^a and A^aa^o both give rise to group A, and a^oa^o gives rise to group O. The situation is, however, rather more complicated since some pigs have blood which does not react with either anti-A or anti-O, and so belong to a third phenotype group i. Moreover, some pigs may have cellular A substance even though it does not occur in either parent. These facts lead to the hypothesis that the A–O system is affected by the epistatic action of other genes as has been described for sheep (Rendel *et al.*, 1954). This second set of alleles S and s provides that animals with genes SS or Ss express the A–O system, but that an animal with genes ss supresses the A–O system. Thus a pig with genes ss and A^aA^a would be group i (Rasmussen, 1964). A further complicating factor is that A-positive material can be found in the saliva (Goodwin and Coombs, 1956). It appears that the soluble A substance is primary material which is later adsorbed onto the red cells, possibly when the necessary enzymes are present (Andresen, 1963).

The use in experimental cardiovascular studies of mixed blood collected from pigs killed in a slaughter-house has been described by Guiney (1965).

The B to O systems

In the remaining systems any naturally occurring antibodies are weak, and in blood transfusions they would therefore prove troublesome only if a second transfusion was made from the same animal. The remaining systems, however, provide useful labels with which to study the genetics of pig populations. Moustgaard and Hesselholt (1966) in a review of the subject gave a table which listed the various blood group systems, the blood factors, and the genes which control them. The subject advances so rapidly, however, that the table was made obsolete in the same year.

HAEMOLYTIC DISEASE OF THE NEW-BORN

Incompatibility between mother and offspring resulting in haemolytic disease has been reported in the pig by Bruner *et al.* (1949). According to Goodwin *et al.* (1955), Goodwin and Saison (1957) and Joysey *et al.* (1959), this is probably caused by vaccination with material from pooled pig blood, although Andresen *et al.* (1965) and Andresen and Baker (1963) believe it may also be caused by anti-B^a.

LYMPH

According to Binns and Hall (1966), the efferent lymph of pigs is apparently unique among species studied hitherto in containing very few cells (less than 10–250 per mm^3), although flow rates and protein and lipid concentrations are at usual levels. Cell types in efferent lymph are similar to those in other species. The lymphocytes of the pig show the typical proportionate increase and marked transformation which are seen in the lymph of other species following antigenic stimulation of the draining lymph node. Intestinal lymph contains slightly larger numbers of cells than efferent lymph, and has a milky appearance due to absorbed intestinal lipids. Thoracic duct lymph contains large numbers of erythrocytes and blood leucocytes and is apparently contaminated through a lymphatic-venous anastomosis. This contamination probably accounts for the tendency for clots to form in chronic thoracic duct fistulae.

RESPIRATION

The pig does not sweat (Ingram, 1964b, 1967) and therefore depends on an increased respiratory frequency in order to increase evaporative heat loss. Information about respiration must therefore be related to the environmental temperature at which the measurements are made. At temperatures below the critical temperature the resting pig aged 3 months takes 12–18 breaths per min but with activity or slight excitement this may increase to 30/min. Above the upper limit of the thermoneutral zone the frequency increases and under very hot and humid conditions may approach 400/min (Ingram, 1964a, 1965a).

Figures for tidal volume within the thermoneutral zone for the 90 kg pig are given by Morrison *et al.* (1967) as 790 ml; smaller pigs of 25 kg have a tidal volume of about 420 ml (Ingram and Legge, 1970). The quantity of air taken at each breath again depends on the environmental temperature; in a cold environment it increases slightly to meet the extra demand for oxygen, and when hot it decreases with an increase in frequency (Fig. 7). At high body temperatures there is no tendency for the pattern of breathing to change to deeper slower breathing as has been reputed for other panting species, but as in other panting animals the pH of the blood rises as ventilation rate increases.

KIDNEY

The renal physiology of the pig has been reviewed and compared with that of man and dog by Nielsen *et al.* (1966), who concluded that in many respects the pig's kidney is similar to man's, although there are important differences with respect to creatinine which in the pig is reabsorbed by the proximal tubule.

The principal anatomical features of the kidney are that it has both long

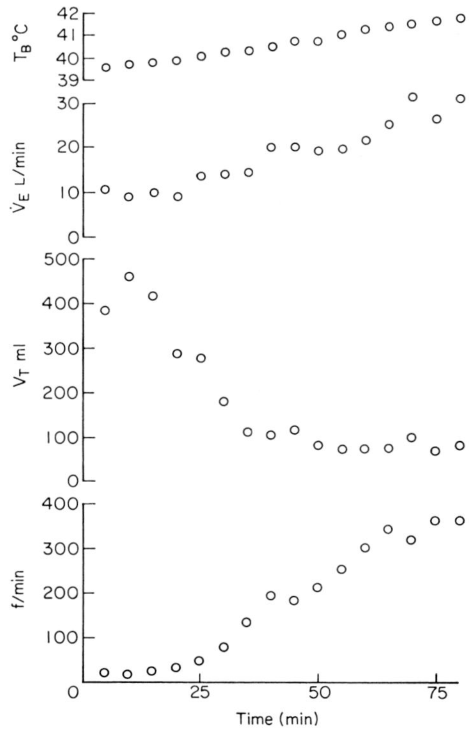

Fig. 7 The relation of frequency of breathing to tidal volume (by permission from Ingram and Legge, 1970).

and short nephrons although most (97%) are short; the relative medullary thickness is only about half that of man but nevertheless the pig is able to concentrate urine to the same extent (Schmidt-Nielsen and O'Dell, 1961). Published figures for the physiological characteristics of the pig's kidney give the maximal osmolal concentration of urine as 1060–1080 mOsm/kg, and the maximal urine/plasma osmolal ratio as 3·7; both values are comparable with those found for man (Nielsen *et al.*, 1966). According to Schmidt-Nielsen *et al.* (1961), loading with urea has no effect on renal concentrating ability in the pig. Inulin clearance has been reported as 5·0 ml/kg.min while renal plasma flow is 19·5 ml/kg.min according to Munsick *et al.* (1958). Nielsen *et al.* (1966) suggest that this latter figure, which was obtained with para-aminohippuric acid, may have to be amended since this substance is acetylated by the pig.

METABOLISM

The rate at which a resting animal uses oxygen depends on its size, state of nutrition, time since feeding, time of day, environmental temperature and on the presence of other pigs or bedding. Measurements can be made either by collecting expired air from a mask over the pig's mouth, or by using a respiration chamber.

For new-born pigs the minimum metabolic rate is observed at about 35°C (Mount, 1968) and this critical temperature falls to around 25°C by the time the animal is 3 months old (Ingram, 1964b).

Variations in metabolic rate over a 24 hour period have been reported by Cairnie and Pullar (1959), while the effects of severe under-nutrition have been studied by McCance and Mount (1960). For a full discussion of metabolic rate in the pig see Mount (1968). The deep body temperature of the pig is normally in the region of 39°C.

DIGESTIVE SYSTEM

Investigations into the digestive enzymes of growing pigs have been given impetus by the problems involved in the early weaning of animals on a commercial scale. Proteinase, amylase, maltase and sucrase activities are low during the first 2 weeks after birth and then rise with age; a consequence of the lack of invertase, or sucrase, is that sucrose administered to the piglet acts as a purgative, causing diarrhoea (see Chapter 5).

Pancreatic tributyrinase is relatively high at birth but nevertheless increases with time. By contrast lactase is high in activity at birth and declines with age to reach a constant level at 3 weeks (Bailey *et al.*, 1956; Kitts *et al.*, 1956; Braude *et al.*, 1958; Hartman *et al.*, 1961).

Experiments on the digestibility of carbohydrates and sugars and on the effect of adding enzymes have been carried out by Cunningham and Brisson (1957a,b) and Cunningham (1959). They found that the rate of digestion of starch was not improved by the addition of amylase in the new-born pig probably because the factor limiting starch digestion is the rupture of the granule itself. Soluble starch, on the other hand, is digested at about the same rate as maltose. A useful technique often employed in digestibility studies in pigs is that in which chromium oxide is added to the meal. Clawson *et al.* (1955) have shown that results obtained with this method are close to those in which a total faeces collection is made.

The rate of passage of food through the gut has been studied using stained non-digestible particles in the food and noting the time taken for this to be cleared from the body. For growing pigs the mean clearance time of half the coloured material was about 25 hours and was not affected by the live weight of the animal. Breeding sows on the other hand retained food in the gut for longer periods. The time taken for the food to pass through the gut varies with the size of meal, the larger meals moving more quickly. This relation is true whether the bulk is made up by solid meal or water (Castle and Castle, 1956, 1957). Kidder *et al.* (1961) made radiographic studies on the alimentary tracts of piglets using barium meal. The meal passed through the intestines of young pigs weaned at 3 days of age more slowly than in sow-reared piglets of the same age.

CHAPTER 8

Anaesthesia

The chief problems of anaesthesia in the pig arise from the size of the animal and the difficulties associated with restraint and handling. As Vaughan (1961) points out, the economic value of the average bacon pig does not warrant anything beyond simple surgical operations for therapeutic purposes, and as a consequence the veterinary information available on anaesthesia and surgery in pigs is less than that for the horse or the ox. In spite of this, much attention has been given to pigs in recent years and general accounts of anaesthesia are available (Hill and Perry, 1959; Vaughan, 1961; Westhues and Fritsch, 1964; Hall, 1966).

Restraint and Handling

Animals of up to about 25–30 kg can usually be restrained manually, for the purpose of giving a gaseous anaesthetic or an intraperitoneal injection, providing that two people are available. One method of restraint for the administration of gaseous anaesthesia is to hold the pig under one arm while the forelimbs are restrained by the other hand. A second person further restrains the head and pushes the mask over the snout. As soon as the anaesthetic begins to take effect, usually in less than a minute, the animal can be placed on a table. Larger animals may be restrained if an assistant grasps the animal's forelimbs from behind and holds its body between his legs. Again a second person holds the mask over the snout while restraining the animal's head. The noisy protest which such handling inevitably evokes is the common accompaniment of almost any manipulation of the pig.

For the administration of an intraperitoneal injection, an assistant should hold the pig off the ground by the hind limbs with the belly away from his body; the animal's head may be held between the assistant's legs although this should not be necessary. The needle can then be inserted to one side of the mid-line about midway between the diaphragm and the navel.

Intramuscular injections can be administered while the assistant holds the animal off the ground by the hind limbs with the animal's belly towards his body. The injection can then be given into the rump. As an alternative if the animal is in a cage or trolley with an open top the needle can be introduced into the rump if the rump is first tapped three or four times sharply with the ends of the fingers. The needle can then be inserted without undue reaction on the part of the animal.

Restraint of large animals provides a much greater problem. It is of considerable advantage to have the assistance of the attendant who feeds and tends the animals, and who will often be able to manoeuvre the animal into the cage or trolley with the minimum of disturbance. Use can be made of the fact that if a pig is held by a loop of rope round the upper jaw it will always pull backwards and so tend to tighten the loop. Miller and Robertson (1952) describe various "twitches" for the pig which consist in principle of a loop of rope on the end of a rod of up to 1 metre in length. Once the upper jaw has been secured in a loop the other end of the rope can be passed through a ring fixed to the wall, or round a firmly fixed pipe. Intramuscular injections can be given as described above or into the muscle behind the ear; Paulick *et al.* (1967) give a series of dissections in which sites for injection are indicated. Gaseous anaesthetics can be administered while animals are held by this method by threading a specially made mask on to the rope before the rope is secured.

Miller and Robertson (1952) describe a method of casting sows and boars by means of ropes round the upper jaw and round the feet. For the administration of anaesthesia, however, this should not be necessary, especially if the animal has been treated previously with a tranquillizer (see below).

In experiments dealing with large animals, full attention should be given to the problem of handling the pig once it is anaesthetized. Considerable saving of effort can be obtained by ensuring that as the animal becomes unconscious it falls on to a low trolley or stretcher. One such trolley, which is described by Hill and Perry (1959), is illustrated in use in Fig. 1. It is also useful to remove the animal from the group a day or two before surgery, to allow it to become accustomed to the surroundings in which it will recover after the operation.

PREMEDICATION

When dealing with large animals some premedication is to be recommended. This makes the process of anaesthetizing an animal much easier, and avoids a struggle which is particularly undesirable for the investigator

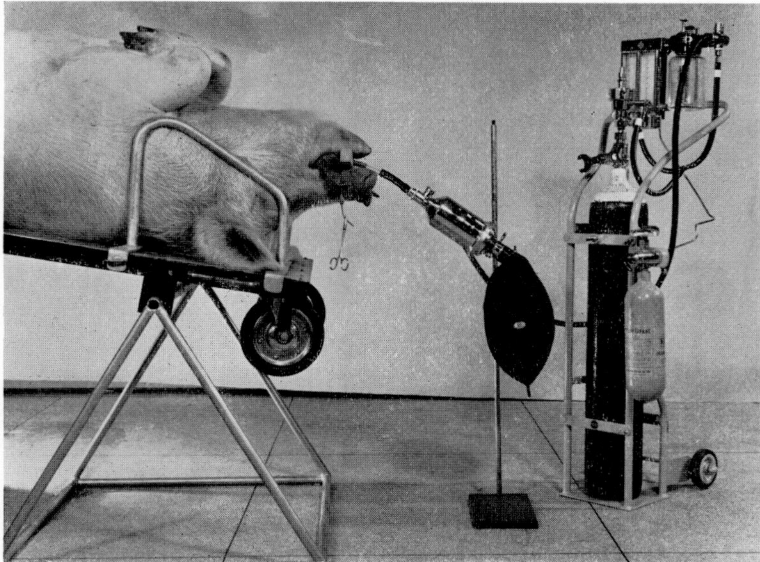

Fig. 1 Large pig in position on a low trolley which has been raised onto a stand (after Hill and Perry, 1959).

before he undertakes any surgery. Intravenous injections into the ear are very difficult without the use of a tranquillizer since the slightest movement will bring the needle out of the vein.

Tranquillizers

These are not in themselves anaesthetics but simply produce sedation without drowsiness. The method of action appears to be a specific depression of the reticular system in the brain and hence various stimuli which are normally capable of causing arousal are not relayed to the cortex. In addition these drugs have a variety of side-effects such as hypothermia and hypotension, and some have anti-histamine and anti-adrenaline activity, although the degree to which these side-effects are manifest varies considerably.

Phencyclidine

A full account of the use of this substance in the pig has been published by Tavernor (1963); it has been used with success in this Institute. The drug is injected intramuscularly in a dose of 2 mg/kg for pigs aged 8 weeks to 2 years, and the animals can be expected to fall over in about 5 min. A positive effect is also achieved by lower doses down to 0·25 mg/kg. The principal side-effect of this drug is stimulation of the salivary gland. The drug is reported to have no adverse side-effects, although when given in high doses respiratory rate and body temperature may increase. Tavernor (1963) has used it on the pregnant sow without ill effects.

Chlorpromazine

A full account of the use of this substance in the pig has been published by Ritchie (1957). The drug is useful in sedating animals before surgery, but it does not render them insensitive to peripheral stimuli. Vasodilatation occurs on administration, and for this reason care should be taken if Fluothane is later used as an anaesthetic as this also causes a fall in blood pressure. Chlorpromazine also has anti-adrenaline, anti-emetic and vagolytic properties.

The doses recommended for intramuscular injection vary considerably from one authority to another; Ritchie (1957) gives 2–4·4 mg/kg, Hill and Perry (1959) used 2–2·5 mg/kg, Vaughan (1961) suggests 4 mg/kg, and Hall (1966) recommends 1·0 mg/kg. The dose used must therefore depend on the degree to which the animal is to be sedated, and whether or not the animal is accustomed to being handled. Half to three-quarters of an hour should be allowed for the full effects to occur; little or no effect is noted on respiratory rate after intramuscular injection.

When given intravenously in the pig in doses of 0·5–3·3 mg/kg for animals weighing 15–227 kg, chlorpromazine may increase respiratory rate to 60–100/min irrespective of whether the dose is given slowly or quickly. The full effect of sedation occurs after 10–20 min; in the majority of instances, however, the intravenous route is not recommended, partly on account of the difficulty of injecting into the ear vein of an untranquillized pig, and partly because unless very dilute solutions are used thrombosis of the ear may occur.

In most instances, pigs are simply sedated by chlorpromazine and are more easily handled, but according to both Ritchie (1957) and Vaughan (1961) a state of general anaesthesia can be produced in some animals. One of Ritchie's (1957) pigs, on the other hand, went into a state of violent excitement after about 4·5 hours. The effect of the drug may not disappear completely for 24–48 hours. There is little or no potentiating action with barbiturates, but the induction time with ether and chloroform is reduced.

Promazine hydrochloride

This drug is similar to chlorpromazine, but for its full effect in the pig it must be given intravenously. The doses suggested by Vaughan (1961) are 2 mg/kg for intravenous injection or 4 mg/kg intramuscularly, but because full effects are obtained only by the intravenous route, it is not the best choice for the pig. Accounts of the use of this drug have been published by Sinclair (1961) and by Johnson (1961).

Trimeprazine tartrate

As with promazine hydrochloride, the best results are obtained after intravenous administration. Trimeprazine produces a less marked fall in blood pressure and less prolonged effects than chlorpromazine. According to Hall (1966), it is generally more effective and preferable to chlorpromazine, but since intravenous injection is not easy in the unsedated pig it probably loses its advantages in this species. Tavernor (1960) reports that after the intravenous injection of 1·0 mg/lb (2·2 mg/kg) sedation is produced in 5–10 min, while 1·5–3·0 mg/lb (3·3–6·6 mg/kg) are needed to produce sedation by the intramuscular route which is less reliable and takes some 45 min to produce the full effect.

Promethazine hydrochloride

This drug has long-acting anti-histamine properties, and is more often thought of in this connection, but it also has a sedative action. According to Hall (1966), the drug is irritant to the tissues, and should be injected deeply in a large muscle mass.

Perphenazine

This is used in doses of 0·2 mg/kg for sedation in the pig.

Steroid anaesthesia

Huhn and Schulze (1961) have reported the use of 21-hydroxy-pregnon-3,20 dion as an anaesthetic agent in pigs. They found that although satisfactory narcosis was produced there was the disadvantage of acute irritation of the tissues.

From this account it will be apparent that the premedication of choice lies between chlorpromazine, which sedates the pig, and phencyclidine, which brings the animal down with a minimum of delay.

Anaesthesia

INTRAPERITONEAL ANAESTHESIA

For small animals which can easily be held up by the hind legs without premedication an intraperitoneal injection of pentobarbitone is in some instances quite satisfactory. A dose of about 20 mg/kg is suggested, but since the level of anaesthesia is difficult to control it is preferable not to use this route for general surgery.

INTRAVENOUS ANAESTHESIA

The only convenient veins for intravenous injections in the pig are those in the ear. In a cool environment it is useful to warm the ear with hot water in order to increase the circulation, and if no assistance is available the venous return can be reduced by placing an elastic band around the base of the ear. The simpler procedure, however, is to have an assistant occlude the vessels at the base of the ear with the finger. As soon as the needle is in the vessel (as can be shown by withdrawing blood into the syringe) the assistant takes his fingers off the vein, or if an elastic band has been used this must be cut. The injection should then be given slowly. It is important to have enough anaesthetic in the syringe since during re-filling the needle may inadvertently leave the vein; for the same reason, the intravenous procedure can prove very difficult in an unsedated animal.

For operations which are likely to last long enough for a second injection of anaesthetic to become necessary, it is best to insert a catheter into the jugular vein, or into the superior vena cava. Although the jugular vein is not readily visible in the pig, an intravenous catheter can be inserted into this vein in young animals (see pp. 99–103). For the administration of anaesthetics during a long operation it is best to fill the catheter with heparinized saline in order to avoid clotting.

Pentobarbitone sodium

The dose needed for intravenous injection depends on the age of the animal. Vaughan (1961) suggests the following guide:

up to 50 kg	24 mg/kg
50–100 kg	20 mg/kg
100–300 kg	10 mg/kg

Individual pigs vary quite considerably, and for this reason it is best to give the injection slowly and to judge the depth of anaesthesia. Since the pig, unlike some animals, is not subject to an excitement phase during the administration of pentobarbitone, the slowness of administration presents no special difficulties. As the animal loses consciousness the musculature becomes relaxed and the animal can be laid down and the reflexes tested before more anaesthetic is given. If an overdose is given, breathing stops before the heart and it is thus possible to save the animal by artificial respiration. Full surgical anaesthesia lasts only 15–45 min, after which a further injection of about half the original dose will be needed. For operations from which the animal is intended to survive it is better to supplement with a gaseous anaesthetic and to avoid too lengthy a recovery period, which in any event will last 3–8 hours.

Thiopentone sodium

This drug is shorter acting than pentobarbitone, and therefore has an advantage if it is to be used in preparation for the administration of a gaseous anaesthetic or for a short operation. As pointed out by Vaughan (1961) and Kubin (1952), the pig has a very variable tolerance to this drug. Doses of between 10 and 15 mg/kg are suggested. If given very slowly, anaesthesia will last 10–20 min, while faster administration will require less anaesthetic, although the duration of anaesthesia will be shortened. Since this drug is very alkaline and can thus produce necrosis of tissue, care must be taken while making the injection, especially into the ear vein of a conscious animal which may move and bring the needle out of the vessel.

INHALATION ANAESTHESIA

Gaseous anaesthetics may be administered either by means of a mask over the snout or through a cuffed endotracheal tube from a standard Boyle's anaesthetic apparatus. When the anaesthetic is given by mask an alternative to the commercial mask can be made conveniently from a polyethylene beaker. A hole is made in the bottom of the beaker and fitted with a brass tube of appropriate size for connection to the anaesthetic machine; the open end of the beaker is then covered by a piece of rubber sheet with a central circular hole, large enough for the snout to pass through. By making two or three sizes of mask, pigs of up to 40 or 50 kg can be fitted. For a larger animal the mask needs to be specially made. A respiratory mask with a leak-proof facial seal with an inflatable liner has been described (Routledge *et al.*, 1968).

The use of the endotracheal tube avoids any leaks of anaesthetic, but since its insertion is not simple it is dealt with in a later section.

Halothane

The use of this anaesthetic in the pig has been described by various workers including Vaughan (1961), Dziuk *et al.* (1964) and Hall (1966). Its use in the pig is highly satisfactory. For the induction of anaesthesia the air stream is bubbled through the liquid in order to give a high concentration of anaesthetic. Using this method, anaesthesia can be induced in about 3 min. After the pig becomes unconscious, it is necessary only to pass the air stream from the anaesthetic machine over the surface of the liquid. A special container which can be heated and thermostatically controlled is obtainable for halothane, but its use is not essential. Recovery is rapid and uneventful.

This anaesthetic causes a fall in blood pressure, and for some surgery involving the insertion of catheters into blood vessels this may be a disadvantage, although it is not a serious one. A considerable advantage of this anaesthetic is that the vapour is not inflammable, so that cautery or drills and other apparatus employing an electric brush motor may safely be used.

Metafane

This anaesthetic appears to cause less peripheral vasodilation in pigs than halothane. Induction is slow and for large animals some premedication is essential.

Cyclopropane

According to Hall (1966), this anaesthetic is satisfactory in pigs, but it is inclined to make the animal vomit and for this reason pigs recovering from anaesthesia should not be left unattended. The gas is highly inflammable and should be used only on a closed circuit system; preferably it should be given through an endotracheal tube.

Chloroform

This agent can be given to pigs through a mask strapped to the head, and it is useful for short periods of anaesthesia. Its use for long periods of anaesthesia is not recommended.

Ether

This anaesthetic is satisfactory for pigs but may cause excessive salivation.

Nitrous oxide

Hall (1966) states that the use of this gas after thiopentone is very satisfactory, even for some lengthy experimental surgery, producing a light anaesthesia from which recovery is rapid.

SPINAL ANAESTHESIA

For experimental purposes it is not very probable that spinal anaesthesia will be necessary. Accounts are, however, given by Westhues and Fritsch (1964) and by Hall (1966).

ELECTRO-ANAESTHESIA

An additional method of anaesthesia in the pig has been used by Short (1964). In this method electrodes are placed on the head and a current of 100 mA is passed at a frequency of 700 cyc/sec. In the pig, however, the setting required is not uniform for all animals. Nevertheless Short has used this method for hernia repair and ovariectomy, and reports that consciousness returns as soon as the current is turned off.

Control of Anaesthesia

The tests by which the level of anaesthesia can be judged are complex, and vary somewhat from one anaesthetic to another and also with different combinations of agents. It is intended here to give only an outline of the methods which can be used in most experimental laboratories. In the initial stages, before sterile drapes have been used, the carpo-pedal reflex is a convenient indicator; the thumb is pushed between the trotters of the hind limb and the grip tightened, upon which the leg will be withdrawn by a lightly anaesthetized subject. During the operation it is not always convenient to use this test, but, if it is possible to observe the eye, the cornea may be touched with a twist of cotton wool, when a blink will indicate that anaesthesia is light. A very useful index of anaesthesia is the position of the eye. As the animal becomes anaesthetized the eyes roll downwards exposing the whites. As anaesthesia becomes deeper more white is exposed, but in very deeply anaesthetized animals the eye rolls back again to expose the pupil. For most surgical procedures, the level of anaesthesia which corresponds to an eye position with the white visible and regular steady ventilation will be found satisfactory. The depth of anaesthesia required will, however, vary with the type of operation; for example, while some slight movement or chewing on the part of the pig would not be serious in some instances, it can not be tolerated when the animal's head is in a stereotaxic instrument.

If muscle relaxant drugs have been used some of the above reflexes will not function; great care will then have to be exercised in gauging the depth of anaesthesia.

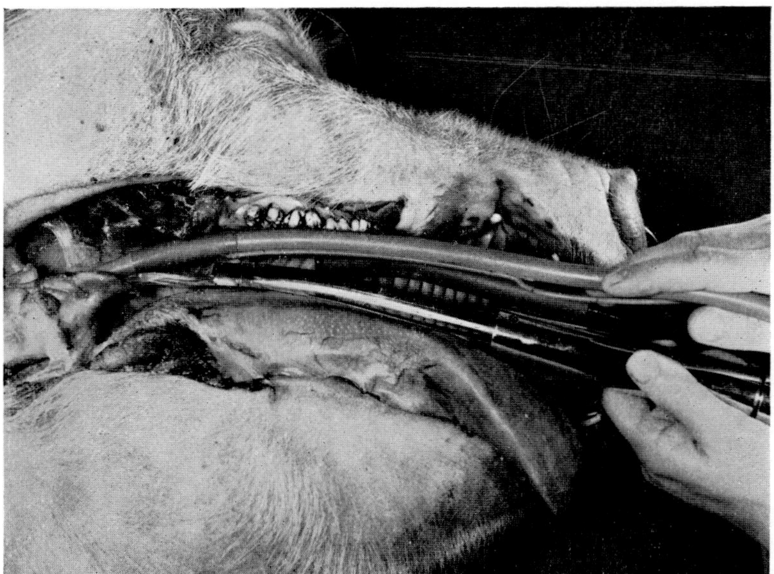

Fig. 2 Dissection of the head of a dead pig to show the relations of the parts to the insertion of an endotracheal tube. The lower hand holds a laryngoscope consisting of a brass battery and lamp holder with a perspex rod which holds the tongue down and transmits light to the larynx (after Hill and Perry, 1959).

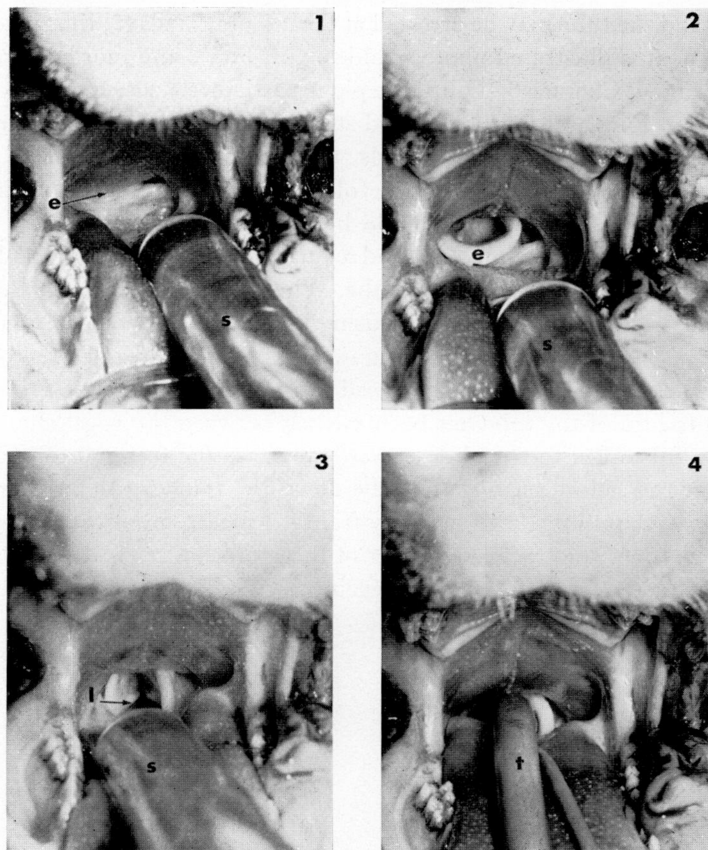

Fig. 3 The steps involved in intubation of the trachea demonstrated in a dead pig.

1. Depression of tongue with laryngoscope reveals closed epiglottis.
2. Epiglottis is opened with end of laryngoscope.
3. Epiglottis is held down with laryngoscope to reveal larynx which in this photograph is relaxed enough to allow endotracheal tube to enter.
4. Endotracheal tube in position.

e, epiglottis; l, larynx; s, laryngoscope; t, endotracheal tube.

Intubation

In pigs the insertion of a tracheal tube is difficult, and although several accounts are available (Hill and Perry, 1959; Rowson, 1965; Hall, 1966) there is no single solution to the problem. First, the vocal chords tend to obstruct the passage of the tube. Second, the larynx is set at an angle to the trachea and the tube may be arrested at the cricoid cartilage; the anatomical arrangement is illustrated photographically in Fig. 2 and more diagramatically in Fig. 6, Chapter 6. Third, the pig is particularly susceptible to laryngeal spasm. For these reasons several attempts may be needed to introduce the tube, especially in young animals.

Attention should be given to the following points in order to minimize these difficulties. The animal should be very deeply anaesthetized, to the point when breathing is slow, deep and regular and has almost stopped, before any attempt is made to insert the tube. When the pig is anaesthetized, the mouth is held open by an assistant using two lengths of cord or a gag, and the tongue is depressed with a laryngoscope. A rod is passed down the tube to the tip and the assembly introduced under direct vision (see Fig. 3). As soon as the tip of the tube has been pushed between the vocal chords it is often useful to flex the head downwards towards the feet and to withdraw the inner rod before passing the tube down the trachea. As an alternative some workers pull the head up away from the feet in order to insert the lip of the tube into the trachea. Neither of these procedures is essential, however, once some experience has been gained.

CHAPTER 9

Surgery

The pig has advantages as an animal for use in experiments involving surgery as the tissue heals well and the animal is resistant to infection. Against this are the large size of the full grown animal and the difficulties associated with intubation of the trachea. The special techniques used in experimental surgery are described by Markowitz *et al.* (1959), and it is intended here to confine comment to work which relates specifically to the pig.

Theatre Sterility and Recovery from Surgery

The pig is not especially subject to post-operative infection, but the cost of experimental work demands that full aseptic surgical techniques should be used whenever possible. The extra trouble which has to be taken is very little compared with the possibility of a valuable animal becoming infected.

The operating theatre should have an annexe or anaesthetics room in which the animal can be prepared. Once the pig has been anaesthetized, the body and particularly the legs and feet can be cleaned, and the feet then covered with plastic or linen bags, before the animal is taken into the theatre. Fully grown animals are very heavy and difficult to manoeuvre once they are anaesthetized, and some type of hoist or lifting gear is very helpful. Some theatres built for large animal work are equipped with an overhead hoist which can be moved along a rail and used to lift the animal. As an alternative, the operating table may be lowered to the floor on a car hoist until the table top is at floor level, when the animal can be moved into position. The hoist is then raised and the table wheeled into the theatre. The pig can also be anaesthetized on a detachable table top which can then be lifted by hand, or with the aid of a hoist, onto the operating table.

RECOVERY

Animals are best left to recover by themselves in a pen, preferably on a thick rubber mattress, with the walls covered by a resilient layer to prevent self-damage to the pig through incoordinated movements during recovery. Equipment in the pen should be kept to a minimum, and no sawdust, shavings or straw which may enter wounds should be used. It has been found that very young animals can sometimes be returned to the mother after surgery; thus Ingram and Ślebodziński (1965) found that the sow always accepted the 2-week-old piglet back after it had been thyroidectomized, and recovery was uneventful. Similar experiences have been reported by K. J. Hill (personal communication), who has been able to carry out surgical procedures on animals within minutes after birth.

Where possible animals which are to be used in special cages or stands are best trained before the operation and accustomed to handling by the experimenter. The pig has one very great advantage at this stage in that it is always interested in food. With the aid of a handful of pellets the experimenter can rapidly accustom the animal to the handling, and after surgery the inspection of wounds can be carried out with the minimum of disturbance. Animals which are taken directly from a farm pig house and subsequently recover from surgery in a strange cage, may on the other hand take some time to adjust to the new conditions. (See also Chapter 3 on handling and cages.)

Digestive Tract

There is considerable literature dealing with the techniques used to insert cannulae and fistulae in dogs, sheep and cattle, but until very recently the pig has been little used in this type of work. The Russian literature, however, shows that these techniques present no problems in pigs of 20–25 kg body weight. Protasenya (1961) has studied digestion and metabolism in pigs using a fistula which was placed in the ileum by an adaptation of the methods developed for the dog. The cannula, which was adapted from the London cannula, was made of stainless steel and polyethylene, and proved to be reliable and stable. The only material difference was that the size had to be increased for use in weaner pigs. Protasenya (1961) also illustrates a simple frame in which the pig was held while samples were taken from the digestive tract.

Techniques being developed in other countries are also based principally on the adaptation of those methods which have proved successful in dogs and sheep. Redman *et al.* (1964) have performed caecal fistulation in sixty pigs of 7–20 kg using the technique described by Markowitz. An opening was made in the abdominal wall, and by separating each layer of muscle in the direction of the fibres the number of sutures needed was reduced because contraction of muscle held the cannula in place. The animals were housed afterwards in smooth-walled pens, and in some instances two animals were housed together for a month without complications. This technique has been used to study the absorption of Vitamin B_{12} by Henderick *et al.* (1964).

Cunningham *et al.* (1962) have used a re-entrant fistula for digestion studies on pigs of 32 kg weight. The animals were fed a diet of skimmed milk for a week before the operation in order to exclude from the digestive tract coarse particles of food, which might block the cannula, and all food and water were withheld for 24 hours before the operation. A 10 cm incision was made in the abdomen to the left of the mid-line just caudal to the navel. The intestine was transected 12·5 cm from the caecum, and the cannulae inserted. The cannulae were then exteriorized through small stab incisions on either side of the mid-line. The cannulae, which are illustrated in the article, were made from polyethylene tubing of 1·27 cm inside diameter, with small semi-circular collars heat-fused to the outer ends to aid in connecting them together with rubber tubing. These cannulae were used over a 5 month period and became blocked only when coarsely ground rations were fed. The technique of gastrectomy in the pig has also been described by Cunningham (1967).

Endocrine Organs

REPRODUCTIVE TRACT

The procedure for the removal of the ovaries and testes may be found in texts on veterinary surgery, and is similar in principle to that for other animals. For experimental purposes, however, it may be necessary to make frequent observations of the ovary in order to follow the development of follicles and corpora lutea. Betteridge and Raeside (1962) have devised a system by which observations can be carried out frequently. A peritoneal cannula large enough to allow the ovary to be drawn through the shaft is fixed into the abdominal wall. The ovary is located and a thread tied to the mesovarium. The thread is then passed through the cannula and the incision closed. A plug of Plexiglass is pushed into the cannula so that it is flush on the inside, and the whole is covered by a screw cap lubricated by antiseptic cream. When required, the ovary can be drawn through the shaft for photography.

Serial observations of the sow's uterus have been made by J. S. Perry (personal communication) by means of a series of four or five laparotomies.

In these experiments, in addition to taking care to avoid adhesions to the body wall, it was also necessary to make each successive incision in a different place, because of the tendency of the wound to calcify. Thus after several operations the abdominal wall may develop a calcified plate which can not be cut with a scalpel.

Sub-total and total hysterectomy have been described by several authors, especially in connection with the effects of removal of the uterus on the survival of the corpora lutea. A ventral mid-line incision is made and the uterine blood supply and cervix ligated. The uterus may then be removed from the mid-cervix to the tubo-uterine junction while avoiding damage to the ovarian blood supply (Spies *et al.*, 1960); the operation is straightforward and without complications (Anderson *et al.*, 1961). Transplantation of the ovary, complete with blood vessels, to an orthotopic site has been described by Binns *et al.* (1967).

THYROID GLAND

The exposure of the pig's thyroid presents no difficulty. The gland is not markedly divided into lobes, but is triangular in shape (see Chapter 6). The posterior end of the gland tends to be drawn out into a long thread and in thyroidectomy it is important to ensure that all this tissue is removed. In animals of a few days of age no ligation of the vessels is necessary. Accessory tissue, however, is widespread and regeneration is sometimes substantial after 35 days, particularly at the junction of the jugular veins (Schmidt *et al.*, 1964). After comparing the effectiveness of surgical thyroidectomy and that produced after radioiodine treatment, Stromlund *et al.* (1960) favoured the latter because their surgical methods never removed all the tissue. As an alternative, Ingram and Slebodzinski (1965) removed the thyroid gland and then treated animals with thiouracil.

ADRENAL GLAND

The operation of adrenalectomy is not easy in the pig on account of the close proximity of the gland to the posterior vena cava blood vessels. The degree to which this occurs varies between animals, and the investigator must be prepared to select his animal at operation. The technique has been described in pigs of 42–60 kg body weight by Kolb *et al.* (1962). The glands are removed in two stages: the right adrenal is first removed, and a second operation is performed at least 14 days later. An incision is made 4 cm lateral to the transverse processes of the lumbar vertebrae, extending from the last rib to the tuber coxae. The kidney is mobilized by blunt dissection and pushed aside centro-medially using large swabs and a retractor, and other retractors are needed to hold back the gut. The adipose tissue is then dissected away towards the spine until the adrenal is visible, the topography varying very much between animals. The gland is now mobilized cranio-caudally. This process is complicated by the ventral surface of the gland adhering closely to the peritoneum, and by the presence of several supra-renal arteries and veins which must be tied off.

Complications may occur from the adrenal gland being situated so far cranially that part of the gland disappears under the diaphragm and manipulation is possible only during inspiration. Strong movements of the intestine may sometimes prove troublesome. Lastly, considerable bleeding may occur from the vena cava when the gland adheres closely to this vessel.

Unilateral adrenalectomy had no ill effect on the pig, but after the removal of both adrenals all pigs died within 10 days, and five of nine animals died within 5 hours of the second operation.

HYPOPHYSIS

Two techniques have been described for hypophysectomy, the parapharyngeal approach and the lateral approach. The parapharyngeal method is essentially the same as used on the rat, but it can also be used on small pigs of up to 15 kg (St. Clair, 1945). The posterior wall of the pharynx is pushed forward and stripped of its attachment to the skull. The insertions of the two ventral straight muscles on the tuberosity at the spheno-occipital junction are thus exposed and the bullae of the petrous temporal bone can be felt laterally. A hole is then made anterior to the insertion of the ventral straight muscles and the pituitary sucked out.

More recently, the lateral approach has been used on pigs of 100–150 kg, and appears to be the preferred method. The technique has been described by Du Mesnil du Buisson *et al.*, (1964a, b). A triangular-shaped incision running from the mid-line down to the posterior limit of the orbit is used and a whole section of bone is removed. The brain can then be lifted to expose the supra-optic artery which is clipped and sectioned and any bleeding stopped with the aid of cautery. The stalk of the pituitary is then cut, the gland removed, and the space packed with gelatin foam after bleeding has been stopped by cautery. Full details of the method are available in the papers mentioned above along with comments on post-operative care.

SPLEEN

The technique of splenectomy has been described by Seamer and Walker (1960) on the basis of a study of 204 animals. A laparotomy is performed through an abdominal left paracostal incision. The principal site of haemorrhage has been found to be the origin of the external oblique muscle. A finger is inserted and directed forward under the costal arch, and the spleen is trapped against the inside of the rib cage and drawn into the incision. The gastro-splenic artery can then be ligated and the spleen pulled out and the splenic artery ligated. It is always necessary to dissect the hilum of the spleen from the left extremity.

PANCREAS

The pancreas of the pig occupies a dorsal position in the abdominal cavity and extends from the duodenum on the right side to the left kidney. In order to remove the pancreas it is essential to obtain adequate exposure of the organ and this is best achieved through a mid-line laparotomy. An assistant holds the duodenum while the pancreatic duct is cut between ligatures. The organ is carefully removed by blunt dissection from the mesoduodenum so that half the mesoduodenum is left intact. This thin film of mesentery appears to be sufficient to prevent a coil of intestine from passing through the space occupied by the pancreas and becoming strangulated. Further, the pancreas may totally or partially surround the portal vein and careful dissection in this region is necessary in order to remove the organ without causing a severe haemorrhage (R. W. Ash, personal communication). A brief account of the maintenance of the pancreatectomized pig is given by Anderson and Ash (1970).

Skin Grafting and Wound Healing

The technique of skin grafting in the pig has been studied by various authors. Baker and Andresen (1963, 1964) used the dorsal surface of the ear as a graft bed, taking pieces of skin from both the dorsal and ventral ear surfaces; the grafts were held in place by means of an adhesive patch. Homografts were lost after 32–37 days. In a further study, Baker (1964) found a mean survival time of 13 days. The shorter time by comparison with the earlier study was thought to be due to the fact that the grafts were covered continuously in the second experiment, preventing scab formation. Saison and Ingram (1963), on the other hand, found in their study that grafts to the shoulder were lost between 14 and 20 days; antibodies to red cells were produced in some animals. The time taken to establish a blood flow through the vessels of grafts has been studied by Clemmessen (1964) by means of Indian ink injections. Restoration begins in about 48 hours, and is complete at the end of 4 or 5 days.

Bromberg *et al.* (1964) have shown that methyl-2-cyanoacrylate is useful in attaching the graft, and that it may have some advantages over the use of sutures. The grafts were made in pigs 20–60 kg in weight; the skin was prepared by clipping free of hair, scrubbing and bathing with tincture of iodine and alcohol in the usual way.

Thymectomy

The technique of thymectomy in the new-born pig has been described by Pestana *et al.* (1965). The gland in the pig extends from the floor of the mouth to the thoracic inlet, but it has been successfully removed in experiments on the effect of the thymus on the immune response. Pestana *et al.* conclude that in contrast to rodents and chickens the thymus of the pig has no immunological function.

Implantation of Probes into the Brain

The shape of the skull varies from breed to breed and with the age of the animal. In the adult the face becomes dish-shaped while in the weaner the nasal, frontal and parietal bones lie in almost a straight line. At the posterior end of the skull is the nuchal crest, a thick ridge of the parietal bone which provides for the attachment of the well-developed ligamentum nuchae. Behind the crest the bone falls away sharply, giving the skull its characteristic shape (see Fig. 5, Chapter 6). The ligamentum nuchae and the muscles of the neck may make any approach to the spinal column at this point very difficult by comparison with the ease with which the vertebral canal can be cannulated in the cat.

A feature of the skull which makes it difficult to fix the head in a stereotaxic frame is that the external auditory meatus slopes downwards as in the rabbit, making the use of ordinary straight ear bars impossible. In this laboratory an instrument has been developed (Fig. 1) in which pressure plates are pushed against the zygomatic arch and the head is supported under the jaw. In addition the usual mouth and ear bars are used.

The brain occupies a relatively small region of the skull, particularly in the adult, and is protected from above by the frontal bone. This bone is

Fig. 1 A stereotaxic frame for use with the pig.

traversed by the frontal sinus and in older animals is virtually in two layers. In the 3 month pig, however, the sinuses are confined to the anterior region, while the posterior portion is about 7 mm thick. This fact makes it difficult to expose a large area of the brain in an operation from which the animal is to recover, but on the other hand the very thickness of the bone makes it easy to insert bone screws for the attachment of electrodes, or mounds of dental cement.

The procedure which has been developed for the insertion of probes into the brain has been adapted from that used in cats and dogs. The scalp is opened from just in front of the nuchal crest to a point in front of the level of the eyes; the periosteum is scraped from the bone and any bleeding is stopped with diathermy. The site of the trephine hole is then determined from an X-ray photograph. Before the hole is made, it is useful to drill two small holes in front and two behind the trephine site and to insert short self tapping stainless steel screws into the skull. These screws are later used to anchor the dental cement in which electrodes or probes are embedded. The insertion of the screws at this point is recommended because of the danger of using a drill and screwdriver in the region of electrodes which have been implanted into the brain. Once the electrodes are in position the hole is covered with gel foam (Sterispon, Allen and Hanbury Ltd.) and a little bone wax is placed on top and around the electrode. The bone surface must be made perfectly dry; the use of a hair dryer is recommended, but not a hot air blower. A layer of acrylic cement can now be applied to the bone. As soon as the cement is dry a further layer is applied until enough material has been built up to hold the electrode. The electrode carrier can now be removed, and the plug used to make connection with the electrode embedded in cement. The mound of cement is then built up to an oval shape and the skin closed around it. The use of antibiotics is important, and no sawdust or powdered food should be allowed in the pen on account of the danger that this may find its way into the wound.

Usenik *et al.* (1962) have described a technique for the insertion of stainless steel discs on to the dura for the recording of the electrocorticogram of adult pigs. A subminiature radio tube socket to which wires have been attached is used. Each wire terminates in a stainless steel dish which is inserted through a drilled hole in the skull to rest on the dura. The radio tube socket is then embedded in dental cement and external connections made to it. The authors maintained some of their pigs for a year, but they mention the problem of infection round the mound of dental cement which may become loose if the infected area is not treated with antibiotics.

Vascular Surgery

The pig's blood vessels tend to be rather more friable than those of the dog, which in turn are more friable than those of man. The difficulties, however, are not so great as to make anastomoses difficult, and the pig has been found to be quite suitable for peripheral vascular surgery (R. Y. Calne, personal communication). Similarly, the pig's heart is more difficult to cope with than that of dog or man, but again the animal has proved to be satisfactory for use in experimental cardiovascular surgery.

INSERTION OF CATHETERS AND WIRES

For many experiments it is necessary to insert catheters into blood vessels or lymph ducts in order to take serial samples or to record pressures.

A second requirement is to implant wire for thermocouples or thermistors to record temperatures in various parts of the body. A common problem in all these procedures is to end the catheter at some point where the animal is unable to damage it by rubbing or scratching. A suitable position is on the back, a little anterior to the scapulae. Catheters and wires which emerge at this point are least likely to be damaged by the animal, and will undergo a minimum of distortion when the animal moves. This last point is of particular importance where wires are to be implanted, since constant movement by the animal can break a thin (40 s.w.g.) wire within a few hours. A further advantage of this position is that if the animal is provided with a harness (Fig. 2), a small box can be carried on the animal's back and connections can be made in it to the indwelling catheter. The illustration shows an animal with a harness which holds a radio transmitter for the telemetry of body temperature.

Probes and catheters which have been implanted into the carotid arteries or jugular veins can be led up to the neck through a guide tube made from stainless steel tube; the tube is passed through from the operation site and pulled out at the other side. The occlusion of one carotid artery or jugular vein is almost inevitable in small animals, but no ill effects are observed. Catheters have been maintained patent for periods up to 47 days when introduced into the superior vena cava by way of the external jugular vein (Christison and Curtin, 1969). For temperature measurements, however, a blind-ended catheter can be attached to the wall of the carotid artery. This arrangement has the double advantage of not occluding the vessel and enabling the measuring probe to be withdrawn for inspection or replacement.

An alternative to the use of the blood vessels of the neck for catheterization is to pass a cannula down either a vein, or in larger animals an artery in the ear (Anderson and Elsley, 1969). This procedure is not always easy, but it is reported that the catheter can be left in place for several weeks and does not trouble the animal. Some difficulty may be experienced in passing the catheter through the junction between the ear and head when a marginal ear vein is used; in this respect, use of the central vein is to be preferred.

Catheters and wires implanted in the body become enclosed with fibrous tissue, and at the end of 2 or 3 weeks they can be freed only with difficulty. All catheters leaving the body must be long enough to allow movement of the animal. Thus a catheter leaving the skin at the back of the neck will be drawn 3 or 4 cm into the body when the animal puts it head down, and will protrude again when the head is held up. This movement makes com-

Fig. 2 A simple harness which may be used either for restraint of the animal or for the attachment of apparatus such as a transmitter for telemetry.

plete healing around the wound impossible, and although the movement causes the pig no irritation it is liable to carry infection into the body. Such infection is likely to be completely localized, and can be controlled by the local application of antibiotics.

J. L. Linzell (personal communication) has also cannulated the internal saphenous artery and the mammary vein in the sow. He stresses the tendency of the arteries of the pig to spasm, and of the layers of the wall to strip as the catheter is passed down. For external connections Linzell uses a metal plate over which a protective box can be fitted (Linzell *et al.*, 1969).

Replacement Surgery

It has been shown that the stomach can be used as a substitute for the bladder, and Morelle (1963) has carried out experiments in pigs in which a segment of the stomach has been used to replace the ureters. The abdomen is opened over its entire length and a segment of the stomach 15 cm long folded into a tube; its blood supply is derived from the left gastroepiploic artery. A piece of bladder 1·5–2 cm in diameter is then removed, and the distal end of the "gastro-ureter" sutured to the bladder. A catheter with many lateral holes is placed in the "gastro-ureter" and led outside through the bladder and urethra. This catheter invests a thinner ureteral catheter. The proximal end of the left ureter is grafted in the top of the stomach tube, and the thin ureteral catheter passes through this anastomosis into the ureter.

Homotransplantation of Organs

The pig has been used successfully in studies involving the homotransplantation of organs including the liver (Calne *et al.*, 1967a, b; Riddell *et al.*, 1967), the kidney (Calne *et al.*, 1967b), and the ovary (Binns *et al.*, 1967). The animal has been regarded favourably as a subject for experimental surgical techniques, and has displayed a remarkable ability to accept liver homografts which survive even though kidney and skin grafts are rejected.

Withdrawal of Blood Samples Without Implantation of Catheter

VENOUS BLOOD FROM THE ANTERIOR VENA CAVA

Blood may be taken from the superior vena cava by direct puncture (Carle and Dewhirst, 1942). In the case of animals up to about 30 kg, the animal is placed on its back in a V-shaped holder with its head hanging downwards (Fig. 3). An assistant holds the forelimbs and head, while another holds the hind limbs. Once the pig is on its back it remains surprisingly still, and with small pigs only one assistant is needed. With larger animals, the pig is allowed to stand; a slip-noose over the upper jaw causes the animal to pull back, so exposing the required area.

The operator should run his finger down the side of the trachea to a point just in front of the sternum; the finger will now be in a hollow formed by the angle of the ribs with the trachea, and it is into this hollow that the needle is placed. If a point is aimed for just below the anterior end of the sternum a 5–7 cm needle can be inserted into the anterior vena cava. If the needle is inserted more cranially it is possible to cannulate the jugular vein. By chance the needle may enter the carotid artery, and in this event care must be taken in case damage is done to the nerves which pass alongside this artery. Once the needle has been inserted a blood sample can be taken.

If a catheter is to be inserted it may be threaded through the needle, and the needle withdrawn. The catheter may then be strapped to the body by means of adhesive tape and the animal allowed to move about in a cage. By

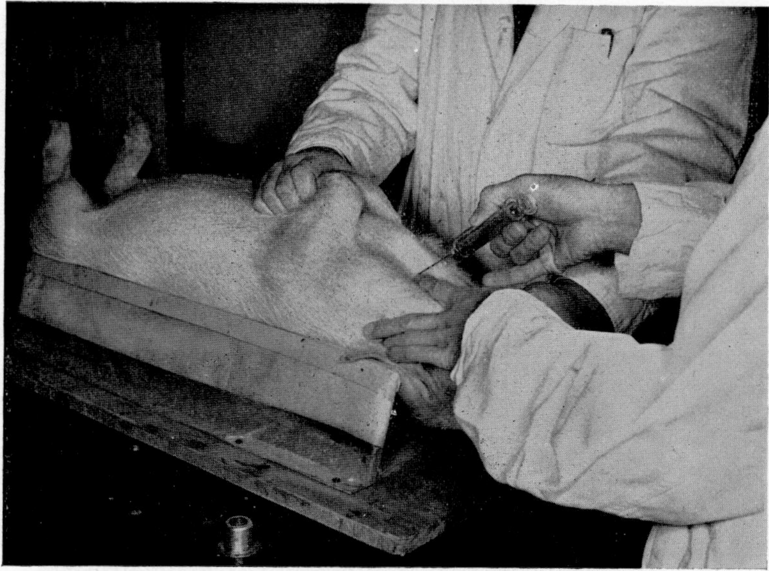

Fig. 3 The approach to withdrawal of blood from the superior vena cava while the pig is supported in the supine position in a V-shaped stand. The needle is shown in position prior to insertion.

using a long catheter it has been possible in this laboratory to take blood samples, or to make injections from a sufficient distance not to disturb the animal, a factor which can be of considerable importance.

HEART

Blood may be withdrawn from either the right or left ventricle by direct puncture. The left ventricle lies beneath the 5th rib and may be punctured from either side. The risk of the needle entering the right ventricle is however less if the needle is inserted between the 5th and 6th ribs. To withdraw blood from the right ventricle the needle should be inserted between ribs 3 and 4. In the anaesthetized relaxed 3 month old animal the left ventricle will lie about 2 fingers width posterior to the "elbow".

VENOUS SINUSES

Venous blood from the sinus around the pituitary gland may be collected by means of a needle inserted through the skull under the guidance of radiographs and fixed to the skull with dental cement (B. A. Baldwin, personal communication). A technique has also been described for obtaining blood samples from the orbital sinus (Huhn *et al.*, 1969).

OTHER BLOOD VESSELS

The technique for surgical cannulation of the portal vein has been described by Arsac and Rérat (1962).

Collection of Lymph in Unanaesthetized Pigs

The problems are essentially similar to those in other species, and the techniques used are modifications of these. Yoffey and Courtice (1956), Lascelles and Morris (1961), Binns and Hall (1966) and R. Binns (personal communication) have cannulated the thoracic duct by right thoracotomy at the seventh rib, and the intestinal and lumbar ducts, cranially and caudally respectively to the right renal vessels, at laparotomy on the right side. They found that the efferent lymphatic ducts of the lymph nodes were frequently multiple and difficult to cannulate, but they collected lymph from the efferent ducts of the retropharyngeal, dorsal superficial cervical, supramammary, lateral iliac and subiliac nodes.

Lymph from the thoracic duct was found to be contaminated with blood, and the catheter was difficult to maintain. Flow rates and protein and lipid contents were similar to those of other species, but lymphocyte counts were very low.

Sampling of Bone-marrow

R. Binns (1967 and personal communication) has devised a method for the collection of large amounts of bone marrow for transplantation experiments. The proximal antero-medial surface of the tibia is drilled, and the femoral artery is temporarily occluded. The majority of the tibial marrow is then removed with minimal blood contamination, by combined gouging and suction using sharpened curved siliconized stainless steel tubes connected to a suction reservoir containing sterile citrated Ringer saline. Intermittent uptake of diluent facilities the passage of bony fragments along the 3 and 7 mm steel tubes which are found to be satisfactory for young pigs. No post-operative complications have been observed, and the pigs walk normally afterwards.

Injection and Marking of the Foetal Pig

Foetal pigs from 60 days of gestation to term have been injected by the intraperitoneal, subcutaneous and intramuscular routes, and through the uterine wall at laparotomy on the sows. At these times, they have been satisfactorily marked for post-natal identification by the shallow subcutaneous injection of a mixture of 10 mg Evans' Blue and 0·1 ml non-toxic carbon suspension, C11/1431A (Gunther Wagner, Pelican Werke, Hanover) per kg of estimated foetal body weight (R. Binns, 1967 and personal communication).

Intramuscular injections in the foetus have been made into the ham or deep neck tissue, intraperitoneal injections through the mid-flank between the palpated hip, knee and costal angle, and shallow subcutaneous injections over the sacrum. The needle is directed cranially at an acute angle to the skin. Penetration of the skin is verified by gently moving the foetus caudally, when the needle point also moves and in so doing moves the uterine wall.

CHAPTER 10

Behaviour

Behavioural studies may be considered broadly under two main headings: (i) observations made on animals in situations which occur normally during the course of their life history, and (ii) experiments in which the pig learns to perform some action, usually under strictly controlled conditions. These will be considered in turn.

Observational Studies

The types of observation which can be made on undisturbed pigs in their natural or usual husbandry environment have been reviewed by Hafez *et al.* (1962). These observations can be classified under a number of headings.

FEEDING

Pigs are truly omnivorous animals which will eat or attempt to eat almost anything, and except under very hot conditions they will always display considerable interest in food.

Under natural conditions the pig roots in the ground, and this propensity for finding food which is buried in the soil has been used in France where pigs have been trained to search for truffles. Observations of any field where pigs have been kept reveals that rooting behaviour can result in considerable damage to plants even when the animals have been fitted with a ring in the nose to discourage this activity. The pig is also fond of nuts, and in certain areas of England grazing rights are still held in forest land; indeed certain old documents describe land areas in terms of keep for given numbers of pigs.

The rooting behaviour of the pig is very marked; it can be used in training pigs to perform tasks, e.g. opening a food container, and can be considerably increased by scattering the feed over the floor. Particular note should be taken of this fact when feeding animals under experimental conditions, for while it is very useful to give an animal a few pellets of food while it is being trained to stand in a stall, the procedure may excite the animal so much that it will continue to root for some time for odd scraps of food.

Under both farm and laboratory conditions pigs may be fed on either a pelleted or a powdered diet which is high in cereal content. The feed may be provided either *ad libitum* or on a restricted basis (see Chapter 5). Water is usually provided in a drinking bowl, but under some commercial systems it is mixed with the food. The amount of food taken by the pigs can be influenced by the constituents of the diet, and by a variety of other factors, including competition between members of a group; pigs tend to eat more when there is some competition. On the other hand, the variation in growth rate in a group of pigs fed together is likely to be greater than that among the same pigs fed separately. Where food intake is to be measured, some device such as a bar across the front of the feeder is essential in order to reduce the spilling of food due to rooting and playing activity.

Information about the preferences of young pigs for sugar or saccharin has been derived from two types of experiment. Diaz *et al.* (1956) fed different pigs the same type of rations to which had been added refined sugar cane, unrefined sugar cane or molasses, and found that more food was taken by the groups receiving sugar, whether it was refined or not. Aldinger *et al.* (1959) carried out a similar experiment from which it was concluded that pigs prefer food treated with saccharin to that without saccharin. Experiments of a different kind have been carried out by Kare *et al.* (1965). In these studies the pigs were given a choice of several solutions containing saccharin, sugar, distilled water and quinine. The pigs preferred saccharin or sugar. These experiments, and others by Fowler and Ensminger (1960), suggested that there may be some genetic basis for appetite and preferences.

An interesting preference for certain metals has been reported by Braude (1948), who found that pigs prefer to lick a copper plate rather than other metals such as tin or aluminium.

Points which arise in experiments in which the animals have a choice are that if the particles of food in a fixed diet are too small the animal can not sort them out (Hafez *et al.*, 1962); further, pigs take more food from the feeder nearest to the drinking bowl, regardless of the type of food it contains (Facto *et al.*, 1959).

Temperature also influences food intake. Heitman and Hughes (1949), have studied pigs under hot conditions, and have shown that large animals, particularly, decrease food intake at high temperatures. Similar findings have also been reported by Ingram and Mount (1965) for young pigs. Other environmental conditions can influence the effect of temperature on food intake; for example the provision of a water wallow (Bray and Singletary, 1948), or even a drinking bowl from which water can be splashed, may enable the pig to wet itself, and so to lose heat by evaporation, with the result that food intake is increased.

DRINKING

When given food and water in separate containers, pigs may either eat first and then drink or alternate between the two. Individual pigs can often be seen to move from water bowl to feeder, sometimes very frequently. Even young pigs of a few weeks of age will learn to operate a self-watering bowl. The amounts of water taken will depend on age, environmental temperature and lactation (see Chapter 5).

ELIMINATION

Pigs tend to reserve a section of their living space for defaecation. When housed inside, the animals will rapidly learn to use a dunging passage, and when provided with an outside run a corner of this is usually reserved for defaecation. Even when restricted to a cage some animals will reserve a corner for elimination, and use can be made of this fact for the collection of faeces. The animals may be further encouraged to use a particular region for defaecation by the location of the water bowl, since pigs tend to defaecate near the source of water. On this account, however, the water may often become contaminated with faeces, so that it is advisable to provide a flap cover to the bowl. According to Hafez *et al.* (1962) there is some evidence that the practice of reserving a particular area for defaecation may be learnt by young piglets from the sow.

Under very warm conditions when pigs are deprived of a water wallow they will urinate and roll in their urine and in this way achieve some evaporative cooling (Heitman and Hughes, 1949).

FIGHTING

Fighting occurs chiefly between adult boars, but females and immature pigs may also indulge in some aggressive behaviour towards each other.

New-born pigs will snap at each other or slash with their temporary canine teeth when establishing their teat order, but after this vicious attacks are not common, and such behaviour is confined to play.

When pigs from two litters are mixed at weaning some fighting usually occurs especially if one set of pigs is allowed into the pen some time before the other set, and a single pig introduced into a group of pigs may be attacked by several pigs at once. Young pigs are frequently seen to be covered with a mass of small scratches made by the teeth of their opponents, but this can sometimes be avoided by spraying newly mixed animals with a strongly smelling substance which appears to prevent the pigs from recognizing the newcomers.

The fighting behaviour of boars has been described in some detail by Hafez *et al.* (1962). At the beginning of a conflict the two boars grunt, grind the teeth, foam at the mouth and may paw the ground. They then take up a typical fighting stance standing shoulder to shoulder. The boar applies pressure to the opponent's flank, and brings the head upwards with the mouth open. These slashing movements are extremely dangerous, and, if the tusks are intact, may result in severe wounds. The ears and legs may also be bitten, and on occasion an agressor may charge his opponent. Such encounters may last for up to an hour, but once the dominance of one animal has been established subsequent encounters consist of no more than a grunt or threatening posture. The boar may also make attacks on man, and in this event the consequences can be serious particularly if the victim is knocked down.

Females are not as a rule agressive unless with a young litter. On these occasions great care should be taken on entering the farrowing pen since confrontation with an angry sow can lead to a serious encounter. Particular attention should be given to handling the litter, since the removal of one or two members to the accompaniment of considerable squealing may arouse the sow. In this event she will usually give a warning grunt and threaten to bite, and if provoked further she will attack.

MATING

Females reach sexual maturity at seven months of age or a little later, according to the breed, and before this time display no sexual interest in the boar. Once the animals have become mature, oestrous cycles are repeated every 21 days until pregnancy takes place, and, in domestic animals at least, there appears to be no breeding season. As in other species, however, the rhythm of the cycles may be disturbed by changes in management, or other environmental conditions.

During di-oestrus the female is not only unreceptive but antagonistic to any boar which displays attention, and will bite, or run away to hide or escape. Pro-oestrus is characterized by an increase in activity and the ease with which the female can be disturbed. This phase of the cycle is relatively short (14 hours) and increasing interest is shown in the boar, or even in other females which she may attempt to mount although she will not allow herself to be mounted by the boar. This type of behaviour increases in intensity as oestrus approaches, and the sow makes more frequent attempts to mount other animals, urinates more often and increases spontaneous activity to twice the normal level (Altmann, 1941). During oestrus the boar is accepted, and the female also receives attention from other females particularly during the first days. If pregnancy ensues, a faint oestrus may again be displayed at an interval of another sexual cycle, while animals which fail to become pregnant pass into di-oestrus and the cycle begins again.

Males reach full sexual maturity at about eight months of age, but may begin to show interest in oestrous females before this time.

The mating of pigs is usually preceded by some courtship behaviour, the duration of which depends on several factors. If the boar approaches a female which stands still, he presses his snout against her head or side and gradually reaches the genital-anal region where the nuzzling becomes intense; on occasion he may lift the female's hind-quarters off the ground by pushing his snout between her legs and raising his head suddenly. The boar urinates frequently at this stage and grinds the teeth and foams at the mouth (cf. fighting behaviour). If the female runs away from the boar he will pursue her and attempt to bring her to a stand-still. The chase and the ensuing courtship is accompanied by the so-called "mating song", which is a series of grunts at 6–8 per second at 85–95 decibels, but this song is almost absent when the female remains stationary. Sometimes the mating response can be elicited in the oestrous female by a touch on the back; the sow then stands still and lowers the middle of her back and in this state it is very difficult to make her move. The response can, moreover, be more easily obtained when a recording of the mating song is played. Olfactory stimuli have also been shown to be of importance.

Some females may display a preference for a particular boar, or alternatively they may reject an animal particularly if it has previously wounded the female with its tusks during the nuzzling process or with the dew claw on mounting. A second mating by a boar is usually more prolonged; thus while a vigorous male may mate the first time with hardly any foreplay, on the second occasion there is almost always some courtship behaviour.

When the boar mounts the female he rests his forefeet on her back, and then if he has approached from the rear he may step forward to lay his belly on her back; on some occasions several mountings may take place before intromission. Mounting of the female from the front is more common among young males with little experience. Once intromission into the vagina has occurred the mating usually continues until ejaculation has been completed, but if the penis has been introduced into the anus it is usually withdrawn. In most instances more than one ejaculation occurs at each intromission, the total time spent being about 10 min, although some animals may spend as long as half an hour. Although a second mating with the same female will usually occur later, if the male is presented with a fresh animal in oestrus a response is elicited much sooner. When a succession of females in oestrus is presented, however, the time taken to reach ejaculation increases, and the duration of the ejaculation decreases with each successive female (Hafez *et al.*, 1962).

When two boars compete for the same female, fighting does not usually occur although the second boar may interfere with the mating and if possible mate her himself.

Homosexual relationships between boars can be established, even when both males copulate with females.

MATERNAL BEHAVIOUR

Signs of maternal behaviour begin to appear towards the end of pregnancy when females find a site for their nest. At this time the gilt or sow will make a hollow in the ground and carry dried grass or straw to it. This site is then defended with the same type of threatening behaviour seen in sows defending their young. Even when the sow is placed in a concrete pen, some attempt at nest building may occur.

Most pigs farrow on their side, but some individuals may stand or lie on their belly. Restless and nervous animals tend to move about between the delivery of each foetus and this can lead to some of the new-born pigs being crushed. At other times a nervous sow which has been disturbed may kill the new-born piglets and eat them.

When the pig is born it is still covered by the foetal membranes, and has to free itself by kicking. The sow pays little attention until the litter has been born and then she herds the piglets into a corner of the nest. The sow will defend her young against intrusion and upon the utterance of the warning grunt the small piglets crouch and remain motionless. The sow may then either attack the intruder or lead it away from the litter.

SUCKLING

As soon as the piglet is born it begins to display interest in seeking milk from the sow. Animals make their way to the teats and compete for them, a strong preference being shown for the anterior nipple. After an initial period the piglets settle down to a particular teat and use it irrespective of the side on which the sow lies. The factors which help the piglet to locate its own teat have been investigated by several workers (Donald, 1937; Gill and Thompson, 1956; McBride, 1963). The most powerful clues appear to be derived from body conformation, since more errors are made in finding teats in the centre of the udder than at either end. The ability to locate the right teat is not, on the other hand, influenced by the presence of mud or scented substances on the udder. When the piglets change teats, they most frequently attempt to move towards the pectoral region. This apparently innate preference may confer some advantages in that piglets at the front of the sow are less likely to be damaged by kicks or to be trampled when the sow rises. In addition it has been shown that the larger pigs tend to obtain the anterior teats (Wyeth and McBride, 1964) and that the anterior teats are more productive.

When nursing their young, sows usually lie on one side and turn over for the next session, but some animals display a preference for lying on one side and most sows will occasionally nurse the litter while standing.

The sow may call the litter to feed by a series of grunts, or if the piglets are hungry they may themselves take the initiative. The sound of other litters which have been aroused will usually arouse the rest of the litters in a house. If the sow is aroused by some shock such as a loud bang, suckling may be initiated; the piglets may be woken from sleep by an imitation of suckling sounds (Hartman and Pond, 1960).

Barber *et al.* (1955) have divided suckling into three phases: (a) nosing, during which the piglets push against the teat; this phase increases in duration as the litter grows; (b) a short, quiet phase when the animals lie down with ears drawn back, but when little milk is obtained; (c) true suckling which lasts from 13 to 37 seconds.

The young piglets fall asleep at the end of the suckling period, but in older animals the sow may move away. As the litter begins to eat solid food, the piglets first go to the teats, then eat food before finally sleeping.

The frequency of suckling is greater by day than by night, but on average piglets are suckled once per hour (Braude, 1954; Barber *et al.*, 1955). After the 7th day post partum the frequency declines.

Sows usually accept other pigs for fostering, especially if they are presented within 48 hours of birth. With older piglets it is best to eliminate olfactory clues by which sows recognize their own litter. This may be done by mixing the foster piglets with the original litter for 2 or 3 hours, by smearing the piglets with the sow's excreta, or by using a spray of some strongly smelling substance.

Experimental Investigations

The types of behavioural investigation which are carried out with pigs in the laboratory, as opposed to observations made under conditions akin to those in the field, fall into two groups. The first group includes classical conditioning experiments, of the sort associated with Pavlov, and the second group comprises the operant conditioning studies usually associated with the name of Skinner.

CLASSICAL CONDITIONING

In a classical conditioning experiment a stimulus such as a bell is sounded in the animal's presence. After a while the animal ignores this sound and at this stage the bell is sounded just before the presentation of food (the unconditioned stimulus). On presentation of the food the animal salivates (unconditioned response), and after a series of training periods the sound of the bell alone will elicit salivation, which is then termed the conditioned response. Thus the bell has become the conditioned stimulus and can now elicit the conditioned response.

Experiments such as these have many variations. The pig has been very little used in this type of study, although Pavlov himself used them, but according to Marcuse and Moore (1944) with little or no success. Marcuse and Moore (1942) have, however, succeeded in using the pig in experiments which involve some classical conditioning. Three pigs were prepared with parotid fistulae and were trained using a sound tone as the conditioned stimulus, and food as the unconditioned stimulus. In their experiment the pigs also had to learn to lift the lid of a box in order to obtain the food. Two animals developed the unconditioned response of salivation in response to a tone of 480 cyc/sec, while the third animal which did not salivate nonetheless learned to distinguish between 480 and 550 cyc/sec, using the lifting of the lid of the food box and the pig's vocal response as criteria. In later experiments, Marcuse and Moore (1946) found that pigs learned to respond to the sound of a metronome by lifting a lid for food, and the authors were able to study various indices of conditioning including heart rate and motor activity. These workers stress the need to train animals at an early age.

OPERANT CONDITIONING

As may be expected from the fact that pigs learn to use automatic feeders and watering bowls, the animals rapidly learn to operate a panel-type switch, and this faculty has been used in experiments involving operant conditioning. In experiments carried out in this laboratory pigs have learned to push a panel (Fig. 1) in order to obtain a short burst of infra-red heat (Baldwin and Ingram, 1967a, b; 1968a b). The pigs learn this response both when

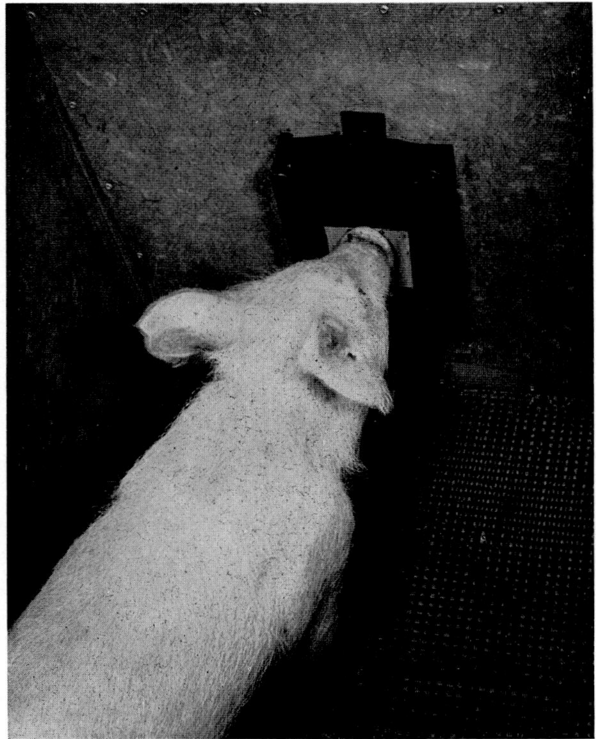

Fig. 1 Pig operating a panel-type switch. The central white area on the panel is hinged and has a micro-switch behind it.

exposed singly in cages, and in groups under normal farm conditions. Experiments of this kind have also been made by Klopfer (1966), who has been able to show that within the wavelength range of light from 465 to 680mμ pigs can distinguish differences in wavelength as small as 20 mμ. The pig, however, tends to use non-visual stimuli, and, in order to use visual stimuli, Wesley (1955) found it necessary to halt the animal for a few seconds in front of the stimulus card. Moreover, Konyukhova (1955) has shown that pigs develop conditioned responses to sound much earlier than to light.

In order to perform experiments of this kind it is essential to take care in the handling of the animal, and, to a lesser extent, in the choice of individual pigs. In this laboratory it has been found an advantage to choose those members of a litter which most readily approach the experimenter. These pigs tend to be both more ready to explore their environment and to have the least fear of man. From the time of selection the animals are best housed if possible in the same type of cage as that in which the experiment is to be performed (Fig. 2). Thus in the experiments of Baldwin and Ingram (1967a) the cage in which the experiments were performed was a modified cage of the type used for housing experimental animals. By ensuring that the animals were accustomed to being handled and were used to their surroundings, pigs exposed to a cold environment learned within half an hour, and in exceptional circumstances within minutes, to operate a simple switch (Fig. 1) in order to obtain infra-red heat. Under less favourable conditions, using pigs which had not been specially selected, the animal's attention could be drawn to the switch by sticking some brightly coloured tape on the panel with a loose end sticking out into the cage. In chewing this the pig came to associate the panel with the reward (or more correctly the reinforcement) for correct operation. Alternatively a mixture of food and water was rubbed over the panel switch.

For experiments in which it is necessary to attach wires or tubes to the animal, some type of stand in which the animal can be held is useful (Moore and Marcuse, 1945). The type illustrated in Fig. 3 was used to investigate the effects of cooling the hypothalamus on the rate at which the pig responded for infra-red heat. In this study it was essential to connect tubes to connections on the animal's head in order to perfuse with cold alcohol the thermode implanted into the hypothalamus. The restriction of the animal in this frame considerably altered the time taken to learn the response, and the animals which had learned to respond when free to move about in a cage took some little time to respond regularly when they were

Fig. 2 Cage equipped with panel switch and infra-red heaters. A similar cage without the switch and heaters can be used to house the pig before and after experiments, so that habituation is established.

in the stand. The best procedure was found to be one in which the animal was first trained while free to move about in a cage, and then placed in the stand in a warm room and given some food. Only after the animal had become accustomed to the stand was the switch displayed and the room temperature lowered. Failure to take care over the training procedure, or haste in taking two steps simultaneously, often led to difficulties and rarely saved time. Animals which are to be subjected to some surgical procedure are best trained before the operation.

Mount (1963) investigated the pig's behaviour in relation to the choice of temperature and heat loss. Piglets were given the choice of a range of temperatures between 23 and 37°C in adjoining compartments, care being taken to ensure that they were introduced at random into a given compartment in the apparatus; the preferred temperature proved to be close to

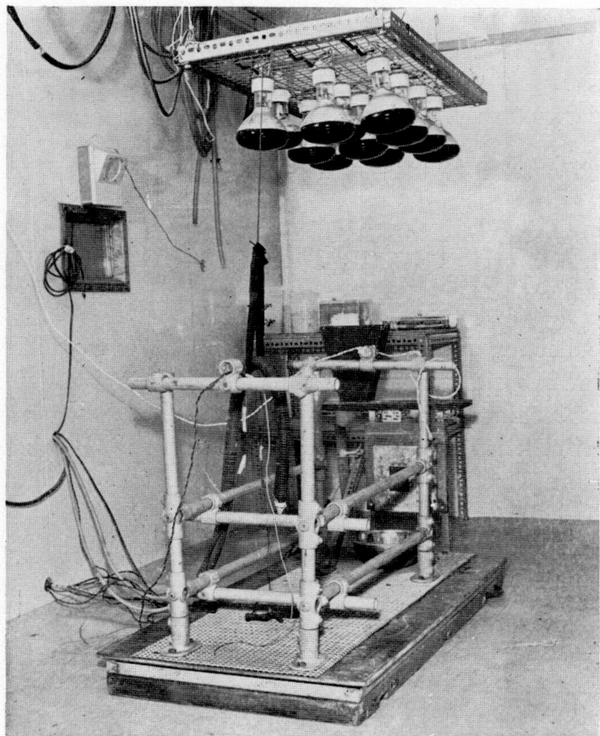

Fig. 3 Stand in which a pig can be restrained for experiments involving the attachment of leads and other apparatus. In the illustration the equipment includes a panel switch which can be used either to turn on the infra-red heaters, or to deliver food into the bowl.

30°C. In other experiments the effect of huddling on oxygen consumption and the effect of posture on heat loss to the floor have been investigated. The behavioural responses involved had the effect of reducing heat loss to the environment by as much as 40% (Mount, 1960, 1967). Other adaptations to the environment have been reported by Hafez *et al.* (1962); while pigs are diurnal in moderate temperatures, they are nocturnal during the hot months of the year.

CHAPTER 11

Uses of the Pig as a Laboratory Animal

The task of collecting information on the diverse ways in which the pig has been used as an experimental animal is made easier as a result of the publication of the proceedings of a symposium held in the United States in 1965 (Bustad and McClellan, 1966). In addition to this, ways in which the pig is of use in medical research have also been summarized in tabular form by Doyle *et al.* (1968).

The uses made of the pig in biomedical research will be considered in this chapter under the following headings: Cardiovascular Research; The Digestive System; Metabolism and Nutrition; The New-born Pig; The Skin; Nervous System and Behaviour; Immunological Studies; Skeletal and Dental Research; Radiobiology; Genetics; Miniature Pigs.

Cardiovascular Research

The use of pigs in cardiovascular research has been reviewed by Detweiler (1966) and by Engelhardt (1966). Baro- and chemo-receptor mechanisms have been investigated (Booth *et al.*, 1966), and cardiac output monitored by implanted sensing devices (Sawyer and Stone, 1966; Stone and Sawyer, 1966). Cardiac failure has been induced experimentally in pigs.

The distribution of the coronary arteries is similar to that in man; the coronary vessels supplying the atrioventricular node and the bundle of His have been ligated to study the ensuing collateral circulation and the effects of drugs (Lumb, 1966). Maaske *et al.* (1966) partially occluded the main pulmonary artery and produced a slow progressive right heart failure accompanied by signs of clinical heart failure; ventricular dilatation, enlarged liver and ascites were found at autopsy. Atherosclerosis in pigs has been studied by a number of workers, and the lesions have been found to be comparable with those occurring in man (Luginbühl, 1966; Greer *et al.*, 1966; Rowsell *et al.*, 1966). Congenital heart malformations in the pig have been described, and flow through the ductus arteriosus and mitral regurgitation in new-born pigs have been investigated (Evans *et al.*, 1963; Rowe *et al.*, 1964).

Guiney (1965) has discussed the use of the pig as an experimental animal, particularly in cardiovascular surgery. Other aspects of cardiovascular physiology in the pig have been reviewed by Hörnicke (1966a).

The Digestive System

Work by Huber and Wallin (1966) has included the analysis of gastric secretion from Heidenhain pouches or gastric fistulae, and pancreatic secretion has been studied by Magee and Hong (1966). Oesophagogastric ulcers which occur in pigs are similar to those seen in man, and their incidence is influenced by the nature of the diet (Perry *et al.*, 1966; Reese *et al.*, 1966). The clinical and pathological features of these ulcers have been described by several workers (Curtin *et al.*, 1963; Orfeur, 1965; Rothenbacher, 1965; Curtin and Goetsch, 1966).

McCance and Wilkinson (1967) have successfully carried out resection of the whole of the lower half of the small intestine, with the exception of the last few cm of the ileum, in pigs weighing 3–4 kg. The exposure and handling of the intestines may produce crippling adhesions.

Metabolism and Nutrition

There has been a great deal of investigation of the type of food appropriate to rapid growth and maintenance of health in the pig. The dietary requirements of the animal have been summarized by the Agricultural Research Council (1967) in the United Kingdom and by Cunha (1966) in the United States. The subject of energy exchanges between the animal and its environment has been reviewed by Mount (1968).

Pigs weaned on to low protein diets at 3 weeks of age show changes resembling those seen in kwashiorkor in human infants (Pond *et al.*, 1966). Filer *et al.* (1966) have suggested that the growing pig could be used as a model for comparison with the human infant in respect of the effects of age, sex and diet on the rate of growth and body composition during infancy. The human infant, however, takes about 6 months to double its birth weight, whereas the pig achieves the same proportionate increase in about 1 week. Consequently the new-born pig provides a fore-shortened and accelerated model of human development, insofar as the animal can be used for this purpose.

Johnson (1966) found apparently irreversible damage in the myocardium, arteries and arterioles as a result of refeeding following starvation. Thiamine-deficient pigs also show cardiovascular changes, in the form of bradycardia, sinus arrhythmia and atrioventricular blocks (Miller and Ullrey, 1966).

The New-born Pig

In some ways the new-born pig is similar to the new-born human infant, for example in respect of metabolic rate in relation to body temperature and environmental temperature (Mount, 1966). The new-born pig allows an experimental study of the neonatal cold injury which occurs in the human (Mann and Elliott, 1957). One of the features of this condition in the baby is hypoglycaemia. When the new-born pig is fasted, particularly under cold conditions, its blood sugar level falls and the animal dies in hypoglycaemia unless it is given glucose (Goodwin, 1957; McCance and Widdowson, 1959). During the course of the first week after birth the pig becomes increasingly resistant to hypoglycaemia; this and other aspects of metabolism in the new-born pig are discussed elsewhere (Mount, 1968).

The Skin

The skin of the pig has often been regarded as similar to human skin, but appearances are to some degree misleading (Montagna and Yun, 1964; Montagna, 1966), as pointed out in Chapter 1. The skins of both species, however, contain similar proteins (Weinstein, 1966).

Cutaneous evaporative loss is small in the pig as compared with other mammals. Ingram (1967) found no evidence of thermal sweating from the apocrine glands in the animal's skin, and using a ventilated capsule on the skin he found only a small rise in evaporative loss under hot conditions (Ingram, 1965a). At high temperatures the animal can make up for the lack of endogenously derived cutaneous evaporation by its wallowing behaviour. Mud smeared on the animal's surface leads to evaporative heat loss at a level corresponding with that in man (Ingram, 1965b). It is clear that the big difference in sweating ability between man and pig must not be overlooked when the pig is used as subject in obtaining experimental results which are to be applied to the human.

Nervous System and Behaviour

The pig has been used in behavioural studies by some workers. Klopfer (1966) has investigated visual learning in pigs, and as a result has been able to demonstrate wavelength discrimination and to determine the photopic and scotopic visibility curves. In this connection, the eye of the pig shows certain similarities to the human eye; the general vascular structure is similar and there is no tapetum (Bloodworth *et al.*, 1965).

The pig has been used in operant conditioning experiments in which the animal learns to operate a switch or lever in order to obtain a reward (Baldwin and Ingram, 1967a). Thus, the animal quickly learns to switch on heaters in a cold environment or to turn off an unwelcome draught of air. The success achieved in these experiments is haply at variance with the experience of Pavlov (quoted by Engelhardt, 1966) who once said: "As soon as a swine was lifted onto a stand, it squealed at the top of its voice, and all work in the laboratory was impossible . . . we spent a month without obtaining any results . . . All pigs are hysterical."

Dellmann and McClure (1966) have been concerned with the development of a stereotaxic atlas for the pig brain.

Immunological Studies

Germ-free, colostrum-deprived pigs, taken from the sow at term by hysterectomy under sterile conditions, have proved to be very useful for immunological studies (Betts *et al.*, 1960; Landy and Ledbetter, 1966; Kim *et al.*, 1966). Studies have been made of the absorption by the newborn pig of intact proteins from colostrum and other sources (Ullrey *et al.*, 1966; Pierce and Smith, 1967), and of "gut closure" to the absorption of gamma-globulin-sized particles.

Immunological problems arise in the transplantation of organs, which has been referred to in Chapter 9. Blood groups in the pig have been discussed in Chapter 7.

Skeletal and Dental Research

The pig has lent itself to a number of studies on bone, joints and teeth. Pratt and McCance (1966) investigated the structure of bones in pigs subjected to severe undernutrition; when the animals were undernourished and then refed, all the bones ultimately reached lengths close to the expected values. Arthritis occurs in pigs; hypersensitivity may be involved (Neher and Carter, 1966). The experimental production of pannus in a rheumatoid-like arthritis in pigs has been described by Sikes *et al.* (1966) and Ajmal (1970) has discussed chronic proliferative arthritis in the pig in relation to human rheumatoid arthritis.

The pig is unique among common laboratory animals in having a long period of deciduous and transitional dentition, which permits the experimental study of many dental problems which arise in children (Jump and Weaver, 1966). Miniature pigs are suitable experimental animals for work on restorative dentistry, orthodontics, periodontics and the pathology and therapeutics of the tooth pulp.

Radiobiology

The pig appears to be relatively resistant to radiation injury and Nachtway *et al.* (1966) report the development of radio-resistance in pigs some weeks after exposure to radiation, in that the LD_{50} radiation dose rose considerably. The recovery rate of pigs is greater than that found in other large animals, including man.

In view of the complex of immediate, delayed and genetic effects of irradiation, it is clear that it would be difficult to compare pig, man and other animals in general in respect of sensitivity to irradiation. Specific tissue or functional comparisons are more useful, and in this connection George and Bustad (1966) investigated the comparative effects of beta irradiation of pig, sheep and rabbit skin. They looked for early and late gross changes in the skin following localized exposure to 2000, 8000 or 16,000 rads from Sr^{90} or P^{32} plaques. The most severe overall response occurred in pigs, and the least in sheep, although sheep showed the most intense initial reaction. The pattern of response observed in pigs and rabbits resembled that described for man; complete trans-epidermal necrosis with wet desquamation occurred in pig and rabbit skin following 8000 and 16,000 rads. No hair regrowth took place in any of the three species after exposure to 16,000 rads.

Pigs have been followed for a period of 2 years subsequent to exposure to the radiation flux of a nuclear explosion, and some of the animals were re-irradiated with X-rays 4 months after the initial irradiation (Shively *et al.*, 1964). All the test animals gained less weight than the controls and showed a prolonged leucopenia. There was no reduction in reproductive efficiency, or undue morbidity or mortality. Pigs which survived this brief exposure were capable of efficient reproduction when mated 8 months or more after the irradiation.

In the pig, the period of maximum sensitivity of the male germ-cells to irradiation extends from 50 days pre-natally to 70 days post-natally. Sub-lethal doses do not sterilize, as in the rat, but only diminish the sperm-producing capacity (Erickson, 1965). In the female mammal there are wide species differences in oocyte sensitivity. Erickson found that gilts exposed to 200 or 400 r gamma radiation at 1–40 days following birth showed irregular oestrous cycles and did not become pregnant; the ovaries contained cystic corpora lutea and few vesicular follicles. However, six sows exposed to 500 r had each produced four successive litters of pigs of normal size. Erickson and Murphree (1964) found the foetal pig to be markedly refractory to gamma irradiation.

Genetics

The domestic pig provides unique opportunities for research in this field on account of the considerable genetic variation found between and within the different breeds (Cox, 1966). In addition, the pig is suited for work in cytogenetics since it has only 38 chromosomes which can be readily paired and grouped in a manner similar to human chromosomes (McFeely and Hare, 1966).

No evidence of monozygotic or chimaerical twins was found in a study on 40 pigs (Baker and Andresen, 1964); there was also no such evidence in a second study in which pigs of the same blood group and sex were selected from 1384 animals from 172 litters (Baker, 1964).

Miniature Pigs

Strains of miniature pigs have been used for experimental work in reproduction, dental research, bone development, radiobiology and other fields. Haematological and biochemical values similar to those found in the usual domestic pig have been described for miniature pigs (Tegeris *et al.*, 1966).

The miniature pig was developed in an attempt to provide an animal of mature size small enough to be easily handled and maintained, and with normal physiological characteristics (Weaver and McKean, 1965; Bustad *et al.*, 1966). The Hormel and Pitman-Moore strains of miniature pig were developed in the United States by selection from wild pigs; there are naturally occurring small pigs in Indonesia, Bhutan, Nepal and Vietnam.

Haring *et al.* (1966) have described the successful production of miniature pigs, at Göttingen in Germany. The following account is based on discussions with Professor R. Gruhn and Professor D. Smidt.

The development of the miniature pig at Göttingen is currently at the stage of intensive selection for small body size coupled with high fertility. The animals are of two types, coloured and white, and they are derived from crosses between the Minnesota miniature pig (Rempel and Dettmers, 1966), the Vietnamese pig and the German Landrace. The breeding plan on which the development has taken place has involved the production of the coloured miniature pig from Minnesota–Vietnamese crosses (about 60% Minnesota and 40% Vietnamese), and of the white miniature pig from a three-way cross between the Minnesota, Vietnamese and German Landrace breeds. The white miniature pig combines small size with the white colour and high fertility of the Landrace; one of the disadvantages of some miniature pigs has been the small litter size.

Table I

Numbers and weights of the coloured and white strains of the Göttingen miniature pig at the beginning and end of a 5-year period in which selection for small size was carried out

	Coloured		White	
	1962/63	1967/68	1962/63	1967/68
Number of pigs born	6·2	5·8	6·0	7·1
Birth weight (kg)	0·51	0·50	0·59	0·51
8 weeks weight (kg)	5·7	5·0	7·3	5·3
20 weeks weight (kg)	16·8	11·5	25·3	13·0
1 year weight (kg)	46·0	29·7	61·7	35·9

Table I indicates the progress which has been made in the selection, with reference for example to the weight at 20 weeks of age. This is now 11·5 kg for the coloured pig and 13·0 kg for the white pig, reductions of 31 and 49% respectively over 5 years in the corresponding weight for age. The animals are fed twice daily, as much feed as they can consume in about half an hour on each occasion. It has been found necessary to restrict the feed intake of older animals in order to prevent them becoming excessively fat.

Over the same period of selection, however, the birth weight has changed little, from 0·51 to 0·50 kg for the coloured pig and from 0·59 to 0·51 kg for the white. The range of birth weight extends from approximately 0·25 to 0·7 kg; pigs which weigh less than 0·3 kg at birth do not survive. The mean number of pigs born in the litter has remained close to six in the coloured strain, and increased from six to seven in the white pigs; the number in an individual litter falls between three and ten.

Management of the animal is similar to that for ordinary farm pigs. The miniature pigs are normally weaned at 8 weeks of age by removing the sow. During the suckling period the pigs have had access to feed pellets of the "creep feed" (see Chapter 5) type; at weaning, the pigs are initially allowed only a limited quantity of these, in order to prevent gastro-intestinal disturbances which could follow excessive intake; the amount is increased slowly over several days from about 40 g per pig to the 50–80 g which represents the *ad libitum* intake. There is about 80% survival between birth and weaning.

The weights of the breeding animals are indicated in Table I, where 1-year-old coloured pigs are seen to weigh about 30 kg and the white pigs about 36 kg. The gilt reaches sexual maturity at about 5 months of age, and is used for breeding at about 6–7 months of age. The gestation period is close to 112 days, as with the farm pig. The boar reaches sexual maturity at an early age; living sperm can be obtained at 8–12 weeks, and spermatogenesis has been detected at 4 weeks of age. By comparison, in the German Landrace spermatogenesis is observed at 15 weeks of age at the earliest, and the usual time is 17–19 weeks. Artificial insemination has been practised with 90% success in the unit at Göttingen. Also at Göttingen, work has now begun in a new unit with the miniature pig as a specific-pathogen-free animal.

The types of miniature pig, and the research uses to which the animals have been put, have been reviewed by Bustad and McClellan (1968).

CHAPTER 12

Health and Legislation

by

G. A. EMBLETON

In all livestock enterprises, the health of the animals is a function of the level of management, and in the pig herd this depends upon the competence and conscientiousness of the stockman. The standard of husbandry is reflected in the temperament of the animals, and the herdsman who does not need protection when moving amongst his animals is rarely lacking in knowledge or concern for the health of his herd.

Experience with pigs free of enzootic pneumonia supports the view that to start building up a pig herd from hysterectomy-derived, S.P.F. or minimal disease pigs is a worthwhile investment. The terminology which is applied to gnotobiotic animals (in which association with micro-organisms is defined) is confused, but the following definitions are generally acceptable. *Germ-free pigs* are those removed aseptically from the uterus and reared under sterile conditions. *Specific pathogen-free* (*S.P.F.*) *animals* are removed from the uterus aseptically and are reared free from pathogens, although they are not necessarily free of other micro-organisms. *Minimal disease animals* are obtained in the same way, but they are reared in a conventional manner so that the microflora are normal without pathogens or other organisms which are considered to be undesirable. *Hysterectomy-derived* is a self-explanatory term, and bears no relation to the animals' microflora.

By using these animals, known pathogens are excluded at the outset, and experimental results are not influenced by subclinical disease. There is also less likelihood of scientists using animals which, although affected by disorders producing only mild signs, are markedly abnormal in the physiological sense.

Anyone contemplating the establishment of a pig unit, particularly in buildings which have not previously housed pigs, should consider using gnotobiotic animals from the outset, since restocking with S.P.F. or minimal disease pigs in buildings which have held conventionally reared pigs is a hazardous venture. Techniques for producing "clean" pigs have been described (Betts *et al.*, 1960, Waxler *et al.*, 1966) and information is accumulating regarding their commercial performance and value in research.

Disease in Relation to Age

It is accepted that the period of highest morbidity and mortality extends from birth to 4 months of age in commercial herds, and this is also the case in the experimental herd at Babraham. Surveys and single herd studies (Veterinary Investigation Service 1959, 1960; Sharpe, 1966) indicate that the highest mortality rates are the result of physical factors—trauma, starvation, congenital abnormalities—in the neonatal period and generalized infections and enteritis in pigs up to 4 months old. Clinical records at Babraham show that the two main troubles arise from traumatic lesions of the skin and sub-cutis in the first few days of life and from dysfunctions of the alimentary tract. Although the sows are retained in the herd rather longer than in commercial undertakings, fertility does not decline to a significant degree though litter numbers tend to diminish with increasing age of the sow. Occasional cases of injury, sepsis, pneumonia and generalized bacterial infection occur within all age groups, but, in general, certain diseases are most common to particular age groups as outlined in the following pages.

Disease in the New-born Pig

Deaths from cold are no longer acceptable in the new-born pig since, apart from any use of nests or bedding, infra-red heating is an inexpensive item. If a farrowing crate is used, an infra-red lamp in the creep attracts the piglets away from the sow; this lowers the incidence of the crushing of piglets by the sow, which is the cause of most neonatal deaths.

SURGICAL CONDITIONS

An occasional structural defect such as cleft palate, umbilical hernia, scrotal hernia or atresia ani can occur in any population as a simple accident of development and is not necessarily an inherited defect. Genetically determined defects are not uncommon in pigs, and herd records will show which common ancestor should be culled if the incidence of affected pigs is above average. Hydrocephalus and atresia ani ultimately prove fatal, although it is sometimes possible to correct the latter by surgery. If correction of any genetic defect is successful, care should be taken that the animal is not used for breeding. Umbilical hernia, despite the large hernial sac which occasionally results, is rarely an indication for surgery.

Scrotal hernia is usually unilateral, rarely becomes complicated, and is best corrected under general anaesthesia when castration is carried out. The two sides of the scrotum of affected pigs differ markedly in size and consistency due to fat or small intestine entering the peritoneal sac, and this disparity is corrected if the pig is suspended by the hind limbs allowing the ectopic viscera to return to the abdomen. Should the presence of scrotal hernia be signalled by protrusion of a viscus after castration, return of the viscus to the abdomen followed by skin suturing may be adequate, if cosmetically imperfect; speedy action is the keynote of success.

INFECTIONS

The aetiology of a severe circumoral and facial cellulitis is obscure, but the jostling for teats which occurs at feeding time, and the exposure of the animals to the sharp canine teeth of their litter mates are certainly contributory factors. The removal of the canine teeth with specially designed pincers is a task requiring some degree of skill, but it can be performed by the herdsman with little discomfort to the piglets.

Contusions of the anterior surface of the carpus and all joints below it in the forelimb, and corresponding areas in the hind limb, appear to arise as a result of the jostling and nuzzling activity at the teats during suckling. Almost all piglets sustain these lesions, but on concrete floors with an abrasive surface the lesions extend into the deeper layers. On inadequately cleaned floors infection supervenes, progressing in some cases to synovitis, peri-arthritis and suppurative arthritis, any or all of which may appear long after the initial lesion has healed. A non-abrasive surface in the farrowing pen and strict hygiene are the only effective preventive measures (Penny *et al.*, 1965). If infection does occur, parenteral antibiotic treatment is essential to avoid complications.

The significance and mode of action of *E. coli* in swine enteric disorders are discussed briefly but informatively by Stevens (1963), and the various syndromes described by him are depressingly familiar to all clinicians. The most common feature is diarrhoea, or scour as it is termed in agricultural parlance. Raised body temperature is not a constant feature, nor are dehydration and anorexia except in the later stages of the disease.

Neonatal scour occurs between birth and the 7th day, with a peak at the 3rd day. When *E. coli* is involved, infection appears to be acute, causing scour or sudden death in a number of animals simultaneously. A proportion of animals which recover from the acute phase will remain chronically unthrifty, feeding less, drinking more and suffering intermittent bouts of diarrhoea. Isolation of haemolytic *E. coli* or a strain known to be pathogenic in pigs should be followed up by treatment started at the first sign of disease. Scour at this time and at the 3-week period can usually be ignored as being non-infectious if the sow is seen to be eating the faeces from affected pigs (H. Williams-Smith, personal communication). Though not an edifying sight this can be reassuring.

Whilst opinions differ as to the efficacy of *E. coli* vaccines as a prophylactic measure, in a herd trial it was found that vaccinated sows reared 2% more piglets to weaning than the unvaccinated sows.

OTHER CONDITIONS

Hypoglycaemia is the terminal phase of a number of syndromes (Goodwin, 1955), including agalactia in the sow. Hypoglycaemic piglets are usually weak, have a subnormal temperature, rapid heart rate and exhibit mouth breathing and cyclic convulsions. The treatment of uncomplicated hypoglycaemia is discussed under diseases associated with parturition.

Haemolytic disease of the new-born has become rare since vaccination with crystal violet swine fever vaccine (which contained whole blood) has been discontinued. The visible signs are the classic ones of haemolysis, with jaundice being particularly obvious in the white pig; the diagnosis is described by Goodwin (1957).

Disease in the Suckled Pig

Diagnostic signs in the 3–8 week old pig are normally better defined than in the new-born. At this age the healthy pig closely resembles his counterpart in the nursery picture book, being rotund, pink, sparsely covered by fine hair and inquisitive except when asleep or feeding. Experience and observation eventually enable one to pick out the animal which, although not thin, excessively hairy, shivering or in a state of rapid decline is nonetheless certainly not normal.

DIGESTIVE TRACT

Diseases and disorders of the digestive system are the cause of the greatest loss of condition and of deaths of pigs during this period. Coliform enteritis at 3 weeks of age may produce scour lasting 2 or 3 days; it often affects the whole litter and leaves some animals in poor health and in an unthrifty condition for weeks or for the remainder of their lives. Treatment with a specific drug, accompanied by a high standard of hygiene, is more efficient and more profitable than blanket medication with one or more antibiotic drugs. Indiscriminate treatment is followed sooner or later by the emergence of a resistant strain of organism.

BLOOD

Piglets reared indoors from birth are especially prone to iron deficiency anaemia since at this period of rapid growth the daily iron requirement is approximately 7 mg, only one-sixth of which is normally present in the sow's milk. Injections of iron as an iron-dextran complex when the piglets are 3 days old compensate for the deficiency, but such treatment introduces those hazards which are inherent in any injection, including needle breakage, infection and nerve damage, the first two of these being of particular concern where surplus pigs are sold for bacon. For some time a powder containing iron fumarate and copper has been in use at Babraham; the piglets take it readily, and it introduces them to solid food at an early age. The cost per pig is significantly below that of the injection, and there is the added advantage that the herdsman automatically inspects the litter as he replenishes the powder in the creep daily.

Thrombocytopenic purpura, believed to be an iso-immunization reaction, is not uncommon in unweaned pigs, and is characterized by petechial haemorrhages in the skin and other organs (Nordstoga, 1965). It is important for two reasons; first its resemblance to swine fever, and second the fact that it is a disease in which the sow rather than the boar should be culled, since once the sow is sensitized succeeding litters will be affected.

INFECTIONS

Swine pox is most likely in pigs at the late pre-weaning stage. It follows the typical pattern seen in other pock diseases, although the intermediate stages may occur without being noted. The lesions, initially raised red areas on the skin of the belly, thighs and flanks, progress to scabby patches, sometimes fairly extensive, and occasionally death occurs. Isolation of the sow and the infected litter is advisable, although many cases occur in only a single piglet from a litter and semi-isolation prevents further spread.

Enzootic pneumonia, attributed now to infection with a mycoplasma, is noticed from 3 weeks of age onwards and gives rise to coughing and, later, in the absence of secondary bacterial infection, the symptoms associated with a mild pneumonia which are laboured breathing, elevated temperature and roughness of the coat. Macroscopically the pathognomonic lesion is said to be consolidation of parts of the anterior lobes of the lung, and treatment is directed towards the prevention or elimination of secondary infections. Elimination of this disease is difficult and expensive, and it is preferable to begin with pigs from a herd certified free of enzootic pneumonia.

Leptospiral infection is possible where rats are abundant or where they are able to contaminate the food or water. Vermin control is the essential preventive measure, and this is important since human infection—the organism gaining entry through unnoticed fissures in the skin—is often fatal despite treatment. Normally more than one member of the litter is affected, and a febrile illness results with jaundice appearing 3–4 days after the infection, which is usually fatal. Treatment involves dosing with the appropriate antibiotic.

Infections acquired in the immediately post-natal period through the navel or through lesions on the skin often reach a climax during the next month. Swelling of one or more joints with marked lameness or anorexia are a sequel to infection by either route whilst serositis (pericarditis, pleurisy and peritonitis alone or in combination) with massive effusion may cause sudden death, though the lesions are of a chronic inflammatory nature.

PARASITES

Ascarid infection may produce pneumonia due to the larval forms of *A. lumbricoides* migrating through the lungs. Jaundice and enteritis may result from migrations in the digestive tract. The examination of faeces for worms and worm eggs assists diagnosis, but the worms are best eliminated at source (the sow) by the McLean County system of washing and dosing the sows before parturition (British Veterinary Association, 1967).

Disease in Weaned Pigs

ENTERITIS ASSOCIATED WITH *E. COLI*

This commonly occurs 7–14 days after weaning, irrespective of the age at which weaning is carried out. An epidemiological study based on accurate records and clinical histories will show if the occurrence of the disease follows a pattern with respect to weaning or subsequent movements. Bacteriological examination of affected animals and treatment with a specific drug in the litters which are at risk can greatly reduce the incidence.

SALMONELLA ENTERITIS

Salmonella enteritis has scouring as its cardinal feature. It is followed by chronic unthriftiness due to progressive necrosis of the bowel wall. Bacteriological examination is necessary to detect the affected animals, to eliminate carriers and to prevent the spread of infection. Treatment and prevention are the province of the veterinary surgeon who will also advise on the Public Health aspect of the disease, and the precautions necessary to prevent human infection.

TRANSMISSIBLE GASTRO-ENTERITIS

A viral disease, affecting pigs of all ages. It is usually fatal in the newborn and less severe in adult pigs. The feeding of whole blood or serum from recovered animals appears at present to be the only effective treatment (Noble, 1964). Pigs which recover appear to develop a degree of immunity which lasts for some years. Should an outbreak occur, pregnant sows should be mixed with the infected animals in the hope that by contracting the disease and developing immunity before farrowing, the sow and litter will not become infected.

SWINE FEVER

This may appear or be noticed first as scouring in a number of animals in a group. In the early stages the pigs are dejected, body temperature may range from 40–42·4°C and although they refuse food the animals may be excessively thirsty. Some cases die within 2–3 days, whilst others develop oculo-nasal discharge, pin-point areas of skin haemorrhage and reddish discoloration of the ears and skin of the belly, and sometimes pneumonia. The existence of this disease should be borne in mind whenever there are a number of pigs present with scour, cough or other signs of systematic disease, and advice should be sought in the early stages.

ERYSIPELAS

Erysipelas may affect pigs of all ages; in unvaccinated herds it is most common in spring and autumn. Acute septicaemic, sub-acute cutaneous and chronic forms of the disease are recognized. The cutaneous form is characterized by raised reddened areas of skin, approximately diamond-shaped, raised body temperature and inappetance. The acute septicaemic form is often fatal, but sub-acute cases after recovery may later exhibit arthritic symptoms, and death may result from lesions on the heart valves following endothelial damage. All pigs which are in contact should receive antiserum. Vaccination should be carried out twice each year; it is virtually 100% effective in preventing the disease. The organism (*Erysipelothrix insidiosa*) is capable of causing disease in man.

OTHER CONDITIONS

A syndrome characterized by massive haemorrhage into the lumen of the gut followed by sudden death has been described by Jones (1967). He suggests that this may be the result of a hypersensitive reaction since inflammatory lesions are absent. Commonly occurring in whey-fed pigs, it is not confined to animals on such a diet but appears to occur only in fattening pigs.

A common sequel to enteritis is rectal prolapse, usually confined to one or two members of a group. The prolapsed portion of rectum is often attacked by others in a litter, but most simple cases and some cases with extensive lacerations and congestion will recover after reduction, which may require anaesthesia, or amputation. Predisposing causes should be eliminated.

Intussusception occurs under the same conditions as rectal prolapse, but is rarely diagnosed until it has been in existence for a length of time which has allowed peritonitis to develop; this effectively precludes operation.

Apart from lesions subsequent to neonatal infection of the skin, abscesses occur in and around the vertebral column as a sequel to tail biting by penmates; lameness, paraplegia or paralysis result. Treatment is rarely successful, but the initial trauma may be avoided by suspending a chain or spreading some straw in the pen for the pigs to play with, thus relieving the boredom which is said to result in this type of behaviour.

Disease in the Pregnant and Parturient Sow

PREGNANCY

There are few conditions which are specifically associated with pregnancy, but one of these is infection of the urinary tract by *Corynebacterium suis*, usually concurrently with a streptococcus. The aetiology of the condition is not certain, but pressure on the ureters from the pregnant uterus appears to be a contributory factor (Soltys and Spratling, 1957). Symptoms are most common in the first third of pregnancy when turbid and sometimes bloody urine is voided, a mucoid vaginal discharge is produced, and there may be pain on pressure in the lumbar region. The temperature rarely rises above normal. Experience has proved that if the organism can be recovered from the urine treatment is valueless, although most cases will live until the last few days of pregnancy. In the absence of bacteriological evidence, animals exhibiting signs of urinary disease should be treated with high therapeutic doses of chloramphenicol.

PARTURITION

The pregnant sow frequently refuses all food in the 48 hours preceding parturition. Although it is unusual for labour to start later than 24 hours after colostrum is secreted, in an otherwise healthy sow delay need cause no concern.

Uterine inertia usually develops as the result of a very large or small number of foetuses. The signs are loss of foetal fluids followed by weak or ineffectual expulsive efforts. Vaginal exploration allows the presenting foetuses to be delivered after correcting any malpresentations or postures. The intramuscular injection of pituitrin may be repeated, if necessary, at 20 min intervals; it usually has the desired effect.

Caesarean section in the sow is no more hazardous than in other species though conduct of the operation under pentobarbitone sodium alone can lead to prolonged (up to 12 hours) recumbency of mother and offspring. Such recumbency is effectively terminated by intravenous bemegride, and this treatment is essential to prevent primary hypoglycaemia in the piglets.

AFTER PARTURITION

Probably the most common syndrome seen in the newly farrowed sow is that collection of signs known as farrowing fever, the mastitis–metritis–

agalactia syndrome (Martin *et al.*, 1967), or porcine agalactia (Swarbrick, 1968). Agalactia is the one constant feature, and is usually discovered when the piglets are seen to be sucking ineffectually, squeaking, losing the sheen on their bodies and losing weight. A watery or white, mucoid vaginal discharge may be present and the mammary glands—particularly at the front or hind end—may be tense and swollen and fever is sometimes present. A body temperature of 38·2–39·5°C precludes inflammatory lesions; the mammary turgidity results from congestion rather than from inflammation. No parturient sow should be inspected less often than once every 8 hours, and if treatment of the afebrile case is instituted at the onset of the disease, a single dose of 40 i.u. A.C.T.H., regardless of the nature or quantity of the vaginal discharge which is exhibited by 94% of affected sows, will effect a cure. Prompt diagnosis and treatment are essential if the piglets are to be saved. Piglets can be carried over the critical period by (i) administration of 5–20 ml of 40% glucose by mouth every 4 to 6 hours using a pipette teat attached to a syringe, and (ii) allowing them access to a watery solution of sow milk substitute in a shallow tray. Once they have tasted this from the attendant's finger, through falling into the tray, or licking their companions, they will drink it readily, and the strength of the mixture can than be increased according to the maker's instructions. Patience is essential in these manoeuvres which can easily be 90% successful.

Mastitis and metritis undoubtedly exist as clinical entities, and are frequently accompanied by, or induce, agalactia. In these cases treatment usually depends on antibiotic, anti-inflammatory and oxytocic drugs and it is important to identify the organism present in the secretions or discharge.

Hypocalcaemia presents as an initially uneasy then recumbent post-parturient sow, showing little or no rise in temperature, with a low serum calcium level. The slow intravenous injection of calcium borogluconate is a specific cure.

Occasionally a sow or gilt will savage her litter at birth. This is a deliberate process, and differs entirely from the behaviour of the animal with mastitis who lies on her abdomen or sits up, presumably to prevent the litter nudging a tender abdomen, and occasionally pushes piglets away with her snout. In savaging, the piglets are bitten at birth or as soon as they crawl within reach of the sow's jaws. Whilst it is usually unwise to upset the parturient sow to avoid mismothering, in this instance no further harm will result from snaring the sow by a noose round the upper jaw to facilitate injection into an ear vein. Chlorpromazine should be given at a dose rate which abolishes aggressive reactions when a piglet is placed near to or on the sow's head. When the effect of the drug has worn off, mother and offspring are normally at peace; in the interim the piglets have fed. Succeeding farrowings are usually uneventful.

Oral dosing is difficult except in the very young or with substances palatable enough to be included in the food. This route of administration is especially useful in enteric infections, and the "Varidoser" gun with cartridges of the required drug is most useful.

Notifiable Diseases (Great Britain)

These diseases are important because of their infectivity and their effect on animals in other herds or on man.

Foot-and-mouth disease is not fatal in all cases, and is more benign in pigs than in other susceptible animals, so that on a number of occasions pigs have been the cause of widespread outbreaks. Any lame pigs should be thoroughly examined, and where lameness is accompanied by blisters around the hoof or on the lips or tongue, a high temperature and reluctance to feed, expert advice should be obtained at once. Doubtful cases, or a number of cases of lameness occurring at one time, should be regarded with suspicion. Preceding lameness by 24–48 hours is a period of inappetance and raised body temperature, when the animals are reluctant to stand and, if forced to rise, quickly lie down again. If the infection is confirmed, the entire herd, and other susceptible animals in contact, are slaughtered by the authorities.

Anthrax, caused by *Bacillus anthracis*, is another disease which is less dramatic in the pig than in other species although any case of sudden death should be regarded as anthrax until proved otherwise. Whilst it is an offence to fail to report a case of the disease it is also an offence to open the carcase of a suspected case. Diagnosis is best left to the veterinarian, who may take samples from the oedematous tissue or body cavities. Sub-mandibular oedema is the most common sign, due to the ingestion of contaminated food. Involvement of the regional lymph nodes also occurs, and affected pigs commonly have a temperature exceeding 41°C.

Swine fever is specific to pigs, and in its various forms may present as the unexplained death of a number of pigs simultaneously, fever accompanied by constipation and followed by intractable diarrhoea, or large areas of skin reddening and progressing to necrosis. Again, suspicious cases should be referred for veterinary advice, and compulsory slaughter of the entire herd is carried out by the Ministry of Agriculture if the disease is confirmed.

It is not an offence to be unable to diagnose these diseases but it is an offence to fail to suspect the disease and act accordingly.

Other Legislation (Great Britain)

Under the Diseases of Animals Act there are a number of Orders relating to matters other than the Notifiable Diseases, and pig keepers in particular should be aware of these.

The Movement of Animals (Records) Order (1960) requires all pig owners to record, in a specified form, movement of animals on and off the premises.

The Regulation of Movement of Swine Order (1959) requires that all purchased pigs shall remain in isolation on the premises of the buyer for 28 days. The only exception is in the case of pigs which are to be sent direct to a slaughter house from the buyer's premises, for which a licence can be obtained from the Local Authority.

The Diseases of Animals (Waste Foods) Order (1957) makes it an offence to feed any waste food to pigs without first boiling it.

The Foot and Mouth Disease (Packing Material Amendment) Order (1926) prohibits the use as bedding of hay and straw originating as packing material.

The Licensing of Boars (England and Wales) Order (1956) requires all entire (uncastrated) male pigs to be licensed before they attain the age of 6 months.

Information on these and other relevant sections of the Act can be obtained from the offices of the Ministry of Agriculture.

Legislation relating to the disposal of waste and effluent must also be borne in mind when designing a pig unit, and where a suitable incinerator is not included on the site carcass disposal may have to be arranged with the Local Authority.

The position of the pig unit in relation to other buildings, especially dwelling houses, must comply with certain regulations which are the subject of local authority bye-laws. These may vary between districts, and should be consulted in the early planning stages.

If animals from which tissues are to be removed by the owner are sent for slaughter for human consumption, the Meat Inspector of the Public Health department should be advised in good time. If any part of an animal is removed from the slaughterhouse before it has been passed by the inspector, the entire carcass will be condemned.

Salmonellosis from pigs sent for slaughter from the herd will result in investigations by the appropriate departments concerned with human health, since food poisoning due to this group of organisms is quite common and often serious.

Zoonoses

In common with other species, the pig is subject to a number of diseases transmissible to man. It is wise to inform one's medical attendant of the nature of the work on which one is engaged so that those physicians used to dealing with urban populations are reminded of the conditions commonly seen in rural areas but which are almost unknown in towns.

Anthrax is normally acquired through abrasions of the skin, where a local "boil" is produced, the "malignant pustule" of cutaneous anthrax. Any person coming into contact with a case of anthrax is well advised to see his own doctor at once.

Erysipelothrix insidiosa, the organism causing swine erysipelas, produces erysipeloid in man. This causes skin lesions—which have been mistakenly diagnosed as shingles—accompanied by systemic upsets which vary in severity.

Listeria monocytogenes, an organism associated with infections of the central nervous system in pigs, can cause a fatal meningitis or meningo-encephalitis in man, and has been isolated from premature infants.

Enteritis or dysentery due to infection by species of salmonella requires prompt treatment. Persons working with pigs should not only have adequate toilet facilities, but should be encouraged to use them, especially before eating and drinking, since infection is normally conveyed by the hands to the mouth.

As mentioned earlier, Leptospirosis (Swineherds' disease, or Weil's disease) and Canicola Fever can be contracted from rats or their excreta via the mouth or broken skin, and gloves should always be worn if rat carcases or contaminated food have to be handled. In any illness, recent contact with rats should be included in the medical history.

Mange mites may cause a transient skin irritation but as the mite is unable to survive away from the pig the condition is self limiting, provided re-infection is prevented.

Ringworm is rare in pigs in this country but it is probable that in common with ringworm fungi affecting other species, transfer between pig and man is possible in either direction.

Opportunities for contact with the tetanus organism are always present when dealing with pigs and other farm animals, and people engaged in this type of work are well advised to undergo a course of anti-tetanus vaccinations rather than wait until lacerations necessitate treatment with antiserum.

Finally, in recent years it has been shown that a pulmonary condition described as "Farmer's lung" is due to the inhalation of mould spores present in spoilt hay. Workers dealing with such material or frequently engaged in mixing food should wear face masks of an approved design.

Preventive Medicine

The efficient running of the pig unit is greatly facilitated by accurate records. At Babraham each breeding animal has an individual record card on which is entered its name, herd number, breed, sex, date of birth, weights at birth and weaning, date of purchase, sire's and dam's numbers and the date of and reason for disposal. The sow's card has spaces for service dates and the boar's number, farrowing date with the number, sex and weight of the piglet's born and their ear numbers and weaning weights. The reverse side of the card shows the date and nature of all illnesses, vaccinations or other prophylactic measures, and the results of diagnostic tests. The movement of animals is also recorded in the official movements book.

The following scheme has been found valuable in keeping disease to a minimum, and it requires little conscious effort for its implementation.

Approximately 1 week before farrowing, the pregnant sow is brought into the concrete yard adjoining the farrowing house where she is scrubbed with a pig shampoo, and, after the final rinse, with a wash containing gamma-benzene hexachloride, which is effective against mange mites. She is then transferred to a pen in the farrowing house which has been scrubbed and disinfected, and with her first feed is given a dose of an anthelminthic to remove the common roundworms of pigs.

At farrowing time the sow is left undisturbed until farrowing is complete and after the piglets have settled down the total litter weight and other information is recorded. The day after farrowing the sow is offered the usual food instead of the wet bran which has been offered during the 3 days before parturition.

When the piglets are 3 days old, iron fumarate powder is placed in the creep at the level advised by the manufacturer. Castration is performed at 3 weeks of age. Before weaning, when the sow (never the litter) is removed from the pen, sow and litter are vaccinated against erysipelas.

All breeding animals are tuberculin tested at 6 monthly intervals, and boars, sows which have not farrowed during the previous 6 months and gilts intended for breeding are vaccinated against erysipelas in March and September. Any animal reacting to tuberculin should be disposed of and the herd tested again after an interval of 90 days.

Pigs are purchased only from herds registered with the Pig Health Control Association, and the lungs of all pigs dying or sent for slaughter are examined for evidence of enzootic pneumonia.

Since only adult animals have access to pasture, routine worming of young stock is not practised, and random samples have shown no evidence of helminth infection in weaned or fattening pigs.

All purchased pigs are isolated for 60 days in a unit separated from the main Institute, where they are supervised by the technician on the veterinary surgeon's staff. During this time he has no contact with other animals in the Institute, and a separate set of garments is used and remains in the isolation unit until laundered.

The most common diseases of pigs, and the best way to avoid them, have been indicated in this chapter. Two factors contributing to healthy growth and development are the minimum number of changes in housing and the gradual introduction of new food. The unit should be planned in such a way that if experimental methods militate against health, as they do for example in experiments at unusual temperatures, the animals at risk should be isolated permanently from the rest of the herd from the start of the experiment. Experiments which involve housing pigs at unusually high or low ambient temperature can lead to digestive disorders and respiratory disease. Pigs on weaner type food at the start of an experiment may be unable to assimilate this ration for the duration of the experiment, which nevertheless precludes a change of diet. Since latent gut infection may be stimulated under these conditions, a sudden change in feeding on returning to the main herd will exacerbate an already unhealthy situation. If they must be returned to the main herd, such pigs should be housed separately until they are established in the routine of the herd and are otherwise normal. In this way pathogens are not passaged and then transmitted throughout the herd.

If a veterinarian is not included on the staff it is wise to establish a system of advisory visits with a veterinary surgeon in the area, who is acquainted with local conditions and who may well be able to pinpoint a potential source of trouble before experiments are ruined, or before disease becomes established in the herd.

Veterinary advice is essential before the use of antibiotics as food additives is decided upon. Supplementation of foods with antibiotics is used to prevent disease under modern husbandry systems which are designed to increase productivity. Unless the pig unit is intended to study pig husbandry techniques, intensive units should not be used, and antibiotic supplements should not be necessary.

The publications on pig diseases published by Field (1964) and the British Veterinary Association (1967) are excellent sources of information on the more common pig diseases, and a comprehensive reference volume has been edited by Dunne (1970).

References

Aamdal, J. and Högset, I. (1957). Artificial insemination in swine. *J. Am. vet. med. Ass.* **131**, 59–64.

Adams, W. E. (1958). "The Comparative Morphology of the Carotid Body and Carotid Sinus." Charles C. Thomas, Springfield, U.S.A.

Adrian, E. D. (1943a). Sensory areas of the brain. *Lancet* ii, 33–36.

Adrian, E. D. (1943b). Afferent areas in the brain of ungulates. *Brain* **66**, 89–103.

Agricultural Research Council (1967). The nutrient requirements of farm livestock. No. 3 Pigs: technical reviews and summaries. HMSO, London.

Ajmal, M. (1970). Chronic proliferative arthritis in swine in relation to human rheumatoid arthritis. *Vet. Bull.* **40**, 1–8.

Aldinger, S. M., Speer, V. C., Hays, V. W. and Catron, D. V. (1959). Effect of saccharin on consumption of starter rations by baby pigs. *J. Anim. Sci.* **18**, 1350–1355.

Altmann, M. (1939). Behaviour of the sow in relation to sex cycle. *Am. J. Physiol.* **126**, 421.

Altmann, M. (1941). A study of patterns of activity and neighbourly relations in swine. *J. comp. Psychol.* **31**, 473–479.

Anderson, D. M. and Ash, R. W. (1970). Experimental diabetes mellitus in the pig. *Proc. Nutr. Soc.* **29**, 27A.

Anderson, D. M. and Elsley, F. W. H. (1969). A note on the use of indwelling catheters in conscious adult pigs. *J. agric. Sci., Camb.* **72**, 415–477.

Anderson, L. L., Butcher, R. L. and Melampy, R. M. (1961). Subtotal hysterectomy and ovarian function in gilts. *Endocrinology* **69**, 571–580.

Andresen, E. (1962). Blood groups in pigs. *Ann. N.Y. Acad. Sci.* **97**, 205–225.

Andresen, E. (1963). "A Study of the Blood Groups of the Pig". Munksgaard, Copenhagen.

Andresen, E. and Baker, L. N. (1963). Haemolytic disease in pigs caused by anti-B_a1. *J. Anim. Sci.* **22**, 720–725.

Andresen, E., Preston, K. S., Ramsey, F. K. and Baker, L. N. (1965). Further studies on haemolytic disease in pigs caused by anti-B_a. *Am. J. vet. Res.* **26**, 303–309.

Arsac, M. and Rérat, A. (1962). Technique de fistulation de la veine porte chez le porc. *Annls Biol. anim. Biochim. Biophys.* **2**, 335–343.

Asdell, S. A. (1964). "Patterns of Mammalian Reproduction" (2nd edition). Constable, London.

Asplund, J. M., Grummer, R. H. and Phillips, P. H. (1962). Absorption of colostral gamma-globulins and insulin by the new-born pig. *J. Anim. Sci.* **21**, 412–413.

Attinger, E. O. and Cahill, J. M. (1960). Cardiopulmonary mechanics in anesthetized pigs and dogs. *Am. J. Physiol.* **198**, 346–348.

Bailey, C. B., Kitts, W. D. and Wood, A. J. (1956). The development of the digestive enzyme system of the pig during its preweaning phase of growth. B, intestinal lactase, sucrase and maltase. *Can. J. agric. Sci.* **36**, 51–55.

Baker, L. N. (1964). Skin grafting in pigs: a search for monozygotic twins. *Transplantation* **2**, 434–436.

Baker, L. N. and Andresen, E. (1963). Full thickness skin grafting in pigs. "Genetics Today". Proc. XI Int. Conf. Genetics. 1. Iowa State University.

Baker, L. N. and Andresen, E. (1964). Skin grafting in pigs: evidence for a histoincompatibility mechanism. *Transplantation* **2**, 118–119.

Baldwin, B. A. and Ingram, D. L. (1967a). Behavioural thermoregulation in pigs. *Physiol. Behav.* **2**, 15–21.

Baldwin, B. A. and Ingram, D. L. (1967b). The effect of heating and cooling the hypothalamus on behavioural thermoregulation in the pig. *J. Physiol., Lond.* **191**, 375–392.

Baldwin, B. A. and Ingram, D. L. (1968a). Factors influencing behavioural thermoregulation in the pig. *Physiol. Behav.* **3**, 409–415.

Baldwin, B. A. and Ingram, D. L. (1968b). The effects of food intake and acclimatization to temperature on behavioural thermoregulation in pigs and mice. *Physiol. Behav.* **3**, 395–400.

Barber, R. S., Braude, R. and Mitchell, K. G. (1955). Studies on milk production of Large White pigs. *J. agric. Sci., Camb.* **46**, 99–118.

Barker, C. A. V. (1967). Control of estrus in pigs fed ICI Compound 33828. *Can. vet. J.* **8**, 39–46.

Barone, R. (1952). Topographie des viscères, chez le porc et la truie. (Topography of the organs of the pig and sow). *Revue Méd. vét.* **103**, 688–697.

Bartels, H. and Harms, H. (1959). Oxygen dissociation curves in the blood of man, rabbit, guinea pig, dog, cat, pig, ox and sheep. *Pflügers Arch. ges. Physiol.* **268**, 334–365.

Becker, D. E., Ullrey, D. E., Terrill, S. W. and Notzold, R. A. (1954). Failure of the new-born pig to utilize dietary sucrose. *Science, N.Y.* **120**, 345–346.

Berg, R. (1964a). Über den Entwicklungsgrad des Koronarge fässmusters beim Hausschwein (Sus scrofa domesticus), (Degree of development of the pattern of coronary vessels in the domestic pig). *Anat. Anz.* **115**, 193–204.

Berg, R. (1964b). Beitrag zur Phylogenese des Verhaltens der Koronararterien zum Myokard beim Hausschwein (Sus scrofa domesticus), (Phylogenesis of the disposition of the coronary arteries to the myocardium in the domestic pig). *Anat. Anz.* **115**, 184–192.

Berg, R. (1965). Zur Morphologie der Koronargefässe des Schweines unter besonderer Berücksichtigung ihres Verhaltens zum Myokard (Morphology of the coronary blood vessels of swine with special reference to their relations to the myocardium). *Arch. exp. VetMed.* **19**, 1145–1307.

Betteridge, K. J. and Raeside, J. I. (1962). Observation of the ovary by peritoneal cannulation in pigs. *Res. vet. Sci.* **3**, 390–398.

Betts, A. O., Lamont, P. H. and Littlewort, M. C. G. (1960). The production by hysterectomy of pathogen-free, colostrum-deprived pigs and the foundation of a minimal-disease herd. *Vet. Rec.* **72**, 461–468.

Binns, R. M. (1967). Bone marrow and lymphoid cell injection of the pig foetus resulting in transplantation tolerance or immunity, and immunoglobulin production. *Nature, Lond.* **214**, 179–181.

Binns, R. M. and Hall, J. G. (1966). The paucity of lymphocytes in the lymph of unanaesthetised pigs. *Br. J. exp. Path.* **47**, 275–280.

Binns, R. M., Harrison, F. A. and Heap, R. B. (1967). Transplantation of the ovary in the pig and in the pregnant sheep. *Acta endocr., Copnh.* Suppl. 119, 193.

Bloodworth, J. M. B., Jr., Gutgesell, H. P., Jr and Engerman, R. L. (1965). Retinal vasculature of the pig. Light and electron microscope studies. *Expl Eye Res.* **4**, 174–178.

Boda, J. M. (1959). The Estrous Cycle of the Sow. *In* "Reproduction in Domestic Animals", Vol. 1. Academic Press, London and New York.

Booth, N. H., Bredeck, H. E. and Herin, R. A. (1960). Baroceptor reflex mechanisms in swine. *Am. J. Physiol.* **199**, 1189–1191.

Booth, N. H., Bredeck, H. E. and Herin, R. A. (1966). Baroceptor and chemoceptor reflex mechanisms in swine. *In* "Swine in Biochemical Research" (L. K. Bustad and R. O. McClellan, eds), pp. 331–346. Battelle Memorial Institute.

Bowland, J. P. (1966). Swine milk composition—a summary. *In* Swine in Biomedical Research" (L. K. Bustad and R. O. McClellan, eds), pp. 97–107. Battelle Memorial Institute.

Boylan, W. J., Rempel, W. E. and Comstock, R. E. (1961). Heritability of litter size in swine. *J. Anim. Sci.* **20**, 506–568.

Braude, R. (1948). Some observations on the behaviour of pigs in an experimental piggery. *Bull. Anim. Behav.* **6**, 17–25.

Braude, R. (1954). Pig Nutrition. *In* "Progress in the Physiology of Farm Animals", Vol. 1. (J. Hammond, ed.). Butterworth, London.

Braude, R., Clarke, P. M. and Mitchell, K. G. (1955). Analysis of the breeding records of a herd of pigs. *J. agric. Sci., Camb.* **45**, 19–27.

Braude, R., Dollar, A. M., Mitchell, K. G., Porter, J. W. G. and Wallser, D. M. (1958). Proteolytic enzymes and the clotting of milk in the stomach of the young pig. *Proc. Nutr. Soc.* **17**, xlix.

Bray, C. I. and Singletary, C. B. (1948). Effect of hog wallows on gains of fattening swine. *J. Anim. Sci.* **7**, 521–522.

Breazile, J. E. (1967). The cytoarchitecture of the brain stem of the domestic pig. *J. comp. Neurol.* **129**, 169–188.

Breazile, J. E., Swafford, B. C. and Thompson, W. D. (1966). Study of the motor cortex of the domestic pig. *Am. J. vet. Res.* **27**, 1369–1373.

Briggs, H. M. (1969). "Modern Breeds of Livestock", 3rd edition. Macmillan, London.

British Veterinary Association (1967). "Husbandry and Diseases of Pigs" 4th edition. British Veterinary Association, London.

Bromberg, B. E., Song, I. C., Koehnlein, E. and Mohn, M. P. (1964). Non-suture fixation of split-thickness skin grafts. *Surgery* **55**, 846–853.

Brooks, C. C. and Davis, J. W. (1969). Changes in hematology of the perinatal pig. *J. Anim. Sci.* **28**, 517–522.

Bruner, D. W., Brown, R. G., Hull, F. E. and Kinkaid, A. S. (1949). Blood factors and baby pig anaemia. *J. Am. vet. med. Ass.* **115**, 94–96.

Bryden, W. (1933). The chromosomes of the pig. *Cytologia* **5**, 149–153.

Burger, J. F. (1952). Sex physiology of pigs. *Onderstepoort. J. vet. Sci. Anim. Ind.* Suppl. 1, 3–218.

Bush, J. A., Jensen, W. N., Cartwright, G. E. and Wintrobe, M. M. (1955a). Blood volume studies in normal and anemic swine. *Am. J. Physiol.* **181**, 9–14.

Bush, J. A., Berlin, N. I., Jensen, W. N., Brill, A. B., Cartwright, G. E. and Wintrobe, M. M. (1955b). Erythrocyte life span in growing swine as determined by glycine-2-C^{14}. *J. exp. Med.* **101**, 451–459.

Bustad, L. K. and McClellan, R. O. (1966). "Swine in Biomedical Research". Battelle Memorial Institute.

Bustad, L. K. and McClellan, R. O. (1968). Miniature swine: development, management and utilization. *Lab. Anim. Care* **18**, 280–287.

Bustad, L. K., Horstman, V. G. and England, D. C. (1966). Development of Hanford miniature swine. *In* "Swine in Biomedical Research", (L. K. Bustad and R. O. McClellan, eds), pp. 769–774. Battelle Memorial Institute.

Cairnie, A. B. and Pullar, J. D. (1959). An investigation into the efficient use of time in the calorimetric measurement of heat output. *Br. J. Nutr.* **13**, 431–439.

Calne, R. Y., White, H. J. O., Yoffa, D. E., Maginn, R. R., Binns, R. M., Samuel, J. R. and Molina, V. P. (1967a). Observations of orthotopic liver transplantation in the pig. *Br. med. J.* **2**, 478–480.

Calne, R. Y., White, H. J. O., Yoffa, D. E., Binns, R. M., Maginn, R. R., Herbertson, R. M., Millard, P. R., Molina, V. P. and Davis, D. R. (1967b). Prolonged survival of liver transplants in the pig. *Br. med. J.* **4**, 645–648.

Carle, B. N. and Dewhirst, W. H. (1942). A method for bleeding swine. *J. Am. vet. med. Ass.* **101**, 495–496.

Carmichael, W. J. and Rice, J. B. (1920). Variations in farrow, with special reference to the birth weight of pigs. *Bull. Ill. agric. Exp. Stn* **226**, 67–95.

Casida, L. E. (1935). Prepubertal development of the pig ovary and its relation to stimulation with gonadotrophic hormones. *Anat. Rec.* **61**, 389–396.

Castle, E. J. and Castle, M. E. (1956). The rate of passage of food through the alimentary tract of pigs. *J. agric. Sci., Camb.* **47**, 196–203.

Castle, E. J. and Castle, M. E. (1957). Further studies of the rate of passage of food through the alimentary tract of pigs. *J. agric. Sci., Camb.* **49**, 106–112.

Christensen, G. C. and Campeti, F. L. (1959). Anatomic and functional studies of the coronary circulation in the dog and pig. *Am. J. vet. Res.* **20**, 18–26.

Christison, G. I. and Curtin, T. M. (1969). A simple venous catheter for sequential blood sampling from unrestrained pigs. *Lab. Anim. Care* **19**, 259–262.

Clawson, A. J., Reid, J. T., Sheffy, B. E. and Willmam, J. P. (1955). Use of chromium oxide in digestion studies with swine. *J. Anim. Sci.* **14**, 700–709.

Clemmessen, Th. (1964). The early circulation in split skin grafts. *Acta chir. scand.* **127**, 1–8.

Cole, H. H. and Cupps, P. T. (1959). "Reproduction in Domestic Animals", 1st edition. Academic Press, New York and London.

Corner, G. W. (1921). Cyclic changes in the ovaries and uterus of the sow, and their relation to the mechanism of implantation. *Contr. Embryol.* **13**, 117–146.

Cox, D. F. (1964a). Genetic variation in the gestation period of swine. *J. Anim. Sci.* **23**, 746–751.

Cox, D. F. (1964b). Relation of litter size and other factors to the duration of gestation in the pig. *J. Reprod. Fert.* **7**, 405–407.

Cox, D. F. (1966). Swine genetics and biomedical research. *In* "Swine in Biomedical Research" (L. K. Bustad and R. O. McClellan, eds), pp. 1–12. Battelle Memorial Institute.

Craft, W. A. and Moe, L. H. (1932). Statistical observations involving weight, hemoglobin and the proportion of white blood cells in pigs. *J. Am. vet. med. Ass.* **58**, 405–407.

Crew, F. A. E. and Kollner, P. C. (1939). Cytogenetical analysis of the chromosomes of the pig. *Proc. R. Soc. Edinb.* **59**, 163–179.

Crighton, D. B. (1970). The induction of pregnancy during lactation in the sow: the effects of a treatment imposed at 21 days of lactation. *Anim. Prod.* **12**, 611–617.

Cunha, T. J. (1966). Nutritional requirements of the pig. *In* "Swine in Biomedical Research" (L. K. Bustad and R. O. McClellan, eds), pp. 681–695. Battelle Memorial Institute.

Cunningham, H. M. (1959). Digestion of starch and some of its degradation products by newborn pigs. *J. Anim. Sci.* **18**, 964–975.

Cunningham, H. M. (1967). The digestibility, rate of passage and rate of gain in the gastrectomized pig. *J. Anim. Sci.* **26**, 500–503.

Cunningham, H. M. and Brisson, G. J. (1957a). The utilization of maltose by newborn pigs. *J. Anim. Sci.* **16**, 574–577.

Cunningham, H. M. and Brisson, G. J. (1957b). The effect of proteolytic enzymes on the utilization of animal and plant proteins by new-born pigs and the response to predigested protein. *J. Anim. Sci.* **16**, 568–573.

Cunningham, H. M., Friend, D. W. and Nicholson, J. W. G. (1962). Note on re-entrant fistula for digestion studies with pigs. *Can. J. anim. Sci.* **42**, 112–113.

Curtin, T. M. and Goetsch, G. D. (1966). Some altered gastric mucins associated with oesophagogastric ulcers in swine. *Am. J. vet. Res.* **27**, 1013–1016.

Curtin, T. M., Goetsch, G. D. and Hollandbeck, R. (1963). Clinical and pathologic characterization of esophagogastric ulcers in swine. *J. Am. vet. med. Ass.* **143**, 854–860.

Dahlqvist, A. (1961). Intestinal carbohydrases of the new-born pig. *Nature, Lond.* **190**, 31–32.

Darwin, C. (1882). *In* "The Variation in Animals and Plants Under Domestication", Vol. I, Chap. III. J. Murray, London.

Dellmann, H.-D. and McClure, R. C. (1966). Skull measurements for the establishment of a coordinate system for stereotaxic placement in the brain of swine (Sus scrofa). *In* "Swine in Biomedical Research" (L. K. Bustad and R. O. McClellan, eds), pp. 537–542. Battelle Memorial Institute.

Derivaux, J. (1959). Artificial insemination in pigs. *Annls. Méd. vét.* **103**, 369–393.

Detweiler, D. K. (1966). Swine in comparative cardiovascular research. *In*

"Swine in Biomedical Research" (L. K. Bustad and R. O. McClellan, eds), pp. 301–306. Battelle Memorial Institute.

Diaz, F., Speer, V. C., Ashton, G. C., Liu, C. H. and Catron, D. V. (1956). Comparison of refined cane sugar, invert cane molasses and unrefined cane sugar in starter rations for early weaned pigs. *J. Anim. Sci.* **15**, 315–319.

Donald, H. P. (1937). Suckling and suckling preference in pigs. *Emp. J. exp. Agric.* **5**, 361–368.

Doornenbal, H., Asdell, S. A. and Wellington, G. H. (1962). Chromium-51 determined red cell volume as an index of "Lean Body Mass" in pigs. *J. Anim. Sci.* **21**, 461–463.

Doyle, R. E., Garb, S., Davis, L. E., Meyer, D. K. and Clayton, F. W. (1968). Domesticated farm animals in medical research. *Ann. N.Y. Acad. Sci.* **147**, 129–204.

Dukes, H. H. (1955). "The Physiology of Domestic Animals", 7th edition. Comstock, New York.

Dukes, H. H. and Schwarte, L. H. (1931). The blood pressure of the pig and the influence of non-nervous and nervous factors on the cardiovascular apparatus. *J. Am. vet. med. Ass.* **79**, 37–62.

Dunne, H. W. (1963). Field and laboratory diagnosis of hog cholera. *Vet. Med.* **53**, 222–239.

Dunne, H. W. (1970). "Diseases of Swine", 3rd edition. Ames Iowa State University Press.

Dziuk, P. J. (1960). Influence of orally administered progestins on estrus and ovulation in swine. *J. Anim. Sci.* **19**, 1319–1320.

Dziuk, P. J. and Baker, R. D. (1962). Induction and control of ovulation in swine. *J. Anim. Sci.* **21**, 697–699.

Dziuk, P. J. and Polge, C. (1965). Fertility in gilts following induced ovulation. *Vet. Rec.* **77**, 236–238.

Dziuk, P. J., Phillips, T. N. and Graber, J. W. (1964). Halothane closed-circuit anaesthesia in the pig. *Am. J. vet. Res.* **109**, 1773–1775.

Eibl, K. and Wettke, K. (1964). Artificial insemination of pigs using CO_2 in activated boar semen. *Vet. Rec.* **76**, 856–858.

Ellinger, T. (1921). The influence of age on fertility in swine. *Proc. natn. Acad. Sci. USA* **7**, 134–138.

Elsley, F. W. H., Bannerman, Mary, Bathurst, E. V. J., Bracewell, A. G., Cunningham, J. M. M., Dodsworth, T. L., Dodds, P. A., Forbes, T. J. and Laird, R. (1969). The effect of level of feed intake in pregnancy and in lactation upon the productivity of sows. *Anim. Prod.* **11**, 225–241.

Engelhardt, W. von (1963). Untersuchungen am Schwein ueber die systolen-und Diastolendauer des Herzens und ueber den Blutdruck in der Ruhe und waehrend der Erholung nach koerperlicher Beluslung. *Zentbl. VetMed.* **10**, 39–50.

Engelhardt, W. von (1966). Swine cardiovascular physiology—a review. *In* "Swine in Biomedical Research" (L. K. Bustad and R. O. McClellan, eds), pp. 307–329. Battelle Memorial Institute.

Engelhardt, W. von, Hörnicke, H. and Hampel, K.-H. (1961). Blutdruckmessung beim Schwein. *Berl. Münch. tierärzte Wschr.* **74**, 349–352.

Erickson, B. H. (1965). Symposium on atomic energy in animal science: radiation effects on gonadal development in farm animals. *J. Anim. Sci.* **24**, 568–583.

Erickson, B. H. and Murphree, R. L. (1964). Limb development in prenatally irradiated cattle, sheep and swine. *J. Anim. Sci.* **23**, 1066–1071.

Evans, J. R., Rowe, R. D., Downie, H. G. and Rowsell, H. C. (1963). Murmurs arising from ductus arteriosus in normal new-born swine. *Circulation Res.* **12**, 85–93.

Facto, L. A., Diaz, F., Hays, V. W. and Catron, D. V. (1959). Time lapse cinematography as a method of studying animal behaviour and ration preferences. *J. Anim. Sci.* **18**, 1488.

Field, H. I. (1964). Disease of pigs. Ministry of Agriculture, Fisheries and Food Bulletin 171, H.M.S.O., London.

Filer, L. J. Jr., Owen, G. M. and Fomon, S. J. (1966). Effect of age, sex, and diet on carcass composition of infant pigs. *In* "Swine in Biomedical Research" (L. K. Bustad and R. O. McClellan, eds), pp. 141–150. Battelle Memorial Institute.

Foote, W. C., Waldorf, D. P., Chapman, A. B., Self, H. L., Grummer, R. H. and Casida, L. E. (1956). Age at puberty of gilts produced by different systems of mating. *J. Anim. Sci.* **15**, 959–969.

Fowler, E. H. and Calhoun, M. L. (1964). The microscopic anatomy of developing foetal pig skin. *Am. J. vet. Res.* **25**, 156–164.

Fowler, S. H. and Ensminger, M. E. (1960). Interactions between genotype and plane of nutrition in selection for rate of gain in swine. *J. Anim. Sci.* **19**, 434–449.

Fraser, A. C. (1938). A study of the blood of pigs. *Br. vet. J.* **94**, 3–21.

Gardiner, M. R., Sippel, W. L. and McCormick, W. C. (1953). The blood picture in new-born pigs. *Am. J. vet. Res.* **14**, 68–71.

George, L. A. and Bustad, L. K. (1966). Comparative effects of beta irradiation of swine, sheep and rabbit skin. *In* "Swine in Biomedical Research" (L. K. Bustad and R. O. McClellan, eds), pp. 491–500. Battelle Memorial Institute.

Ghoshal, N. G. and Getty, R. (1968). The arterial blood supply to the appendages of the domestic pig (Sus scrofa domesticus). *Iowa St. J. Sci.* **43**, 125–152.

Gilbert, S. G. (1963). "Pictorial Anatomy of the Fetal Pig", 2nd edition. University of Washington Press, Seattle.

Gill, J. C. and Thompson, W. (1956). Observations on the behaviour of suckling pigs. *Br. J. Anim. Behav.* **4**, 46–51.

Gimenez-Martin, G., Lopez-Saez, J. F. and Monge, E. G. (1962). Somatic chromosomes of the pig. *J. Hered.* **53**, 281.

Glover, T. D. (1955). The semen of the pig. *Vet. Rec.* **67**, 36–40.

Goodwin, R. F. W. (1955). Some common factors in the pathology of the new-born pig. *Br. vet. J.* **111**, 361–372.

Goodwin, R. F. W. (1957). The clinical diagnosis of haemolytic disease in the new-born pig. *Vet. Rec.* **69**, 505–507.

Goodwin, R. F. W. and Coombs, R. R. A. (1956). The blood groups of the pig. IV. The A antigen-antibody system and haemolytic disease in new-born piglets. *J. comp. Path. Ther.* **66**, 317–331.

Goodwin, R. F. W. and Saison, R. (1957). The blood groups of the pig. V.

Further observations on the epidemiology of haemolytic disease in the newborn. *J. comp. Path. Ther.* **67**, 126–144.

Goodwin, R. F. W., Saison, R. and Coombs, R. R. A. (1955). The blood groups of the Pig. II Red cell iso-antibodies in the sera of pigs injected with crystal violet swine fever vaccine. *J. comp. Path. Ther.* **65**, 79–92.

Grahame, T. and Morris, P. G. D. (1957). Comparison of the vascular supply to the virgin and post-gravid uterus of the pig, ox and sheep. *Br. vet. J.* **113**, 498–501.

Greer, S. A. N., Hays, V. W., Speer, V. C. and McCall, J. T. (1966). Effect of dietary fat, protein and cholesterol on atherosclerosis in swine. *J. Nutr.* **90**, 183–190.

Groves, T. W. (1966). Controlled oestrus and ovulation in pigs. *Vet. Rec.* **79**, 24–25.

Guiney, E. J. (1965). The pig as an experimental animal with particular reference to cardiovascular surgery. *Ir. J. med. Sci.* No. 474.

Hadek, R. and Getty. R. (1959a). Age change studies of the ovary of the domesticated pig. *Am. J. vet. Res.* **20**, 578–584.

Hadek, R. and Getty, R. (1959b). The changing morphology in the uterus of the growing pig. *Am. J. vet. Res.* **20**, 573–577.

Hafez, E. S. E., Sumption, L. J. and Jakway, J. S. (1962). Behaviour of swine. *In* "The Behaviour of Domestic Animals", (Hafez, ed.). Ballière Tindall and Cox, London.

Hall, L. W. (1966). Wright's "Veterinary Anaesthesia and Analgesia". Ballière, Tindall and Cox, London.

Hansard, S. L. (1956). Residual organ blood volume of cattle, sheep and swine. *Proc. Soc. exp. Biol. Med.* **91**, 31–34.

Hansard, S. L., Sauberlich, H. E. and Comar, C. L. (1951a). Blood volume of swine. *Proc. Soc. exp. Biol. Med.* **78**, 544–545.

Hansard, S. L., Plumlee, M. P., Hobbs, C. S. and Comar, C. L. (1951b). The design and operation of metabolism units for nutritional studies with swine. *J. Anim. Sci.* **10**, 88–96.

Haring, R., Gruhn, R., Smidt, D. and Scheven, B. (1966). Miniature swine development for laboratory purposes. *In* "Swine in Biomedical Research" (L. K. Bustad and R. O. McClellan, eds), pp. 789–796. Battelle Memorial Institute.

Hartman, D. A. and Pond, W. G. (1960). Design and use of a milking machine for sows. *J. Anim. Sci.* **19**, 780–785.

Hartman, P. A., Hays, V. W., Baker, R. O., Neagle, L. H. and Catron, D. V. (1961). Digestive enzyme development in the young pig. *J. Anim. Sci.* **20**, 114–123.

Hauser, E. R., Dickerson, G. E. and Mayer, D. T. (1952). Reproductive development and performance of inbred and crossbred boars. *Res. Bull. Mo. agric. Exp. Stn*, 503.

Heap, F. C. and Lodge, G. A. (1967). Changes in body composition of the sow during pregnancy. *Anim. Prod.* **9**, 237–246.

Heinze, W. (1961). Die Kaumuskulatur des Schweines in anatomischer und funktioneller Hinsicht. (Antomical and functional aspects of the jaw musculature of the pig). *Anat. Anz.* **109**, 269–291.

Heitman, H., Jr., and Hughes, E. H. (1949). The effects of air temperature and relative humidity on the physiological well being of swine. *J. Anim. Sci.* **8**, 171–181.

Hemmings, W. A. and Brambell, F. W. R. (1961). Protein transfer across the foetal membranes. *Br. med. Bull.* **17**, 96–101.

Henderick, H. K., Teague, H. S., Redman, D. R. and Grifo, A. P., Jr. (1964). Absorption of vitamin B_{12} from the colon of the pig. *J. Anim. Sci.* **23**, 1036–1038.

Hetzer, H. O. (1945). Inheritance of coat colour in swine. 1. General survey of major colour variations in swine. *J. Hered.* **36**. 121–128.

Hicks, N. (1965). "Pigs, Their Breeding, Feeding and Management" (Fishwick). Crosby Lockwood, London.

Hill, K. J. and Perry, J. S. (1959). A method for closed circuit anaesthesia in the pig. *Vet. Rec.* **71**, 296–299.

Hörnicke, H. (1966a). Review of the Berlin symposium on swine circulatory system. *In* "Swine in Biomedical Research" (L. K. Bustad and R. O. McClellan, eds), pp. 419–424. Battelle Memorial Institute.

Hörnicke, H. (1966b). Blutdruck des Schweines unter verschiedenen Einfluessen Eine Uebersicht. *Arch. exp. VetMed.* **20**, 1035–1047.

Hŏjny, J. and Hăla, K. (1964). Blood group systems O in pigs. In Blood groups of animals: Proceedings of the 9th European Animal Blood Group Conference. Prague, August, 1964, Czechoslovak Academy of Sciences.

Holt, J. P., Rhode, E. A., Peoples, S. A. and Kines, H. (1962). Left ventricular function in mammals of greatly different size. *Circulation Res.* **10**, 798–806.

Huber, W. G. and Wallin, R. F. (1966). Gastric secretion and ulcer formation in the pig. *In* "Swine in Biomedical Research" (L. K. Bustad and R. O. McClellan, eds), pp. 121–127. Battelle Memorial Institute.

Hudson, A. E. A. (1955). Fragility of erythrocytes in blood from swine of two age groups. *Am. J. vet. Res.* **16**, 120–122.

Huhn, J. E. and Schulze, H. U. (1961). Steroid anaesthesia in young pigs. *Berl. Münch. tierärztl. Wschr.* **74**, 369–374.

Huhn, R. G., Osweiler, G. D. and Switzer, W. P. (1969). Application of the orbital sinus bleeding technique to swine. *Lab. Anim. Care* **19**, 403–405.

Hunt, A. C. (1968). Micro-anatomy of the lymph nodes of the pig. *Br. J. exp. Path.* **49**, 338–339.

Ingram, D. L. (1964a). The effect of environmental temperature on body temperatures, respiratory frequency and pulse rate in the young pig. *Res. vet. Sci.* **5**, 348–356.

Ingram, D. L. (1964b). The effect of environmental temperature on heat loss and thermal insulation in the young pig. *Res. vet. Sci.* **5**, 357–364.

Ingram, D. L. (1965a). The effect of humidity on temperature regulation and cutaneous water loss in the young pig. *Res. vet. Sci.* **6**, 9–17.

Ingram, D. L. (1965b). Evaporative cooling in the pig. *Nature, Lond.* **207**, 415–416.

Ingram, D. L. (1967). Stimulation of cutaneous glands in the pig. *J. comp. Path. Ther.* **77**, 93–98.

Ingram, D. L. and Legge, K. F. (1970). The effect of environmental temperature on respiratory ventilation in the pig. *Resp. Physiol.* **8**, 1–12.

Ingram, D. L. and Mount, L. E. (1965). Metabolic rates of young pigs living at high ambient temperatures. *Res. vet. Sci.* **6**, 300–306.

Ingram, D. L. and Ślebodziński, A. (1965). Oxygen consumption and thyroid gland activity during adaptation to high ambient temperatures in young pigs. *Res. vet. Sci.* **6**, 522–530.

Ito, S., Kudo, S. and Niwa, T. (1959). Studies on the normal oestrus in swine with special reference to proper time for service. *Annls Zootech.* Ser D. 8. Suppl. pp. 105–107.

Jarmoluk, L. and Fredeen, H. T. (1965). Application of radiography in carcass research with swine. *Med. Radiogr. Photogr.* **41**, 63–65.

Ježková, D. (1960). Blood pressure studies on piglets. *Sb. čsl. Akad. zeměd. Věd. Veterinárni medicina.* **5**, 93–100.

Johnson, B. C. (1966). Some enzymatic and cardiovascular effects of starvation-refeeding stress. *In* "Swine in Biomedical Research" (L. K. Bustad and R. O. McClellan, eds), pp. 193–211. Battelle Memorial Institute.

Johnson, R. F. (1961). Promazine hydrochloride as a tranquillizer for sows. *Vet. Rec.* **73**, 588.

Jones, J. E. T. (1967). An intestinal haemorrhage syndrome. *Br. vet. J.* **123**, 286–293.

Joysey, V. C., Goodwin, R. F. W. and Coombs, R. R. A. (1959). The blood groups of the pig. VIII. The distribution of twelve red cell antigens in seven breeds. *J. comp. Path. Ther.* **69**, 292–299.

Jump, E. B. and Weaver, M. E. (1966). The miniature pig in dental research. *In* "Swine in Biomedical Research" (L. K. Bustad and R. O. McClellan, eds), pp. 543–558. Battelle Memorial Institute.

Kaman, J. (1966). Die Grossramifikation der Leberblutgefässe des Schweines. (The gross ramifications of the blood vessels of the pig liver). *Zentbl. VetMed. Reihe A* **13**, 719–745.

Kaman, J. and Červený, Č. (1968). Akzessorische Lungenarterien beim Schweine (Sus scrofa f. domestica). (Accessory pulmonary arteries in pigs). Anat. Anz. 122, 60–67.

Karagianes, M. T. (1968). An instrumentation harness assembly for long-term use on miniature swine. *J. appl. Physiol.* **25**, 641–643.

Kare, M. R., Pond, W. C. and Campbell, J. (1965). Individual variations in taste reactions of pigs. *Anim. Behav.* **13**, 265–269.

Kelly, W. A. and Eckstein, P. (1969). Implantation, development of the uterus and foetal membranes, *In* "Reproduction in Domestic Animals", 2nd edition (H. H. Cole, and P. T. Cupps, eds), pp. 385–413. Academic Press: New York and London.

Kidder, D. E., Manners, M. J. and McCrea, M. T. (1961). The passage of food through the alimentary tract of the piglet. *Res. vet. Sci.* **2**, 227–231.

Kidder, D. E., Manners, M. J. and McCrea, M. R. (1963). The digestion of sucrose by the piglet. *Res. vet. Sci.* **4**, 131–144.

Kim, Y. B., Bradley, S. G. and Watson, D. W. (1966). Antibody synthesis in germ-free colostrum-deprived miniature piglets. *In* "Swine in Biomedical

Research" (L. K. Bustad and R. O. McClellan, eds), pp. 273–284. Battelle Memorial Institute.

Kitts, W. D., Bailey, C. B. and Wood, A. J. (1956). The development of the digestive enzyme system of the pig during its preweaning phase of growth. A. Pancreatic amylase and lipase. *Can. J. agric. Sci.* **36**, 45.

Klopfer, F. D. (1966). Visual learning in swine. *In* "Swine in Biomedical Research" (L. K. Bustad and R. O. McClellan, eds), pp. 559–574. Battelle Memorial Institute.

Kolb, E., Müller, I., Seidel, H. and Ellrich, T. (1962). Studies on pigs II. Thorn's test after adrenalectomy with reference to sodium and potassium in serum and muscle. *Zentbl. VetMed.* **9**, 664–692 (also *Vet. Bull.* **33**, 1343, 1963).

Konyukhova, V. A. (1955). Conditioned reflexes in pigs. *Fiziol. zh. SSSR.* **41**, 326–333.

Kornegay, E. T., Brent, B. E., Long, C. H., Miller, E. R., Ullrey, D. E. and Hoefer, J. A. (1963). *J. Anim. Sci.* **22**, 1125.

Kubin, G. (1952). Pentothal anaesthesia for swine. *Wien. tierärztl. Mschr.* **39**, 542–545.

Laing, W., Rowsell, H. C. and Mustard, J. F. (1963). Morphological studies of the pig's aorta. *Anat. Rec.* **145**, 251 (Abstr.).

Landy, J. J. and Ledbetter, R. K. (1966). Delivery and maintenance of the germ-free pig. *In* "Swine in Biomedical Research" (L. K. Bustad and R. O. McClellan, eds), pp. 619–632. Battelle Memorial Institute.

Lascelles, A. K. and Morris, B. (1961). Surgical techniques for the collection of lymph from unanaesthetized sheep. *Q. Jl exp. Physiol.* **46**, 199–205.

Lasley, E. L. (1957). Ovulation, prenatal mortality and litter size in swine. *J. Anim. Sci.* **16**, 335–340.

Lie, H. (1968). Thrombocytes, leucocytes and packed red cell volume in piglets during the first two weeks of life. *Acta. vet scand.* **9**, 105–111.

Linzell, J. L., Mepham, T. B., Annison, E. F. and West, C. E. (1969). Mammary metabolism in lactating sows: arteriovenous differences of milk precursors and the mammary metabolism of [^{14}C] glucose and [^{14}C] actetate. *Br. J. Nutr.* **23**, 319–332.

Lodge, G. A. (1969). The effects of pattern of feed distribution during the reproductive cycle on the performance of sows. *Anim. Prod.* **11**, 133–143.

Luginbühl, H. (1966). Spontaneous atherosclerosis in swine, *In* "Swine in Biomedical Research" (L. K. Bustad and R. O. McClellan, eds), pp. 347–363. Battelle Memorial Institute.

Luke, D. (1953). The differential leucocyte count in the normal pig. *J. comp. Path. Ther.* **63**, 346–354.

Lumb, G. D. (1966). Experimentally induced cardiac failure in swine: pathological changes. *In* "Swine in Biomedical Research" (L. K. Bustad and R. O. McClellan, eds), pp. 389–404. Battelle Memorial Institute.

Lush, J. L. and Mollin, A. E. (1942). Litter size and weight as permanent characteristics of sows. *U.S.D.A. Techn. Bull.* 836.

McBride, G. (1963). The "teat order" and communication in young pigs. *Anim. Behav.* **11**, 53–56.

McCance, R. A. (1960). Severe undernutrition in growing and adult animals. 1. Production and general effects. *Br. J. Nutr.* **14**, 59–73.

McCance, R. A. and Mount, L. E. (1960). Severe undernutrition in growing and adult animals. 5. Metabolic rate and body temperature in the pig. *Br. J. Nutr.* **14**, 509–518.

McCance, R. A. and Widdowson, E. M. (1959). The effect of lowering the ambient temperature on the metabolism of the new-born pig. *J. Physiol., Lond.* **147**, 124–134.

McCance, R. A. and Widdowson, E. M. (1959). The effect of colostrum on the composition and volume of the plasma of new-born piglets. *J. Physiol., Lond.* **145**, 547–550.

McCance, R. A. and Wilkinson, A. W. (1967). Experimental resection of the intestine in new-born pigs. *Br. J. Nutr.* **21**, 731–740.

McConnell, J., Fechheimer, N. S. and Gilmore, L. O. (1963). Somatic chromosomes of the domestic pig. *J. Anim. Sci.* **22**, 374–379.

McFee, A. F., Banner, M. W. and Rary, J. M. (1966). Variation in chromosome number among European wild pigs. *Cytogenetics* **5**, 75–81.

McFeely, R. A. and Hare, W. C. D. (1966). Cytogenetic studies of the domestic pig. *In* "Swine in Biomedical Research" (L. K. Bustad and R. O. McClellan, eds), pp. 13–23. Battelle Memorial Institute.

McKenzie, F. F. (1926). The normal oestrous cycle in the sow. *Mo. agric. Exp. Stn Bull.* **86**, 5–41.

McKenzie, F. F. and Miller, J. C. (1930). Studies on reproduction of swine at the Missouri station *Mo. agric. Exp. Stn Bull.* **285**, 43–46.

McKibben, J. and Getty, R. (1969). Innervation of heart of domesticated animals: pig. *Am. J. vet. Res.* **30**, 779–789.

Maaske, C. A., Booth, N. H. and Nielsen, T. W. (1966). Experimental right heart failure in swine. *In* "Swine in Biomedical Research" (L. K. Bustad and R. O. McClellan, eds), pp. 377–387. Battelle Memorial Institute.

Magee, D. F. and Hong, S. S. (1966). Studies on pancreatic physiology in pigs. *In* "Swine in Biomedical Research" (L. K. Bustad and R. O. McClellan, eds), pp. 109–120. Battelle Memorial Institute.

Makino, S. (1944). The chromosome complex of the pig. *Cytologia*, **13**, 170–178.

Mann, T. (1959). Biochemistry of semen and secretions of male accessory organs. *In* "Reproduction in Domestic Animals", 1st edition (H. H. Cole and P. T. Cupps, eds), pp. 51–73. Academic Press: New York and London.

Mann, T. P. and Elliott, R. I. K. (1957). Neonatal cold injury due to accidental exposure to cold. *Lancet* i, 229–234.

Marcarian, H. Q. and Calhoun, M. L. (1966). Microscopic anatomy of the integument of adult swine. *Am. J. vet. Res.* **27**, 765–772.

Marcuse, F. L. and Moore, A. U. (1942). Conditioned reflexes in the pig. *Bull. Can. Psychol. Ass.* **2**, 13–14.

Marcuse, F. L. and Moore, A. U. (1944). Tantrum behaviour in the pig. *J. comp. Psychol.* **37**, 235–241.

Marcuse, F. L. and Moore, A. U. (1946). Motor criteria of discrimination. *J. comp. Psychol.* **39**, 25–27.

Markowitz, J., Archibald, J. and Downie, H. G. (1959). "Experimental Surgery". Baillière Tindall and Cox, London.

Martin, C. E., Hooper, B. E., Armstrong, C. H. and Amstutz, H. E. (1967). A clinical and pathologocal study of the mastitis-metritis-agalactia syndrome of sows. *J. Am. vet. med. Ass.* **151**, 1629–1634.

Mayo, R. H. J. (1961). Swine metabolism unit. *J. Anim. Sci.* **20**, 71–74.

Du Mesnil du Buisson, F., Léglise, P. C. and Anderson, L. L. (1964a). Hypophysectomy in pigs. *J. Anim. Sci.* **23**, 1226–1227.

Du Mesnil du Buisson, F., Léglise, P. C., and Chodkiewicz, J. P. (1964b). Technique for hypophysectomy by the transfrontal supra-orbital approach in the pig. *Ann. Biol. Anim. Biochem. Biophys.* **4**, 229–237.

Miller, E. R. and Ullrey, D. E. (1966). Electrocardiographic changes in normal and thiamine-deficient baby pigs. *In* "Swine in Biomedical Research", (L. K. Bustad and R. O. McClellan, eds), pp. 225–234. Battelle Memorial Institute.

Miller, E. R., Ullrey, D. E., Ackerman, I., Schmidt, D. A., Hoefer, J. A. and Luecke, R. W. (1961a). Swine hematology from birth to maternity. I. Serum proteins. *J. Anim. Sci.* **20**, 31–35.

Miller, E. R., Ullrey, D. E., Ackerman, I., Schmidt, D. A., Luecke, R. W. and Hoefer, J. A. (1961b). Swine hemetology from birth to maternity. II. Erythrocyte population size and hemoglobin concentration. *J. Anim. Sci.* **20**, 890–897.

Miller, W. L. and Robertson, E. D. S. (1952). Practical Animal Husbandry. Oliver and Boyd: Edinburgh.

Montagna, W. (1966). The microscopic anatomy of the skin of swine and man. *In* "Swine in Biomedical Research" (L. K. Bustad and R. O. McClellan, eds), pp. 285–286. Battelle Memorial Institute.

Montagna, W. and Yun, J. S. (1964). The skin of the domestic pig. *J. invest. Derm.* **43**, 11–21.

Moody, W. G. and Zobrisky, S. E. (1966). Study of backfat layers of swine. *J. Anim. Sci.* **25**, 809–813.

Moore, A. U. and Marcuse, F. L. (1945). Salivary, cardiac and motor indices of conditioning in two sows. *J. comp. Psychol.* **38**, 1–16.

Morelle, V. R. (1963). Replacement of the ureter by a segment of the stomach, in pigs and dogs. *Arch. Chirurg. Neerland.* **15**, 293–301.

Morgan, D. P. and Davey, R. J. (1965). Adjustable metabolism units for barrows. *J. Anim. Sci.* **24**, 13–15.

Morris, D. (1965). "The Mammals". Hodder and Stoughton, London.

Morrison, S. R., Bond, T. E. and Heitman, H., Jr. (1967). Skin and lung moisture loss from swine. Trans. *Am. Soc. agric. Engrs* **10**, 691–696.

Mount, L. E. (1959). The metabolic rate of the new-born pig in relation to environmental temperature and to age. *J. Physiol., Lond.* **147**, 333–345.

Mount, L. E. (1960). The influence of huddling and body size on the metabolic rate of the young pig. *J. agric. Sci., Camb.* **55**. 101–105.

Mount, L. E. (1963). The environmental temperature preferred by the young pig. *Nature, Lond.* **199**, 122–123.

Mount, L. E. (1964). The tissue and air components of thermal insulation in the new-born pig. *J. Physiol., Lond.* **170**, 286–295.

Mount, L. E. (1966). Thermal and metabolic comparisons between the new-born pig and human infant. *In* "Swine in Biomedical Research" (L. K. Bustad and R. O. McClellan, eds), pp. 501–509. Battelle Memorial Institute.

Mount, L. E. (1967). The heat loss from new-born pigs to the floor. *Res. vet. Sci.* **8**. 175–186.

Mount L.E.(1968)."The Climatic Physiology of the Pig". Edward Arnold, London.

Mount, L. E., Holmes, C. W., Start, I. B. and Legge, A. J. (1967). A direct calorimeter for the continuous recording of heat loss from groups of growing pigs over long periods. *J. agric. Sci., Camb.* **68**, 47–55.

Moustgaard, J. and Hesselholt, M. (1966). Blood groups and serum protein polymorphism in pigs. *In* "Swine in Biomedical Research" (L. K. Bustad and R. O. McClellan, eds), pp. 25–44. Battelle Memorial Institute.

Munsick, C. R., Sawyer, W. H. and Van Dyke, H. B. (1958). The antidiuretic potency of arginine and lysine vasopressins in the pig with observations on porcine renal function. *Endocrinology*, **63**, 688–693.

Nachtway, D. S., Ainsworth, E. J., Leong, G. F. and Page, N. P. (1966). Acute mortality response and recovery from radiation injury in swine. *In* "Swine in Biomedical Research" (L. K. Bustad and R. O. McClellan, eds), pp. 429–434. Battelle Memorial Institute.

Neher, G. M. and Carter, J. M. (1966). Experimental rheumatoid-like arthritis in swine. *In* "Swine in Biomedical Research", (L. K. Bustad and R. O. McClellan, eds), pp. 235–248. Battelle Memorial Institute.

Nielsen, T. W., Maaske, C. A. and Booth, N. H. (1966). Some comparative aspects of porcine renal function. *In* "Swine in Biomedical Research" (L. K. Bustad and R. O. McClellan, eds), pp. 529–536. Battelle Memorial Institute.

National Academy of Sciences (1968), "Nutrient Requirements of Swine", 6th ed. Publication 1599. Washington, D.C.

Niwa, T. and Hizuho, A. (1954). Studies on the age of sexual maturity in the boar. II on the Berkshire breed. *Bull. natn. Inst. agric. Sci., Tokyo* Series G (Animal Husbandry), **9**, 141–159.

Noble, W. A. (1964). Methods used to combat transmissible gastroenteritis. *Vet. Rec.* **76**. 1497–1498.

Nordstoga, K. (1965). Thrombocytopenic purpura in baby pigs caused by maternal isoimmunization. *Pathologia vet.* **2**, 601–610.

Orfeur, N. B. (1965). Gastric ulcers in fattening pigs. *Vet. Rec.* **77**, 781–788.

Oxenreider, S. L., McClure, R. C. and Day, B. N. (1965). Arteries and veins of the internal genitalia of female swine. *J. Reprod. Fert.* **9**, 19–27.

Palmer, W. M., Teague, H. S. and Venzke, W. G. (1964). Post-partum changes in the reproductive tract of the lactating sow. *J. Anim. Sci.* **23**, 1227.

Parkes, A. S. (1956). The biology of spermatozoa and artificial insemination. *In* "Marshall's Physiology of Reproduction" (A. S. Parkes, ed.), pp. 161–263. Longmans Green, London.

Paulick, Ch., Neurand, K. and Wilkens, H. (1967). Topographical anatomy of injection sites in pigs. *Deut. Tierärtzl. Wochenschr.* **74**, 519–524.

Pekas, J. C. (1968). Versatile swine laboratory apparatus for physiologic and metabolic studies. *J. Anim. Sci.* **27**, 1303–1306.

Penny, R. H. C., Osborne, A. D., Wright, A. I. and Stephens, T. K. (1965). Foot-rot in pigs: observations on the clinical disease. *Vet. Rec.* **77**, 1101–1108.

Perk, K., Frei, Y. F. and Herz, A. (1964). Osmotic fragility of red blood cells of young and mature domestic and laboratory animals. *Am. J. vet. Res.* **25**, 1241–1248.

Perry, G. C. and Watson, J. H. (1967). A note on the selective uptake of serum proteins in piglets following dosage with porcine serum, *Anim. Prod.* **9**, 557–560.

Perry, J. S. (1956). Observations on reproduction in a pedigree herd of Large White pigs. *J. agric. Sci., Camb.* **47**, 332–343.

Perry, T. W., Pickett, R. A , Curtin, T. M., Beeson, W. M.. and Nuwer, A. J. (1966). Studies on esophagogastric ulcers in swine. *In* "Swine in Biomedical Research" (L. K. Bustad and R. O. McClellan, eds.), pp. 129–140. Battelle Memorial Institute.

Pestana, C., Hellenbeck, G. A. and Shorter, R. G. (1965). Thymectomy in new-born pigs. *J. Surg. Res.* **5**, 306–312.

Phillips, R. W. and Zeller, J. H. (1943). Sexual development in small and large types of swine. *Anat. Rec.* **85**, 387–400.

Pierce, A. E. and Smith, M. W. (1967). The intestinal absorption of pig and bovine lactoglobulin and human serum albumin by the new-born pig. *J. Physiol., Lond.* **190**, 1–18.

Polge, C. (1956). Artificial insemination in pigs. *Vet. Rec.* **68**, 62–76.

Polge, C. (1965). Effective synchronisation of oestrus in pigs after treatment with ICI compound 3328. *Vet. Rec.* **77**, 232–236.

Pond, W. G., Barnes, R. H., Reid, I., Krook, L., Kwong, E. and Moore, A. U. (1966). Protein deficiency in baby pigs. *In* "Swine in Biomedical Research" (L. K. Bustad and R. O. McClellan, eds), pp. 213–224. Battelle Memorial Institute.

Pope, C. H. (1962). On the natural superiority of pigs. *Turtox News* **40**, 190–193, 230–231, 258–261.

Pratt, C. W. M. and McCance, R. A. (1966). The structure of the long bones in pigs stunted by severe undernutrition. *In* "Swine in Biomedical Research" (L. K. Bustad and R. O. McClellan, eds), pp. 169–172. Battelle Memorial Institute.

Protasenya, T. P. (1961). Fistula-angiostomy method for studying digestion and metabolism in pigs. *Trudy Novocherkassk. zoovet. Inst.* **13**, 199–206.

Pugh, O. L. and Penny, R. H. C. (1966). A crate for the restraint of large pigs. *Vet. Rec.* **79**, 390–392.

Radford, P. (1965). Synchronization of ovulation in sows at the first post weaning oestrus. *Vet. Rec.* **77**, 239–240.

Ramirez, C. G., Miller, E. R., Ullrey, D. E. and Hoefer, J. A. (1962). Blood volume of the nursing pig. *J. Anim. Sci.* **21**, 1028–1029.

Ramirez, C. G., Miller, E. R., Ullrey, D. E. and Hoefer, J. A. (1963). Swine hematology from birth to maturity. III. Blood volume of the nursing pig. *J. Anim. Sci.* **22**, 1068–1074.

Rasmussen, B. A. (1964). Gene interaction and the A–O blood group system in pigs. *Genetics, Princeton.* **50**, 191–198.

Redman, D. R., Teague, H. S., Henderick, H. K. and King, N. B. (1964). Cecal fistulation of the pig using two forms of indwelling cannulas. *J. Anim. Sci.* **23**, 1032–1035.

Reese, N. A., Muggenburg, B. A., Kowalczyk, T., Grummer, R. H. and Hoekstra, W. G. (1966). Nutritional and environmental factors influencing gastric ulcers in swine. *J. Anim. Sci.* **25**, 14–20; 21–24.

Reiner, L., Vrbanovic, D. and Madrazo, A. (1961). Inter-arterial coronary anastomoses in neonatal pigs. *Proc. Soc. exp. Biol. Med.* **106**, 732–734.

Rempel, W. E. and Dettmers, A. E. (1966). Minnesota's miniature pigs. *In* "Swine in Biomedical Research" (L. K. Bustad and R. O. McClellan, eds), pp. 781–787. Battelle Memorial Institute.

Rendel, J., Neimann Sørensen, A. and Irwin, M. R. (1954). Evidence for epistatic action of genes for antigenic substances in sheep. *Genetics, Princeton,* **39**, 396–409.

Riddell, A. G., Terblanche, J., Peacock, J. H., Tierris, E. J. and Hunt, A. C. (1967). Experimental liver homotransplantation in pigs. *In* "Advances in Transplantation", pp. 639–641. Munksgaard.

Ripke, E. (1964). Beitrag zue Kenntnis des Schweinegebisses (The teeth of the pig). *Anat. Anz.* **114**, 181–211.

Ritchie, H. E. (1957). Chlorpromazine sedation in the pig. *Vet. Rec.* **69**, 895–900.

Robertson, G. L., Grummer, R. H., Casida, L. E. and Chapman, A. B. (1951). Age at puberty and related phenomena in outbred Chester White and Poland China gilts. *J. Anim. Sci.* **10**, 647–656.

Robinson, K. and Lee, D. H. K. (1941). Reactions of the pig to hot atmospheres. *Proc. Roy. Soc., Queensland* **53**, 145–158.

Robinson, K. L. and Coey, W. E. (1957). "The Pig". The UFAW handbook on the care and management of laboratory animals. UFAW, London.

Romack, F. E., Turner, C. W. and Day, B. N. (1964). Anatomy of the porcine thyroid. *Res. Bull. Mo. Agric. Exp. Sta.* No. 838, pp. 10.

Rothe, K. (1963). Artificial insemination in pigs. *Arch. exp. VetMed.* **17**, 957–1017.

Rothenbacher, H. (1965). Esophagogastric ulcer syndrome in young pigs. *Am. J. vet. Res.* **26**, 1214–1217.

Rothschild, N. M. B. (1965). "A Classification of Living Animals". Longmans, London.

Routledge, F. G., Topliff, E. D. L. and Livingstone, S. D. (1968). Animal respiratory mask. *Can. J. Physiol. Pharmac.* **46**, 700–701.

Rowe, R. D., Sinclair, J. D., Kerr, A. R. and Gage, P. W. (1964). Duct flow and mitral reguritation during changes of oxygenation in new-born swine. *J. appl. Physiol.* **19**, 1157–1163.

Rowsell, H. C., Mustard, J. F., Packham, M. A. and Dodds, W. J. (1966). The hemostatic mechanism and its role in cardiovascular disease of swine. *In* "Swine in Biomedical Research" (L. K. Bustad and R. O. McClellan, eds), pp. 365–376. Battelle Memorial Institute.

Rowson, L. E. A. (1956). Artificial insemination of pigs. *Agriculture, Lond.* **62**, 571–573.

Rowson, L. E. A. (1958). Techniques of collection, dilution and storage of semen. *In* "Reproduction in Domestic Animals", 1st edition (H. H. Cole, and P. T. Cupps, eds), pp. 95–134. Academic Press: New York and London.

Rowson, L. E. A. (1965). Endotracheal intubation in the pig. *Vet. Rec.* **77**, 1465.

Saar, L. I. and Getty, R. (1964). The interrelationship of the lymph vessel connections of the lymph nodes of the head, neck and shoulder regions of swine *Am. J. vet. Res.* **25**, 618–636.

Sainsbury, D. W. B. (1963). "Pig Housing". Farming Press: Ipswich.

Sainsbury, D. W. B. (1967). "Animal Health and Housing". Baillière, Tindall and Cassell, London.

Saison, R. and Ingram, D. G. (1963). Production of specific haemagglutinins in pigs after receiving skin homografts. *Nature, Lond.* **197**, 296–297.

Sanders, D. P., Heidenreich, C. J. and Jones, H. W. (1964). Body temperature and the oestrous cycle in swine. *Am. J. vet. Res.* **25**, 851–853.

Sawyer, D. C. and Stone, H. L. (1966). Thoracic surgery for electromagnetic flow sensor implantation on miniature swine. *In* "Swine in Biomedical Research (L. K. Bustad and R. O. McClellan, eds), pp. 405–409. Battelle Memorial Institute.

Scarborough, R. A. (1930). The blood picture of normal laboratory animals. *Yale J. Biol. Med.* **3**, 547–552.

Schalm, O. W. (1965). "Veterinary Hematology". 2nd edition. Lea and Febiger: Philadelphia.

Schilling, E. and Röstel, W. (1964). Methodische Untersuchungen zur Brunstfeststellung beim Schwein. *Dt–öst. tierärztl. Wschr.* **71**, 429–436.

Schmidt, E. M. (1968). Blood pressure response to aortic nerve stimulation in swine. *Am. J. Physiol.* **215**, 1488–1492.

Schmidt, J. E., St. Clair, L. E. and Reber, E. F. (1964). Thyroid gland activity in normal and surgically thyroidectomized pigs. *Fedn Proc. Fedn Am. Socs exp. Biol.* 23, 108.

Schmidt-Nielsen, B. and O'Dell, R. (1961). Structure and concentrating mechanisms in the mammalian kidney. *Am. J. Physiol.* **200**, 1119–1124.

Schmidt-Nielsen, B., O'Dell, R. and Osaki, H. (1961). Interdependence of urea and electrolytes in the production of a concentrated urine. *Am. J. Physiol.* **200**, 1125–1132.

Seamer, J. and Walker, R. G. (1960). Splenectomy and the spleen weight of young pigs. *Res. vet. Sci.* **1**, 125–128.

Self, H. L., Grummer, R. H. and Casida, L. E. (1955). The effects of various sequences of full and limited feeding on the reproductive phenomena in Chester White and Poland China gilts. *J. Anim. Sci.* **14**, 573–592.

Sharpe, H. B. A. (1966). Pre-weaning mortality in a herd of Large White pigs. *Br. vet. J.* **122**, 99–111.

Shively, J. N., Warner, A. R., Jr., Miller, H. P., Kurtz, H. J., Andrews, H. L. and Woodward, K. T. (1964). Study of swine surviving exposure to the gamma-neutron flux of a nuclear detonation. *Am. J. vet. Res.* **25**, 1341–1346.

Short, C. E. (1964). The application of electro-anaesthesia on large animals. A report of 100 administrations. *J. Am. vet. med. Ass.* **145**, 1104–1106.

Sidney, S. (1871). "The Pig". Routledge and Sons, London.

Sikes, D., Fletcher, O., Jr. and Papp, E. (1966). Experimental production of pannus in a rheumatoid-like arthritis of swine. *Am. J. vet. Res.* **27**, 1017–1025.

Sillar, F. C. and Meyler, R. M. (1961). "The Symbolic Pig". Oliver and Boyd, Edinburgh.

Sinclair, D. V. (1961). Promazine hydrochloride as a tranquillizer for sows. *Vet. Rec.* **73**, 561.

Sisson, S. and Grossman, J. D. (1953). "The Anatomy of the Domestic Animals", 4th edition. Saunders: Philadelphia and London.

Smidt, D. (1963). Possibilities of heat diagnosis in pigs. *Dt.-öst. tierärztl. Wschr.* **70**, 517–522.

Smidt, W. J. (1962). The practical application of artificial insemination in pigs. *Proc. IVth Int. Congr. Anim. Reprod. The Hague* **4**, 864–867.

Smith, G. S., Smith, J. L., Mameesh, M. S., Simon, J. and Johnson, B. C. (1964). Hypertension and cardiovascular abnormalities in starved-refed swine. *J. Nutr.* **82**, 173–182.

Smith, J. L. and Calhoun, M. L. (1964). The microscopic anatomy of the integument of new-born swine. *Am. J. vet. Res.* **25**, 165–173.

Soltys, M. A. and Spratling, R. F. (1957). Infectious cystitis and pyelonephritis of pigs: a preliminary communication. *Vet. Rec.* **69**, 500–504.

Spalding, J. F. and Berry, R. O. (1956). A chromosome study of the wild pig (Peccari angulatus) and the domestic pig (Sus scrofa). *Cytologia* **21**, 81–84.

Spencer, S. (1921). "The Pig". Arthur Pearson, London.

Spies, H. G., Zimmerman, D. R., Self, H. L. and Casida, L. E. (1960). Effect of exogenous progesterone on the corpora lutea of hysterectomized gilts. *J. Anim. Sci.* **19**, 101–108.

Sprague, L. M. (1958). On the recognition and inheritance of the soluble blood groups property "O_c" of cattle. *Genetics, Princeton* **43**, 906–912.

Squiers, C. D., Dickerson, G. E. and Mayer, D. T. (1952). Influence of inbreeding, age and growth rate of sows on sexual maturity, rate of ovulation, fertilization and embryonic survival. *Res. Bull. Mo. agric. Exp. Stn.* 494.

St. Clair, L. E. (1945). Hypophysectomy and its physiologic effects in the pig (Sus scrofa domestica). Ph.D. Thesis, Ames, Iowa, Iowa State University.

Stevens, A. J. (1963). Enteritis in pigs—a working hypothesis. *Br. vet. J.* **119**, 520–526.

Stone, H. L. and Sawyer, D. C. (1966). Cardiac output and related measurements in unanaesthetized miniature swine. *In* "Swine in Biomedical Research" (L. K. Bustad and R. O. McClellan, eds), pp. 411–418. Battelle Memorial Institute.

Stowe, C. M. and Good, A. L. (1961). Estimation of cardiac output by the direct Fick technique in domestic animals, with observations on a case of traumatic pericarditis. *Am. J. vet. Res.* **22**, 1093–1096.

Stromlund, E. V., Reber, E. F., St. Clair, L. E. and Kodras, R. (1960). Response of swine to surgical and radioiodine thyroidectomies. *J. Anim. Sci.* **19**, 1337–1338.

Swarbrick, O. (1968). The porcine agalactia syndrome, clinical and histological observations. *Vet. Rec.* **82**, 241–252.

Talanti, S. (1959). Studies on the lungs in the pig. *Anat. Anz.* **106**, 68–75.

Talbot, R. B. and Swenson, M. J. (1963a). Measurement of porcine plasma volume using T1842 dye. *Am. J. vet. Res.* **24**, 467–471.

Talbot, R. B. and Swenson, M. J. (1963b). Survival of Cr^{51} labelled erythrocytes in swine. *Proc. Soc. exp. Biol. Med.* **112**, 573–576.

Tavernor, W. D. (1960). Clinical observation on the use of trimeprazine tartrate as a sedative and premedicant in horses, cattle, pigs and dogs. *Vet. Rec.* **72**, 317–321.

Tavernor, W. D. (1963). A study of the effect of phencyclidine in the pig. *Vet. Rec* **75**, 1377–1382 and 1383.

Tegeris, A. S., Earl, F. L. and Curtis, J. M. (1966). Normal hematological and biochemical parameters of young miniature swine. *In* "Swine in Biomedical Research" (L. K. Bustad and R. O. McClellan, eds), pp. 575–596. Battelle Memorial Institute.

Thomas, C. E. (1957). The muscular architecture of the ventricles of hog and dog hearts. *Am. J. Anat.* **101**, 17–57.

Thomas, C. E. (1959). The muscular architecture of the atria of hog and dog hearts. *Am. J. Anat.* **104**, 207–236.

Tribe, D. E. (1957). The baby pig. In UFAW handbook on the care and management of laboratory animals. UFAW: London.

Ullrey, D. E., Long, C. H. and Miller, E. R. (1966). Absorption of intact protein from the intestinal lumen of the neonatal pig. *In* "Swine in Biomedical Research" (L. K. Bustad and R. O. McClellan, eds), pp. 249–262. Battelle Memorial Institute.

Usenik, E. A., Kitchell, R. L. and Herschler, R. C. (1962). A surgical technique for permanent implantation of electrocorticographic electrodes in the Burro and pig. *Am. J. vet. Res.* **23**, 70–73.

Vaughan, L. C. (1961). Anaesthesia in the pig. *Br. vet. J.* **117**, 383–391.

Venn, J. A. J. (1944). Variations in the leukocyte count of the pig during the first twelve weeks of life. *J. comp. Path. Ther.* **54**, 172–178.

Veterinary Investigation Service, Ministry of Agriculture, Fisheries and Food (1959). A survey of the incidence and causes of mortality in pigs. I. Sow survey. *Vet. Rec.* **71**, 777–786.

Veterinary Investigation Service, Ministry of Agriculture, Fisheries and Food. (1960). A survey of the incidence and causes of mortality in pigs. II. Findings at post-mortem examination of pigs. *Vet. Rec.* **71**, 1240–1247.

Wachtel, W. (1963). Untersuchungen über Herzminutenvolumen, arteriovenöse Sauerstoffdifferenz, Haemoglobingehalt und Erythrozytenzahlen bei Haus-und Wildschweinen. *Arch. exp. VetMed.* **17**, 787–798.

Wachtel, W., Lyhs, L and Lehmann, E. (1963). Blutdruck messung beim Schwein. *Arch. exp. VetMed.* **17**, 355–360.

Walker, D. M. (1959). The development of the digestive system of the young animal. *J. agric. Sci., Camb.* **52**, 357–363.

Wallace, C. (1949). The effects of castration and stilboestrol treatment on the semen production of the boar. *J. Endocr.* **6**, 205–217.

Warwick B. L. (1926). Intra-uterine migration of ova in the sow. *Anat. Rec.* **33**, 29–33.

Waxler, G. L., Schmidt, D. A. and Whitehair, C. K. (1966). Technique for rearing gnotobiotic pigs. *Am. J. Vet. Res.* **27**, 300–307.

Weaver, M. E. and McKean, C. F. (1965). Miniature swine as laboratory animals. *Lab. Anim. Care* **15**, 49–56.

Weinstein, G. D. (1966). Comparison of turnover time and of keratinous protein fractions in swine and human epidermis. *In* "Swine in Biomedical Research" (L. K. Bustad and R. O. McCelllan, eds), pp. 287–297. Battelle Memorail Institute.

Wesley, F. (1955). Visual discrimination learning in swine. MSc thesis in Psychology, Dept. of Psychology, Washington State Univ., Pullman, Washington, U.S.A.

Westhues, H. and Fritsch, R. (1964). "Animal Anaesthesia". Oliver and Boyd, London.

Widdowson, E. M. and McCance, R. A. (1955). The effect of suckling anaemia on the pig's heart. *Br. J. exp. Path.* **36**, 175–178.

Wiggins, E. L., Casida, L. E. and Grummer, R. H. (1950). The effect of season of birth on sexual development in gilts. *J. Anim. Sci.* **9**, 277–280.

Wilson, K. M. (1926). Histological changes in the vaginal mucosa of the sow in relation to the oestrous cycle. *Am. J. Anat.* **37**, 417–432.

Woolsey, C. N. and Fairman, D. (1946). Contralateral, ipsilateral and bilateral representation of cutaneous receptors in somatic areas I and II of the cerebral cortex of pig, sheep and other mammals. *Surgery* **19**, 684–702.

Wyeth, G. S. F. and McBride, G. (1964). Social behaviour of domestic animals. V. A note on suckling behaviour in young pigs. *Anim. Prod.* **6**, 245–247.

Yamashita, T. (1961). Histological studies of the ovaries of sows. IV. Stereographical study of the vascular arrangement in the various structures of ovaries by use of neoprene latex casting specimens. *Jap. J. vet. Res.* **9**, 31–40.

Yoffey, J. M. and Courtice, F. C. (1956). "Lymphatics, Lymph and Lymphoid Tissue". Edward Arnold, London.

Zeller, J. H. (1964). Breeds of Swine. U.S.D.A. Bulletin 1263.

Zimmerman, D. R., Self, H. L. and Casida, L. E. (1957). The effect of flushing for various lengths of time on the ovulation rate of Chester White and Chester White Poland China crossbred gilts. *J. Anim. Sci.* **16**, 1099–1100.

Zintzsch, I. and Flechsig, G. (1966). Anatomy of the sella turcica of the pig. *Anat. Anz.* **119**, 147–161.

Ziolo, Irena (1965). Myelinization of the nerve fibres of pig spinal cord. *Acta. Anat.* **61**, 297–320.

Author Index

Numbers in *italics* indicate the pages on which references are listed in full

Subject Index

STAT GEEK BASEBALL
the BEST EVER book
from baseballevaluation.com

First Edition: Paperback and Ebook, 2010

Book jacket photo courtesy of Jeff Hire.

Printed/Digital Copy in the United States of America by JDP ECON Publications & Books, 2217 Bruce Drive, Pottstown, Pa., U.S.A.19464-1514

ISBN: 09745338-58, 978-09745338-58

Disclaimer: The Baseball Evaluation system was developed independently of Major League Baseball and the Major League Baseball Players Association and is not endorsed by or associated in any way with either organization.

Note: Historic baseball statistics were provided courtesy of baseball-databank.org and the Sean Lahman Database. Other statistic sources used include Baseball-reference.com & the Chuck Roscium Catcher Database.

All Baseball Evaluation Stats, including PEVA, RAVE, EXPEQ, etc., were developed by JDP Econ & are the proprietary property of JDP ECON. All rights reserved. If any Baseball Evaluation stat is used in articles, etc., please credit baseballevaluation.com.

The Baseball Evaluation System, Stats, Rankings, and other information is the intellectual property of JDP Econ and the data herein is provided for entertainment and information purposes only.

STAT GEEK BASEBALL
About the Best Ever Book 2011.

Essentially it's a book of lists, or rankings, or Player Ratings, and how your favorite players of today and those of yesterday rank against each other, all updated through the 2010 season. What batter had the best season ever? We've got that subject covered. Which pitcher dominated the regular season at a higher level than any other pitcher in history? The ranking list is there. But we have not stopped at the best seasons by the top players, there are hundreds of players listed, plus explanations of the Top Ten, and the new additions to the list from the current season. And it's more than the top seasons of individual players, we've compiled the Best Ever list of Top Players by every Team and Franchise in baseball history, from the long forgotten ones liked the Keokuk Westerns to your favorites from coast to coast, yes, the New York Yankees to the Minnesota Twins and the Los Angeles Angels of Anaheim. We've got your favorite teams covered.

Is there More Than the Top Seasons Ranked?

You betcha, as a Tea Party maven might say. We've got the Top Careers of the Regular Season for both batters and pitchers and the Top Postseason Careers Best Ever ranks, too.

It's list after list and stats next to all.

Included Best Ever Ranking Lists

- Best Ever Seasons by Batters
- Best Ever Seasons by Pitchers
- Best Ever Seasons by Relief Pitchers
- Best Ever Seasons Ranked by Every Team and Franchise
- Best Ever Careers by Batters
- Best Ever Careers by Pitchers
- Best Ever Careers by Every Team and Franchise
- Best Ever Postseason Years in History for Batters
- Best Ever Postseason Years in History for Pitchers
- Best Ever Postseason Batting Careers by Total Ratings
- Best Ever Postseason Batting Careers by Average Year
- Best Ever Postseason Pitching Careers by Total Ratings
- Best Ever Postseason Pitching Careers by Average Year
- Hall of Fame Ranks

And bonus coverage for the upcoming 2011 season

- Fantasy Baseball Cheatsheet

So has Stat Geek Baseball just made these rankings up off the top of their head?

No. The best ever lists in Stat Geek Baseball, the Best Ever Book, come from the baseballevaluation.com Player Rating decision model, a research study that took five thousand hours to complete and rated every player in baseball history, the millions of stats (over 4 million) and came up with nearly 2 million new ones of their own. And it's not a quick one, two, plus three system. It tracks correlations between how baseball statistics of all eras compare and are valued by real baseball. There's no fantasy addition here, although you might find it an interesting system for your fantasy baseball team, too.

At the heart of that system is PEVA, a player rating or player performance grade for every player in history for every season of their career. And yes, it does account for each era and can be used to rather accurately compare a player or pitcher no matter whether they played in a time of limited home runs or steroid induced power. How can it do that? It tracks domination. How well the player dominated his time, the season, or the postseason involved.

So sit back and enjoy Stat Geek Baseball, the Best Ever Book, its rankings, ratings, and stats. See how they compare to your own lists. See how your favorite player makes out compared to others. See which one of your current favorites still playing the game is moving up the career list for their team or the whole of baseball and track what that means to their Hall of Fame potential.

Raise the office debate or baseball forum to another level, or just add these numbers and rankings to the heap of baseball opinion and sabermetric gobble already out there. We hope you enjoy it. We hope you debate it. We thank you for taking a look.

STAT GEEK BASEBALL and baseballevaluation.com About Our Stats

Just like you, we were always fascinated with the game of baseball, the way the game moved, its history, and yes, it's stats. But just how can you determine whether a player from the distant past is better than a player from today. Or even how can you tell just whether a certain player of today is better than another? So the staff at baseballevaluation.com set about on that task. Now, it's a task taken by others in the past. We are not saying we are the first, or the best, or that likely location, somewhere in between. But after 5,000 hours of research, we think we've come up with a Player Rating system that mirrors the way real baseball values its players, and here, as well as on the website of baseballevaluation.com, we give you stats you can get nowhere else!

What stats might you say?

How about PEVA?

PEVA is the acronym for Stat Geek Baseball's New Player Rating value. This grade is given to each player and pitcher each season, rating their performance on a peer to peer review. Six components for pitchers and batters are melded together into the PEVA Rating, which ranges each year from 0.200 to 64.000. PEVA ratings are available for every pitcher and hitter in baseball history and are valid no matter whether you or your favorite played in the dead ball era, the Babe Ruth era, or the steroid era.

PEVA Player Rating Boxscore		
64.000	Maximum	
32.000	Fantastic	MVP or CY Young Candidate
20.000	Great	All-League
15.000	Very Good	All-Star Caliber
10.000	Good	Plus Starter
3.500	Average	Bench Player
0.200	Minimum	

What are the PEVA Components?

Which stats go into the Player Rating components that form the PEVA rating and how are they measured? PEVA components are measured on a comparative statistical basis within the year in question, scaling the value between the Minimum, Average, and Maximum values for the year. There are exceptions to these rules, however, for components that test the boundaries of the maximum, for example, as well as for components that have more value and impact on other components. Yes, it's complicated, but that's how things are measured in real baseball, too.

Position Player Components
Games Played
Plate Appearances
Run Production (Runs Scored plus Runs Batted In)
On Base Percentage
Slugging Percentage
Field Value

Pitcher Components
Games or Games Started
Wins or Saves or Wins plus Saves
Earned Run Average
WHIP9 (Walks and Hits Per 9 Innings Pitched)
Innings Pitched

So what is RAVE?

Rave or Rolling Average is a graded composite of the last three (RAVE) PEVA values. The traditional RAVE value is measured by taking the most recent PEVA year at a 50% level, 2nd year at 30%, and the 3rd year at 20%. In the Best Ever ranking lists, RAVE does not show up much, but it does in other factors and projections within the baseballevaluation.com statistics.

What is EXPEQ and how does it factor in the Best Ever ranking equation?

EXPEQ or Experience Equivalent becomes important when trying to compare the Best Ever Career ranks and how to put them into perspective. It is calculated using a correlation between the maximum number of games played, plate appearances, or innings pitched in a current season against the individual player and whether a player reached a certain percentage of those factors. It is formatted in a number of years and percentage of that year fashion, instead of MLSTs use of years.days.

For the Best Ever Career ranks you will see PEVA per EQ year listed beside each player and it will be different from the total amount of seasons played or the comparative PEVA per Year figure. An EQ year essentially gives you an MLST type approximation for players of any era and allows you to figure out how a player's PEVA rating looks over an average full season.

We know the above explanation really does not go anywhere near explaining enough about the system, although you can argue whether that matters or not when viewing the Best Ever rankings in this Best Ever Book of baseball players. There's more at baseballevaluation.com for those that are interested.

Now, onto the fun and the rankings. Again, we thank you for your interest, and hope we've added one small measure to the stats and best ever rankings of all Stat Geeks out there.

Stat Geek Baseball, The Best Ever Book Chapter Index

Definitions/Abbreviations

Regular Statistics

R	Runs	
RBI	Runs Batted In	
HR	Home Runs	
RPR	Run Production	Runs Plus Runs Batted In
OBP	On-Base Percentage	Hits and Walks plus Hit By Pitch divided by the total of At Bats plus Walks plus Hit By Pitch plus Sacrifice Flies.
SLG	Slugging Percentage	Total Bases divided by At Bats.
OPS	On Base Plus Slugging	OBP plus SLG percentages.
AVE	Batting Average	Hits Divided by At Bats.
PA	Plate Appearances	At Bats, Walks, Hit By Pitch, Sacrifice Flies, Sacrifices, and Defensive Interference.
AB	At Bats	
W	WIns	
L	Losses	
SV	Saves	Defined as a pitcher who a) finished a game & b) does not get the win and c) pitches at least one inning with a lead of no more than three runs or pitches with the tying run at bat, on base, or on deck, or pitches 3 innings with the lead.
IP	Innings Pitched	
ERA	Earned Run Average	The amount of earned runs divided by Innings Pitched times 9.
WHIP	Walks/Hits per Inning	Walks plus hits divided by Innings Pitched.
WHIP9	Walks/Hits per 9 Innings	Walks plus hits over 9 Innings Pitched
SO/W	Strikeout to Walk Ratio	Strikeouts divided by Walks
HR9IP	Home Runs per 9 IP	Home Runs divided by Innings Pitched multipied by 9.
MLST	Major League Service Time	Years.days a player was on the 25 man roster.
New Stats		
PEVA-B	PEVA Batting Rating	Overall Player Rating for Batting
PEVA-P	PEVA Pitching Rating	Overall Player Rating for Pitching
PEVA-T	PEVA Total Rating	Overall Player Rating for Regular Season, including both Pitching and Batting.
FV	Field Value	Stat measuring a player's fielding value over the season compared to other players. Stats used within this value include Innings/Games Played, Fielding Percentage, Caught Stealing Percentage, Outfield Assists Per Game/9 IP, and Range Factor. The maximum, average, and minimum value for Field Value differs with each position. Maximums range from 2.10 for Catchers, 1.75 for Shortstops, 1.70 for OF/3B, 1.50 for 2B, and 1.40 for 1B.
EQ	Equivalent Years	Equivalent of Full Season Years by a Player based on Games, Plate Appearances, and IP compared to total # of games in a season.
Per EQ	Per Equivalent Years	PEVA divided by EQ Year
RAVE	PEVA Rolling Average	Rolling Average of PEVA values based on 50% of the most recent year, 30% of one year back, and 20% of two years prior.

Chapter 1 - BEST SEASONS EVER, POSITION PLAYERS

From Stat Geek Baseball and baseballevaluation.com

PEVA Player Rating Boxscore		
64.000	Maximum	Maximum Player Rating
32.000	Fantastic	MVP/CY Young Candidate
20.000	Great	All-League
15.000	Very Good	All-Star Caliber
10.000	Good	Plus Starter
3.500	Average	Bench Player
0.200	Minimum	Minimum Player Rating

Baseball fans have debated the best ever players over the century and one-half of the major leagues. Who were the best players ever, and what player had the best season ever? It made for interesting water cooler or bottled water debate and continues to interest baseball fans from the casual watcher to sabermetrics gurus everywhere. And there were always factors that made it so very hard to measure. How do you account for the dead ball era when almost nobody hit a home run? What about the days when the ball got livelier? Were the players better then just because their stats began to climb? And boy, how about those yesteryear heros who played a schedule that did not approach that 154 game schedule or the 162 game schedule, but sat at about half of that number? Then there's the steroid era, perhaps the hardest of them all. How do you figure that out, not only not knowing how it affected the game overall, but not knowing who or how many players were involved? And now that we're past that time (at least we hope we are) how do the players of the last couple years compare?

Well, all those questions are the ones we tried to find answers to, albeit likely imperfect ones. But after 5,000 hours of trying to figure out the correlations in all those many stats, baseballevaluation.com has come up with a system that allows for some of those answers to come into view. At least in our statistical opinion. So sit back and enjoy this debate one more time. I'm sure it will spur just as many questions as it will answer them. Hope you enjoy the ride. It will be one that deals with players and how dominant they were in the time period they played in, because in the end, that's how you truly measure greatness, isn't it?

So who will be judged as having the best season ever and just why did it come out that way? Will it be someone you thought would be way up the list or a surprise? Well, just a little hint in the best season ever front before we get any further down the line. There's not a whole lot of surprises on the batting front, with the top ten being dominated by very few players. But oh, just wait for Chapter 2 and the Best Seasons Ever by a Pitcher list, nine different players show up on that list. So without further consternation, we will unveil the player who had the best season ever for a position player, and later in the chapter we'll chronicle a bit about the players from the 2010 season who cracked the Top 200 of this list.

1. Babe Ruth, New York Yankees, 1923

No surprise whatsoever that the Bambino topped the list, although some will be surprised that his 1927 season did not come in at Number One. But do not fret too much, there will be more than one additional season by the great Babe Ruth in the top ten seasons of all-time. And more than two or three as well. Sometimes it's hard for fans of the game today to realize just how dominant Babe Ruth was as a player in the era from 1920-1930. But before we get into the stat details, just think about the cultural context. Even today, you can play in one of his "Babe Ruth" leagues; over 1 million kids do each year. You can go into a supermarket and buy a Baby Ruth bar. Yes, we know there's controversy about the name. Even though there's now Yankees Stadium II, most acknowledge that it is still the house that Ruth built. 1923 was the first year the Yankees played in the first house that Babe constructed. Ruth was so dominant in his time, well, he hit more home runs in some seasons than entire teams did. But let's get to the number one season itself, 1923.

Babe Ruth put so much fear into the opposition that he walked 170 times in 1923, 69 times more than the hitter in second place. While this was not one of his most prodigious home run seasons, he still tied for the Major League lead at 41. But where you really see the domination is in the area of Slugging and On Base

Percentage. And even though the folks from 1923 likely thought of OPS in other terms, there's no doubt to its import. Just think what an OPS of 1.309 would look like in any park in 2010. Just awesome! And BTW, he stole 17 bases in 1923, 9th place in the AL. Of course, he was thrown out 21 times to get there. He wasn't perfect. Not only in that stat, but in the PEVA Rating system. There has never been a player in the history of baseball who was perfect, 64.000, in all system stat categories for a full regular season, but oh, boy, he sure was close. And it's not like he suffered from having nobody else play in this era when the likes of Gehrig was plying his trade at first base for Ruth's own Yankees and there was some guy named Hornsby in the other league.

Stat Dominance Chart		Babe Ruth 1923		58.931 PEVA RATING	
Stats for Year	**Ruth**	**Rank**	**% Of Year Ave.**		**Yr. Ave.**
RUNS	151	1			
RBI	131	1	RPR	525.4%	53.67
OBP	0.545	1		157.1%	0.347
SLG	0.764	1		195.4%	0.391
Field Value	1.57				

*Note: Maximum FV for OF 1.70. RPR = Runs plus RBI.

2. Babe Ruth, New York Yankees, 1920

Many people think, and it's perfectly understandable, that Ruth's 1920 season was his best, and from an offensive standpoint, that may be so. It was the year that proved just how dominant Ruth was, and what was to come over the next decade. It was the first season of Babe Ruth as a Yankee after that trade from the Red Sox that catapulted the dislike between the two teams and fans to the level we see today. And it was the first year that Babe Ruth was no longer really a pitcher, logging only 4 innings in 1920 after hurling more than 100 each season from 1915 to 1919. So here goes the dominance. Babe Ruth hit 54 Home Runs in 1920; second place was 19. Babe Ruth slugged at a rate of 0.847, the average SLG% in MLB in 1920 was 0.372. So then why does the baseballevaluation PEVA system rank him slightly lower than in 1923. Well, for one thing, Ruth was less durable in 1920 than in 1923, so when he was not on the field, either for reason of rest or injury, Ruth could reek less havoc on opposing pitchers. Durability is important. It helps teams win games. Ruth played in 142 contests in 1920 and got to the plate 616 times compared to 152 and 699 respectively in 1923. And Ruth was a much better outfielder after several more seasons of full-time outfield play, rating a 1.57 out of 1.70 for his defense in 1923, but only 1.29 in 1920. His defensive abilities, which were better than most give him credit for, were just rounding into shape in his first Yankee year when Ruth was still only 25 years old.

But if you want to consider Ruth in 1920 as the better player because you discount defense in the equation, that's fine with us. The PEVA rating system does not, however, penalize a prodigious offensive player because of his poor defense, but does reward a great offensive player who is also a good defensive player. Ruth benefitted from good defense in 1923 by gaining approximately 11% in his overall rating over 1920. So when you add in all those factors, the 158 Runs Scored and the 137 RBI's, too, all first place totals just like in 1923, he comes in at Number Two on our list of best seasons ever.

Stat Dominance Chart		Babe Ruth 1920		55.754 PEVA RATING	
Stats for Year	**Ruth**	**Rank**	**% Of Year Ave.**		**Yr. Ave**
RUNS	158	1			
RBI	137	1	RPR	600.3%	49.14
OBP	0.532	1		158.8%	0.335
SLG	0.847	1		227.7%	0.372
Field Value	1.29				

*Note: Maximum FV for OF 1.70. RPR = Runs plus RBI.

3. Barry Bonds, San Francisco Giants, 2001

Oh, boy. Now comes the Bonds. Not since Babe Ruth banged out those seasons in the 1920s had baseball seen such a fantastic hitting season. It was truly Ruthian in its exploits. Now we know, or think we know, that those exploits are somewhat tainted in today's PED allegation environment, but how tainted, we can't say for sure. So we will go forward with the acknowledgement that Bonds played in the steroid era, and that some players participated in PEDs that makes it difficult to know who, how many, and what affect it had on the statistics of the day. But we will judge, until we know with more certainty, the era within the era itself and will judge domination in that way. And boy, Barry Bonds was certainly dominant in 2001.

Managers did not want to pitch to him, he walked 177 times. There were 73 home runs hit, still the Major League record, even though most of us think of it with an asterick. His OPS was up in the stratosphere, 1.378, which was miles ahead of the man in second place that year, Sammy Sosa, at 1.174. So yes, there's no doubt, he dominated this steroid era year, although not quite as much as Ruth did in the two years above. Bonds did not lead baseball in Runs or RBI, partly due to only appearing in 153 games of the 162 game schedule. Although, to be fair, and to flesh out what we do know, he did accomplish this in PacBell Park, a notorious pitcher's park.

I know, for many, including us, it's hard to accept the fact that Bonds would appear high on any list with the allegations that are in the air. And this will be the problem with ascertaining where his, and others, accomplishments should sit. But here it is, a player from the steroid era, sitting high on another list, as we measured his accomplishments within the era he played in. Now we won't even get into the whole Hall of Fame worthiness debate now will we.

Stat Dominance Chart		**B. Bonds 2001**		**55.207 PEVA RATING**	
Stats for Year	**Bonds**	**Rank**	**% Of Year Ave.**		**Yr. Ave**
RUNS	129	4			
RBI	137	6	RPR	444.1%	59.9
OBP	0.515	1		155.1%	0.332
SLG	0.863	1		202.1%	0.427
Field Value	1.31				

*Note: Maximum FV for OF 1.70. RPR = Runs plus RBI.

4. Babe Ruth, New York Yankees, 1921

Come on. Really. How do you make the case for one Ruth year over the other? It's really a parsing out of fabulous stats against other fabulous stats. Sure, the three that stand out here versus 1923 is a slightly lower On Base Percentage, Batting Average, and a Fielding Value that still hadn't improved to the level it would a couple years hence. But we were in the 2nd year of Ruth domination in the Yankee world while still playing in the Polo Grounds they shared with the Giants who were none too pleased with the success and even suggested they move. So plans were made, spurred on by Ruth's popularity and the Yankees finally becoming the Yankees we know of today.

With 177 Runs Scored and 171 Runs Batted In, there was very little opponents could do to minimize the damage Ruth could cause during a game. And the fact that he was a darling of the New York and national media, both with his bat and exploits outside the diamond made for one great combination. And it was just what baseball ordered after the 1919 Black Sox scandal. The New York baseball public clamored for as much Ruth as possible, and the Yankee stadium that would be started after the 1921 season would now warrant nearly 60,000 seats, double the amount in the usual stadium at the time. Babe Ruth was 6 times as likely to produce a run as the average major league player in 1921, and at 26 years of age in 1921, there were going to be a decade or more of such statistics in the future. It would bode well for the Yankees, and even though it's hard to believe today, 1921 was the first year the Yankees won the pennant, eventually losing the World Series to the Giants, their co-tenant at those soon-to-be vacated Polo Grounds.

Stat Dominance Chart		Babe Ruth 1921	54.876 PEVA RATING		
Stats for Year	**Ruth**	**Rank**	**% Of Year Ave.**		**Yr. Ave.**
RUNS	177	1			
RBI	171	1	RPR	616.8%	56.42
OBP	0.512	1		147.1%	0.348
SLG	0.846	1		209.9%	0.403
Field Value	1.47				

*Note: Maximum FV for OF 1.70. RPR = Runs plus RBI.

5. Barry Bonds, San Francisco Giants, 2002

Once again, Barry Bonds dominated Major League Baseball in 2002, albeit in a PED allegation year. His OBP was at 0.582 a record in the history of the game, only to be broken just two years later by, you guessed it, Barry Bonds. Bonds was way down in homers in 2002 with 46 compared to his record of the year before; managers just plain walked him. They walked him 198 times, another record, to be broken two years later by, you guessed it, Barry Bonds again, with 232. They were walking him with men on first base. They were planning to walk him even before the game began. One reason Bonds' numbers weren't even higher, the production numbers and not rate stats, were those walks, plus the lack of games played, only 143. He ranked only #8 in Runs Scored and #14 in Runs Batted In for the season. But those rate stats of OBP and SLG were enormous, scaling the average player at a rate 175.8 percent and 191.6% of the norm.

There's not a whole lot more to say. It's a year and an era too many of us remember well.

Stat Dominance Chart		B. Bonds 2002	54.848 PEVA RATING		
Stats for Year	**Bonds**	**Rank**	**% Of Year Ave.**		**Yr. Ave.**
RUNS	117	8			
RBI	110	14	RPR	379.9%	59.75
OBP	0.582	1		175.8%	0.331
SLG	0.799	1		191.6%	0.417
Field Value	1.19				

*Note: Maximum FV for OF 1.70. RPR = Runs plus RBI.

6. Ted Williams, Boston Red Sox, 1942

The Splendid Splinter jumps into the #6 position in the best seasons ever list with a spectacular campaign in 1942 that saw the fans at Fenway Park see a marvelous year. Only the second player season on this list that had a Run Production multiplier over the average player of 6, Williams was a run producing machine. He was on base often, a 0.499 clip, and slugged his way around the league to the tune of 0.648. Williams fielded his position well, too, with a 1.55 Field Value.

But this would be the last season for Ted Williams until 1946, as his commitment to serve the nation during World War II would begin after the 1942 season. With the two seasons he'd just had, (1941 comes in at #8), it's easy to imagine that some of those missed years would have wound up pretty high on the best ever list, too, particularly when his returning season in 1946 comes in at #26.

Williams was a marvel who deserves to be on this list above some above, and perhaps include more seasons, too. But everyone knows that, or at least should. Certainly the folks from Boston are aware that the Splendid Splinter is one of the best players in the history of the game, with two of the top ten seasons of all-time.

Stat Dominance Chart		T. Williams 1942	52.075 PEVA RATING		
Stats for Year	**Williams**	**Rank**	**% Of Year Ave.**		**Yr. Ave.**
RUNS	141	1			
RBI	137	1	RPR	647.9%	42.91

OBP	0.499	1		154.5%	0.323
SLG	0.648	1		185.1%	0.350
Field Value	1.55				

*Note: Maximum FV for OF 1.70. RPR = Runs plus RBI.

7. Babe Ruth, New York Yankees, 1927

Yes, it made the top ten, and for most, still includes the most important number in the history of the game. 60. Babe Ruth hit sixty home runs in 1927 in a 154 game season. But when you look at some of his other numbers, no matter how spectacular they are, the magnificent season of 1927 ranks #7 on the All-Time Batting Seasons list, and only #4 in the best seasons ever by a man named Babe.

To justify this is again a quibble. His On Base Percentage had begun to drop, from its high of 0.545 in 1923 to the rate of 0.486. He wouldn't have an OBP above 0.500 for the rest of his career. Probably because of that lack of selectivity, Ruth hit those Home Runs and had Run Production over the roof that Ruth built, again topping a 6 multiple compared to both players in the American and National Leagues.

The Yankees won 110 games in 1927 and many regard this team as the best team in baseball history. They won the World Series in four straight games over the Pittsburgh Pirates. It was a wonderful season, led by the Big Bambino.

Stat Dominance Chart		**Babe Ruth 1927**		**51.850 PEVA RATING**	
Stats for Year	**Ruth**	**Rank**		**% Of Year Ave.**	**Yr. Ave.**
RUNS	158	1			
RBI	164	2	RPR	639.5%	50.35
OBP	0.486	1		140.9%	0.345
SLG	0.772	1		196.4%	0.393
Field Value	1.43				

*Note: Maximum FV for OF 1.70. RPR = Runs plus RBI.

8. Ted Williams, Boston Red Sox, 1941

The world was at war, but the USA still remained on the outside, not yet attacked at Pearl Harbor, and baseball reigned throughout the summer of 1941 as a diversion to a nation that knew it likely wouldn't be long until they, too, were involved. But in this season, there was always Ted to lead the way. Williams was only 23 years old in 1941 and had already become the most dominant player in the game. Taking nothing away from the other great players at the time, including the great Joe DiMaggio, but this was the season when the last man hit 0.400, actually 0.406. There's a lot of debate as to whether we'll ever see that again. Well, all we know for sure, is we haven't seen it yet.

Ted Williams could just plain hit. He hit for that average. He hit for power, knocking 37 homers out of the park in 1941 while knocking in 120. Nobody did it better than the splinter in 1941; only seven seasons were better than his 1941 campaign in the batting history of the game.

Stat Dominance Chart		**T. Williams 1941**		**51.730 PEVA RATING**	
Stats for Year	**Williams**	**Rank**		**% Of Year Ave.**	**Yr. Ave.**
RUNS	135	1			
RBI	120	4	RPR	535.0%	47.66
OBP	0.553	1		165.6%	0.334
SLG	0.735	1		196.0%	0.375
Field Value	1.33				

*Note: Maximum FV for OF 1.70. RPR = Runs plus RBI.

9. Babe Ruth, New York Yankees, 1926

Ho, hum, just another year when Babe Ruth led baseball in Runs Scored, Runs Batted In, On Base and Slugging Percentage while leading his team to the World Series. They'd lose to the Cardinals there. At 31 years of age, Ruth still played the field well, and hit the ground running on the bases, and the plate swinging to the tune of 47 Home Runs and 146 Runs Batted In.

Stat Dominance Chart		Babe Ruth 1926	51.603 PEVA RATING		
Stats for Year	**Ruth**	**Rank**	**% Of Year Ave.**		**Yr. Ave.**
RUNS	139	1			
RBI	146	1	RPR	552.1%	51.62
OBP	0.516	1		149.6%	0.345
SLG	0.737	1		189.5%	0.389
Field Value	1.47				

*Note: Maximum FV for OF 1.70. RPR = Runs plus RBI.

10. Babe Ruth, New York Yankees, 1924

There's a tendency for people to think that Babe Ruth was solely a power hitter, somewhat incapable of hitting the ball up the middle or to get a base hit, but once again his campaign in 1924 proved that to be false. Ruth hit 0.378 in 1924 while hitting those prodigious shots, all 46 of them. He also knocked out 39 doubles, 7 triples, and 108 singles, to the total of 200 hits. Add that to his 142 walks and you can see how 342 times on base can lead to 143 runs scored.

Coming in as the 6th entrant of Babe Ruth on the all-time best seasons list, it may bring up the rear, however, it once again added to the legacy of the man who built a Yankee Stadium, filled it to the rafters each night, and charmed, in a roguesh manner, the public and media throughout his career.

Stat Dominance Chart		Babe Ruth 1924	51.324 PEVA RATING		
Stats for Year	**Ruth**	**Rank**	**% Of Year Ave.**		**Yr. Ave.**
RUNS	143	1			
RBI	121	3	RPR	522.3%	50.55
OBP	0.513	1		147.4%	0.348
SLG	0.739	1		187.6%	0.394
Field Value	1.53				

*Note: Maximum FV for OF 1.70. RPR = Runs plus RBI.

Who Jumped Onto the Best Ever Seasons List in 2010

Well, as most scribes have noted, this was the year of the pitcher. No hitters abounded, even one in the postseason, and ERA's were down so far, you'd have thought that PED meant Pitcher's Extra Dominance and not the Performance Enhancing Drugs of the decade or so in the past. But that didn't mean that some of today's best sluggers did not perform very well, a number of them did. And three did so well, in fact, that they jumped into the Best Ever Season lists for position players; yes, Albert Pujols again, and Miguel Cabrera, too, plus the man who may win the MVP over Pujols, but probably should not, Joey Votto.

Joining the List

Name	**Team**	**Rank**
Albert Pujols	St. Louis Cardinals	100

Miguel Cabrera	Detroit Tigers	166
Joey Votto	Cincinnati Reds	224

Best Position Player Years Ever (1871-2010) - Regular Season

Rank	Name	First	Year	Team	Lg	HR	RBI	AVG	Age	PEVA-B
1	Ruth	Babe	1923	NYA	AL	41	131	0.393	28	**58.931**
2	Ruth	Babe	1920	NYA	AL	54	137	0.376	25	**55.754**
3	Bonds	Barry	2001	SFN	NL	73	137	0.328	37	**55.207**
4	Ruth	Babe	1921	NYA	AL	59	171	0.378	26	**54.876**
5	Bonds	Barry	2002	SFN	NL	46	110	0.370	38	**54.848**
6	Williams	Ted	1942	BOS	AL	36	137	0.356	24	**52.075**
7	Ruth	Babe	1927	NYA	AL	60	164	0.356	32	**51.850**
8	Williams	Ted	1941	BOS	AL	37	120	0.406	23	**51.730**
9	Ruth	Babe	1926	NYA	AL	47	146	0.372	31	**51.603**
10	Ruth	Babe	1924	NYA	AL	46	121	0.378	29	**51.324**
11	Gehrig	Lou	1927	NYA	AL	47	175	0.373	24	**50.953**
12	Shaffer	Orator	1878	IN1	NL	0	30	0.338	27	**50.240**
13	Gehrig	Lou	1934	NYA	AL	49	165	0.363	31	**50.136**
14	Cobb	Ty	1917	DET	AL	6	102	0.383	31	**48.911**
15	Wagner	Honus	1908	PIT	NL	10	109	0.354	34	**48.837**
16	Bonds	Barry	2004	SFN	NL	45	101	0.362	40	**48.632**
17	Rosen	Al	1953	CLE	AL	43	145	0.336	29	**48.401**
18	Mantle	Mickey	1956	NYA	AL	52	130	0.353	25	**47.236**
19	Barnes	Ross	1876	CHN	NL	1	59	0.429	26	**46.554**
20	Gehrig	Lou	1936	NYA	AL	49	152	0.354	33	**45.995**
21	Yount	Robin	1982	ML4	AL	29	114	0.331	27	**45.717**
22	Ruth	Babe	1919	BOS	AL	29	114	0.322	24	**45.484**
23	Hornsby	Rogers	1922	SLN	NL	42	152	0.401	26	**45.149**
24	Robinson	Frank	1962	CIN	NL	39	136	0.342	27	**44.982**
25	Jones	Charley	1879	BSN	NL	9	62	0.315	29	**44.728**
26	Williams	Ted	1946	BOS	AL	38	123	0.342	28	**44.043**
27	Pujols	Albert	2009	SLN	NL	47	135	0.327	29	**43.955**
28	Hines	Paul	1879	PRO	NL	2	52	0.357	24	**43.731**
29	Cobb	Ty	1915	DET	AL	3	99	0.369	29	**43.671**
30	Bonds	Barry	1993	SFN	NL	46	123	0.336	29	**43.404**
31	Speaker	Tris	1912	BOS	AL	10	90	0.383	24	**43.198**
32	Mays	Willie	1965	SFN	NL	52	112	0.317	34	**43.109**
33	Robinson	Frank	1966	BAL	AL	49	122	0.316	31	**42.918**
34	Stone	George	1906	SLA	AL	6	71	0.358	30	**42.914**
35	Barnes	Ross	1873	BS1	NA	2	62	0.425	23	**42.762**
36	Delahanty	Ed	1899	PHI	NL	9	137	0.410	32	**42.557**
37	Foxx	Jimmie	1933	PHA	AL	48	163	0.356	26	**42.539**
38	Cobb	Ty	1911	DET	AL	8	127	0.420	25	**42.464**
39	Mantle	Mickey	1961	NYA	AL	54	128	0.317	30	**42.398**
40	Mantle	Mickey	1960	NYA	AL	40	94	0.275	29	**42.394**
41	Speaker	Tris	1916	CLE	AL	2	79	0.386	28	**42.388**
42	Bench	Johnny	1972	CIN	NL	40	125	0.270	25	**42.271**
43	Brouthers	Dan	1892	BRO	NL	5	124	0.335	34	**42.196**
44	Ruth	Babe	1931	NYA	AL	46	163	0.373	36	**41.968**
45	Hornsby	Rogers	1929	CHN	NL	39	149	0.380	33	**41.923**
46	Aaron	Hank	1959	ML1	NL	39	123	0.355	25	**41.920**
47	Williams	Ted	1949	BOS	AL	43	159	0.343	31	**41.778**
48	Mathews	Eddie	1953	ML1	NL	47	135	0.302	22	**41.677**
49	Mathews	Eddie	1960	ML1	NL	39	124	0.277	29	**41.382**
50	Musial	Stan	1943	SLN	NL	13	81	0.357	23	**41.278**

Best Position Player Years Ever (1871-2010)

Rank	Name	First	Year	Team	Lg	HR	RBI	AVG	Age	PEVA-B
51	Thomas	Frank	1994	CHA	AL	38	101	0.353	26	**41.278**
52	Foxx	Jimmie	1932	PHA	AL	58	169	0.364	25	**41.252**
53	Piazza	Mike	1997	LAN	NL	40	124	0.362	29	**41.108**
54	Schmidt	Mike	1981	PHI	NL	31	91	0.316	32	**40.963**
55	Mays	Willie	1962	SFN	NL	49	141	0.304	31	**40.947**
56	McGwire	Mark	1998	SLN	NL	70	147	0.299	35	**40.809**
57	Rodriguez	Alex	2007	NYA	AL	54	156	0.314	32	**40.717**
58	Mantle	Mickey	1957	NYA	AL	34	94	0.365	26	**40.637**
59	Musial	Stan	1953	SLN	NL	30	113	0.337	33	**40.502**
60	Yastrzemski	Carl	1967	BOS	AL	44	121	0.326	28	**40.383**
61	Delahanty	Ed	1896	PHI	NL	13	126	0.397	29	**40.317**
62	Aaron	Hank	1963	ML1	NL	44	130	0.319	29	**40.268**
63	Gehrig	Lou	1931	NYA	AL	46	184	0.341	28	**40.131**
64	Holmes	Tommy	1945	BSN	NL	28	117	0.352	28	**40.065**
65	Ruth	Babe	1928	NYA	AL	54	142	0.323	33	**40.061**
66	Gehrig	Lou	1930	NYA	AL	41	174	0.379	27	**39.839**
67	Cobb	Ty	1909	DET	AL	9	107	0.377	23	**39.724**
68	Brouthers	Dan	1883	BFN	NL	3	97	0.374	25	**39.597**
69	Mays	Willie	1955	NY1	NL	51	127	0.319	24	**39.404**
70	Brett	George	1985	KCA	AL	30	112	0.335	32	**39.324**
71	Ruth	Babe	1930	NYA	AL	49	153	0.359	35	**39.019**
72	Sosa	Sammy	2001	CHN	NL	64	160	0.328	33	**38.918**
73	Mays	Willie	1960	SFN	NL	29	103	0.319	29	**38.898**
74	Musial	Stan	1944	SLN	NL	12	94	0.347	24	**38.717**
75	Snider	Duke	1953	BRO	NL	42	126	0.336	27	**38.655**
76	Lajoie	Nap	1901	PHA	AL	14	125	0.426	27	**38.604**
77	Mathews	Eddie	1959	ML1	NL	46	114	0.306	28	**38.592**
78	Hines	Paul	1878	PRO	NL	4	50	0.358	23	**38.405**
79	Campanella	Roy	1953	BRO	NL	41	142	0.312	32	**38.348**
80	Anson	Cap	1881	CHN	NL	1	82	0.399	29	**38.143**
81	Bagwell	Jeff	1994	HOU	NL	39	116	0.368	26	**38.044**
82	Jackson	Joe	1912	CLE	AL	3	90	0.395	23	**37.904**
83	Musial	Stan	1952	SLN	NL	21	91	0.336	32	**37.889**
84	DiMaggio	Joe	1941	NYA	AL	30	125	0.357	27	**37.809**
85	Gehrig	Lou	1937	NYA	AL	37	159	0.351	34	**37.760**
86	Cobb	Ty	1918	DET	AL	3	64	0.382	32	**37.701**
87	Flick	Elmer	1900	PHI	NL	11	110	0.367	24	**37.675**
88	Hornsby	Rogers	1925	SLN	NL	39	143	0.403	29	**37.655**
89	Foxx	Jimmie	1938	BOS	AL	50	175	0.349	31	**37.579**
90	Dunlap	Fred	1884	SLU	UA	13		0.412	25	**37.571**
91	Murray	Eddie	1984	BAL	AL	29	110	0.306	28	**37.406**
92	Musial	Stan	1951	SLN	NL	32	108	0.355	31	**37.339**
93	Speaker	Tris	1914	BOS	AL	4	90	0.338	26	**37.327**
94	Aaron	Hank	1962	ML1	NL	45	128	0.323	28	**37.264**
95	Kiner	Ralph	1951	PIT	NL	42	109	0.309	29	**37.003**
96	Kelley	Joe	1896	BLN	NL	8	100	0.364	25	**36.936**
97	McCovey	Willie	1969	SFN	NL	45	126	0.320	31	**36.924**
98	Delahanty	Ed	1893	PHI	NL	19	146	0.368	26	**36.872**
99	Gehrig	Lou	1933	NYA	AL	32	139	0.334	30	**36.863**
100	Pujols	Albert	2010	SLN	NL	42	118	0.312	30	**36.842**

Best Position Player Years Ever (1871-2010)

Rank	Name	First	Year	Team	Lg	HR	RBI	AVG	Age	PEVA-B
101	Mantle	Mickey	1955	NYA	AL	37	99	0.306	24	**36.784**

102	Mays	Willie	1958	SFN	NL	29	96	0.347	27	**36.606**
103	Baker	Frank	1912	PHA	AL	10	130	0.347	26	**36.589**
104	Ott	Mel	1929	NY1	NL	42	151	0.328	20	**36.437**
105	Mantle	Mickey	1958	NYA	AL	42	97	0.304	27	**36.436**
106	Williams	Ted	1947	BOS	AL	32	114	0.343	29	**36.436**
107	Pujols	Albert	2006	SLN	NL	49	137	0.331	26	**36.422**
108	Mays	Willie	1964	SFN	NL	47	111	0.296	33	**36.258**
109	Magee	Sherry	1910	PHI	NL	6	123	0.331	26	**36.256**
110	Cash	Norm	1961	DET	AL	41	132	0.361	27	**36.206**
111	Cobb	Ty	1912	DET	AL	7	83	0.409	26	**36.188**
112	White	Deacon	1877	BSN	NL	2	49	0.387	30	**36.179**
113	Canseco	Jose	1988	OAK	AL	42	124	0.307	24	**36.135**
114	Bonds	Barry	1990	PIT	NL	33	114	0.301	26	**36.095**
115	Sheckard	Jimmy	1903	BRO	NL	9	75	0.332	25	**36.040**
116	Thomas	Frank	1991	CHA	AL	32	109	0.318	23	**36.023**
117	Bagwell	Jeff	1999	HOU	NL	42	126	0.304	31	**35.960**
118	Ripken Jr.	Cal	1984	BAL	AL	27	86	0.304	24	**35.874**
119	Seymour	Cy	1905	CIN	NL	8	121	0.377	33	**35.870**
120	Ripken Jr.	Cal	1991	BAL	AL	34	114	0.323	31	**35.853**
121	Evans	Dwight	1984	BOS	AL	32	104	0.295	33	**35.847**
122	Wagner	Honus	1907	PIT	NL	6	82	0.350	33	**35.816**
123	Musial	Stan	1948	SLN	NL	39	131	0.376	28	**35.811**
124	DiMaggio	Joe	1937	NYA	AL	46	167	0.346	23	**35.806**
125	Wilson	Hack	1930	CHN	NL	56	191	0.356	30	**35.799**
126	Browning	Pete	1885	LS2	AA	9	73	0.362	24	**35.668**
127	McVey	Cal	1875	BS1	NA	3	87	0.355	26	**35.627**
128	Santo	Ron	1964	CHN	NL	30	114	0.313	24	**35.556**
129	Howard	Ryan	2006	PHI	NL	58	149	0.313	27	**35.507**
130	Wagner	Honus	1905	PIT	NL	6	101	0.363	31	**35.473**
131	Duffy	Hugh	1894	BSN	NL	18	145	0.440	28	**35.452**
132	Bonds	Barry	1992	PIT	NL	34	103	0.311	28	**35.418**
133	Kauff	Benny	1914	IND	FL	8	95	0.370	24	**35.356**
134	Delahanty	Ed	1897	PHI	NL	5	96	0.377	30	**35.320**
135	Mattingly	Don	1986	NYA	AL	31	113	0.352	25	**35.246**
136	Speaker	Tris	1923	CLE	AL	17	130	0.380	35	**35.232**
137	Nicholson	Bill	1943	CHN	NL	29	128	0.309	29	**35.190**
138	Greenberg	Hank	1937	DET	AL	40	183	0.337	26	**35.096**
139	White	Deacon	1875	BS1	NA	1	60	0.367	28	**35.065**
140	Speaker	Tris	1920	CLE	AL	8	107	0.388	32	**35.002**
141	Hamilton	Billy	1894	PHI	NL	4	87	0.404	28	**34.987**
142	Doby	Larry	1952	CLE	AL	32	104	0.276	29	**34.855**
143	Brouthers	Dan	1891	BS2	AA	5	109	0.350	33	**34.762**
144	Nicholson	Bill	1944	CHN	NL	33	122	0.287	30	**34.757**
145	Snider	Duke	1955	BRO	NL	42	136	0.309	29	**34.757**
146	Allen	Dick	1972	CHA	AL	37	113	0.308	30	**34.744**
147	Johnson	Bob	1944	BOS	AL	17	106	0.324	39	**34.713**
148	Delahanty	Ed	1902	WS1	AL	10	93	0.376	35	**34.689**
149	Giambi	Jason	2000	OAK	AL	43	137	0.333	29	**34.676**
150	Martinez	Edgar	1995	SEA	AL	29	113	0.356	32	**34.622**

Best Position Player Years Ever (1871-2010)

Rank	Name	First	Year	Team	Lg	HR	RBI	AVG	Age	PEVA-B
151	Boggs	Wade	1987	BOS	AL	24	89	0.363	29	**34.592**
152	Rodriguez	Alex	2000	SEA	AL	41	132	0.316	25	**34.588**
153	Banks	Ernie	1959	CHN	NL	45	143	0.304	28	**34.546**
154	Torre	Joe	1971	SLN	NL	24	137	0.363	31	**34.530**

155	Cobb	Ty	1916	DET	AL	5	68	0.371	30	**34.488**
156	Keeler	Willie	1897	BLN	NL	0	74	0.424	25	**34.488**
157	Mays	Willie	1963	SFN	NL	38	103	0.314	32	**34.477**
158	Brouthers	Dan	1882	BFN	NL	6	63	0.368	24	**34.359**
159	Hornsby	Rogers	1924	SLN	NL	25	94	0.424	28	**34.349**
160	Stovey	Harry	1889	PH4	AA	19	119	0.308	33	**34.320**
161	Lajoie	Nap	1906	CLE	AL	0	91	0.355	32	**34.293**
162	Gehrig	Lou	1928	NYA	AL	27	142	0.374	25	**34.257**
163	Foster	George	1977	CIN	NL	52	149	0.320	29	**34.233**
164	Rodriguez	Alex	2005	NYA	AL	48	130	0.321	30	**34.230**
165	Jackson	Joe	1911	CLE	AL	7	83	0.408	22	**34.179**
166	Cabrera	Miguel	2010	DET	AL	38	126	0.328	27	**34.146**
167	Delgado	Carlos	2000	TOR	AL	41	137	0.344	28	**34.143**
168	Jennings	Hughie	1896	BLN	NL	0	121	0.401	27	**34.143**
169	Gehrig	Lou	1932	NYA	AL	34	151	0.349	29	**34.089**
170	Fielder	Prince	2009	MIL	NL	46	141	0.299	25	**34.062**
171	Lajoie	Nap	1904	CLE	AL	6	102	0.376	30	**34.027**
172	Schmidt	Mike	1974	PHI	NL	36	116	0.282	25	**34.027**
173	Mays	Willie	1954	NY1	NL	41	110	0.345	23	**33.994**
174	Foxx	Jimmie	1934	PHA	AL	44	130	0.334	27	**33.948**
175	Bonds	Barry	1996	SFN	NL	42	129	0.308	32	**33.905**
176	Jackson	Reggie	1969	OAK	AL	47	118	0.275	23	**33.898**
177	Thomas	Frank	1992	CHA	AL	24	115	0.323	24	**33.764**
178	Killebrew	Harmon	1969	MIN	AL	49	140	0.276	33	**33.634**
179	Banks	Ernie	1958	CHN	NL	47	129	0.313	27	**33.578**
180	Stargell	Willie	1973	PIT	NL	44	119	0.299	33	**33.533**
181	Aaron	Hank	1960	ML1	NL	40	126	0.292	26	**33.522**
182	Carter	Gary	1982	MON	NL	29	97	0.293	28	**33.521**
183	Pujols	Albert	2003	SLN	NL	43	124	0.359	23	**33.510**
184	Jeter	Derek	1999	NYA	AL	24	102	0.349	25	**33.492**
185	Wagner	Honus	1900	PIT	NL	4	100	0.381	26	**33.466**
186	Lynn	Fred	1979	BOS	AL	39	122	0.333	27	**33.455**
187	Pujols	Albert	2008	SLN	NL	37	116	0.357	28	**33.388**
188	Connor	Roger	1885	NY1	NL	1	65	0.371	28	**33.375**
189	Murcer	Bobby	1971	NYA	AL	25	94	0.331	25	**33.373**
190	Mitchell	Kevin	1989	SFN	NL	47	125	0.291	27	**33.233**
191	Musial	Stan	1954	SLN	NL	35	126	0.330	34	**33.159**
192	Rosen	Al	1952	CLE	AL	28	105	0.302	28	**33.117**
193	Klein	Chuck	1930	PHI	NL	40	170	0.386	26	**33.052**
194	Santo	Ron	1966	CHN	NL	30	94	0.312	26	**33.049**
195	Bench	Johnny	1974	CIN	NL	33	129	0.280	27	**33.034**
196	Barnes	Ross	1871	BS1	NA	0	34	0.401	21	**32.973**
197	Morgan	Joe	1976	CIN	NL	27	111	0.320	33	**32.946**
198	Collins	Eddie	1914	PHA	AL	2	85	0.344	27	**32.932**
199	McCovey	Willie	1970	SFN	NL	39	126	0.289	32	**32.785**
200	Burkett	Jesse	1901	SLN	NL	10	75	0.376	33	**32.782**

Best Position Player Years Ever (1871-2010)

Rank	Name	First	Year	Team	Lg	HR	RBI	AVG	Age	PEVA-B
201	Snider	Duke	1954	BRO	NL	40	130	0.341	28	**32.766**
202	Kelly	King	1884	CHN	NL	13	95	0.354	27	**32.756**
203	Berra	Yogi	1950	NYA	AL	28	124	0.322	25	**32.744**
204	Winfield	Dave	1979	SDN	NL	34	118	0.308	28	**32.688**
205	Williams	Ted	1957	BOS	AL	38	87	0.388	39	**32.674**
206	Jones	Chipper	1999	ATL	NL	45	110	0.319	27	**32.668**
207	Sheffield	Gary	1996	FLO	NL	42	120	0.314	28	**32.618**

208	Mauer	Joe	2009	MIN	AL	28	96	0.365	26	**32.590**
209	Belle	Albert	1995	CLE	AL	50	126	0.317	29	**32.570**
210	Kauff	Benny	1915	BRF	FL	12	83	0.342	25	**32.472**
211	Barnes	Ross	1872	BS1	NA	1	44	0.432	22	**32.428**
212	Cobb	Ty	1910	DET	AL	8	91	0.383	24	**32.427**
213	Greenberg	Hank	1935	DET	AL	36	170	0.328	24	**32.405**
214	Pujols	Albert	2005	SLN	NL	41	117	0.330	25	**32.329**
215	Barnes	Ross	1875	BS1	NA	1	58	0.364	25	**32.311**
216	Greenberg	Hank	1940	DET	AL	41	150	0.340	29	**32.198**
217	Dalrymple	Abner	1880	CHN	NL	0	36	0.330	23	**32.143**
218	Mathews	Eddie	1955	ML1	NL	41	101	0.289	24	**32.143**
219	Aaron	Hank	1967	ATL	NL	39	109	0.307	33	**32.133**
220	Cabrera	Miguel	2006	FLO	NL	26	114	0.339	23	**32.126**
221	Wagner	Honus	1903	PIT	NL	5	101	0.355	29	**32.062**
222	Lajoie	Nap	1910	CLE	AL	4	76	0.384	36	**32.056**
223	Bonds	Barry	2000	SFN	NL	49	106	0.306	36	**32.041**
224	Votto	Joey	2010	CIN	NL	37	113	0.324	26	**32.034**
225	Winfield	Dave	1984	NYA	AL	19	100	0.340	33	**32.023**

Note: Age = Player age at the end of the year. All position players above 32 PEVA-B (Peva Batting) Player Ratings listed.

Note: 2010 PEVA = Reflects preliminary final PEVA numbers. Marginal adjustment may be made with final stats and park factors.

Chapter 2 - BEST SEASONS EVER, PITCHERS

From Stat Geek Baseball and baseballevaluation.com

PEVA Player Rating Boxscore		
64.000	Maximum	Maximum Player Rating
32.000	Fantastic	MVP/CY Young Candidate
20.000	Great	All-League
15.000	Very Good	All-Star Caliber
10.000	Good	Plus Starter
3.500	Average	Bench Player
0.200	Minimum	Minimum Player Rating

On the pitching side of the best ever equation, we have seasons that span the history of the game. From the days prior to the turn of the century to seasons only a decade ago. And as diverse as the span of time in this group of spectacular pitching years is, it includes just as diverse a collection of pitchers with differing styles. Unlike the best ever seasons by position players, there is no dominance by several players over the course of the game's history. It includes nine different players. And yes, you will recognize most of them, although there is a surprise or two in the bunch. Well, not so much of a surprise when you start looking through the numbers, but a surprise in names you haven't heard bantered around lately, except in Hall of Fame circles. So let's get started on the Pitching Top, the best Seasons Ever by those men who man the mound. Did any of the pitchers from 2010 push their way into the Top 10, and how many of them made the Best Ever List overall in this great year of pitching? Read below to find out.

1. Greg Maddux, Atlanta Braves, 1994

It may not come as a surprise that the wizard of Fulton County would be included amongst the game's best, or that one of his seasons would be feted, but the 1994 season is one that most fans of the game would not list at the top. Oh, not because it was not dominant. But because it is a year that most baseball fans would like to forget. A year when the players, their union, and Major League Baseball went on strike on August 12, subsequently leading to the owners cancelling the World Series when the strike could not be settled. For most teams, including Atlanta, that meant playing about 114 games. It was that almost unthinkable year, and as much as people do remember, beyond the sabermetrics crowd, they recall that strike and the harm it caused. They recall that end of the season with no playoff baseball, no World Series to crown a glorious champ. Mostly, they'd like to forget.

But for players such as Greg Maddux, it was a season to remember. During the span of two seasons, both marred by a strike, Greg Maddux had mastered the game. In 1994, he did not give up hits, walks, or runs, and although not a strikeout pitcher by any means, he did strike them out, too, 156 times in 1994, 5th most in baseball that year. All done in the cozy confines of a hitter's park. But that didn't matter to Maddux, nobody hit the ball in the air. They were baffled by an unending array of pitches that nipped whatever corner Maddux wanted to nip. In 1994, Maddux only gave up 0.18 HR for every 9 innings he pitched compared to the average in baseball that year of 1.04 HR per 9 IP. His ERA, in that hitter's park, was 34.6% of the average ERA in baseball. That leading ERA was over 1 run better than the man in second place in 1994. Batter's couldn't hit against him, and remember, he didn't walk them either, only 31 times all year. Baseball was witnessing the best pitching season in baseball history and all they could think about was how much more money they might make if they'd just strike, from the union standpoint, or just cancel, from the owner's side. But that shouldn't take anything away from what Maddux accomplished, no matter whether the season was shortened to 114 games or not, or whether his stat line does not look as impressive as it would have if you would prorate those stats to the full 162 game schedule. We're talking 23-9, 1.56 ERA, 222 SO, 287 IP, plus those fantastic rate stats then. It was the best season ever by a pitcher.

Stat Dominance Chart		Greg Maddux 1994	57.974 PEVA RATING	
Stats for Year	**Maddux**	**Rank**	**% Of Year Ave.**	**Year Ave**
W	16	2	470.6%	3.40
IP	202.0	1	332.1%	60.82
ERA	1.56	1	34.6%	4.51
WHIP9	8.06	1	62.6%	12.87
SO/W Ratio	5.03	3	282.6%	1.78

2. Silver King, St. Louis Browns, 1888

A long time ago when the Cardinals were known as the Browns, a pitcher named Silver was the golden arm in the franchise that played in Sportmen's Park and the American Association league. They won their league that year with a 92-43 record, but there was no Championship Series to be played. Such a shame, because in 1888, Silver King won 45 games over 485.7 innings pitched, which warrants him a spot at #2 on the Best Seasons Ever by a Pitcher list.

It's hard to get your head around the type of numbers pitchers put up in an era when they pitched in what was essentially a two or three man rotation. King started 65 of the 135 games the Browns played that year. That was just the way the game was played at the time. But how dominant were they in that era of short rotations, given the circumstances of the time? Was he more dominant that Maddux in 1994 or some great pitching year in between?

King was certainly pretty darn good, but a bit below Maddux in 1994. King won 45 games in a time when the average pitcher won 9.14, compared to Maddux and his 16 wins vs. 3.40, giving an edge to King in department one with his 492.3% of Year Average figure. King also wins a slight edge in the Innings Pitched category when comparing the % of Year Average numbers, too. (Look how similar those numbers are, however.) But Maddux in 1994 was better in the rate stats compared to baseball averages, notching up an ERA that was 34.6% vs. 55.6%, a WHIP9 at 62.6% vs. 74.2%, and a SO/W ratio that was 282.6% of baseball's average in 1994 compared to 194.8% for King. The ERA stat is where Maddux gets a major edge, as there were more more than 10 pitchers in 1888 within one run of King's ERA with 2nd place at 1.74. So, Maddux has a 3-2 edge in the five categories we've highlighted. But that doesn't take anything away from Silver King and his accomplishments. He was king in the year of 1888 and richly deserves his location on this Top Ten list. King did not have any year in his career like this one again, although he did win 203 games over 10 seasons, but for that one season, Silver King was nearly the best ever of the best.

Stat Dominance Chart		Silver King 1888	56.728 PEVA RATING	
Stats for Year	**King**	**Rank**	**% Of Year Ave.**	**Year Ave**
W	45	1	492.3%	9.14
IP	585.7	1	355.8%	164.63
ERA	1.64	1	55.6%	2.95
WHIP9	7.88	1	74.2%	10.62
SO/W Ratio	3.39	5	194.8%	1.74

3. Amos Rusie, New York Giants, 1894

Advance only six years from the standout season of Silver King and to the city of New York where the baseball Giants pitcher, a Hall of Fame pitcher to boot, was putting up numbers that add up to a 3rd place finish in the Best Seasons Ever by a Pitcher rankings. Baseball took a turn for the offensive during the 1890s. Not yet in the Home Run department, but in the amount of hits, walks, and runs that pitchers gave up and hitter's hit. In only the six years since King's great performance, the major league average ERA had risen from 2.95 to 5.31 and the WHIP9 from 10.62 to 15.38. Home Runs were growing, but still accounted for only 0.41 per game.

None of this mattered to Rusie, whose ERA of 2.78 was 52.4% of the league average during the season. The

Giants were a good team in 1894, winning 88 games while losing only 44, but were not good enough to win the National League pennant, losing out to the National League version of the Baltimore Orioles by one game. This was not the same franchise as the Baltimore Orioles of today, which started out in Milwaukee when the American League began play in the next decade.

Rusie was not a strikeout pitcher, with only 195 in 444 innings, but then again, the era was so odd compared to today (he walked 200 batters). A walk had only been considered four balls since 1889, before then it was more, and the mound had been moved back from 50 feet to 60 feet 6 inches only in 1893. I guess they were still getting used to that. But all in all, comparing apples to apples in the decade of the 1890s, Amos Rusie dominated this season just as King and Maddux had theirs. He finished in the Top 2 pitchers in Wins, Innings Pitched, ERA, WHIP9, and Strike Out to Walk Ratio.

Stat Dominance Chart		**Amos Rusie 1894**		**55.357 PEVA RATING**
Stats for Year	**Rusie**	**Rank**	**% Of Year Ave.**	**Year Ave**
W	36	1	525.5%	6.85
IP	444.0	2	368.6%	120.45
ERA	2.78	1	52.4%	5.31
WHIP9	12.69	2	82.5%	15.38
SO/W Ratio	0.98	2	171.9%	0.57

4. Bret Saberhagan, Kansas City Royals, 1989

He's a pitcher people tend to forget outside of the Midwest, even though he pitched barely one generation ago. Yes, his career wasn't filled with as many great seasons as some, particularly in the second half of his career, but for a couple seasons in the late 1980s, Saberhagan was one of the dominant pitchers in the game. So dominant in 1989 in fact, that he ranked #1 in all five of the major stat categories below. He won the most games, 23, pitched the most innings, 262.3, had the lowest ERA, 2.16, allowed the fewest Walks and Hits, 8.65 per 9 innings, and had the best SO/W Ratio at 4.49, that ratio actually poor by some of his career standards. One year it was over 11.

Now those numbers weren't quite as dominant as those of the three seasons ranked above him, but they were the best of that year and way up the scale, enough to push Saberhagan into the #4 pitching season of all pitching seasons in baseball history.

Stat Dominance Chart		**B. Saberhagan 1989**		**53.447 PEVA RATING**
Stats for Year	**Bret**	**Rank**	**% Of Year Ave.**	**Year Ave**
W	23	1	491.5%	4.68
IP	262.3	1	312.3%	84.00
ERA	2.16	1	58.2%	3.71
WHIP9	8.65	1	72.8%	11.89
SO/W Ratio	4.49	1	256.6%	1.75

5. Walter Johnson, Washington Senators, 1912

The Big Train was coming and he was coming strong and fast. In 1912, Walter Johnson ran through batters like an express train. He struck out 303 batters, barely over 8 batters reached base each nine IP, and pitched to an ERA so minuscule, you had to look a long time before finding any crooked innings numbers. And while that ERA and the WHIP numbers weren't quite as good in comparison to major league averages of the season as Greg Maddux was in 1994 (ERA at 41.2% for Johnson while Maddux was at 34.2%; WHIP at 67.0% for Johnson while Maddux was at 62.6%), it was quite fantastic enough to raise him to the #5 pitching season of all-time.

He was the toast of Washington town for the first decades of their franchise, which would later become the Minnesota Twins in 1960, and his dominance might have been at its height in 1912, but it was not the last time Johnson would dominate. Before the end of his Hall of Fame career, he would win 20 games 12 times and amass 417 total victories. And his stellar seasons would command five other places in the Top Pitching seasons of All-Time, coming in at #33, #39, #42, #96, and #98.

Stat Dominance Chart		W. Johnson 1912		52.347 PEVA RATING
Stats for Year	**Johnson**	**Rank**	**% Of Year Ave.**	**Year Ave**
W	33	2	644.5%	5.12
IP	369.0	2	401.4%	91.93
ERA	1.39	1	41.2%	3.37
WHIP9	8.17	1	67.0%	12.20
SO/W Ratio	3.99	1	311.7%	1.28

6. Lefty Grove, Philadelphia Athletics, 1930

When the turn of Lefty Grove came around for the A's and the fans at Philly's Schibe Park, they knew they were in for a treat. And of those 721,663 fans who passed through the turnstiles of the neighborhood park that 1st place season, they would credit Lefty Grove for 28 of their 102 victories. Grove led all of baseball that year in Wins, ERA, WHIP, and SO/W Ratio, just falling short of the 5 tool sweep by pitching a 5th place total of innings, but it was in the dominance over the league average where Grove really stood out. And when it was all said and done, that added up to the #6 spot in the All-Time Pitching Seasons list.

The A's would be the toast of the Philadelphia town for a couple more decades before they took their traveling show to Kansas City and then Oakland in baseball's migration west, but for the season of 1930 when their team led by Grove took the field, they witnessed a stellar pitching season, one which would propel their star to the Hall of Fame in Cooperstown after his career was done.

Stat Dominance Chart		Lefty Grove 1930		52.344 PEVA RATING
Stats for Year	**Grove**	**Rank**	**% Of Year Ave.**	**Year Ave**
W	28	1	452.3%	6.19
IP	291.0	4	264.9%	109.86
ERA	2.54	1	52.8%	4.81
WHIP9	10.30	1	75.2%	13.69
SO/W Ratio	3.48	1	334.6%	1.04

7. Pete (Grover Cleveland) Alexander, Philadelphia Phillies, 1915

There wasn't a whole lot to cheer about for the Phillies franchise in their first one hundred years of existence. By 1915, they still hadn't won a World Series, and wouldn't for 65 more years. But in 1915, they had one of their best seasons ever with a 90-62 record and first place finish, led by the man known as Grover Cleveland Alexander. The Phillies plied their trade at Baker Bowl; Alexander took the ball and mound for 31 of their victories. His ERA was slight, 1.22, at 42.1% of the the baseball average. His WHIP figure was near historic lows at 7.58 and 67.5% of that Major League norm for 1915. Over the next few seasons, Alexander took the mound for the Phillies and put up great numbers, ranking #19 All-Time for them in 1916 and #65 in 1917, but he'd move onto the Cubs by the time another Top 100 season came along, #63 for them in 1920. And it would be a very long time before the denizens of the dugouts or fans in the stands of a Philly park would witness a pitching season anywhere near the caliber of Alexander again. But in the summer of 1915 in the Bowl they called Baker, and for those two great years after, he took the ball sixty feet and six inches away from the batter at home plate and dominated the competition to the tune of three of the best seasons ever, #7 at its pinnacle best of the best ever list in 1915. All hail the President of the Mound, Grover Cleveland Alexander.

Stat Dominance Chart		P. Alexander 1915		52.205 PEVA RATING
Stats for Year	**Pete**	**Rank**	**% Of Year Ave.**	**Year Ave**
W	31	1	482.1%	6.43
IP	376.3	2	321.8%	116.93

ERA	1.22	1	42.1%	2.90
WHIP9	7.58	1	67.5%	11.23
SO/W Ratio	3.77	1	296.9%	1.27

8. Greg Maddux, Atlanta Braves, 1995

They ended the strike before the end of spring training in 1995, but the season would be shortened again, starting April 25 and lasting 144 games. And even though the offseason was much longer than normal, Greg Maddux just kept going, as if there had been no time between seasons. Over the two year span, Maddux would win 35 games and lose only 8; in 1995 he'd win 19. His ERA continued in a fashion unseen since Bob Gibson, only 1.63 in 1995 after the 1.56 in 1994. Fulton County Stadium was once again abuzz with a masterful craftsman making hitters look silly, and all without a 90 MPH fastball.

The ERA of 1.63 was just 36.6% of the league average, just above the 34.6% level of 1994. He didn't pitch as many innings, ranking #6 in baseball, but some stats even got better. Can you imagine pitching almost 210 innings and walking only 23, plus once again, this non-strikeout oriented pitcher had 181 strikeouts in 1995, 7.76 per 9 innings pitched. When people talk about dominance, they somehow push Maddux out of the equation because he wasn't a glowering Clemens or Gibson on the mound. He wasn't going to strike nineteen batters out in a game. All he was going to do was get you out.

So let's prorate those stats to the full 162 game schedule again just for fun. We're talking 21-2, 1.63 ERA, 204 SO, 236 IP, plus those dizzy WHIP and SO/W ratio numbers. This might not have been Maddux's best season, but it was certainly great, great enough to rank as the great #8 on the All-Time Seasons list.

Stat Dominance Chart		**Greg Maddux 1995**		**52.118 PEVA RATING**
Stats for Year	**Maddux**	**Rank**	**% Of Year Ave.**	**Year Ave**
W	19	1	520.5%	3.65
IP	209.7	6	321.3%	65.27
ERA	1.63	1	36.6%	4.45
WHIP9	7.30	1	57.1%	12.79
SO/W Ratio	7.87	1	439.7%	1.79

9. Randy Johnson, Arizona Diamondbacks, 2001

It had to be scary to look out at that mound sixty feet from home plate and see a basketball player basically handing the ball to the catcher. And when Johnson uncorked that gangly 6'10" frame toward the batter, that's just how it seemed to be. Just ask John Kruk from the indelible video impression at the All-Star game when Randy threw one high and tight. And what made Johnson special, particularly in this 2001 season, was not only the fastball in the high 90s or that towering presence, but the fact that he had both of those things plus an ability to control his balls and strikes. Power and precision, it's the best combination you can have in this game.

Johnson was in his own battle with a teammate this season as both Johnson and Schilling dueled in the desert for the best Diamondback performance. Johnson would win this debate in the PEVA ratings with a 52.009 PEVA Pitching Rating, although Curt was certainly doing his best to compete, coming in at #77 All-Time. But it was Randy Johnson at the head of the class, with the #9 season ever. He led baseball in ERA and WHIP. Remember, this was pitching in the steroid era when Home Runs left the park with the frequency of a busy airport and three run homers dominated the day. But Johnson still posted an ERA of 2.49 in a hitter's ballpark, BankOne, at only 56.3% of the average ERA of the season.

Stat Dominance Chart		**R. Johnson 2001**		**52.009 PEVA RATING**
Stats for Year	**Johnson**	**Rank**	**% Of Year Ave.**	**Year Ave**
W	21	3	510.9%	4.11
IP	249.7	2	340.9%	73.24
ERA	2.49	1	56.3%	4.42

WHIP9	9.081	1	73.2%	12.41
SO/W Ratio	5.239	4	255.6%	2.05

10. Carl Hubbell, New York Giants, 1933

When you pitch to an ERA less than half of the average, you've had one great season. At 43.6% of that average, you're bound to find a spot in the Top Ten, and that's just what Carl Hubbell did for the New York Giants in 1933. Add in baseball bests in WHIP and SO/W ratio and you cement your position. Hubbell is another one of those pitchers who is less remembered than he should be as the years go by. He does not have a best pitcher award named after him like Cy Young or a Spalding glove. However, this does not diminish in any fashion a Hall of Fame career and this great season, plus others that were Top 100 worthy either. Think #14 in 1936 when he won 26 games or the #38 season of 1934.

When the Giants moved out of New York and took residence in San Francisco, some of the luster of these seasons seem to have gone with them. Less so for the Dodgers who claimed the borough of Brooklyn, but seemingly so for the Giants. Perhaps it was the trio of Mays, McCovey, and Marichal that gave the franchise a west coast bias in the minds of the fans, but there was no better pitching season in the long history of the Giants franchise, and no better pitcher ever either than the man named Carl Hubbell, with all due respect to Mathewson, Marichal, and Rusie, too.

Stat Dominance Chart		**Carl Hubbell 1933**		**52.001 PEVA RATING**
Stats for Year	**Hubbell**	**Rank**	**% Of Year Ave.**	**Year Ave**
W	23	3	348.0%	6.61
IP	308.7	2	259.4%	119.02
ERA	1.66	1	43.6%	3.81
WHIP9	8.84	1	70.9%	12.46
SO/W Ratio	3.32	1	328.7%	1.01

Note: WHIP9 (Walks/Hits per 9 Innings Pitched)

Who Jumped Onto the Best Ever Seasons List in 2010

Some will think that more pitchers should join the Top 150 seasons in this year of the pitcher than two, but the single most important factor in the PEVA equation is tracking dominance and while those two pitchers certainly did, in Roy Halladay reaching the #42 spot on the best season list, and Adam Wainwright at #119, the whole of baseball saw a resurgence in pitching that didn't quite allow more than two to jump onto the list. Great seasons from Felix Hernandez and Ubaldo Jimenex, for different reasons, fell just short. Felix just didn't compare highly enough in the victory category, the only pure counting stat category within the PEVA calculation, to quite rise to the 32.000 minimum. For Ubaldo, his second half performance pulled down his final season rating below it as well. And what about the relief pitchers of 2010, did any of them make it to the relief pitcher's list? Well, the answer is no. While five pitchers ranked within the Top 40 pitchers of 2010, none rose to the level of 15.000 PEVA, with Rafael Soriano of Tampa Bay, Billy Wagner of Atlanta, Brian WIlson of San Francisco, and Joakim Soria of Kansas City, falling within 1 point of that designation.

Joining the Best Pitcher List

Name	**Team**	**Rank**
Roy Halladay	Philadelphia Phillies	42
Adam Wainwright	St. Louis Cardinals	119

Best Pitcher Years Ever (1871-2010) - Regular Season

Rank	Name	First	Year	Team	Lg	W	L	SV	IP	ERA	Age	PEVA-P
1	Maddux	Greg	1994	ATL	NL	16	6	0	202.0	1.56	28	**57.974**
2	King	Silver	1888	SL4	AA	45	21	0	585.7	1.64	20	**56.728**

3	Rusie	Amos	1894	NY1	NL	36	13	1	444.0	2.78	23	**55.357**
4	Saberhagen	Bret	1989	KCA	AL	23	6	0	262.3	2.16	25	**53.447**
5	Johnson	Walter	1912	WS1	AL	33	12	2	369.0	1.39	25	**52.347**
6	Grove	Lefty	1930	PHA	AL	28	5	9	291.0	2.54	30	**52.344**
7	Alexander	Pete	1915	PHI	NL	31	10	3	376.3	1.22	28	**52.205**
8	Maddux	Greg	1995	ATL	NL	19	2	0	209.7	1.63	29	**52.118**
9	Johnson	Randy	2001	ARI	NL	21	6	0	249.7	2.49	38	**52.009**
10	Hubbell	Carl	1933	NY1	NL	23	12	5	308.7	1.66	30	**52.001**
11	Maddux	Greg	1993	ATL	NL	20	10	0	267.0	2.36	27	**50.920**
12	Koufax	Sandy	1963	LAN	NL	25	5	0	311.0	1.88	28	**50.345**
13	Maddux	Greg	1992	CHN	NL	20	11	0	268.0	2.18	26	**49.969**
14	Hubbell	Carl	1936	NY1	NL	26	6	3	304.0	2.31	33	**48.808**
15	Young	Cy	1901	BOS	AL	33	10	0	371.3	1.62	34	**48.628**
16	Carlton	Steve	1980	PHI	NL	24	9	0	304.0	2.34	36	**48.494**
17	Newhouser	Hal	1945	DET	AL	25	9	2	313.3	1.81	24	**48.483**
18	Cooper	Mort	1942	SLN	NL	22	7	0	278.7	1.78	29	**47.763**
19	Alexander	Pete	1916	PHI	NL	33	12	3	389.0	1.55	29	**47.561**
20	Nichols	Kid	1897	BSN	NL	31	11	3	368.0	2.64	28	**47.498**
21	Lee	Cliff	2008	CLE	AL	22	3	0	223.3	2.54	30	**47.417**
22	Maddux	Greg	1997	ATL	NL	19	4	0	232.7	2.20	31	**47.132**
23	Clemens	Roger	1997	TOR	AL	21	7	0	264.0	2.05	35	**47.079**
24	Maddux	Greg	1998	ATL	NL	18	9	0	251.0	2.22	32	**46.792**
25	Santana	Johan	2006	MIN	AL	19	6	0	233.7	2.77	27	**46.748**
26	Blackwell	Ewell	1947	CIN	NL	22	8	0	273.0	2.47	25	**46.689**
27	Feller	Bob	1940	CLE	AL	27	11	4	320.3	2.61	22	**46.643**
28	Grove	Lefty	1931	PHA	AL	31	4	5	288.7	2.06	31	**46.335**
29	Koufax	Sandy	1965	LAN	NL	26	8	2	335.7	2.04	30	**46.321**
30	Walsh	Ed	1908	CHA	AL	40	15	6	464.0	1.42	27	**46.136**
31	Halladay	Roy	2008	TOR	AL	20	11	0	246.0	2.78	31	**46.107**
32	Martinez	Pedro	1999	BOS	AL	23	4	0	213.3	2.07	28	**45.790**
33	Johnson	Walter	1910	WS1	AL	25	17	1	370.0	1.36	23	**45.752**
34	Gooden	Dwight	1985	NYN	NL	24	4	0	276.7	1.53	21	**45.645**
35	Johnson	Randy	2004	ARI	NL	16	14	0	245.7	2.60	41	**44.961**
36	Scott	Mike	1986	HOU	NL	18	10	0	275.3	2.22	31	**44.955**
37	Vance	Dazzy	1924	BRO	NL	28	6	0	308.3	2.16	33	**44.534**
38	Hubbell	Carl	1934	NY1	NL	21	12	8	313.0	2.30	31	**43.860**
39	Johnson	Walter	1913	WS1	AL	36	7	2	346.0	1.14	26	**43.737**
41	Young	Cy	1893	CL4	NL	34	16	1	422.7	3.36	26	**43.542**
41	Johnson	Randy	2002	ARI	NL	24	5	0	260.0	2.32	39	**43.254**
42	Halladay	Roy	2010	PHI	NL	21	10	0	251.7	2.44	33	**43.340**
43	Johnson	Walter	1918	WS1	AL	23	13	3	326.0	1.27	31	**43.252**
44	Clemens	Roger	1987	BOS	AL	20	9	0	281.7	2.97	25	**43.210**
45	Clemens	Roger	1991	BOS	AL	18	10	0	271.3	2.62	29	**43.004**
46	Wyatt	Whit	1941	BRO	NL	22	10	1	288.3	2.34	34	**42.854**
47	Brown	Kevin	1998	SDN	NL	18	7	0	257.0	2.38	33	**42.712**
48	Luque	Dolf	1923	CIN	NL	27	8	2	322.0	1.93	33	**42.654**
49	Marichal	Juan	1969	SFN	NL	21	11	0	299.7	2.10	32	**42.465**
50	Chandler	Spud	1943	NYA	AL	20	4	0	253.0	1.64	36	**42.384**

Best Pitcher Years Ever (1871-2010)

Rank	Name	First	Year	Team	Lg	W	L	SV	IP	ERA	Age	PEVA-P
51	Greinke	Zack	2009	KCA	AL	16	8	0	229.3	2.16	26	**42.305**
52	Clarkson	John	1885	CHN	NL	53	16	0	623.0	1.85	24	**42.218**
53	Donohue	Pete	1925	CIN	NL	21	14	2	301.0	3.08	25	**42.024**
54	Bond	Tommy	1879	BSN	NL	43	19	0	555.3	1.96	23	**41.857**
55	Clemens	Roger	1990	BOS	AL	21	6	0	228.3	1.93	28	**41.426**
56	Devlin	Jim	1877	LS1	NL	35	25	0	559.0	2.25	28	**41.423**

Rank	Name	First	Year	Team	Lg	W	L	SV	IP	ERA	Age	PEVA-P
57	Walters	Bucky	1939	CIN	NL	27	11	0	319.0	2.29	30	**41.246**
58	Chesbro	Jack	1904	NYA	AL	41	12	0	454.7	1.82	30	**41.233**
59	Smoltz	John	1996	ATL	NL	24	8	0	253.7	2.94	29	**41.196**
60	Peavy	Jake	2007	SDN	NL	19	6	0	223.3	2.54	26	**41.129**
61	Young	Cy	1899	SLN	NL	26	16	1	369.3	2.58	32	**40.907**
62	Brown	Mordecai	1909	CHN	NL	27	9	7	342.7	1.31	33	**40.455**
63	Martinez	Pedro	2000	BOS	AL	18	6	0	217.0	1.74	29	**40.251**
64	Alexander	Pete	1920	CHN	NL	27	14	5	363.3	1.91	33	**40.230**
65	Brown	Kevin	1996	FLO	NL	17	11	0	233.0	1.89	31	**40.004**
66	Alexander	Pete	1917	PHI	NL	30	13	0	388.0	1.83	30	**39.958**
67	Spahn	Warren	1947	BSN	NL	21	10	3	289.7	2.33	26	**39.610**
68	Guidry	Ron	1978	NYA	AL	25	3	0	273.7	1.74	28	**39.576**
69	Sabathia	C.C.	2008	CLE/MI	AL/N	17	10	0	253.0	2.70	28	**39.539**
70	Sabathia	C.C.	2007	CLE	AL	19	7	0	241.0	3.21	27	**39.506**
71	Trout	Dizzy	1944	DET	AL	27	14	0	352.3	2.12	29	**39.350**
72	Hecker	Guy	1884	LS2	AA	52	20	0	670.7	1.80	28	**39.319**
73	Carlton	Steve	1972	PHI	NL	27	10	0	346.3	1.97	28	**39.105**
74	Santana	Johan	2004	MIN	AL	20	6	0	228.0	2.61	25	**39.021**
75	Newhouser	Hal	1946	DET	AL	26	9	1	292.7	1.94	25	**38.936**
76	Tudor	John	1985	SLN	NL	21	8	0	275.0	1.93	31	**38.832**
77	Roberts	Robin	1951	PHI	NL	21	15	2	315.0	3.03	25	**38.640**
78	Schilling	Curt	2001	ARI	NL	22	6	0	256.7	2.98	35	**38.613**
79	Martinez	Pedro	1997	MON	NL	17	8	0	241.3	1.90	26	**38.574**
80	Friend	Bob	1960	PIT	NL	18	12	1	275.7	3.00	30	**38.546**
81	Grove	Lefty	1929	PHA	AL	20	6	4	275.3	2.81	29	**38.544**
82	Young	Cy	1896	CL4	NL	28	15	3	414.3	3.24	29	**38.520**
83	Roberts	Robin	1952	PHI	NL	28	7	2	330.0	2.59	26	**38.501**
84	Roberts	Robin	1953	PHI	NL	23	16	2	346.7	2.75	27	**38.372**
85	Radbourn	Charley	1883	PRO	NL	48	25	1	632.3	2.05	29	**38.272**
86	Nichols	Kid	1898	BSN	NL	31	12	4	388.0	2.13	29	**38.155**
87	Wood	Joe	1912	BOS	AL	34	5	1	344.0	1.91	23	**38.082**
88	Gooden	Dwight	1984	NYN	NL	17	9	0	218.0	2.60	20	**37.956**
89	Rusie	Amos	1893	NY1	NL	33	21	1	482.0	3.23	22	**37.905**
90	Brecheen	Harry	1948	SLN	NL	20	7	1	233.3	2.24	34	**37.896**
91	Martinez	Pedro	2002	BOS	AL	20	4	0	199.3	2.26	31	**37.759**
92	Lincecum	Tim	2009	SFN	NL	15	7	0	225.3	2.48	25	**37.546**
93	Koufax	Sandy	1966	LAN	NL	27	9	0	323.0	1.73	31	**37.199**
94	Schilling	Curt	2002	ARI	NL	23	7	0	259.3	3.23	36	**37.153**
95	Willis	Vic	1899	BSN	NL	27	8	2	342.7	2.50	23	**37.022**
96	Faber	Red	1921	CHA	AL	25	15	1	330.7	2.48	33	**37.013**
97	Johnson	Walter	1914	WS1	AL	28	18	1	371.7	1.72	27	**36.936**
98	Rogers	Steve	1982	MON	NL	19	8	0	277.0	2.40	33	**36.926**
99	Johnson	Walter	1915	WS1	AL	27	13	4	336.7	1.55	28	**36.871**
100	Halladay	Roy	2003	TOR	AL	22	7	0	266.0	3.25	26	**36.858**

Best Pitcher Years Ever (1871-2010)

Rank	Name	First	Year	Team	Lg	W	L	SV	IP	ERA	Age	PEVA-P
101	Radbourn	Charley	1884	PRO	NL	59	12	1	678.7	1.38	30	**36.790**
102	Bradley	George	1876	SL3	NL	45	19	0	573.0	1.23	24	**36.677**
103	Walters	Bucky	1940	CIN	NL	22	10	0	305.0	2.48	31	**36.538**
104	Gibson	Bob	1968	SLN	NL	22	9	0	304.7	1.12	33	**36.485**
105	Appier	Kevin	1993	KCA	AL	18	8	0	238.7	2.56	26	**36.409**
106	Walsh	Ed	1911	CHA	AL	27	18	4	368.7	2.22	30	**36.402**
107	Palmer	Jim	1975	BAL	AL	23	11	1	323.0	2.09	30	**36.388**
108	Faber	Red	1922	CHA	AL	21	17	2	352.0	2.81	34	**36.363**
109	Shocker	Urban	1922	SLA	AL	24	17	3	348.0	2.97	32	**36.261**

110	Webb	Brandon	2006	ARI	NL	16	8	0	235.0	3.10	27	**36.142**
111	Key	Jimmy	1987	TOR	AL	17	8	0	261.0	2.76	26	**36.007**
112	Hunter	Catfish	1974	OAK	AL	25	12	0	318.3	2.49	28	**35.949**
113	Cicotte	Eddie	1917	CHA	AL	28	12	4	346.7	1.53	33	**35.909**
114	Maddux	Greg	2001	ATL	NL	17	11	0	233.0	3.05	35	**35.904**
115	Luque	Dolf	1925	CIN	NL	16	18	0	291.0	2.63	35	**35.862**
116	Gomez	Lefty	1937	NYA	AL	21	11	0	278.3	2.33	29	**35.814**
117	Passeau	Claude	1940	CHN	NL	20	13	5	280.7	2.50	31	**35.806**
118	Roberts	Robin	1954	PHI	NL	23	15	4	336.7	2.97	28	**35.774**
119	Wainwright	Adam	2010	SLN	NL	20	11	0	230.3	2.42	29	**35.709**
120	Dean	Dizzy	1934	SLN	NL	30	7	7	311.7	2.66	24	**35.666**
121	Maglie	Sal	1951	NY1	NL	23	6	4	298.0	2.93	34	**35.603**
122	Richard	J.R.	1979	HOU	NL	18	13	0	292.3	2.71	29	**35.177**
123	Webb	Brandon	2007	ARI	NL	18	10	0	236.3	3.01	28	**35.172**
124	Maddux	Greg	1996	ATL	NL	15	11	0	245.0	2.72	30	**35.136**
125	Glavine	Tom	1991	ATL	NL	20	11	0	246.7	2.55	25	**35.128**
126	Carpenter	Chris	2009	SLN	NL	17	4	0	192.7	2.24	34	**35.125**
127	Wood	Wilbur	1971	CHA	AL	22	13	1	334.0	1.91	30	**35.043**
128	Clemens	Roger	1986	BOS	AL	24	4	0	254.0	2.48	24	**35.030**
129	Carlton	Steve	1982	PHI	NL	23	11	0	295.7	3.10	38	**35.010**
130	Johnson	Randy	1999	ARI	NL	17	9	0	271.7	2.48	36	**34.970**
131	Feller	Bob	1946	CLE	AL	26	15	4	371.3	2.18	28	**34.939**
132	Vance	Dazzy	1928	BRO	NL	22	10	2	280.3	2.09	37	**34.912**
133	Lincecum	Tim	2008	SFN	NL	18	5	0	227.0	2.62	24	**34.883**
134	Lee	Bill	1938	CHN	NL	22	9	2	291.0	2.66	29	**34.833**
135	Mathewson	Christy	1909	NY1	NL	25	6	2	275.3	1.14	29	**34.761**
136	Clemens	Roger	1992	BOS	AL	18	11	0	246.7	2.41	30	**34.707**
137	Jenkins	Fergie	1970	CHN	NL	22	16	0	313.0	3.39	28	**34.695**
138	Walsh	Ed	1910	CHA	AL	18	20	5	369.7	1.27	29	**34.646**
139	Rixey	Eppa	1925	CIN	NL	21	11	1	287.3	2.88	34	**34.428**
140	Derringer	Paul	1939	CIN	NL	25	7	0	301.0	2.93	33	**34.402**
141	Jenkins	Fergie	1974	TEX	AL	25	12	0	328.3	2.82	32	**34.395**
142	Andujar	Joaquin	1982	SLN	NL	15	10	0	265.7	2.47	30	**34.352**
143	Dierker	Larry	1969	HOU	NL	20	13	0	305.3	2.33	23	**34.302**
144	McGinnity	Joe	1899	BLN	NL	28	16	2	366.3	2.68	28	**34.184**
145	Blue	Vida	1971	OAK	AL	24	8	0	312.0	1.82	22	**34.160**
146	Derringer	Paul	1938	CIN	NL	21	14	3	307.0	2.93	32	**34.132**
147	Johnson	Randy	1995	SEA	AL	18	2	0	214.3	2.48	32	**34.064**
148	Hunter	Catfish	1975	NYA	AL	23	14	0	328.0	2.58	29	**34.002**
149	Hernandez	Felix	2009	SEA	AL	19	5	0	238.7	2.49	23	**33.947**
150	Marberry	Firpo	1929	WS1	AL	19	12	11	250.3	3.06	31	**33.863**

Best Pitcher Years Ever (1871-2010)

Rank	Name	First	Year	Team	Lg	W	L	SV	IP	ERA	Age	PEVA-P
151	McGinnity	Joe	1904	NY1	NL	35	8	5	408.0	1.61	33	**33.860**
152	Marichal	Juan	1966	SFN	NL	25	6	0	307.3	2.23	29	**33.441**
153	Singer	Bill	1969	LAN	NL	20	12	1	315.7	2.34	25	**33.351**
154	Hahn	Noodles	1899	CIN	NL	23	8	0	309.0	2.68	20	**33.291**
155	Morris	Ed	1886	PT1	AA	41	20	1	555.3	2.45	24	**33.289**
156	Hutchison	Bill	1890	CHN	NL	42	25	2	603.0	2.70	31	**33.191**
157	Clemens	Roger	1998	TOR	AL	20	6	0	234.7	2.65	36	**33.173**
158	Drysdale	Don	1962	LAN	NL	25	9	1	314.3	2.83	26	**33.051**
159	Perry	Gaylord	1972	CLE	AL	24	16	1	342.7	1.92	34	**33.015**
160	Coombs	Jack	1910	PHA	AL	31	9	1	353.0	1.30	28	**32.981**
161	Ramsey	Toad	1886	LS2	AA	38	27	0	588.7	2.45	22	**32.963**
162	Clarkson	John	1889	BSN	NL	49	19	1	620.0	2.73	28	**32.959**

163	Young	Cy	1895	CL4	NL	35	10	0	369.7	3.26	28	**32.941**
164	Spahn	Warren	1953	ML1	NL	23	7	3	265.7	2.10	32	**32.929**
165	Melton	Cliff	1937	NY1	NL	20	9	7	248.0	2.61	25	**32.906**
166	Blanton	Cy	1935	PIT	NL	18	13	1	254.3	2.58	27	**32.884**
167	Mathewson	Christy	1905	NY1	NL	31	9	2	338.7	1.28	25	**32.837**
168	Nichols	Kid	1893	BSN	NL	34	14	1	425.0	3.52	24	**32.804**
169	Rijo	Jose	1993	CIN	NL	14	9	0	257.3	2.48	28	**32.798**
170	Stratton	Scott	1890	LS2	AA	34	14	0	431.0	2.36	21	**32.787**
171	Marichal	Juan	1965	SFN	NL	22	13	1	295.3	2.13	28	**32.690**
172	Palmer	Jim	1976	BAL	AL	22	13	0	315.0	2.51	31	**32.669**
173	Tewksbury	Bob	1992	SLN	NL	16	5	0	233.0	2.16	32	**32.658**
174	Haines	Jesse	1927	SLN	NL	24	10	1	300.7	2.72	34	**32.652**
175	Gibson	Bob	1969	SLN	NL	20	13	0	314.0	2.18	34	**32.616**
176	Meekin	Jouett	1894	NY1	NL	33	9	2	409.0	3.70	27	**32.597**
177	Dean	Dizzy	1935	SLN	NL	28	12	5	325.3	3.04	25	**32.595**
178	Newhouser	Hal	1944	DET	AL	29	9	2	312.3	2.22	23	**32.575**
179	Carpenter	Chris	2005	SLN	NL	21	5	0	241.7	2.83	30	**32.538**
180	McLain	Denny	1968	DET	AL	31	6	0	336.0	1.96	24	**32.509**
181	Zettlein	George	1871	CH1	NA	18	9	0	240.7	2.73	27	**32.500**
182	Bond	Tommy	1877	BSN	NL	40	17	0	521.0	2.11	21	**32.408**
183	Cuellar	Mike	1969	BAL	AL	23	11	0	290.7	2.38	32	**32.370**
184	Bunning	Jim	1957	DET	AL	20	8	1	267.3	2.69	26	**32.324**
185	Johnson	Walter	1919	WS1	AL	20	14	2	290.3	1.49	32	**32.304**
186	Mathewson	Christy	1911	NY1	NL	26	13	3	307.0	1.99	31	**32.300**
187	Mathewson	Christy	1908	NY1	NL	37	11	5	390.7	1.43	28	**32.234**
188	Blyleven	Bert	1973	MIN	AL	20	17	0	325.0	2.52	22	**32.194**
189	Jansen	Larry	1951	NY1	NL	23	11	0	278.7	3.04	31	**32.060**
190	Drysdale	Don	1964	LAN	NL	18	16	0	321.3	2.18	28	**32.039**
191	Maddux	Greg	2000	ATL	NL	19	9	0	249.3	3.00	34	**32.027**
192	Clarkson	John	1887	CHN	NL	38	21	0	523.0	3.08	26	**32.014**

Best Relief Pitcher Years Ever (1871-2010)

Rank	Name	First	Year	Team	Lg	W	L	SV	IP	ERA	Age	PEVA-P
1	Quisenberry	Dan	1983	KCA	AL	5	3	45	139.0	1.94	30	**24.535**
2	Gagne	Eric	2003	LAN	NL	2	3	55	82.3	1.20	27	**23.161**
3	McDaniel	Lindy	1960	SLN	NL	12	4	26	116.3	2.09	25	**22.970**
4	Moore	Wilcy	1927	NYA	AL	19	7	13	213.0	2.28	30	**22.864**
5	Hernandez	Willie	1984	DET	AL	9	3	32	140.3	1.92	30	**21.765**
6	Rivera	Mariano	2001	NYA	AL	4	6	50	80.7	2.34	32	**21.526**
7	Konstanty	Jim	1950	PHI	NL	16	7	22	152.0	2.66	33	**20.439**
8	Sutter	Bruce	1979	CHN	NL	6	6	37	101.3	2.22	26	**20.400**
9	Gagne	Eric	2002	LAN	NL	4	1	52	82.3	1.97	26	**20.362**
10	Fingers	Rollie	1981	ML4	AL	6	3	28	78.0	1.04	35	**20.222**
11	Wetteland	John	1993	MON	NL	9	3	43	85.3	1.37	27	**20.171**
12	Rivera	Mariano	2005	NYA	AL	7	4	43	78.3	1.38	36	**20.171**
13	Sutter	Bruce	1977	CHN	NL	7	3	31	107.3	1.34	24	**19.860**
14	Jones	Doug	1992	HOU	NL	11	8	36	111.7	1.85	35	**19.538**
15	Putz	J.J.	2006	SEA	AL	4	1	36	78.3	2.30	29	**19.452**
16	Radatz	Dick	1962	BOS	AL	9	6	24	124.7	2.24	25	**19.399**
17	Eckersley	Dennis	1990	OAK	AL	4	2	48	73.3	0.61	36	**18.752**
18	Gagne	Eric	2004	LAN	NL	7	3	45	82.3	2.19	28	**18.671**
19	Gossage	Rich	1977	PIT	NL	11	9	26	133.0	1.62	26	**18.589**
20	Nathan	Joe	2006	MIN	AL	7	0	36	68.3	1.58	32	**18.535**

21	Quisenberry	Dan	1984	KCA	AL	6	3	44	129.3	2.64	31	**18.502**
22	Jones	Doug	1997	ML4	AL	6	6	36	80.3	2.02	40	**18.476**
23	Montgomery	Jeff	1993	KCA	AL	7	5	45	87.3	2.27	31	**18.356**
24	Hiller	John	1973	DET	AL	10	5	38	125.3	1.44	30	**18.324**
25	Sutter	Bruce	1984	SLN	NL	5	7	45	122.7	1.54	31	**18.275**
26	Hoffman	Trevor	1998	SDN	NL	4	2	53	73.0	1.48	31	**18.234**
27	Eichhorn	Mark	1986	TOR	AL	14	6	10	157.0	1.72	26	**18.200**
28	Wagner	Billy	1999	HOU	NL	4	1	39	74.7	1.57	28	**18.157**
29	Abernathy	Ted	1967	CIN	NL	6	3	28	106.3	1.27	34	**18.113**
30	Kern	Jim	1979	TEX	AL	13	5	29	143.0	1.57	30	**18.109**
31	Shaw	Jeff	1997	CIN	NL	4	2	42	94.7	2.38	31	**18.060**
32	Nen	Robb	1998	SFN	NL	7	7	40	88.7	1.52	29	**18.036**
33	Quisenberry	Dan	1982	KCA	AL	9	7	35	136.7	2.57	29	**18.009**
34	Braxton	Garland	1927	WS1	AL	10	9	13	155.3	2.95	27	**17.897**
35	Putz	J.J.	2007	SEA	AL	6	1	40	71.7	1.38	29	**17.868**
36	Smoltz	John	2003	ATL	NL	0	2	45	64.3	1.12	36	**17.673**
37	Eckersley	Dennis	1992	OAK	AL	7	1	51	80.0	1.91	38	**17.431**
38	Rivera	Mariano	1999	NYA	AL	4	3	45	69.0	1.83	30	**17.373**
39	Harvey	Bryan	1991	CAL	AL	2	4	46	78.7	1.60	28	**17.346**
40	Papelbon	Jonathan	2008	BOS	AL	5	4	41	69.3	2.34	28	**17.149**
41	McDaniel	Lindy	1970	NYA	AL	9	5	29	111.7	2.01	35	**17.038**
42	Rivera	Mariano	2006	NYA	AL	5	5	34	75.0	1.80	37	**16.920**
43	Foulke	Keith	2001	CHA	AL	4	9	42	81.0	2.33	29	**16.823**
44	Rivera	Mariano	2008	NYA	AL	6	5	39	70.7	1.40	39	**16.794**
45	Papelbon	Jonathan	2006	BOS	AL	4	2	35	68.3	0.92	26	**16.791**
46	Jenks	Bobby	2007	CHA	AL	3	5	40	65.0	2.77	26	**16.658**
47	Arroyo	Luis	1961	NYA	AL	15	5	29	119.0	2.19	34	**16.604**
48	Tekulve	Kent	1979	PIT	NL	10	8	31	134.3	2.75	32	**16.556**
49	Ryan	B.J.	2006	TOR	AL	2	2	38	72.3	1.37	31	**16.535**
50	Nen	Robb	2000	SFN	NL	4	3	41	66.0	1.50	31	**16.515**

Best Relief Pitcher Years Ever (1871-2010)

Rank	Name	First	Year	Team	Lg	W	L	SV	IP	ERA	Age	PEVA-P
51	Hoffman	Trevor	1996	SDN	NL	9	5	42	88.0	2.25	29	**16.509**
52	Hrabosky	Al	1975	SLN	NL	13	3	22	97.3	1.66	26	**16.466**
53	Marshall	Mike	1974	LAN	NL	15	12	21	208.3	2.42	31	**16.455**
54	Nathan	Joe	2007	MIN	AL	4	2	37	71.7	1.88	33	**16.389**
55	Garber	Gene	1982	ATL	NL	8	10	30	119.3	2.34	35	**16.356**
56	Thigpen	Bobby	1990	CHA	AL	4	6	57	88.7	1.83	27	**16.291**
57	Kinder	Ellis	1951	BOS	AL	11	2	14	127.0	2.55	37	**16.231**
58	Saito	Takashi	2007	LAN	NL	2	1	39	64.3	1.40	37	**16.224**
59	Rivera	Mariano	2003	NYA	AL	5	2	40	70.7	1.66	34	**16.179**
60	Ward	Duane	1993	TOR	AL	2	3	45	71.7	2.13	29	**16.148**
61	Harvey	Bryan	1993	FLO	NL	1	5	45	69.0	1.70	30	**16.133**
62	Wagner	Billy	2005	PHI	NL	4	3	38	77.7	1.51	34	**16.057**
63	Henke	Tom	1987	TOR	AL	0	6	34	94.0	2.49	30	**16.027**
64	Nathan	Joe	2004	MIN	AL	1	2	44	72.3	1.62	30	**16.025**
65	Rivera	Mariano	2009	NYA	AL	3	3	44	66.3	1.76	40	**15.959**
66	Black	Joe	1952	BRO	NL	15	4	15	142.3	2.15	28	**15.953**
67	Wagner	Billy	2003	HOU	NL	1	4	44	86.0	1.78	32	**15.939**
68	Nen	Robb	1996	FLO	NL	5	1	35	83.0	1.95	27	**15.924**
69	Hoffman	Trevor	1999	SDN	NL	2	3	40	67.3	2.14	32	**15.899**
70	Radatz	Dick	1964	BOS	AL	16	9	29	157.0	2.29	27	**15.866**
71	Wilhelm	Hoyt	1964	CHA	AL	12	9	27	131.3	1.99	42	**15.866**
72	Page	Joe	1947	NYA	AL	14	8	17	141.3	2.48	30	**15.857**
73	Jones	Doug	1988	CLE	AL	3	4	37	83.3	2.27	31	**15.824**

74	Face	Roy	1962	PIT	NL	8	7	28	91.0	1.88	34	**15.603**
75	Kinder	Ellis	1953	BOS	AL	10	6	27	107.0	1.85	39	**15.443**
76	Lidge	Brad	2004	HOU	NL	6	5	29	94.7	1.90	28	**15.433**
77	Foulke	Keith	2003	OAK	AL	9	1	43	86.7	2.08	31	**15.410**
78	James	Bob	1985	CHA	AL	8	7	32	110.0	2.13	27	**15.396**
79	Gossage	Rich	1982	NYA	AL	4	5	30	93.0	2.23	31	**15.395**
80	Rivera	Mariano	2004	NYA	AL	4	2	53	78.7	1.94	35	**15.380**
81	Gossage	Rich	1975	CHA	AL	9	8	26	141.7	1.84	24	**15.358**
82	Nathan	Joe	2009	MIN	AL	2	2	47	68.7	2.10	34	**15.346**
83	Beck	Rod	1993	SFN	NL	3	1	48	79.3	2.16	25	**15.240**
84	Corbett	Doug	1980	MIN	AL	8	6	23	136.3	1.98	28	**15.172**
85	Berry	Joe	1944	PHA	AL	10	8	12	111.3	1.94	40	**15.157**
86	Broxton	Jonathan	2009	LAN	NL	7	2	36	76.0	2.61	25	**15.087**
87	Face	Roy	1960	PIT	NL	10	8	24	114.7	2.90	32	**15.056**

Note: Relief pitchers defined as those with three times the amount of Games Pitched vs. Games Started.

Note: Age = Player age at the end of the year. All pitchers above 32 PEVA-P (Peva Pitching) Player Ratings listed, plus Relief Pitchers above 15 PEVA-P.

Note: 2010 PEVA = Reflects preliminary final PEVA numbers. Marginal adjustment may be made with final stats and park factors.

Chapter 3 - BEST SEASONS EVER - TEAMS/FRANCHISES

PEVA Player Rating Boxscore		
64.000	Maximum	Maximum Player Rating
32.000	Fantastic	MVP/CY Young Candidate
20.000	Great	All-League
15.000	Very Good	All-Star Caliber
10.000	Good	Plus Starter
3.500	Average	Bench Player
0.200	Minimum	Minimum Player Rating

All Franchises (Current and Past) - Top Position Player Seasons

Who Jumped Onto the Best Ever Seasons List for Their Team in 2010

So who became the player who had one of the best batting seasons in your favorite team's history. Well, for two, they jumped all the way up to #2 in their franchises, and even though not yet storied to the extent of a Yankee or Red Sox team, the Toronto Blue Jays and Tampa Bay Rays were quite pleased to add Jose Bautista and Evan Longoria to that spot. For others, it's becoming a case if multiple entries, populating the list of their team with each year they play, such as Adrian Gonzalez coming in at #4 for the Padres after last year's #2 finish. Of course, this could be the last year Adrian gets to ply his first base trade in San Diego. It really starts to get interesting when you see the Hall of Fame company that many of these players are starting to keep. Miguel Cabrera took the #10 spot on the Tigers list, sandwiching himself between two seasons of the great Ty Cobb. For Pujols, his 2010 season ranks between Stan Musial and Joe Torre at #12 on the Cardinal list. Of course, Albert may be the most interesting name on that list after a couple more years and the eventual Hall of Fame nod himself. Others have begun the knockout of famous folks round. Joey Votto knocked Frank Robinson's 1961 campaign off their Top 20 in Cincy. Pretty heady stuff, don't you think. And Mr. Helton is gonna need a good steak dinner from the Gonzalez and Tulowitski pair after they knocked two of his seasons out of the Colorado list.

Joining the Top 20 Best Batting Seasons for Their Team List

Name	Team	Rank
Chris Young	Arizona Diamondbacks	11
Kelly Johnson	Arizona Diamondbacks	13
Joey Votto	Cincinnati Reds	7
Carlos Gonzalez	Colorado Rockies	7
Troy Tulowitski	Colorado Rockies	20
Miguel Cabrera	Detroit Tiger	10
Dan Uggla	Florida Marlins	10
Ryan Braun	Milwaukee Brewers	20
David Wright	New York Mets	19
Adrian Gonzalez	San Diego Padres	6
Albert Pujols	St. Louis Cardinals	10
Evan Longoria	Tampa Bay Rays	2
Carl Crawford	Tampa Bay Rays	4
Josh Hamilton	Texas Rangers	11
Jose Bautista	Toronto Blue Jays	2

Pitching anyone. There was a whole lot of it in the 2010 Major League Baseball season, and it added into the Best Pitching Seasons in Franchise history in a lot of ways. For a newer franchise, the Colorado Rockies, it saw a pitcher with a stellar first half propel him to the top of their list. For another new team, the Tampa Bay Rays, they saw three pitchers come to the fore, including new number one David Price with his 19 wins, relief pitcher Rafael Soriano and his 45 saves at

#5, and Matt Garza with the #9 season in their tenure. For a franchise with a long history, a new pitcher came to town, Roy Halladay, and took their #4 spot. Now there are some in Philadelphia today stating that it was the best year in their history, but despite his great accomplishments in 2010, that just isn't the case, at least as far as the regular season goes. Don't forget, despite all that former losing, there were a couple Hall of Fame pitchers in their past, Grover Cleveland Alexander and Steve Carlton for two who had years that surpassed Roy. Carlton at the #2 spot went 24-9 over 304 Innings with a 2.34 ERA in 1980; Roy was 21-10, 250.7 IP, and a 2.44 ERA. Mr. Alexander had two seasons, in 1915 and 1916 with a combined 62-22 to take spots #1 and #3. Up in the Pacific Northwest, Seattle ace Felix Hernandez tossed himself into the Cy Young race and the #3 spot in Mariner history. This gives Felix the #2 and #3 seasons in a row between last year and this one. In St. Louis, Adam Wainwright wedged himself between two Hall of Famers in Bob Gibson and Dizzy Dean with his season. Who'd think we were talking about Adam in that company, but we are, and an off topic question comes to mind; why didn't the Cardinals win more games with this crew, both on the pitching and batting side?

Joining the Top 20 Best Pitching Seasons for Their Team List

Name	Team	Rank
Ubaldo Jimenez	Colorado Rockies	1
Josh Johnson	Florida Marlins	5
Anibal Sanchez	Florida Marlins	18
Jered Weaver	LA Angels of Anaheim	10
C.C. Sabathia	New York Yankees	15
Roy Halladay	Philadelphia Phillies	4
Mat Latos	San Diego Padres	15
Felix Hernandex	Seattle Mariners	3
Adam Wainwright	St. Louis Cardinals	8
David Price	Tampa Bay Rays	1
Rafael Soriano	Tampa Bay Rays	5
Matt Garza	Tampa Bay Rays	9
C.J. Wilson	Texas Rangers	12

Stat Geek Baseball Franchise Players Best Years

Top 20 Position Players By Team/Franchise (1871-2010)

Rank	Name	First	Year	Team	Lg	HR	RBI	AVG	Age	PEVA-B
	Altoona Mountain City									
1	Smith	Germany	1884	ALT	UA	0		0.315	21	**5.227**
2	Shafer	Taylor	1884	ALT	UA	0		0.284	18	**1.717**
3	Moore	Jerrie	1884	ALT	UA	1		0.313	29	**0.940**
4	Brown	Jim	1884	ALT	UA	1		0.250	24	**0.724**
5	Cross	Clarence	1884	ALT	UA	0		0.571	28	**0.642**

Top Position Players By Franchise (1871-2010)

Rank	Name	First	Year	Team	Lg	HR	RBI	AVG	Age	PEVA-B
	Los Angeles Angels of Anaheim									
1	Salmon	Tim	1995	CAL	AL	34	105	0.330	27	**29.547**
2	DeCinces	Doug	1982	CAL	AL	30	97	0.301	32	**27.834**
3	Baylor	Don	1979	CAL	AL	36	139	0.296	30	**25.905**
4	Downing	Brian	1982	CAL	AL	28	84	0.281	32	**25.802**
5	Glaus	Troy	2000	ANA	AL	47	102	0.284	24	**23.161**
6	Downing	Brian	1979	CAL	AL	12	75	0.326	29	**22.294**
7	Jackson	Reggie	1982	CAL	AL	39	101	0.275	36	**22.210**
8	Guerrero	Vladimir	2004	ANA	AL	39	126	0.337	28	**21.307**
9	Erstad	Darin	2000	ANA	AL	25	100	0.355	26	**21.163**
10	Edmonds	Jim	1995	CAL	AL	33	107	0.290	25	**20.927**
11	Guerrero	Vladimir	2006	LAA	AL	33	116	0.329	30	**20.282**
12	Fregosi	Jim	1964	LAA	AL	18	72	0.277	22	**20.269**

Rank	Name	First	Year	Team	Lg	HR	RBI	AVG	Age	PEVA-B
13	Guerrero	Vladimir	2005	LAA	AL	32	108	0.317	29	**20.134**
14	Salmon	Tim	1997	ANA	AL	33	129	0.296	29	**20.097**
15	Lynn	Fred	1982	CAL	AL	21	86	0.299	30	**19.796**
16	Pearson	Albie	1963	LAA	AL	6	47	0.304	29	**19.259**
17	Robinson	Frank	1973	CAL	AL	30	97	0.266	38	**18.829**
18	Joyner	Wally	1987	CAL	AL	34	117	0.285	25	**18.548**
19	Bonds	Bobby	1977	CAL	AL	37	115	0.264	31	**18.326**
20	Downing	Brian	1984	CAL	AL	23	91	0.275	34	**18.201**

Note: Previous Names of Franchise: Los Angeles Angels (LAA), California Angels (CAL), Anaheim Angels (ANA)

Top Position Players By Franchise (1871-2010)

Rank	Name	First	Year	Team	Lg	HR	RBI	AVG	Age	PEVA-B
	Arizona Diamondbacks									
1	Gonzalez	Luis	2001	ARI	NL	57	142	0.325	34	**30.146**
2	Gonzalez	Luis	1999	ARI	NL	26	111	0.336	32	**20.603**
3	Williams	Matt	1999	ARI	NL	35	142	0.303	34	**17.667**
4	Gonzalez	Luis	2000	ARI	NL	31	114	0.311	33	**17.155**
5	Bell	Jay	1999	ARI	NL	38	112	0.289	34	**16.663**
6	Finley	Steve	1999	ARI	NL	34	103	0.264	34	**14.535**
7	Gonzalez	Luis	2003	ARI	NL	26	104	0.304	36	**14.087**
8	Finley	Steve	2004	ARI	NL	23	48	0.275	39	**13.475**
9	Finley	Steve	2000	ARI	NL	35	96	0.280	35	**13.458**
10	Glaus	Troy	2005	ARI	NL	37	97	0.258	29	**13.277**
11	Young	Chris	2010	ARI	NL	27	91	0.257	27	**13.085**
12	Reynolds	Mark	2009	ARI	NL	44	102	0.260	26	**12.981**
13	Johnson	Kelly	2010	ARI	NL	26	71	0.284	28	**12.520**
14	Gonzalez	Luis	2002	ARI	NL	28	103	0.288	35	**11.697**
15	Green	Shawn	2005	ARI	NL	22	73	0.286	33	**11.513**
16	Young	Chris	2008	ARI	NL	22	85	0.248	25	**11.400**
17	Gonzalez	Luis	2005	ARI	NL	24	79	0.271	38	**11.162**
18	Byrnes	Eric	2007	ARI	NL	21	83	0.286	31	**10.887**
19	Drew	Stephen	2008	ARI	NL	21	67	0.291	25	**10.818**
20	Batista	Tony	1999	ARI	NL	5	21	0.257	26	**10.433**

Top Position Players By Franchise (1871-2010)

Rank	Name	First	Year	Team	Lg	HR	RBI	AVG	Age	PEVA-B
	Atlanta Braves									
1	Jones	Charley	1879	BSN	NL	9	62	0.315	29	**44.728**
2	Aaron	Hank	1959	ML1	NL	39	123	0.355	25	**41.920**
3	Mathews	Eddie	1953	ML1	NL	47	135	0.302	22	**41.677**
4	Mathews	Eddie	1960	ML1	NL	39	124	0.277	29	**41.382**
5	Aaron	Hank	1963	ML1	NL	44	130	0.319	29	**40.268**
6	Holmes	Tommy	1945	BSN	NL	28	117	0.352	28	**40.065**
7	Mathews	Eddie	1959	ML1	NL	46	114	0.306	28	**38.592**
8	Aaron	Hank	1962	ML1	NL	45	128	0.323	28	**37.264**
9	White	Deacon	1877	BSN	NL	2	49	0.387	30	**36.179**
10	Duffy	Hugh	1894	BSN	NL	18	145	0.440	28	**35.452**
11	Aaron	Hank	1960	ML1	NL	40	126	0.292	26	**33.522**
12	Jones	Chipper	1999	ATL	NL	45	110	0.319	27	**32.668**
13	Mathews	Eddie	1955	ML1	NL	41	101	0.289	24	**32.143**
14	Aaron	Hank	1967	ATL	NL	39	109	0.307	33	**32.133**
15	Murphy	Dale	1983	ATL	NL	36	121	0.302	27	**31.161**
16	O'Rourke	Jim	1877	BSN	NL	0	23	0.362	27	**29.542**
17	Murphy	Dale	1987	ATL	NL	44	105	0.295	31	**29.525**
18	Aaron	Hank	1957	ML1	NL	44	132	0.322	23	**29.093**

19	Evans	Darrell	1973	ATL	NL	41	104	0.281	26	**28.985**
20	Aaron	Hank	1958	ML1	NL	30	95	0.326	24	**28.666**

Note: Previous Names of Franchise:
Boston Red Caps, Beaneaters, Doves, Braves (BSN) 1876-1952; Milwaukee Braves (ML1) 1953-1965

Top Position Players By Franchise (1871-2010)

Rank	Name	First	Year	Team	Lg	HR	RBI	AVG	Age	PEVA-B
	Baltimore Orioles									
1	Robinson	Frank	1966	BAL	AL	49	122	0.316	31	**42.918**
2	Stone	George	1906	SLA	AL	6	71	0.358	30	**42.914**
3	Murray	Eddie	1984	BAL	AL	29	110	0.306	28	**37.406**
4	Ripken Jr.	Cal	1984	BAL	AL	27	86	0.304	24	**35.874**
5	Ripken Jr.	Cal	1991	BAL	AL	34	114	0.323	31	**35.853**
6	Ripken Jr.	Cal	1983	BAL	AL	27	102	0.318	23	**31.713**
7	Sisler	George	1920	SLA	AL	19	122	0.407	27	**31.637**
8	Murray	Eddie	1983	BAL	AL	33	111	0.306	27	**28.889**
9	Robinson	Brooks	1964	BAL	AL	28	118	0.317	27	**28.302**
10	Williams	Ken	1922	SLA	AL	39	155	0.332	32	**27.478**
11	Singleton	Ken	1979	BAL	AL	35	111	0.295	32	**26.973**
12	Singleton	Ken	1977	BAL	AL	24	99	0.328	30	**26.828**
13	Murray	Eddie	1982	BAL	AL	32	110	0.316	26	**26.525**
14	Clift	Harlond	1938	SLA	AL	34	118	0.290	26	**26.044**
15	Gentile	Jim	1961	BAL	AL	46	141	0.302	27	**25.625**
16	Murray	Eddie	1985	BAL	AL	31	124	0.297	29	**24.922**
17	Anderson	Brady	1996	BAL	AL	50	110	0.297	32	**24.626**
18	Robinson	Frank	1967	BAL	AL	30	94	0.311	32	**24.064**
19	Clift	Harlond	1937	SLA	AL	29	118	0.306	25	**23.969**
20	Robinson	Frank	1969	BAL	AL	32	100	0.308	34	**23.795**

Note: Previous Names of Franchise: Milwaukee Brewers (MLA) 1901; St. Louis Browns (SLA) 1902-1953

Top Position Players By Franchise (1871-2010)

Rank	Name	First	Year	Team	Lg	HR	RBI	AVG	Age	PEVA-B
	Buffalo Bisons									
1	Brouthers	Dan	1883	BFN	NL	3	97	0.374	25	**39.597**
2	Brouthers	Dan	1882	BFN	NL	6	63	0.368	24	**34.359**
3	O'Rourke	Jim	1884	BFN	NL	5	63	0.347	34	**25.770**
4	Brouthers	Dan	1884	BFN	NL	14	79	0.327	26	**21.208**
5	White	Deacon	1884	BFN	NL	5	74	0.325	37	**20.391**
6	Brouthers	Dan	1885	BFN	NL	7	59	0.359	27	**20.230**
7	O'Rourke	Jim	1883	BFN	NL	1	38	0.328	33	**17.804**
8	Rowe	Jack	1884	BFN	NL	4	61	0.315	28	**16.672**
9	Richardson	Hardy	1881	BFN	NL	2	53	0.291	26	**15.906**
10	Richardson	Hardy	1883	BFN	NL	1	56	0.311	28	**15.819**
11	Foley	Curry	1882	BFN	NL	3	49	0.305	26	**13.418**
12	O'Rourke	Jim	1881	BFN	NL	0	30	0.302	31	**12.725**
13	Richardson	Hardy	1884	BFN	NL	6	60	0.301	29	**12.683**
14	Purcell	Blondie	1882	BFN	NL	2	40	0.276	28	**12.622**
15	Richardson	Hardy	1882	BFN	NL	2	57	0.271	27	**12.574**
16	Shaffer	Orator	1883	BFN	NL	0	41	0.292	32	**12.566**
17	Richardson	Hardy	1885	BFN	NL	6	44	0.319	30	**12.478**
18	Richardson	Hardy	1879	BFN	NL	0	37	0.283	24	**11.797**
19	O'Rourke	Jim	1882	BFN	NL	2	37	0.281	32	**11.548**
20	Brouthers	Dan	1881	BFN	NL	8	45	0.319	23	**11.243**

Top Position Players By Franchise (1871-2010)

Rank	Name	First	Year	Team	Lg	HR	RBI	AVG	Age	PEVA-B
	Buffalo Bisons (Pacific Coast League)									
1	Hoy	Dummy	1890	BFP	PL	1	53	0.298	28	**15.291**
2	Wise	Sam	1890	BFP	PL	6	102	0.293	33	**12.640**
3	Mack	Connie	1890	BFP	PL	0	53	0.266	28	**10.921**
4	Beecher	Ed	1890	BFP	PL	3	90	0.297	30	**9.205**
5	Rowe	Jack	1890	BFP	PL	2	76	0.250	34	**8.439**

Top Position Players By Franchise (1871-2010)

Rank	Name	First	Year	Team	Lg	HR	RBI	AVG	Age	PEVA-B
	Baltimore Canaries									
1	Pike	Lip	1872	BL1	NA	6	60	0.292	27	**17.098**
2	Hall	George	1872	BL1	NA	1	37	0.336	23	**15.242**
3	Radcliff	John	1872	BL1	NA	1	44	0.290	26	**14.113**
4	Higham	Dick	1872	BL1	NA	2	38	0.343	21	**12.788**
5	York	Tom	1872	BL1	NA	1	41	0.266	22	**11.840**
6	Pike	Lip	1873	BL1	NA	4	50	0.315	28	**11.676**
7	York	Tom	1873	BL1	NA	2	49	0.303	23	**11.520**
8	Mills	Everett	1873	BL1	NA	0	57	0.331	28	**10.791**
9	Force	Davy	1873	BL1	NA	0	31	0.368	24	**10.477**
10	Carey	Tom	1873	BL1	NA	1	55	0.334	24	**10.238**

Top Position Players By Franchise (1871-2010)

Rank	Name	First	Year	Team	Lg	HR	RBI	AVG	Age	PEVA-B
	Baltimore Orioles (American Association/National League)									
1	Kelley	Joe	1896	BLN	NL	8	100	0.364	25	**36.936**
2	Keeler	Willie	1897	BLN	NL	0	74	0.424	25	**34.488**
3	Jennings	Hughie	1896	BLN	NL	0	121	0.401	27	**34.143**
4	Kelley	Joe	1894	BLN	NL	6	111	0.393	23	**28.190**
5	Kelley	Joe	1895	BLN	NL	10	134	0.365	24	**28.125**
6	Jennings	Hughie	1895	BLN	NL	4	125	0.386	26	**28.104**
7	Kelley	Joe	1897	BLN	NL	5	118	0.362	26	**27.788**
8	Jennings	Hughie	1897	BLN	NL	2	79	0.355	28	**26.148**
9	Jennings	Hughie	1898	BLN	NL	1	87	0.328	29	**25.255**
10	Keeler	Willie	1896	BLN	NL	4	82	0.386	24	**24.026**
11	Tucker	Tommy	1889	BL2	AA	5	99	0.372	26	**23.829**
12	Burns	Oyster	1887	BL2	AA	9	99	0.341	23	**21.296**
13	Van Haltren	George	1891	BL3	AA	9	83	0.318	25	**21.128**
14	Stenzel	Jake	1897	BLN	NL	4	116	0.353	30	**20.222**
15	Keeler	Willie	1895	BLN	NL	4	78	0.377	23	**20.191**
16	McGraw	John	1898	BLN	NL	0	53	0.342	25	**20.148**
17	McGraw	John	1899	BLN	NL	1	33	0.391	26	**17.875**
18	Burns	Oyster	1888	BL2	AA	4	42	0.298	24	**17.554**
19	Van Haltren	George	1892	BLN	NL	7	57	0.302	26	**16.603**
20	Kelley	Joe	1893	BLN	NL	9	76	0.305	22	**15.463**

Note: Previous Names of Franchise: Baltimore Orioles (BL2) 1882-1889; (BL3) 1890-1891; (BLN) 1892-1899

Top Position Players By Franchise (1871-2010)

Rank	Name	First	Year	Team	Lg	HR	RBI	AVG	Age	PEVA-B
	Baltimore Marylands									
1	Woodhead	Red	1873	BL4	NA	0	0	0.000	22	**0.200**
2	Johns	Tommy	1873	BL4	NA	0	0	0.000	22	**0.200**
3	Smith	Bill	1873	BL4	NA	0	1	0.174	NA	**0.200**
4	Goldsmith	Wally	1873	BL4	NA	0	0	0.000	24	**0.200**
5	Stratton	Ed	1873	BL4	NA	0	0	0.125	NA	**0.200**

Top Position Players By Franchise (1871-2010)

Rank	Name	First	Year	Team	Lg	HR	RBI	AVG	Age	PEVA-B
	Baltimore Terrapins									
1	Duncan	Vern	1914	BLF	FL	2	53	0.287	24	**12.622**
2	Meyer	Benny	1914	BLF	FL	5	40	0.304	29	**10.742**
3	Swacina	Harry	1914	BLF	FL	0	90	0.280	33	**10.267**
4	Jacklitsch	Fred	1914	BLF	FL	2	48	0.276	38	**9.357**
5	Walsh	Jimmy	1914	BLF	FL	10	65	0.308	28	**8.307**

Top Position Players By Franchise (1871-2010)

Rank	Name	First	Year	Team	Lg	HR	RBI	AVG	Age	PEVA-B
	Baltimore Monumentals									
1	Seery	Emmett	1884	BLU	UA	2		0.311	23	**12.602**
2	Robinson	Yank	1884	BLU	UA	2		0.267	25	**7.959**
3	Say	Lou	1884	BLU	UA	2		0.239	30	**4.524**
4	Fusselback	Eddie	1884	BLU	UA	1		0.284	28	**3.994**
5	Phelan	Dick	1884	BLU	UA	3		0.246	30	**3.499**

Top Position Players By Franchise (1871-2010)

Rank	Name	First	Year	Team	Lg	HR	RBI	AVG	Age	PEVA-B
	Boston Red Sox									
1	Williams	Ted	1942	BOS	AL	36	137	0.356	24	**52.075**
2	Williams	Ted	1941	BOS	AL	37	120	0.406	23	**51.730**
3	Ruth	Babe	1919	BOS	AL	29	114	0.322	24	**45.484**
4	Williams	Ted	1946	BOS	AL	38	123	0.342	28	**44.043**
5	Speaker	Tris	1912	BOS	AL	10	90	0.383	24	**43.198**
6	Williams	Ted	1949	BOS	AL	43	159	0.343	31	**41.778**
7	Yastrzemski	Carl	1967	BOS	AL	44	121	0.326	28	**40.383**
8	Foxx	Jimmie	1938	BOS	AL	50	175	0.349	31	**37.579**
9	Speaker	Tris	1914	BOS	AL	4	90	0.338	26	**37.327**
10	Williams	Ted	1947	BOS	AL	32	114	0.343	29	**36.436**
11	Evans	Dwight	1984	BOS	AL	32	104	0.295	33	**35.847**
12	Johnson	Bob	1944	BOS	AL	17	106	0.324	39	**34.713**
13	Boggs	Wade	1987	BOS	AL	24	89	0.363	29	**34.592**
14	Lynn	Fred	1979	BOS	AL	39	122	0.333	27	**33.455**
15	Williams	Ted	1957	BOS	AL	38	87	0.388	39	**32.674**
16	Evans	Dwight	1982	BOS	AL	32	98	0.292	31	**31.949**
17	Boggs	Wade	1988	BOS	AL	5	58	0.366	30	**31.802**
18	Yastrzemski	Carl	1968	BOS	AL	23	74	0.301	29	**31.561**
19	Yastrzemski	Carl	1970	BOS	AL	40	102	0.329	31	**30.942**
20	Williams	Ted	1951	BOS	AL	30	126	0.318	33	**30.939**

Note: Previous Names of Franchise: Boston Americans, Somersets, Pilgrims

Top Position Players By Franchise (1871-2010)

Rank	Name	First	Year	Team	Lg	HR	RBI	AVG	Age	PEVA-B
	Brooklyn Eckfords									
1	Allison	Doug	1872	BR1	NA	0	5	0.342	26	**7.853**
2	Wood	Jimmy	1872	BR1	NA	0	0	0.200	30	**6.086**
3	Martin	Phonney	1872	BR1	NA	0	9	0.154	27	**3.387**
4	Gedney	Count	1872	BR1	NA	0	7	0.183	23	**2.157**
5	Holdsworth	Jim	1872	BR1	NA	0	0	0.286	22	**1.597**

Top Position Players By Franchise (1871-2010)

Rank	Name	First	Year	Team	Lg	HR	RBI	AVG	Age	PEVA-B
	Brooklyn Atlantics									
1	Pabor	Charlie	1873	BR2	NA	0	42	0.360	27	**9.654**

Rank	Name	First	Year	Team	Lg	HR	RBI	AVG	Age	PEVA-B
2	Pearce	Dickey	1874	BR2	NA	0	25	0.294	38	**7.157**
3	Pearce	Dickey	1873	BR2	NA	1	26	0.275	37	**5.297**
4	Barlow	Tom	1873	BR2	NA	1	14	0.273	21	**5.275**
5	Burdock	Jack	1873	BR2	NA	2	36	0.253	21	**4.967**

Top Position Players By Franchise (1871-2010)

Rank	Name	First	Year	Team	Lg	HR	RBI	AVG	Age	PEVA-B
	Brooklyn Gladiators									
1	Simon	Hank	1890	BR4	AA	0	38	0.257	28	**9.569**
2	Daily	Ed	1890	BR4	AA	1	39	0.239	28	**5.020**
3	Peltz	John	1890	BR4	AA	1	33	0.227	29	**4.916**
4	Gerhardt	Joe	1890	BR4	AA	2	40	0.203	35	**4.772**
5	O'Brien	Billy	1890	BR4	AA	4	67	0.278	30	**3.832**

Top Position Players By Franchise (1871-2010)

Rank	Name	First	Year	Team	Lg	HR	RBI	AVG	Age	PEVA-B
	Brooklyn Tip-Tops									
1	Kauff	Benny	1915	BRF	FL	12	83	0.342	25	**32.472**
2	Evans	Steve	1914	BRF	FL	12	96	0.348	29	**28.788**
3	Evans	Steve	1915	BRF	FL	3	30	0.296	30	**16.293**
4	Cooper	Claude	1915	BRF	FL	2	63	0.294	23	**15.763**
5	Shaw	Al	1914	BRF	FL	5	49	0.324	33	**11.926**
6	Hofman	Solly	1914	BRF	FL	5	83	0.287	32	**11.721**

Top Position Players By Franchise (1871-2010)

Rank	Name	First	Year	Team	Lg	HR	RBI	AVG	Age	PEVA-B
	Brooklyn Ward's Wonders									
1	Ward	John	1890	BRP	PL	4	60	0.335	30	**15.619**
2	Orr	Dave	1890	BRP	PL	6	124	0.371	31	**15.509**
3	Bierbauer	Lou	1890	BRP	PL	7	99	0.306	25	**14.454**
4	Joyce	Bill	1890	BRP	PL	1	78	0.252	25	**13.176**
5	Van Haltren	George	1890	BRP	PL	5	54	0.335	24	**6.382**

Top Position Players By Franchise (1871-2010)

Rank	Name	First	Year	Team	Lg	HR	RBI	AVG	Age	PEVA-B
	Boston Red Stockings									
1	Barnes	Ross	1873	BS1	NA	2	62	0.425	23	**42.762**
2	McVey	Cal	1875	BS1	NA	3	87	0.355	26	**35.627**
3	White	Deacon	1875	BS1	NA	1	60	0.367	28	**35.065**
4	Barnes	Ross	1871	BS1	NA	0	34	0.401	21	**32.973**
5	Barnes	Ross	1872	BS1	NA	1	44	0.432	22	**32.428**
6	Barnes	Ross	1875	BS1	NA	1	58	0.364	25	**32.311**
7	Wright	George	1873	BS1	NA	3	50	0.388	26	**30.121**
8	McVey	Cal	1874	BS1	NA	3	71	0.359	25	**28.884**
9	Wright	George	1875	BS1	NA	2	61	0.333	28	**28.148**
10	McVey	Cal	1871	BS1	NA	0	43	0.431	22	**26.529**
11	White	Deacon	1873	BS1	NA	0	66	0.390	26	**24.754**
12	Leonard	Andy	1875	BS1	NA	1	74	0.321	29	**20.750**
13	Wright	George	1874	BS1	NA	2	44	0.329	27	**19.712**
14	Wright	George	1872	BS1	NA	2	32	0.337	25	**19.266**
15	O'Rourke	Jim	1875	BS1	NA	6	72	0.296	25	**18.580**
16	O'Rourke	Jim	1874	BS1	NA	5	61	0.314	24	**17.832**
17	White	Deacon	1874	BS1	NA	3	52	0.300	27	**15.492**
18	Spalding	Al	1874	BS1	NA	0	54	0.329	24	**14.447**
19	Leonard	Andy	1874	BS1	NA	0	51	0.319	28	**13.215**
20	Spalding	Al	1872	BS1	NA	0	47	0.354	22	**12.841**

Top Position Players By Franchise (1871-2010)

Rank	Name	First	Year	Team	Lg	HR	RBI	AVG	Age	PEVA-B
	Boston Reds (Pacific Coast League/American Association)									
1	Brouthers	Dan	1891	BS2	AA	5	109	0.350	33	**34.762**
2	Brown	Tom	1891	BS2	AA	5	72	0.321	31	**27.645**
3	Duffy	Hugh	1891	BS2	AA	9	110	0.336	25	**23.866**
4	Richardson	Hardy	1890	BSP	PL	13	146	0.326	35	**20.434**
5	Farrell	Duke	1891	BS2	AA	12	110	0.302	25	**20.311**
6	Brouthers	Dan	1890	BSP	PL	1	97	0.330	32	**19.035**
7	Stovey	Harry	1890	BSP	PL	12	84	0.299	34	**16.612**
8	Brown	Tom	1890	BSP	PL	4	61	0.274	30	**13.226**
9	Radford	Paul	1891	BS2	AA	0	65	0.259	30	**12.663**
10	Nash	Billy	1890	BSP	PL	5	90	0.266	25	**11.930**

Top Position Players By Franchise (1871-2010)

Rank	Name	First	Year	Team	Lg	HR	RBI	AVG	Age	PEVA-B
	Boston Reds (UA)									
1	Crane	Ed	1884	BSU	UA	12		0.285	22	**10.742**
2	O'Brien	Tom	1884	BSU	UA	4		0.263	24	**6.013**
3	Irwin	John	1884	BSU	UA	1		0.234	23	**5.791**
4	Hackett	Walter	1884	BSU	UA	1		0.243	27	**5.602**
5	Slattery	Mike	1884	BSU	UA	0		0.208	18	**4.100**

Top Position Players By Franchise (1871-2010)

Rank	Name	First	Year	Team	Lg	HR	RBI	AVG	Age	PEVA-B
	Buffalo Buffeds & Blues									
1	Hanford	Charlie	1914	BUF	FL	12	90	0.291	32	**17.564**
2	Chase	Hal	1915	BUF	FL	17	89	0.291	32	**13.941**
3	Louden	Baldy	1914	BUF	FL	6	63	0.313	31	**11.388**
4	Louden	Baldy	1915	BUF	FL	4	48	0.281	32	**8.621**
5	Dalton	Jack	1915	BUF	FL	2	46	0.293	30	**7.691**

Top Position Players By Franchise (1871-2010)

Rank	Name	First	Year	Team	Lg	HR	RBI	AVG	Age	PEVA-B
	Chicago White Stockings (National Association)									
1	Wood	Jimmy	1871	CH1	NA	1	29	0.378	29	**14.911**
2	Meyerle	Levi	1874	CH2	NA	1	47	0.394	29	**14.600**
3	Hines	Paul	1875	CH2	NA	0	36	0.328	20	**11.256**
4	Treacey	Fred	1871	CH1	NA	4	33	0.339	24	**8.368**
5	Force	Davy	1874	CH2	NA	0	26	0.313	25	**7.817**

Top Position Players By Franchise (1871-2010)

Rank	Name	First	Year	Team	Lg	HR	RBI	AVG	Age	PEVA-B
	Chicago White Sox									
1	Thomas	Frank	1994	CHA	AL	38	101	0.353	26	**41.278**
2	Thomas	Frank	1991	CHA	AL	32	109	0.318	23	**36.023**
3	Allen	Dick	1972	CHA	AL	37	113	0.308	30	**34.744**
4	Thomas	Frank	1992	CHA	AL	24	115	0.323	24	**33.764**
5	Thomas	Frank	1995	CHA	AL	40	111	0.308	27	**31.426**
6	Jackson	Joe	1916	CHA	AL	3	78	0.341	27	**30.605**
7	Thomas	Frank	1997	CHA	AL	35	125	0.347	29	**29.480**
8	Thomas	Frank	1996	CHA	AL	40	134	0.349	28	**29.396**
9	Collins	Eddie	1915	CHA	AL	4	77	0.332	28	**29.073**
10	Thomas	Frank	1993	CHA	AL	41	128	0.317	25	**27.858**
11	Belle	Albert	1998	CHA	AL	49	152	0.328	32	**27.673**
12	Jackson	Joe	1920	CHA	AL	12	121	0.382	31	**27.623**
13	Jackson	Joe	1919	CHA	AL	7	96	0.351	30	**26.577**

Rank	Name	First	Year	Team	Lg	HR	RBI	AVG	Age	PEVA-B
14	Thomas	Frank	2000	CHA	AL	43	143	0.328	32	**25.624**
15	Minoso	Minnie	1960	CHA	AL	20	105	0.311	35	**24.110**
16	Minoso	Minnie	1954	CHA	AL	19	116	0.320	29	**23.962**
17	Appling	Luke	1943	CHA	AL	3	80	0.328	36	**23.511**
18	Collins	Eddie	1920	CHA	AL	3	76	0.372	33	**22.630**
19	Appling	Luke	1936	CHA	AL	6	128	0.388	29	**21.923**
20	Thome	Jim	2006	CHA	AL	42	109	0.288	36	**21.392**

Top Position Players By Franchise (1871-2010)

Rank	Name	First	Year	Team	Lg	HR	RBI	AVG	Age	PEVA-B
	Chicago Chi-Feds									
1	Wilson	Art	1914	CHF	FL	10	64	0.291	29	**25.707**
2	Zwilling	Dutch	1914	CHF	FL	16	95	0.313	26	**25.300**
3	Zwilling	Dutch	1915	CHF	FL	13	94	0.286	27	**22.595**
4	Wickland	Al	1914	CHF	FL	6	68	0.276	26	**17.925**
5	Flack	Max	1915	CHF	FL	3	45	0.314	25	**15.520**
6	Mann	Les	1915	CHF	FL	4	58	0.306	23	**14.224**
7	Wilson	Art	1915	CHF	FL	7	31	0.305	30	**12.652**
8	Wickland	Al	1915	CHF	FL	1	5	0.244	27	**11.088**
9	Beck	Fred	1914	CHF	FL	11	77	0.279	28	**10.316**
10	Fischer	William	1915	CHF	FL	4	50	0.329	24	**10.030**

Top Position Players By Franchise (1871-2010)

Rank	Name	First	Year	Team	Lg	HR	RBI	AVG	Age	PEVA-B
	Chicago Cubs									
1	Barnes	Ross	1876	CHN	NL	1	59	0.429	26	**46.554**
2	Hornsby	Rogers	1929	CHN	NL	39	149	0.380	33	**41.923**
3	Sosa	Sammy	2001	CHN	NL	64	160	0.328	33	**38.918**
4	Anson	Cap	1881	CHN	NL	1	82	0.399	29	**38.143**
5	Wilson	Hack	1930	CHN	NL	56	191	0.356	30	**35.799**
6	Santo	Ron	1964	CHN	NL	30	114	0.313	24	**35.556**
7	Nicholson	Bill	1943	CHN	NL	29	128	0.309	29	**35.190**
8	Nicholson	Bill	1944	CHN	NL	33	122	0.287	30	**34.757**
9	Banks	Ernie	1959	CHN	NL	45	143	0.304	28	**34.546**
10	Banks	Ernie	1958	CHN	NL	47	129	0.313	27	**33.578**
11	Santo	Ron	1966	CHN	NL	30	94	0.312	26	**33.049**
12	Kelly	King	1884	CHN	NL	13	95	0.354	27	**32.756**
13	Dalrymple	Abner	1880	CHN	NL	0	36	0.330	23	**32.143**
14	Gore	George	1880	CHN	NL	2	47	0.360	23	**31.318**
15	Lee	Derrek	2005	CHN	NL	46	107	0.335	30	**31.180**
16	Williams	Billy	1965	CHN	NL	34	108	0.315	27	**31.123**
17	Anson	Cap	1882	CHN	NL	1	83	0.362	30	**31.001**
18	Santo	Ron	1967	CHN	NL	31	98	0.300	27	**30.153**
19	Williams	Billy	1972	CHN	NL	37	122	0.333	34	**29.912**
20	Wilson	Hack	1929	CHN	NL	39	159	0.345	29	**29.897**

Note: Previous Names of Franchise: Chicago White Stockings, Colts, Orphans

Top Position Players By Franchise (1871-2010)

Rank	Name	First	Year	Team	Lg	HR	RBI	AVG	Age	PEVA-B
	Chicago Pirates									
1	Duffy	Hugh	1890	CHP	PL	7	82	0.320	24	**23.069**
2	Ryan	Jimmy	1890	CHP	PL	6	89	0.340	27	**16.850**
3	O'Neill	Tip	1890	CHP	PL	3	75	0.302	32	**12.577**
4	Farrell	Duke	1890	CHP	PL	2	84	0.290	24	**9.023**
5	Pfeffer	Fred	1890	CHP	PL	5	80	0.257	30	**7.185**

Top Position Players By Franchise (1871-2010)

Rank	Name	First	Year	Team	Lg	HR	RBI	AVG	Age	PEVA-B
	Chicago/Pittsburgh (Union League)									
1	Schoeneck	Jumbo	1884	CHU	UA	2		0.317	22	**6.118**
2	Ellick	Joe	1884	CHU	UA	0		0.236	30	**4.672**
3	Gross	Emil	1884	CHU	UA	4		0.358	26	**3.217**
4	Krieg	Bill	1884	CHU	UA	0		0.247	25	**3.184**
5	Householder	Charlie	1884	CHU	UA	1		0.239	28	**2.194**

Top Position Players By Franchise (1871-2010)

Rank	Name	First	Year	Team	Lg	HR	RBI	AVG	Age	PEVA-B
	Cincinnati Reds									
1	Robinson	Frank	1962	CIN	NL	39	136	0.342	27	**44.982**
2	Bench	Johnny	1972	CIN	NL	40	125	0.270	25	**42.271**
3	Seymour	Cy	1905	CIN	NL	8	121	0.377	33	**35.870**
4	Foster	George	1977	CIN	NL	52	149	0.320	29	**34.233**
5	Bench	Johnny	1974	CIN	NL	33	129	0.280	27	**33.034**
6	Morgan	Joe	1976	CIN	NL	27	111	0.320	33	**32.946**
7	Votto	Joey	2010	CIN	NL	37	113	0.324	27	**32.034**
8	Jones	Charley	1878	CN1	NL	3	39	0.310	28	**31.723**
9	Morgan	Joe	1975	CIN	NL	17	94	0.327	32	**30.407**
10	Morgan	Joe	1972	CIN	NL	16	73	0.292	29	**30.209**
11	Perez	Tony	1970	CIN	NL	40	129	0.317	28	**29.169**
12	Jones	Charley	1877	CN1	NL	2	38	0.313	27	**28.647**
13	Jones	Charley	1884	CN2	AA	7	71	0.314	34	**28.250**
14	Kelly	King	1879	CN1	NL	2	47	0.348	22	**28.236**
15	Bench	Johnny	1970	CIN	NL	45	148	0.293	23	**27.796**
16	Robinson	Frank	1965	CIN	NL	33	113	0.296	30	**27.669**
17	Robinson	Frank	1964	CIN	NL	29	96	0.306	29	**27.629**
18	Morgan	Joe	1973	CIN	NL	26	82	0.290	30	**27.309**
19	Kluszewski	Ted	1954	CIN	NL	49	141	0.326	30	**27.291**
20	Rose	Pete	1973	CIN	NL	5	64	0.338	32	**27.194**

Note: Previous Names of Franchise:

Cincinnati Reds (CN1) 1876-1880; Cincinnati Red Stockings (CN2) AA 1882-1889; Cincinnati Reds/Red Legs 1890-present

Note: PEVA for Charley Jones 1877 includes two stints with CN1 & one stint with CHN (8 AB.)

Top Position Players By Franchise (1871-2010)

Rank	Name	First	Year	Team	Lg	HR	RBI	AVG	Age	PEVA-B
	Cleveland Forest Citys									
1	White	Deacon	1871	CL1	NA	1	21	0.322	24	**12.018**
2	Sutton	Ezra	1871	CL1	NA	3	23	0.352	21	**11.624**
3	Hastings	Scott	1872	CL1	NA	0	16	0.391	25	**7.720**
4	Allison	Art	1871	CL1	NA	0	19	0.292	22	**4.591**
5	Pratt	Al	1871	CL1	NA	0	20	0.262	23	**3.548**

Top Position Players By Franchise (1871-2010)

Rank	Name	First	Year	Team	Lg	HR	RBI	AVG	Age	PEVA-B
	Cleveland Blues									
1	Glasscock	Jack	1882	CL2	NL	4	46	0.291	25	**21.812**
2	Glasscock	Jack	1884	CL2	NL	1	22	0.249	27	**17.209**
3	Dunlap	Fred	1883	CL2	NL	4	37	0.326	24	**17.162**
4	Dunlap	Fred	1880	CL2	NL	4	30	0.276	21	**15.702**
5	Shaffer	Orator	1880	CL2	NL	0	21	0.266	29	**14.687**
6	Dunlap	Fred	1881	CL2	NL	3	24	0.325	22	**14.446**
7	Dunlap	Fred	1882	CL2	NL	0	28	0.280	23	**13.374**

8	York	Tom	1883	CL2	NL	2	46	0.260	33	**11.469**
9	Glasscock	Jack	1883	CL2	NL	0	46	0.287	26	**11.330**
10	Muldoon	Mike	1882	CL2	NL	6	45	0.246	24	**10.186**

Top Position Players By Franchise (1871-2010)

Rank	Name	First	Year	Team	Lg	HR	RBI	AVG	Age	PEVA-B
	Cleveland Blues & Spiders									
1	Burkett	Jesse	1896	CL4	NL	6	72	0.410	28	**28.983**
2	Burkett	Jesse	1895	CL4	NL	5	83	0.409	27	**23.591**
3	Burkett	Jesse	1897	CL4	NL	2	60	0.383	29	**22.708**
4	Childs	Cupid	1896	CL4	NL	1	106	0.355	29	**21.580**
5	Childs	Cupid	1892	CL4	NL	3	53	0.317	25	**20.676**
6	Burkett	Jesse	1893	CL4	NL	6	82	0.348	25	**18.979**
7	McKean	Ed	1888	CL3	AA	6	68	0.299	24	**18.637**
8	Wallace	Bobby	1897	CL4	NL	4	112	0.335	24	**18.517**
9	Burkett	Jesse	1898	CL4	NL	0	42	0.341	30	**17.717**
10	McKean	Ed	1890	CL4	NL	7	61	0.296	26	**17.400**
11	McKean	Ed	1895	CL4	NL	8	119	0.342	31	**15.760**
12	McKean	Ed	1896	CL4	NL	7	112	0.338	32	**14.719**
13	Davis	George	1891	CL4	NL	3	89	0.289	21	**14.708**
14	Virtue	Jake	1892	CL4	NL	2	89	0.282	27	**14.526**
15	Childs	Cupid	1891	CL4	NL	2	83	0.281	24	**14.432**
16	Wallace	Bobby	1898	CL4	NL	3	99	0.270	25	**14.090**
17	McKean	Ed	1893	CL4	NL	4	133	0.310	29	**13.841**
18	Burkett	Jesse	1894	CL4	NL	8	94	0.358	26	**13.431**
19	McKean	Ed	1889	CL4	NL	4	75	0.318	25	**13.389**
20	Childs	Cupid	1893	CL4	NL	3	65	0.326	26	**13.360**

Top Position Players By Franchise (1871-2010)

Rank	Name	First	Year	Team	Lg	HR	RBI	AVG	Age	PEVA-B
	Columbus Buckeyes									
1	Brown	Tom	1884	CL5	AA	5	32	0.273	24	**13.429**
2	Mann	Fred	1884	CL5	AA	7	0	0.276	26	**12.439**
3	Brown	Tom	1883	CL5	AA	5	32	0.274	23	**11.146**
4	Smith	Pop	1883	CL5	AA	4	0	0.262	27	**9.359**
5	Richmond	John	1883	CL5	AA	0	0	0.283	29	**8.609**

Top Position Players By Franchise (1871-2010)

Rank	Name	First	Year	Team	Lg	HR	RBI	AVG	Age	PEVA-B
	Columbus Solons									
1	Johnson	Spud	1890	CL6	AA	1	113	0.346	30	**23.376**
2	Marr	Lefty	1889	CL6	AA	1	75	0.306	27	**16.611**
3	Duffee	Charlie	1891	CL6	AA	10	90	0.301	25	**16.545**
4	O'Connor	Jack	1890	CL6	AA	2	66	0.324	21	**15.537**
5	McTamany	Jim	1889	CL6	AA	4	52	0.276	26	**15.356**
6	McTamany	Jim	1890	CL6	AA	1	48	0.258	27	**14.589**
7	McTamany	Jim	1891	CL6	AA	3	35	0.250	28	**13.031**
8	Crooks	Jack	1891	CL6	AA	0	46	0.245	26	**12.304**
9	Orr	Dave	1889	CL6	AA	4	87	0.327	30	**10.793**
10	Reilly	Charlie	1890	CL6	AA	4	77	0.266	23	**10.281**

Top Position Players By Franchise (1871-2010)

Rank	Name	First	Year	Team	Lg	HR	RBI	AVG	Age	PEVA-B
	Cleveland Indians									
1	Rosen	Al	1953	CLE	AL	43	145	0.336	29	**48.401**
2	Speaker	Tris	1916	CLE	AL	2	79	0.386	28	**42.388**
3	Jackson	Joe	1912	CLE	AL	3	90	0.395	23	**37.904**

4	Speaker	Tris	1923	CLE	AL	17	130	0.380	35	**35.232**
5	Speaker	Tris	1920	CLE	AL	8	107	0.388	32	**35.002**
6	Doby	Larry	1952	CLE	AL	32	104	0.276	29	**34.855**
7	Lajoie	Nap	1906	CLE	AL	0	91	0.355	32	**34.293**
8	Jackson	Joe	1911	CLE	AL	7	83	0.408	22	**34.179**
9	Lajoie	Nap	1904	CLE	AL	6	102	0.376	30	**34.027**
10	Rosen	Al	1952	CLE	AL	28	105	0.302	28	**33.117**
11	Belle	Albert	1995	CLE	AL	50	126	0.317	29	**32.570**
12	Lajoie	Nap	1910	CLE	AL	4	76	0.384	36	**32.056**
13	Boudreau	Lou	1948	CLE	AL	18	106	0.355	31	**31.926**
14	Ramirez	Manny	1999	CLE	AL	44	165	0.333	27	**30.231**
15	Jackson	Joe	1913	CLE	AL	7	71	0.373	24	**29.387**
16	Bradley	Bill	1902	CLE	AL	11	77	0.340	24	**29.269**
17	Rosen	Al	1950	CLE	AL	37	116	0.287	26	**29.097**
18	Harrah	Toby	1982	CLE	AL	25	78	0.304	34	**28.962**
19	Averill	Earl	1936	CLE	AL	28	126	0.378	34	**28.867**
20	Flick	Elmer	1906	CLE	AL	1	62	0.311	30	**28.141**

Note: Previous Names of Franchise: Cleveland Blues, Broncos, Naps.

Top Position Players By Franchise (1871-2010)

Rank	Name	First	Year	Team	Lg	HR	RBI	AVG	Age	PEVA-B
	Cleveland Infants									
1	Browning	Pete	1890	CLP	PL	5	93	0.373	29	**29.964**
2	Larkin	Henry	1890	CLP	PL	5	112	0.330	30	**23.049**
3	Radford	Paul	1890	CLP	PL	2	62	0.292	29	**13.819**
4	Tebeau	Patsy	1890	CLP	PL	5	74	0.298	26	**10.414**
5	Delahanty	Ed	1890	CLP	PL	3	64	0.296	23	**9.191**

Top Position Players By Franchise (1871-2010)

Rank	Name	First	Year	Team	Lg	HR	RBI	AVG	Age	PEVA-B
	Cincinnati Kelly's Killers									
1	Canavan	Jim	1891	CN3	AA	7	66	0.228	25	**9.146**
2	Seery	Emmett	1891	CN3	AA	4	36	0.285	30	**7.099**
3	Kelly	King	1891	CN3	AA	1	53	0.297	34	**5.852**
4	Carney	John	1891	CN3	AA	3	43	0.278	25	**5.535**
5	Johnston	Dick	1891	CN3	AA	6	51	0.221	28	**3.338**

Top Position Players By Franchise (1871-2010)

Rank	Name	First	Year	Team	Lg	HR	RBI	AVG	Age	PEVA-B
	Cincinnati Outlaw Reds									
1	Burns	Dick	1884	CNU	UA	4		0.306	21	**5.365**
2	Harbidge	Bill	1884	CNU	UA	2		0.279	29	**5.304**
3	Sylvester	Lou	1884	CNU	UA	2		0.267	29	**4.341**
4	Hawes	Bill	1884	CNU	UA	4		0.278	31	**3.537**
5	Crane	Sam	1884	CNU	UA	1		0.233	30	**2.184**

Top Position Players By Franchise (1871-2010)

Rank	Name	First	Year	Team	Lg	HR	RBI	AVG	Age	PEVA-B
	Colorado Rockies									
1	Walker	Larry	1997	COL	NL	49	130	0.366	31	**28.501**
2	Helton	Todd	2003	COL	NL	33	117	0.358	30	**25.643**
3	Holliday	Matt	2007	COL	NL	36	137	0.340	27	**24.699**
4	Helton	Todd	2000	COL	NL	42	147	0.372	27	**22.244**
5	Atkins	Garrett	2006	COL	NL	29	120	0.329	27	**21.810**
6	Helton	Todd	2001	COL	NL	49	146	0.336	28	**20.540**
7	Gonzalez	Carlos	2010	COL	NL	34	117	0.336	25	**19.178**

Rank	Name	First	Year	Team	Lg	HR	RBI	AVG	Age	PEVA-B
8	Helton	Todd	2004	COL	NL	32	96	0.347	31	**19.092**
9	Holliday	Matt	2006	COL	NL	34	114	0.326	26	**18.861**
10	Tulowitzki	Troy	2009	COL	NL	32	92	0.297	25	**17.461**
11	Castilla	Vinny	1998	COL	NL	46	144	0.319	31	**17.100**
12	Burks	Ellis	1996	COL	NL	40	128	0.344	32	**16.456**
13	Walker	Larry	2001	COL	NL	38	123	0.350	35	**15.835**
14	Bichette	Dante	1995	COL	NL	40	128	0.340	32	**15.780**
15	Helton	Todd	2007	COL	NL	17	91	0.320	34	**15.546**
16	Helton	Todd	2002	COL	NL	30	109	0.329	29	**15.523**
17	Holliday	Matt	2008	COL	NL	25	88	0.321	28	**15.333**
18	Walker	Larry	1999	COL	NL	37	115	0.379	33	**14.918**
19	Galarraga	Andres	1997	COL	NL	41	140	0.318	36	**14.761**
20	Tulowitzki	Troy	2010	COL	NL	27	95	0.315	26	**14.536**

Top Position Players By Franchise (1871-2010)

Rank	Name	First	Year	Team	Lg	HR	RBI	AVG	Age	PEVA-B
	Detroit Tigers									
1	Cobb	Ty	1917	DET	AL	6	102	0.383	31	**48.911**
2	Cobb	Ty	1915	DET	AL	3	99	0.369	29	**43.671**
3	Cobb	Ty	1911	DET	AL	8	127	0.420	25	**42.464**
4	Cobb	Ty	1909	DET	AL	9	107	0.377	23	**39.724**
5	Cobb	Ty	1918	DET	AL	3	64	0.382	32	**37.701**
6	Cash	Norm	1961	DET	AL	41	132	0.361	27	**36.206**
7	Cobb	Ty	1912	DET	AL	7	83	0.409	26	**36.188**
8	Greenberg	Hank	1937	DET	AL	40	183	0.337	26	**35.096**
9	Cobb	Ty	1916	DET	AL	5	68	0.371	30	**34.488**
10	Cabrera	Miguel	2010	DET	AL	38	126	0.328	27	**34.146**
11	Cobb	Ty	1910	DET	AL	8	91	0.383	24	**32.427**
12	Greenberg	Hank	1935	DET	AL	36	170	0.328	24	**32.405**
13	Greenberg	Hank	1940	DET	AL	41	150	0.340	29	**32.198**
14	Cobb	Ty	1907	DET	AL	5	119	0.350	21	**31.905**
15	Ordonez	Magglio	2007	DET	AL	28	139	0.363	33	**31.089**
16	Greenberg	Hank	1938	DET	AL	58	146	0.315	27	**30.495**
17	Fielder	Cecil	1990	DET	AL	51	132	0.277	27	**29.877**
18	Kaline	Al	1955	DET	AL	27	102	0.340	21	**29.684**
19	Trammell	Alan	1987	DET	AL	28	105	0.343	29	**29.008**
20	Heilmann	Harry	1923	DET	AL	18	115	0.403	29	**28.483**

Top Position Players By Franchise (1871-2010)

Rank	Name	First	Year	Team	Lg	HR	RBI	AVG	Age	PEVA-B
	Detroit Wolverines									
1	Brouthers	Dan	1888	DTN	NL	9	66	0.307	30	**28.646**
2	Brouthers	Dan	1886	DTN	NL	11	72	0.370	28	**26.852**
3	Thompson	Sam	1887	DTN	NL	11	166	0.372	27	**23.822**
4	Wood	George	1883	DTN	NL	5	47	0.302	25	**22.913**
5	Bennett	Charlie	1882	DTN	NL	5	51	0.301	28	**22.544**
6	Bennett	Charlie	1883	DTN	NL	5	55	0.305	29	**20.547**
7	Brouthers	Dan	1887	DTN	NL	12	101	0.338	29	**20.013**
8	Richardson	Hardy	1886	DTN	NL	11	61	0.351	31	**19.201**
9	Bennett	Charlie	1881	DTN	NL	7	64	0.301	27	**18.960**
10	Hanlon	Ned	1884	DTN	NL	5	39	0.264	27	**16.317**
11	Hanlon	Ned	1885	DTN	NL	1	29	0.302	28	**15.291**
12	Thompson	Sam	1886	DTN	NL	8	89	0.310	26	**14.910**
13	Wood	George	1882	DTN	NL	7	29	0.269	24	**14.263**
14	Bennett	Charlie	1885	DTN	NL	5	60	0.269	31	**14.160**
15	Wood	George	1884	DTN	NL	8	29	0.252	26	**13.984**
16	Rowe	Jack	1887	DTN	NL	6	96	0.318	31	**12.622**

17	Hanlon	Ned	1882	DTN	NL	5	38	0.231	25	**12.267**
18	White	Deacon	1888	DTN	NL	4	71	0.298	41	**12.053**
19	Knight	Lon	1881	DTN	NL	1	52	0.271	28	**11.598**
20	Richardson	Hardy	1887	DTN	NL	8	94	0.328	32	**11.298**

Top Position Players By Franchise (1871-2010)

Rank	Name	First	Year	Team	Lg	HR	RBI	AVG	Age	PEVA-B
	Elizabeth Resolutes									
1	Booth	Eddie	1873	ELI	NA	0	4	0.292	NA	**1.074**
2	Allison	Doug	1873	ELI	NA	0	8	0.289	27	**1.005**
3	Allison	Art	1873	ELI	NA	0	11	0.323	24	**0.869**
4	Austin	Henry	1873	ELI	NA	0	11	0.248	29	**0.441**
5	Fleet	Frank	1873	ELI	NA	0	10	0.256	25	**0.350**

Top Position Players By Franchise (1871-2010)

Rank	Name	First	Year	Team	Lg	HR	RBI	AVG	Age	PEVA-B
	Florida Marlins									
1	Sheffield	Gary	1996	FLO	NL	42	120	0.314	28	**32.618**
2	Cabrera	Miguel	2006	FLO	NL	26	114	0.339	23	**32.126**
3	Cabrera	Miguel	2005	FLO	NL	33	116	0.323	22	**23.720**
4	Ramirez	Hanley	2009	FLO	NL	24	106	0.342	26	**23.588**
5	Cabrera	Miguel	2007	FLO	NL	34	119	0.320	24	**22.708**
6	Delgado	Carlos	2005	FLO	NL	33	115	0.301	33	**21.013**
7	Ramirez	Hanley	2008	FLO	NL	33	67	0.301	25	**20.938**
8	Ramirez	Hanley	2007	FLO	NL	29	81	0.332	24	**19.908**
9	Floyd	Cliff	2001	FLO	NL	31	103	0.317	29	**16.222**
10	Uggla	Dan	2010	FLO	NL	33	105	0.287	30	**16.179**
11	Cabrera	Miguel	2004	FLO	NL	33	112	0.294	21	**14.825**
12	Rodriguez	Ivan	2003	FLO	NL	16	85	0.297	32	**14.811**
13	Alou	Moises	1997	FLO	NL	23	115	0.292	31	**14.645**
14	Lowell	Mike	2004	FLO	NL	27	85	0.293	30	**14.639**
15	Sheffield	Gary	1998	FLO	NL	6	28	0.272	30	**14.301**
16	Wilson	Preston	2000	FLO	NL	31	121	0.264	26	**13.854**
17	Sheffield	Gary	1997	FLO	NL	21	71	0.250	29	**13.817**
18	Lee	Derrek	2003	FLO	NL	31	92	0.271	28	**13.614**
19	Ramirez	Hanley	2006	FLO	NL	17	59	0.292	23	**13.541**
20	Pierre	Juan	2004	FLO	NL	3	49	0.326	27	**13.525**

Top Position Players By Franchise (1871-2010)

Rank	Name	First	Year	Team	Lg	HR	RBI	AVG	Age	PEVA-B
	Fort Wayne Kekiongas									
1	Foran	Jim	1871	FW1	NA	1	18	0.348	23	**1.456**
2	Mathews	Bobby	1871	FW1	NA	0	10	0.270	20	**0.589**
3	Goldsmith	Wally	1871	FW1	NA	0	12	0.205	22	**0.578**
4	Carey	Tom	1871	FW1	NA	0	10	0.230	22	**0.573**
5	Kelly	Bill	1871	FW1	NA	0	7	0.224	NA	**0.484**

Top Position Players By Franchise (1871-2010)

Rank	Name	First	Year	Team	Lg	HR	RBI	AVG	Age	PEVA-B
	Hartford Dark Blues (National League)									
1	Cassidy	John	1877	HAR	NL	0	27	0.378	20	**22.778**
2	Start	Joe	1877	HAR	NL	1	21	0.332	35	**19.559**
3	York	Tom	1877	HAR	NL	1	37	0.283	27	**13.799**
4	Higham	Dick	1876	HAR	NL	0	35	0.327	25	**11.623**
5	Ferguson	Bob	1877	HAR	NL	0	35	0.256	32	**11.164**

Top Position Players By Franchise (1871-2010)

Rank	Name	First	Year	Team	Lg	HR	RBI	AVG	Age	PEVA-B
	Houston Astros									
1	Bagwell	Jeff	1994	HOU	NL	39	116	0.368	26	**38.044**
2	Bagwell	Jeff	1999	HOU	NL	42	126	0.304	31	**35.960**
3	Bagwell	Jeff	1996	HOU	NL	31	120	0.315	28	**31.039**
4	Bagwell	Jeff	1997	HOU	NL	43	135	0.286	29	**29.539**
5	Cruz	Jose	1984	HOU	NL	12	95	0.312	37	**27.975**
6	Cedeno	Cesar	1972	HOU	NL	22	82	0.320	21	**27.803**
7	Wynn	Jimmy	1968	HOU	NL	26	67	0.269	26	**27.640**
8	Berkman	Lance	2006	HOU	NL	45	136	0.315	30	**27.139**
9	Wynn	Jimmy	1969	HOU	NL	33	87	0.269	27	**27.134**
10	Wynn	Jimmy	1972	HOU	NL	24	90	0.273	30	**26.180**
11	Wynn	Jimmy	1965	HOU	NL	22	73	0.275	23	**25.873**
12	Bagwell	Jeff	2000	HOU	NL	47	132	0.310	32	**25.140**
13	Berkman	Lance	2008	HOU	NL	29	106	0.312	32	**24.959**
14	Biggio	Craig	1997	HOU	NL	22	81	0.309	32	**23.961**
15	Berkman	Lance	2004	HOU	NL	30	106	0.316	28	**23.880**
16	Cruz	Jose	1983	HOU	NL	14	92	0.318	36	**22.088**
17	Wynn	Jimmy	1970	HOU	NL	27	88	0.282	28	**21.944**
18	Berkman	Lance	2001	HOU	NL	34	126	0.331	25	**21.405**
19	Watson	Bob	1976	HOU	NL	16	102	0.313	30	**20.889**
20	Cedeno	Cesar	1974	HOU	NL	26	102	0.269	23	**20.852**

Note: Previous Names of Franchise: Houston Colt 45's

Top Position Players By Franchise (1871-2010)

Rank	Name	First	Year	Team	Lg	HR	RBI	AVG	Age	PEVA-B
	Hartford Dark Blues (National Association)									
1	York	Tom	1875	HR1	NA	0	37	0.296	25	**14.808**
2	Pike	Lip	1874	HR1	NA	1	51	0.355	29	**13.501**
3	Remsen	Jack	1875	HR1	NA	0	34	0.268	25	**11.122**
4	Ferguson	Bob	1875	HR1	NA	0	43	0.240	30	**10.643**
5	Burdock	Jack	1875	HR1	NA	0	35	0.294	23	**9.859**

Top Position Players By Franchise (1871-2010)

Rank	Name	First	Year	Team	Lg	HR	RBI	AVG	Age	PEVA-B
	Indianapolis Blues									
1	Shaffer	Orator	1878	IN1	NL	0	30	0.338	27	**50.240**
2	Clapp	John	1878	IN1	NL	0	29	0.304	27	**20.427**
3	McKelvy	Russ	1878	IN1	NL	2	36	0.225	24	**10.612**
4	Quest	Joe	1878	IN1	NL	0	13	0.205	26	**6.698**
5	Williamson	Ned	1878	IN1	NL	1	19	0.232	21	**5.895**

Top Position Players By Franchise (1871-2010)

Rank	Name	First	Year	Team	Lg	HR	RBI	AVG	Age	PEVA-B
	Indianapolis Hoosiers (American Association)									
1	Keenan	Jim	1884	IN2	AA	3	0	0.293	26	**6.327**
2	Phillips	Marr	1884	IN2	AA	0	0	0.269	27	**5.336**
3	Peltz	John	1884	IN2	AA	3	0	0.219	23	**4.633**
4	Kerins	John	1884	IN2	AA	6	0	0.214	26	**3.299**
5	Weihe	Podge	1884	IN2	AA	4	0	0.254	22	**2.474**

Top Position Players By Franchise (1871-2010)

Rank	Name	First	Year	Team	Lg	HR	RBI	AVG	Age	PEVA-B
	Indianapolis Hoosiers (National League)									
1	Glasscock	Jack	1889	IN3	NL	7	85	0.352	32	**24.296**
2	Seery	Emmett	1889	IN3	NL	8	59	0.314	28	**16.268**

Rank	Name	First	Year	Team	Lg	HR	RBI	AVG	Age	PEVA-B
3	Denny	Jerry	1889	IN3	NL	18	112	0.282	30	**14.488**
4	Denny	Jerry	1887	IN3	NL	11	97	0.324	28	**12.665**
5	Hines	Paul	1888	IN3	NL	4	58	0.281	33	**12.664**
6	Denny	Jerry	1888	IN3	NL	12	63	0.261	29	**11.619**
7	Seery	Emmett	1888	IN3	NL	5	50	0.220	27	**10.417**

Top Position Players By Franchise (1871-2010)

Rank	Name	First	Year	Team	Lg	HR	RBI	AVG	Age	PEVA-B
	Indianapolis Hoosiers & Newark Peppers (Federal League)									
1	Kauff	Benny	1914	IND	FL	8	95	0.370	24	**35.356**
2	Roush	Edd	1915	NEW	FL	3	60	0.298	22	**14.470**
3	Scheer	Al	1915	NEW	FL	2	60	0.267	27	**13.949**
4	Rariden	Bill	1915	NEW	FL	0	40	0.270	27	**13.550**
5	McKechnie	Bill	1914	IND	FL	2	38	0.304	28	**12.768**
6	Esmond	Jimmy	1915	NEW	FL	5	62	0.258	26	**12.217**
7	LaPorte	Frank	1914	IND	FL	4	107	0.311	34	**11.995**

Note: Previous Names of Franchise: Indianapolis Hoosiers (IND) 1914; Newark Peppers (NEW) 1915

Top Position Players By Franchise (1871-2010)

Rank	Name	First	Year	Team	Lg	HR	RBI	AVG	Age	PEVA-B
	Kansas City Cowboys (American Association)									
1	Hamilton	Billy	1889	KC2	AA	3	77	0.301	23	**15.432**
2	Long	Herman	1889	KC2	AA	3	60	0.275	23	**13.249**
3	McTamany	Jim	1888	KC2	AA	4	41	0.246	25	**11.901**
4	Burns	Jim	1889	KC2	AA	5	97	0.304	NA	**11.228**
5	Stearns	Ecky	1889	KC2	AA	3	87	0.286	28	**8.475**

Top Position Players By Franchise (1871-2010)

Rank	Name	First	Year	Team	Lg	HR	RBI	AVG	Age	PEVA-B
	Kansas City Royals									
1	Brett	George	1985	KCA	AL	30	112	0.335	32	**39.324**
2	Porter	Darrell	1979	KCA	AL	20	112	0.291	27	**31.378**
3	Mayberry	John	1975	KCA	AL	34	106	0.291	26	**28.634**
4	Brett	George	1980	KCA	AL	24	118	0.390	27	**27.571**
5	McRae	Hal	1982	KCA	AL	27	133	0.308	37	**27.210**
6	Brett	George	1979	KCA	AL	23	107	0.329	26	**25.426**
7	Mayberry	John	1972	KCA	AL	25	100	0.298	23	**22.654**
8	Brett	George	1982	KCA	AL	21	82	0.301	29	**22.248**
9	Brett	George	1988	KCA	AL	24	103	0.306	35	**21.193**
10	Tartabull	Danny	1987	KCA	AL	34	101	0.309	25	**20.246**
11	Brett	George	1976	KCA	AL	7	67	0.333	23	**19.932**
12	Tartabull	Danny	1991	KCA	AL	31	100	0.316	29	**19.129**
13	Beltran	Carlos	2004	KCA	AL	15	51	0.278	27	**19.118**
14	Mayberry	John	1973	KCA	AL	26	100	0.278	24	**18.772**
15	Cowens	Al	1977	KCA	AL	23	112	0.312	26	**18.749**
16	Seitzer	Kevin	1987	KCA	AL	15	83	0.323	25	**18.408**
17	Otis	Amos	1978	KCA	AL	22	96	0.298	31	**18.024**
18	Brett	George	1990	KCA	AL	14	87	0.329	37	**17.381**
19	Brett	George	1977	KCA	AL	22	88	0.312	24	**17.240**
20	Sweeney	Mike	2000	KCA	AL	29	144	0.333	27	**17.106**

Top Position Players By Franchise (1871-2010)

Rank	Name	First	Year	Team	Lg	HR	RBI	AVG	Age	PEVA-B
	Kansas City Packers									
1	Kenworthy	Bill	1914	KCF	FL	15	91	0.317	28	**23.988**
2	Easterly	Ted	1914	KCF	FL	1	67	0.335	29	**17.642**

Rank	Name	First	Year	Team	Lg	HR	RBI	AVG	Age	PEVA-B
3	Chadbourne	Chet	1914	KCF	FL	1	37	0.277	30	**12.792**
4	Perring	George	1914	KCF	FL	2	69	0.278	30	**10.555**
5	Perring	George	1915	KCF	FL	7	67	0.259	31	**9.935**

Top Position Players By Franchise (1871-2010)

Rank	Name	First	Year	Team	Lg	HR	RBI	AVG	Age	PEVA-B
	Kansas City Cowboys (National League)									
1	Myers	Al	1886	KCN	NL	4	51	0.277	23	**5.041**
2	Radford	Paul	1886	KCN	NL	0	20	0.229	25	**4.964**
3	McQuery	Mox	1886	KCN	NL	4	38	0.247	25	**4.197**
4	Bassett	Charley	1886	KCN	NL	2	32	0.260	23	**3.605**
5	Rowe	Dave	1886	KCN	NL	3	57	0.240	32	**3.549**

Top Position Players By Franchise (1871-2010)

Rank	Name	First	Year	Team	Lg	HR	RBI	AVG	Age	PEVA-B
	Kansas City Cowboys (Union League)									
1	Seery	Emmett	1884	KCU	UA	0		0.500	23	**12.602**
2	Ellick	Joe	1884	KCU	UA	0		0.000	30	**4.672**
3	Say	Lou	1884	KCU	UA	1		0.200	30	**4.524**
4	Whitehead	Milt	1884	KCU	UA	0		0.136	22	**3.837**
5	Wheeler	Harry	1884	KCU	UA	0		0.258	26	**2.302**

Note: All players on 2nd stint of year (partial year with club)

Top Position Players By Franchise (1871-2010)

Rank	Name	First	Year	Team	Lg	HR	RBI	AVG	Age	PEVA-B
	Keokuk Westerns									
1	Hallinan	Jimmy	1875	KEO	NA	0	3	0.275	26	**4.411**
2	Golden	Mike	1875	KEO	NA	0	1	0.130	24	**1.613**
3	Quinn	Paddy	1875	KEO	NA	0	5	0.326	26	**0.941**
4	Jones	Charley	1875	KEO	NA	0	10	0.277	25	**0.729**
5	Simmons	Joe	1875	KEO	NA	0	4	0.170	30	**0.200**

Top Position Players By Franchise (1871-2010)

Rank	Name	First	Year	Team	Lg	HR	RBI	AVG	Age	PEVA-B
	Los Angeles Dodgers									
1	Brouthers	Dan	1892	BRO	NL	5	124	0.335	34	**42.196**
2	Piazza	Mike	1997	LAN	NL	40	124	0.362	29	**41.108**
3	Snider	Duke	1953	BRO	NL	42	126	0.336	27	**38.655**
4	Campanella	Roy	1953	BRO	NL	41	142	0.312	32	**38.348**
5	Sheckard	Jimmy	1903	BRO	NL	9	75	0.332	25	**36.040**
6	Snider	Duke	1955	BRO	NL	42	136	0.309	29	**34.757**
7	Snider	Duke	1954	BRO	NL	40	130	0.341	28	**32.766**
8	Galan	Augie	1944	BRO	NL	12	93	0.318	32	**31.002**
9	Campanella	Roy	1951	BRO	NL	33	108	0.325	30	**30.592**
10	Herman	Babe	1930	BRO	NL	35	130	0.393	27	**30.341**
11	Davis	Tommy	1962	LAN	NL	27	153	0.346	23	**29.550**
12	Piazza	Mike	1996	LAN	NL	36	105	0.336	28	**29.100**
13	Wynn	Jimmy	1974	LAN	NL	32	108	0.271	32	**28.948**
14	Walker	Dixie	1944	BRO	NL	13	91	0.357	34	**28.465**
15	Robinson	Jackie	1952	BRO	NL	19	75	0.308	33	**25.988**
16	Beltre	Adrian	2004	LAN	NL	48	121	0.334	25	**25.807**
17	Murray	Eddie	1990	LAN	NL	26	95	0.330	34	**25.700**
18	Robinson	Jackie	1951	BRO	NL	19	88	0.338	32	**25.424**
19	Sheffield	Gary	2000	LAN	NL	43	109	0.325	32	**25.043**
20	Guerrero	Pedro	1982	LAN	NL	32	100	0.304	26	**24.759**

Note: Previous Names of Franchise: Brooklyn Atlantics (BR3) 1884, Grays (BR3) 1885-1887, Bridegrooms (BR3) 1888-9; Bridegrooms, Grooms, Superbas, Dodgers, Robins (BRO) 1890-1957

Top Position Players By Franchise (1871-2010)

Rank	Name	First	Year	Team	Lg	HR	RBI	AVG	Age	PEVA-B
	Louisville Grays									
1	Hall	George	1877	LS1	NL	0	26	0.323	28	**14.527**
2	Shaffer	Orator	1877	LS1	NL	3	34	0.285	26	**10.281**
3	Gerhardt	Joe	1877	LS1	NL	1	35	0.304	22	**8.914**
4	Snyder	Pop	1877	LS1	NL	2	28	0.258	23	**6.368**
5	Latham	Juice	1877	LS1	NL	0	22	0.291	25	**6.334**

Top Position Players By Franchise (1871-2010)

Rank	Name	First	Year	Team	Lg	HR	RBI	AVG	Age	PEVA-B
	Louisville Eclipse & Colonels									
1	Browning	Pete	1885	LS2	AA	9	73	0.362	24	**35.668**
2	Clarke	Fred	1897	LS3	NL	6	67	0.390	25	**31.409**
3	Browning	Pete	1887	LS2	AA	4	118	0.402	26	**27.217**
4	Wolf	Jimmy	1890	LS2	AA	4	98	0.363	28	**22.630**
5	Collins	Hub	1888	LS2	AA	2	50	0.307	24	**22.619**
6	Browning	Pete	1884	LS2	AA	4	47	0.336	23	**18.743**
7	Clarke	Fred	1899	LS3	NL	5	70	0.342	27	**17.629**
8	Hoy	Dummy	1898	LS3	NL	6	66	0.304	36	**17.456**
9	Wolf	Jimmy	1884	LS2	AA	3	73	0.300	22	**17.063**
10	Wagner	Honus	1899	LS3	NL	7	113	0.336	25	**16.590**
11	Clarke	Fred	1898	LS3	NL	3	47	0.307	26	**16.279**
12	Browning	Pete	1882	LS2	AA	5		0.378	21	**15.646**
13	Clarke	Fred	1896	LS3	NL	9	79	0.325	24	**15.479**
14	McCreery	Tom	1896	LS3	NL	7	65	0.351	22	**14.077**
15	Wolf	Jimmy	1885	LS2	AA	1	52	0.292	23	**13.649**
16	Browning	Pete	1883	LS2	AA	4	0	0.338	22	**13.647**
17	Wagner	Honus	1898	LS3	NL	10	105	0.299	24	**13.502**
18	Cline	Monk	1884	LS2	AA	2	39	0.290	26	**12.997**
19	Clarke	Fred	1895	LS3	NL	4	82	0.347	23	**12.740**
20	Browning	Pete	1888	LS2	AA	3	72	0.313	27	**12.258**

Top Position Players By Franchise (1871-2010)

Rank	Name	First	Year	Team	Lg	HR	RBI	AVG	Age	PEVA-B
	Middletown Mansfields									
1	Booth	Eddie	1872	MID	NA	0	12	0.325	NA	**3.992**
2	Murnane	Tim	1872	MID	NA	0	13	0.359	20	**2.056**
3	Clapp	John	1872	MID	NA	1	10	0.289	21	**1.747**
4	O'Rourke	Jim	1872	MID	NA	0	12	0.307	22	**1.704**
5	McCarton	Frank	1872	MID	NA	0	10	0.329	18	**1.395**

Top Position Players By Franchise (1871-2010)

Rank	Name	First	Year	Team	Lg	HR	RBI	AVG	Age	PEVA-B
	Minnesota Twins									
1	Delahanty	Ed	1902	WS1	AL	10	93	0.376	35	**34.689**
2	Killebrew	Harmon	1969	MIN	AL	49	140	0.276	33	**33.634**
3	Mauer	Joe	2009	MIN	AL	28	96	0.365	26	**32.590**
4	Carew	Rod	1977	MIN	AL	14	100	0.388	32	**31.528**
5	Spence	Stan	1944	WS1	AL	18	100	0.316	29	**30.663**
6	Puckett	Kirby	1988	MIN	AL	24	121	0.356	28	**30.641**
7	Killebrew	Harmon	1967	MIN	AL	44	113	0.269	31	**29.312**
8	Oliva	Tony	1964	MIN	AL	32	94	0.323	26	**26.344**
9	Puckett	Kirby	1986	MIN	AL	31	96	0.328	26	**25.975**

Rank	Name	First	Year	Team	Lg	HR	RBI	AVG	Age	PEVA-B
10	Travis	Cecil	1941	WS1	AL	7	101	0.359	28	**25.776**
11	Mauer	Joe	2006	MIN	AL	13	84	0.347	23	**25.655**
12	Killebrew	Harmon	1964	MIN	AL	49	111	0.270	28	**25.257**
13	Goslin	Goose	1926	WS1	AL	17	108	0.354	26	**24.499**
14	Vernon	Mickey	1953	WS1	AL	15	115	0.337	35	**23.556**
15	Killebrew	Harmon	1970	MIN	AL	41	113	0.271	34	**23.140**
16	Killebrew	Harmon	1966	MIN	AL	39	110	0.281	30	**23.060**
17	Allison	Bob	1963	MIN	AL	35	91	0.271	29	**22.876**
18	Allison	Bob	1964	MIN	AL	32	86	0.287	30	**22.600**
19	Mauer	Joe	2008	MIN	AL	9	85	0.328	25	**22.034**
20	Goslin	Goose	1924	WS1	AL	12	129	0.344	24	**21.994**

Note: Previous Names of Franchise: Washington Senators (1901-1960)

Top Position Players By Franchise (1871-2010)

Rank	Name	First	Year	Team	Lg	HR	RBI	AVG	Age	PEVA-B
	Milwaukee Grays									
1	Dalrymple	Abner	1878	ML2	NL	0	15	0.354	21	**21.930**
2	Peters	John	1878	ML2	NL	0	22	0.309	28	**4.881**
3	Foley	Will	1878	ML2	NL	0	22	0.271	23	**3.871**
4	Goodman	Jake	1878	ML2	NL	1	27	0.246	25	**3.847**
5	Golden	Mike	1878	ML2	NL	0	20	0.206	27	**1.479**

Top Position Players By Franchise (1871-2010)

Rank	Name	First	Year	Team	Lg	HR	RBI	AVG	Age	PEVA-B
	Milwaukee Brewers (American Association)									
1	Canavan	Jim	1891	ML3	AA	3	21	0.268	25	**9.146**
2	Carney	John	1891	ML3	AA	3	23	0.300	25	**5.535**
3	Vaughn	Farmer	1891	ML3	AA	0	9	0.333	27	**1.755**
4	Shoch	George	1891	ML3	AA	1	16	0.315	32	**1.361**
5	Dalrymple	Abner	1891	ML3	AA	1	22	0.311	34	**0.887**

Top Position Players By Franchise (1871-2010)

Rank	Name	First	Year	Team	Lg	HR	RBI	AVG	Age	PEVA-B
	Milwaukee Brewers									
1	Yount	Robin	1982	ML4	AL	29	114	0.331	27	**45.717**
2	Fielder	Prince	2009	MIL	NL	46	141	0.299	25	**34.062**
3	Yount	Robin	1983	ML4	AL	17	80	0.308	28	**28.042**
4	Thomas	Gorman	1982	ML4	AL	39	112	0.245	32	**26.667**
5	Braun	Ryan	2009	MIL	NL	32	114	0.320	26	**25.835**
6	Yount	Robin	1989	ML4	AL	21	103	0.318	34	**25.703**
7	Molitor	Paul	1982	ML4	AL	19	71	0.302	26	**25.602**
8	Cooper	Cecil	1982	ML4	AL	32	121	0.313	33	**24.686**
9	Molitor	Paul	1991	ML4	AL	17	75	0.325	35	**24.360**
10	Cooper	Cecil	1983	ML4	AL	30	126	0.307	34	**24.257**
11	Oglivie	Ben	1980	ML4	AL	41	118	0.304	31	**23.936**
12	Cooper	Cecil	1980	ML4	AL	25	122	0.352	31	**23.765**
13	Yount	Robin	1984	ML4	AL	16	80	0.298	29	**23.370**
14	Lezcano	Sixto	1979	ML4	AL	28	101	0.321	26	**22.688**
15	Fielder	Prince	2007	MIL	NL	50	119	0.288	23	**22.583**
16	Yount	Robin	1988	ML4	AL	13	91	0.306	33	**21.619**
17	Thomas	Gorman	1979	ML4	AL	45	123	0.244	29	**20.520**
18	Scott	George	1975	ML4	AL	36	109	0.285	31	**19.992**
19	Scott	George	1973	ML4	AL	24	107	0.306	29	**19.860**
20	Braun	Ryan	2010	MIL	NL	25	103	0.304	27	**19.028**

Note: Previous Names of Franchise: Seattle Pilots (SE1) 1969

Top Position Players By Franchise (1871-2010)

Rank	Name	First	Year	Team	Lg	HR	RBI	AVG	Age	PEVA-B
	Milwaukee Brewers (Union League)									
1	Sexton	Tom	1884	MLU	UA	0		0.234	19	**0.804**
2	Cushman	Ed	1884	MLU	UA	0		0.091	32	**0.200**
3	Morrissey	Tom	1884	MLU	UA	0		0.170	24	**0.200**
4	Bignell	George	1884	MLU	UA	0		0.222	26	**0.200**
5	Myers	Al	1884	MLU	UA	0		0.326	21	**0.200**

Top Position Players By Franchise (1871-2010)

Rank	Name	First	Year	Team	Lg	HR	RBI	AVG	Age	PEVA-B
	New Haven Elm Citys									
1	McGinley	Tim	1875	NH1	NA	0	10	0.275	NA	**2.581**
2	Luff	Henry	1875	NH1	NA	2	18	0.271	19	**2.444**
3	Pabor	Charlie	1875	NH1	NA	0	2	0.348	29	**2.302**
4	Somerville	Ed	1875	NH1	NA	0	7	0.213	22	**1.558**
5	McKelvey	John	1875	NH1	NA	0	10	0.229	28	**1.415**

Note: McGinley, Pabor, Somerville 2nd stint players (partial year)

Top Position Players By Franchise (1871-2010)

Rank	Name	First	Year	Team	Lg	HR	RBI	AVG	Age	PEVA-B
	New York Mutuals (National Association)									
1	Eggler	Dave	1872	NY2	NA	0	20	0.338	23	**25.075**
2	Wolters	Rynie	1871	NY2	NA	0	44	0.370	29	**21.178**
3	Hatfield	John	1872	NY2	NA	1	45	0.319	25	**18.990**
4	Hicks	Nat	1872	NY2	NA	0	33	0.306	27	**16.985**
5	Start	Joe	1872	NY2	NA	0	50	0.270	30	**12.845**
6	Start	Joe	1871	NY2	NA	1	34	0.360	29	**12.116**
7	Eggler	Dave	1873	NY2	NA	0	34	0.336	24	**11.226**
8	Bechtel	George	1872	NY2	NA	0	41	0.298	24	**11.150**
9	Start	Joe	1874	NY2	NA	2	45	0.314	32	**11.139**
10	Eggler	Dave	1871	NY2	NA	0	18	0.320	22	**10.199**

Top Position Players By Franchise (1871-2010)

Rank	Name	First	Year	Team	Lg	HR	RBI	AVG	Age	PEVA-B
	New York Mutuals (National League)									
1	Force	Davy	1876	NY3	NL	0	0	0.000	27	**5.219**
2	Hallinan	Jimmy	1876	NY3	NL	2	36	0.279	27	**4.913**
3	Treacey	Fred	1876	NY3	NL	0	18	0.211	29	**4.126**
4	Start	Joe	1876	NY3	NL	0	21	0.277	34	**3.497**
5	Holdsworth	Jim	1876	NY3	NL	0	19	0.266	26	**3.157**

Top Position Players By Franchise (1871-2010)

Rank	Name	First	Year	Team	Lg	HR	RBI	AVG	Age	PEVA-B
	New York Metropolitans									
1	Orr	Dave	1884	NY4	AA	9	112	0.354	25	**28.656**
2	Orr	Dave	1885	NY4	AA	6	77	0.342	26	**25.379**
3	Orr	Dave	1886	NY4	AA	7	91	0.338	27	**21.791**
4	Esterbrook	Dude	1884	NY4	AA	1	0	0.314	27	**18.169**
5	Nelson	Candy	1884	NY4	AA	1	0	0.255	35	**16.469**
6	Nelson	Candy	1885	NY4	AA	1	30	0.255	36	**15.627**
7	Roseman	Chief	1884	NY4	AA	4	0	0.298	28	**13.761**
8	Brady	Steve	1885	NY4	AA	3	58	0.295	34	**13.032**
9	Roseman	Chief	1885	NY4	AA	4	46	0.278	29	**12.838**

Rank	Name	First	Year	Team	Lg	HR	RBI	AVG	Age	PEVA-B
10	Nelson	Candy	1883	NY4	AA	0	0	0.305	34	**10.308**

Top Position Players By Franchise (1871-2010)

Rank	Name	First	Year	Team	Lg	HR	RBI	AVG	Age	PEVA-B
	New York Yankees									
1	Ruth	Babe	1923	NYA	AL	41	131	0.393	28	**58.931**
2	Ruth	Babe	1920	NYA	AL	54	137	0.376	25	**55.754**
3	Ruth	Babe	1921	NYA	AL	59	171	0.378	26	**54.876**
4	Ruth	Babe	1927	NYA	AL	60	164	0.356	32	**51.850**
5	Ruth	Babe	1926	NYA	AL	47	146	0.372	31	**51.603**
6	Ruth	Babe	1924	NYA	AL	46	121	0.378	29	**51.324**
7	Gehrig	Lou	1927	NYA	AL	47	175	0.373	24	**50.953**
8	Gehrig	Lou	1934	NYA	AL	49	165	0.363	31	**50.136**
9	Mantle	Mickey	1956	NYA	AL	52	130	0.353	25	**47.236**
10	Gehrig	Lou	1936	NYA	AL	49	152	0.354	33	**45.995**
11	Mantle	Mickey	1961	NYA	AL	54	128	0.317	30	**42.398**
12	Mantle	Mickey	1960	NYA	AL	40	94	0.275	29	**42.394**
13	Ruth	Babe	1931	NYA	AL	46	163	0.373	36	**41.968**
14	Rodriguez	Alex	2007	NYA	AL	54	156	0.314	32	**40.717**
15	Mantle	Mickey	1957	NYA	AL	34	94	0.365	26	**40.637**
16	Gehrig	Lou	1931	NYA	AL	46	184	0.341	28	**40.131**
17	Ruth	Babe	1928	NYA	AL	54	142	0.323	33	**40.061**
18	Gehrig	Lou	1930	NYA	AL	41	174	0.379	27	**39.839**
19	Ruth	Babe	1930	NYA	AL	49	153	0.359	35	**39.019**
20	DiMaggio	Joe	1941	NYA	AL	30	125	0.357	27	**37.809**

Note: Previous Names of Franchise: Baltimore Orioles (BLA) 1901-1902; New York Highlanders 1903-1912

Top Position Players By Franchise (1871-2010)

Rank	Name	First	Year	Team	Lg	HR	RBI	AVG	Age	PEVA-B
	New York Mets									
1	Beltran	Carlos	2006	NYN	NL	41	116	0.275	29	**29.035**
2	Wright	David	2007	NYN	NL	30	107	0.325	25	**28.574**
3	Strawberry	Darryl	1987	NYN	NL	39	104	0.284	25	**26.929**
4	Johnson	Howard	1989	NYN	NL	36	101	0.287	29	**25.048**
5	Piazza	Mike	2000	NYN	NL	38	113	0.324	32	**25.007**
6	Wright	David	2008	NYN	NL	33	124	0.302	26	**24.501**
7	Strawberry	Darryl	1988	NYN	NL	39	101	0.269	26	**23.327**
8	Carter	Gary	1985	NYN	NL	32	100	0.281	31	**23.097**
9	Gilkey	Bernard	1996	NYN	NL	30	117	0.317	30	**22.529**
10	Piazza	Mike	1998	NYN	NL	23	76	0.348	30	**22.368**
11	Hernandez	Keith	1984	NYN	NL	15	94	0.311	31	**22.367**
12	Beltran	Carlos	2008	NYN	NL	27	112	0.284	31	**21.453**
13	Staub	Rusty	1975	NYN	NL	19	105	0.282	31	**21.008**
14	Johnson	Howard	1991	NYN	NL	38	117	0.259	31	**20.769**
15	Wright	David	2006	NYN	NL	26	116	0.311	24	**20.464**
16	Strawberry	Darryl	1990	NYN	NL	37	108	0.277	28	**20.252**
17	Ventura	Robin	1999	NYN	NL	32	120	0.301	32	**19.984**
18	Wright	David	2005	NYN	NL	27	102	0.306	23	**19.979**
19	Wright	David	2010	NYN	NL	29	103	0.283	28	**19.724**
20	Olerud	John	1998	NYN	NL	22	93	0.354	30	**19.289**

Top Position Players By Franchise (1871-2010)

Rank	Name	First	Year	Team	Lg	HR	RBI	AVG	Age	PEVA-B
	New York Giants (Pacific League)									
1	Connor	Roger	1890	NYP	PL	14	103	0.349	33	**25.866**
2	O'Rourke	Jim	1890	NYP	PL	9	115	0.360	40	**16.853**

Rank	Name	First	Year	Team	Lg	HR	RBI	AVG	Age	PEVA-B
3	Ewing	Buck	1890	NYP	PL	8	72	0.338	31	**12.724**
4	Gore	George	1890	NYP	PL	10	55	0.318	33	**10.430**
5	Richardson	Danny	1890	NYP	PL	4	80	0.256	27	**7.003**

Top Position Players By Franchise (1871-2010)

Rank	Name	First	Year	Team	Lg	HR	RBI	AVG	Age	PEVA-B
	Oakland Athletics									
1	Foxx	Jimmie	1933	PHA	AL	48	163	0.356	26	**42.539**
2	Foxx	Jimmie	1932	PHA	AL	58	169	0.364	25	**41.252**
3	Lajoie	Nap	1901	PHA	AL	14	125	0.426	27	**38.604**
4	Baker	Frank	1912	PHA	AL	10	130	0.347	26	**36.589**
5	Canseco	Jose	1988	OAK	AL	42	124	0.307	24	**36.135**
6	Giambi	Jason	2000	OAK	AL	43	137	0.333	29	**34.676**
7	Foxx	Jimmie	1934	PHA	AL	44	130	0.334	27	**33.948**
8	Jackson	Reggie	1969	OAK	AL	47	118	0.275	23	**33.898**
9	Collins	Eddie	1914	PHA	AL	2	85	0.344	27	**32.932**
10	Baker	Frank	1913	PHA	AL	12	117	0.337	27	**31.860**
11	Henderson	Rickey	1990	OAK	AL	28	61	0.325	32	**30.896**
12	Foxx	Jimmie	1935	PHA	AL	36	115	0.346	28	**29.569**
13	Giambi	Jason	2001	OAK	AL	38	120	0.342	30	**29.030**
14	Collins	Eddie	1909	PHA	AL	3	56	0.347	22	**28.689**
15	Bando	Sal	1969	OAK	AL	31	113	0.281	25	**27.934**
16	Johnson	Bob	1939	PHA	AL	23	114	0.338	34	**27.839**
17	Collins	Eddie	1913	PHA	AL	3	73	0.345	26	**27.774**
18	McGwire	Mark	1996	OAK	AL	52	113	0.312	33	**27.171**
19	McGwire	Mark	1987	OAK	AL	49	118	0.289	24	**26.438**
20	Canseco	Jose	1991	OAK	AL	44	122	0.266	27	**26.059**

Note: Previous Names of Franchise: Philadelphia Athletics (PHA) 1901-1954; Kansas City Athletics (KC1) 1955-1967

Top Position Players By Franchise (1871-2010)

Rank	Name	First	Year	Team	Lg	HR	RBI	AVG	Age	PEVA-B
	Philadelphia Athletics (National Association)									
1	Anson	Cap	1872	PH1	NA	0	50	0.415	20	**27.329**
2	Meyerle	Levi	1871	PH1	NA	4	40	0.492	26	**24.682**
3	Force	Davy	1875	PH1	NA	0	49	0.311	26	**17.860**
4	Sutton	Ezra	1875	PH1	NA	1	59	0.324	25	**16.597**
5	Hall	George	1875	PH1	NA	4	62	0.299	26	**16.453**
6	Cuthbert	Ned	1872	PH1	NA	1	47	0.338	27	**16.386**
7	McGeary	Mike	1872	PH1	NA	0	35	0.360	21	**15.043**
8	Malone	Fergy	1871	PH1	NA	1	33	0.343	29	**13.373**
9	Craver	Bill	1875	PH1	NA	2	40	0.319	31	**13.098**
10	Fisler	Wes	1872	PH1	NA	0	48	0.350	31	**12.846**
11	Anson	Cap	1875	PH1	NA	0	58	0.325	23	**12.033**
12	Mack	Denny	1872	PH1	NA	0	34	0.288	21	**10.386**

Top Position Players By Franchise (1871-2010)

Rank	Name	First	Year	Team	Lg	HR	RBI	AVG	Age	PEVA-B
	Philadelphia Athletics (American Association)									
1	Stovey	Harry	1889	PH4	AA	19	119	0.308	33	**34.320**
2	Larkin	Henry	1885	PH4	AA	8	88	0.329	25	**29.794**
3	Stovey	Harry	1885	PH4	AA	13	75	0.315	29	**27.677**
4	Stovey	Harry	1884	PH4	AA	10	83	0.326	28	**26.587**
5	Stovey	Harry	1888	PH4	AA	9	65	0.287	32	**26.048**
6	Lyons	Denny	1889	PH4	AA	9	82	0.329	23	**25.426**
7	Lyons	Denny	1887	PH4	AA	6	102	0.367	21	**23.589**

Rank	Name	First	Year	Team	Lg	HR	RBI	AVG	Age	PEVA-B
8	Larkin	Henry	1886	PH4	AA	2	74	0.319	26	**23.297**
9	Stovey	Harry	1883	PH4	AA	14	66	0.304	27	**21.352**
10	Milligan	Jocko	1891	PH4	AA	11	106	0.303	30	**21.264**
11	Welch	Curt	1888	PH4	AA	1	61	0.282	26	**20.921**
12	Larkin	Henry	1888	PH4	AA	7	101	0.269	28	**17.228**
13	Lyons	Denny	1890	PH4	AA	7	73	0.354	24	**16.142**
14	Moynahan	Mike	1883	PH4	AA	1	67	0.310	27	**16.073**
15	Larkin	Henry	1889	PH4	AA	3	74	0.318	29	**15.376**
16	Lyons	Denny	1888	PH4	AA	6	83	0.296	22	**15.309**
17	Wood	George	1891	PH4	AA	3	61	0.309	33	**15.229**
18	Larkin	Henry	1891	PH4	AA	10	93	0.279	31	**14.140**
19	McTamany	Jim	1891	PH4	AA	3	21	0.225	28	**13.031**
20	Stovey	Harry	1886	PH4	AA	7	59	0.294	30	**12.617**

Top Position Players By Franchise (1871-2010)

Rank	Name	First	Year	Team	Lg	HR	RBI	AVG	Age	PEVA-B
	Philadelphia Whites									
1	Craver	Bill	1874	PH2	NA	0	56	0.343	30	**15.032**
2	Eggler	Dave	1874	PH2	NA	0	31	0.318	25	**12.715**
3	Malone	Fergy	1873	PH2	NA	0	43	0.290	31	**9.974**
4	Meyerle	Levi	1875	PH2	NA	1	54	0.316	30	**9.652**
5	Meyerle	Levi	1873	PH2	NA	3	58	0.349	28	**9.621**

Top Position Players By Franchise (1871-2010)

Rank	Name	First	Year	Team	Lg	HR	RBI	AVG	Age	PEVA-B
	Philadelphia Centennials									
1	Craver	Bill	1875	PH3	NA	0	5	0.277	31	**13.098**
2	Bechtel	George	1875	PH3	NA	0	7	0.279	27	**3.286**
3	Treacey	Fred	1875	PH3	NA	0	2	0.261	28	**2.806**
4	McGinley	Tim	1875	PH3	NA	0	5	0.231	NA	**2.581**
5	Somerville	Ed	1875	PH3	NA	0	6	0.228	22	**1.558**

Top Position Players By Franchise (1871-2010)

Rank	Name	First	Year	Team	Lg	HR	RBI	AVG	Age	PEVA-B
	Philadelphia Phillies									
1	Delahanty	Ed	1899	PHI	NL	9	137	0.410	32	**42.557**
2	Schmidt	Mike	1981	PHI	NL	31	91	0.316	32	**40.963**
3	Delahanty	Ed	1896	PHI	NL	13	126	0.397	29	**40.317**
4	Flick	Elmer	1900	PHI	NL	11	110	0.367	24	**37.675**
5	Delahanty	Ed	1893	PHI	NL	19	146	0.368	26	**36.872**
6	Magee	Sherry	1910	PHI	NL	6	123	0.331	26	**36.256**
7	Howard	Ryan	2006	PHI	NL	58	149	0.313	27	**35.507**
8	Delahanty	Ed	1897	PHI	NL	5	96	0.377	30	**35.320**
9	Hamilton	Billy	1894	PHI	NL	4	87	0.404	28	**34.987**
10	Schmidt	Mike	1974	PHI	NL	36	116	0.282	25	**34.027**
11	Klein	Chuck	1930	PHI	NL	40	170	0.386	26	**33.052**
12	O'Doul	Lefty	1929	PHI	NL	32	122	0.398	32	**31.823**
13	Delahanty	Ed	1895	PHI	NL	11	106	0.404	28	**31.733**
14	Delahanty	Ed	1898	PHI	NL	4	92	0.334	31	**30.919**
15	Allen	Dick	1964	PHI	NL	29	91	0.318	22	**30.770**
16	Schmidt	Mike	1983	PHI	NL	40	109	0.255	34	**30.565**
17	Schmidt	Mike	1982	PHI	NL	35	87	0.280	33	**30.250**
18	Allen	Dick	1966	PHI	NL	40	110	0.317	24	**29.910**
19	Thompson	Sam	1895	PHI	NL	18	165	0.392	35	**29.754**
20	Schmidt	Mike	1984	PHI	NL	36	106	0.277	35	**29.652**

Note: Previous Names of Franchise: Philadelphia Quakers 1883-1889; Philadelphia Blue Jays 1943-1944

Top Position Players By Franchise (1871-2010)

Rank	Name	First	Year	Team	Lg	HR	RBI	AVG	Age	PEVA-B
	Philadelphia Athletics (National League)									
1	Hall	George	1876	PHN	NL	5	45	0.366	27	**19.290**
2	Meyerle	Levi	1876	PHN	NL	0	34	0.340	31	**8.147**
3	Force	Davy	1876	PHN	NL	0	17	0.232	27	**5.219**
4	Fisler	Wes	1876	PHN	NL	1	30	0.288	35	**4.809**
5	Sutton	Ezra	1876	PHN	NL	1	31	0.297	26	**4.636**

Top Position Players By Franchise (1871-2010)

Rank	Name	First	Year	Team	Lg	HR	RBI	AVG	Age	PEVA-B
	Philadelphia Athletics (Pacific League)									
1	Shindle	Billy	1890	PHP	PL	10	90	0.324	30	**19.068**
2	Wood	George	1890	PHP	PL	9	102	0.289	32	**15.312**
3	Griffin	Mike	1890	PHP	PL	6	54	0.286	25	**12.624**
4	Mulvey	Joe	1890	PHP	PL	6	87	0.287	32	**9.179**
5	Farrar	Sid	1890	PHP	PL	1	69	0.256	31	**5.644**

Top Position Players By Franchise (1871-2010)

Rank	Name	First	Year	Team	Lg	HR	RBI	AVG	Age	PEVA-B
	Philadelphia Keystones									
1	Hoover	Buster	1884	PHU	UA	0		0.364	21	**9.374**
2	Geer	Billy	1884	PHU	UA	0		0.250	25	**7.793**
3	McCormick	Jerry	1884	PHU	UA	0		0.285	NA	**5.672**
4	Kienzle	Bill	1884	PHU	UA	0		0.254	NA	**4.253**
5	Clements	Jack	1884	PHU	UA	3		0.282	20	**3.395**

Top Position Players By Franchise (1871-2010)

Rank	Name	First	Year	Team	Lg	HR	RBI	AVG	Age	PEVA-B
	Pittsburgh Pirates									
1	Wagner	Honus	1908	PIT	NL	10	109	0.354	34	**48.837**
2	Kiner	Ralph	1951	PIT	NL	42	109	0.309	29	**37.003**
3	Bonds	Barry	1990	PIT	NL	33	114	0.301	26	**36.095**
4	Wagner	Honus	1907	PIT	NL	6	82	0.350	33	**35.816**
5	Wagner	Honus	1905	PIT	NL	6	101	0.363	31	**35.473**
6	Bonds	Barry	1992	PIT	NL	34	103	0.311	28	**35.418**
7	Stargell	Willie	1973	PIT	NL	44	119	0.299	33	**33.533**
8	Wagner	Honus	1900	PIT	NL	4	100	0.381	26	**33.466**
9	Wagner	Honus	1903	PIT	NL	5	101	0.355	29	**32.062**
10	Wagner	Honus	1906	PIT	NL	2	71	0.339	32	**31.968**
11	Wagner	Honus	1909	PIT	NL	5	100	0.339	35	**31.677**
12	Clemente	Roberto	1967	PIT	NL	23	110	0.357	33	**31.045**
13	Stargell	Willie	1971	PIT	NL	48	125	0.295	31	**30.420**
14	Kiner	Ralph	1947	PIT	NL	51	127	0.313	25	**30.272**
15	Williams	Jimmy	1899	PIT	NL	9	116	0.355	23	**28.883**
16	Kiner	Ralph	1949	PIT	NL	54	127	0.310	27	**28.368**
17	Bonds	Barry	1991	PIT	NL	25	116	0.292	27	**28.310**
18	Vaughan	Arky	1935	PIT	NL	19	99	0.385	23	**26.511**
19	Smith	Elmer	1896	PIT	NL	6	94	0.362	28	**26.333**
20	Wagner	Honus	1912	PIT	NL	7	102	0.324	38	**26.228**

Note: Previous Names of Franchise: Pittsburgh Alleghenys (PT1) 1882-1886; Pittsburgh Alleghenys (PIT) 1887-1890

Top Position Players By Franchise (1871-2010)

Rank	Name	First	Year	Team	Lg	HR	RBI	AVG	Age	PEVA-B
	Providence Grays									
1	Hines	Paul	1879	PRO	NL	2	52	0.357	24	**43.731**

2	Hines	Paul	1878	PRO	NL	4	50	0.358	23	**38.405**
3	York	Tom	1878	PRO	NL	1	26	0.309	28	**27.661**
4	Higham	Dick	1878	PRO	NL	1	29	0.320	27	**27.094**
5	O'Rourke	Jim	1879	PRO	NL	1	46	0.348	29	**25.376**
6	Hines	Paul	1880	PRO	NL	3	35	0.307	25	**24.429**
7	Hines	Paul	1884	PRO	NL	3	41	0.302	29	**23.403**
8	York	Tom	1879	PRO	NL	1	50	0.310	29	**22.474**
9	Brown	Lew	1878	PRO	NL	1	43	0.305	20	**20.812**
10	Hines	Paul	1882	PRO	NL	4	34	0.309	27	**20.338**
11	Wright	George	1879	PRO	NL	1	42	0.276	32	**19.248**
12	York	Tom	1881	PRO	NL	2	47	0.304	31	**17.945**
13	Farrell	Jack	1883	PRO	NL	3	61	0.305	26	**16.797**
14	Hines	Paul	1883	PRO	NL	4	45	0.299	28	**16.189**
15	Start	Joe	1882	PRO	NL	0	48	0.329	40	**16.185**
16	Gross	Emil	1880	PRO	NL	1	34	0.259	22	**13.655**
17	Carroll	Cliff	1884	PRO	NL	3	54	0.261	25	**13.068**
18	Denny	Jerry	1883	PRO	NL	8	55	0.275	24	**13.004**
19	Hines	Paul	1881	PRO	NL	2	31	0.285	26	**12.117**
20	Start	Joe	1879	PRO	NL	2	37	0.319	37	**11.872**

Top Position Players By Franchise (1871-2010)

Rank	Name	First	Year	Team	Lg	HR	RBI	AVG	Age	PEVA-B
	Pittsburgh Rebels									
1	Konetchy	Ed	1915	PTF	FL	10	93	0.314	30	**19.572**
2	Lennox	Ed	1914	PTF	FL	11	84	0.312	31	**18.903**
3	Oakes	Rebel	1914	PTF	FL	7	75	0.312	31	**15.609**
4	Kelly	Jim	1915	PTF	FL	4	50	0.294	31	**11.361**
5	Mowrey	Mike	1915	PTF	FL	1	49	0.280	31	**11.159**
6	Wickland	Al	1915	PTF	FL	1	30	0.301	27	**11.088**
7	Oakes	Rebel	1915	PTF	FL	0	82	0.278	32	**10.565**

Top Position Players By Franchise (1871-2010)

Rank	Name	First	Year	Team	Lg	HR	RBI	AVG	Age	PEVA-B
	Pittsburgh Burghers									
1	Beckley	Jake	1890	PTP	PL	9	120	0.324	23	**23.357**
2	Fields	Jocko	1890	PTP	PL	9	86	0.281	26	**13.610**
3	Visner	Joe	1890	PTP	PL	3	71	0.267	31	**12.981**
4	Carroll	Fred	1890	PTP	PL	2	71	0.298	26	**12.170**
5	Hanlon	Ned	1890	PTP	PL	1	44	0.278	33	**10.532**

Top Position Players By Franchise (1871-2010)

Rank	Name	First	Year	Team	Lg	HR	RBI	AVG	Age	PEVA-B
	Rockford Forest Citys									
1	Anson	Cap	1871	RC1	NA	0	16	0.325	19	**5.003**
2	Stires	Gat	1871	RC1	NA	2	24	0.273	22	**3.837**
3	Hastings	Scott	1871	RC1	NA	0	20	0.254	24	**3.285**
4	Mack	Denny	1871	RC1	NA	0	17	0.246	20	**2.652**
5	Addy	Bob	1871	RC1	NA	0	13	0.271	26	**2.182**

Top Position Players By Franchise (1871-2010)

Rank	Name	First	Year	Team	Lg	HR	RBI	AVG	Age	PEVA-B
	Rochester Broncos									
1	Knowles	Jimmy	1890	RC2	AA	5	84	0.281	34	**10.628**
2	Scheffler	Ted	1890	RC2	AA	3	34	0.245	26	**9.074**
3	Lyons	Harry	1890	RC2	AA	3	58	0.260	24	**8.265**
4	Griffin	Sandy	1890	RC2	AA	5	53	0.307	32	**8.181**

5	McGuire	Deacon	1890	RC2	AA	4	53	0.299	27	**5.883**

Top Position Players By Franchise (1871-2010)

Rank	Name	First	Year	Team	Lg	HR	RBI	AVG	Age	PEVA-B
	Richmond Virginians									
1	Mansell	Mike	1884	RIC	AA	0	0	0.301	26	**2.362**
2	Johnston	Dick	1884	RIC	AA	2	0	0.281	21	**1.770**
3	Nash	Billy	1884	RIC	AA	1	0	0.199	19	**1.547**
4	Glenn	Ed	1884	RIC	AA	1	0	0.246	24	**1.121**
5	Powell	Jim	1884	RIC	AA	0	0	0.245	25	**1.058**

Top Position Players By Franchise (1871-2010)

Rank	Name	First	Year	Team	Lg	HR	RBI	AVG	Age	PEVA-B
	San Diego Padres									
1	Winfield	Dave	1979	SDN	NL	34	118	0.308	28	**32.688**
2	Gonzalez	Adrian	2009	SDN	NL	40	99	0.277	27	**27.865**
3	Gwynn	Tony	1987	SDN	NL	7	54	0.370	27	**27.822**
4	Caminiti	Ken	1996	SDN	NL	40	130	0.326	33	**26.999**
5	Gwynn	Tony	1984	SDN	NL	5	71	0.351	24	**24.767**
6	Gonzalez	Adrian	2010	SDN	NL	31	101	0.298	28	**24.739**
7	Sheffield	Gary	1992	SDN	NL	33	100	0.330	24	**23.846**
8	Giles	Brian	2005	SDN	NL	15	83	0.301	34	**23.632**
9	Gwynn	Tony	1997	SDN	NL	17	119	0.372	37	**22.316**
10	Winfield	Dave	1978	SDN	NL	24	97	0.308	27	**21.529**
11	Colbert	Nate	1972	SDN	NL	38	111	0.250	26	**21.351**
12	Gwynn	Tony	1986	SDN	NL	14	59	0.329	26	**21.280**
13	Gwynn	Tony	1994	SDN	NL	12	64	0.394	34	**21.228**
14	Lezcano	Sixto	1982	SDN	NL	16	84	0.289	29	**21.143**
15	Gonzalez	Adrian	2008	SDN	NL	36	119	0.279	26	**20.657**
16	Vaughn	Greg	1998	SDN	NL	50	119	0.272	33	**20.269**
17	McGriff	Fred	1992	SDN	NL	35	104	0.286	29	**20.225**
18	Hendrick	George	1977	SDN	NL	23	81	0.311	28	**19.141**
19	Nevin	Phil	2001	SDN	NL	41	126	0.306	30	**19.119**
20	Gaston	Cito	1970	SDN	NL	29	93	0.318	26	**19.009**

Top Position Players By Franchise (1871-2010)

Rank	Name	First	Year	Team	Lg	HR	RBI	AVG	Age	PEVA-B
	Seattle Mariners									
1	Martinez	Edgar	1995	SEA	AL	29	113	0.356	32	**34.622**
2	Rodriguez	Alex	2000	SEA	AL	41	132	0.316	25	**34.588**
3	Griffey Jr.	Ken	1997	SEA	AL	56	147	0.304	28	**30.208**
4	Rodriguez	Alex	1996	SEA	AL	36	123	0.358	21	**29.207**
5	Griffey Jr.	Ken	1993	SEA	AL	45	109	0.309	24	**28.976**
6	Davis	Alvin	1984	SEA	AL	27	116	0.284	24	**26.893**
7	Martinez	Edgar	2000	SEA	AL	37	145	0.324	37	**26.156**
8	Griffey Jr.	Ken	1994	SEA	AL	40	90	0.323	25	**24.873**
9	Griffey Jr.	Ken	1991	SEA	AL	22	100	0.327	22	**24.792**
10	Martinez	Edgar	1997	SEA	AL	28	108	0.330	34	**23.979**
11	Griffey Jr.	Ken	1996	SEA	AL	49	140	0.303	27	**23.477**
12	Martinez	Edgar	1996	SEA	AL	26	103	0.327	33	**22.840**
13	Griffey Jr.	Ken	1998	SEA	AL	56	146	0.284	29	**22.701**
14	Rodriguez	Alex	1998	SEA	AL	42	124	0.310	23	**20.656**
15	Griffey Jr.	Ken	1999	SEA	AL	48	134	0.285	30	**20.133**
16	Ibanez	Raul	2006	SEA	AL	33	123	0.289	34	**19.804**
17	Boone	Bret	2001	SEA	AL	37	141	0.331	32	**19.697**
18	Suzuki	Ichiro	2004	SEA	AL	8	60	0.372	31	**19.568**
19	Martinez	Edgar	1992	SEA	AL	18	73	0.343	29	**19.232**

20	Davis	Alvin	1989	SEA	AL	21	95	0.305	29	**19.231**

Top Position Players By Franchise (1871-2010)

Rank	Name	First	Year	Team	Lg	HR	RBI	AVG	Age	PEVA-B
	San Francisco Giants									
1	Bonds	Barry	2001	SFN	NL	73	137	0.328	37	**55.207**
2	Bonds	Barry	2002	SFN	NL	46	110	0.370	38	**54.848**
3	Bonds	Barry	2004	SFN	NL	45	101	0.362	40	**48.632**
4	Bonds	Barry	1993	SFN	NL	46	123	0.336	29	**43.404**
5	Mays	Willie	1965	SFN	NL	52	112	0.317	34	**43.109**
6	Mays	Willie	1962	SFN	NL	49	141	0.304	31	**40.947**
7	Mays	Willie	1955	NY1	NL	51	127	0.319	24	**39.404**
8	Mays	Willie	1960	SFN	NL	29	103	0.319	29	**38.898**
9	McCovey	Willie	1969	SFN	NL	45	126	0.320	31	**36.924**
10	Mays	Willie	1958	SFN	NL	29	96	0.347	27	**36.606**
11	Ott	Mel	1929	NY1	NL	42	151	0.328	20	**36.437**
12	Mays	Willie	1964	SFN	NL	47	111	0.296	33	**36.258**
13	Mays	Willie	1963	SFN	NL	38	103	0.314	32	**34.477**
14	Mays	Willie	1954	NY1	NL	41	110	0.345	23	**33.994**
15	Bonds	Barry	1996	SFN	NL	42	129	0.308	32	**33.905**
16	Connor	Roger	1885	NY1	NL	1	65	0.371	28	**33.375**
17	Mitchell	Kevin	1989	SFN	NL	47	125	0.291	27	**33.233**
18	McCovey	Willie	1970	SFN	NL	39	126	0.289	32	**32.785**
19	Bonds	Barry	2000	SFN	NL	49	106	0.306	36	**32.041**
20	Bonds	Barry	2003	SFN	NL	45	90	0.341	39	**31.974**

Note: Previous Names of Franchise: New York Gothams 1883-1884; New York Giants 1885-1957

Top Position Players By Franchise (1871-2010)

Rank	Name	First	Year	Team	Lg	HR	RBI	AVG	Age	PEVA-B
	St. Louis Red Stockings									
1	Morgan	Pidgey	1875	SL1	NA	0	1	0.261	22	**1.023**
2	Hautz	Charlie	1875	SL1	NA	0	4	0.301	23	**1.022**
3	Redmon	Billy	1875	SL1	NA	0	1	0.195	NA	**0.365**
4	Croft	Art	1875	SL1	NA	0	2	0.200	20	**0.228**
5	McSorley	Trick	1875	SL1	NA	0	2	0.212	23	**0.221**

Top Position Players By Franchise (1871-2010)

Rank	Name	First	Year	Team	Lg	HR	RBI	AVG	Age	PEVA-B
	St. Louis Brown Stockings (National Association)									
1	Pike	Lip	1875	SL2	NA	0	44	0.346	30	**23.989**
2	Pearce	Dickey	1875	SL2	NA	0	29	0.248	39	**7.012**
3	Cuthbert	Ned	1875	SL2	NA	0	17	0.245	30	**6.283**
4	Battin	Joe	1875	SL2	NA	0	33	0.250	24	**4.445**
5	Dehlman	Herman	1875	SL2	NA	0	14	0.224	23	**4.138**

Top Position Players By Franchise (1871-2010)

Rank	Name	First	Year	Team	Lg	HR	RBI	AVG	Age	PEVA-B
	St. Louis Brown Stockings (National League)									
1	Clapp	John	1877	SL3	NL	0	34	0.318	26	**19.615**
2	Pike	Lip	1876	SL3	NL	1	50	0.323	31	**17.351**
3	Clapp	John	1876	SL3	NL	0	29	0.305	25	**13.420**
4	Dorgan	Mike	1877	SL3	NL	0	23	0.308	24	**13.349**
5	Battin	Joe	1876	SL3	NL	0	46	0.300	25	**10.703**

Top Position Players By Franchise (1871-2010)

Rank	Name	First	Year	Team	Lg	HR	RBI	AVG	Age	PEVA-B

Rank	Name	First	Year	Team	Lg	HR	RBI	AVG	Age	PEVA-B
	St. Louis Maroons									
1	Dunlap	Fred	1884	SLU	UA	13		0.412	25	**37.571**
2	Shaffer	Orator	1884	SLU	UA	2		0.360	33	**23.763**
3	Glasscock	Jack	1886	SL5	NL	3	40	0.325	29	**16.318**
4	Glasscock	Jack	1885	SL5	NL	1	40	0.280	28	**12.849**
5	Rowe	Dave	1884	SLU	UA	4		0.293	30	**11.370**
6	Gleason	Jack	1884	SLU	UA	4		0.324	30	**10.214**

Top Position Players By Franchise (1871-2010)

Rank	Name	First	Year	Team	Lg	HR	RBI	AVG	Age	PEVA-B
	St. Louis Terriers									
1	Tobin	Jack	1915	SLF	FL	6	51	0.294	23	**15.463**
2	Borton	Babe	1915	SLF	FL	3	83	0.286	27	**15.217**
3	Miller	Ward	1915	SLF	FL	1	63	0.306	31	**14.797**
4	Johnson	Ernie	1915	SLF	FL	7	67	0.240	27	**8.859**
5	Miller	Ward	1914	SLF	FL	4	50	0.294	30	**8.829**

Top Position Players By Franchise (1871-2010)

Rank	Name	First	Year	Team	Lg	HR	RBI	AVG	Age	PEVA-B
	St. Louis Cardinals									
1	Hornsby	Rogers	1922	SLN	NL	42	152	0.401	26	**45.149**
2	Pujols	Albert	2009	SLN	NL	47	135	0.327	29	**43.955**
3	Musial	Stan	1943	SLN	NL	13	81	0.357	23	**41.278**
4	McGwire	Mark	1998	SLN	NL	70	147	0.299	35	**40.809**
5	Musial	Stan	1953	SLN	NL	30	113	0.337	33	**40.502**
6	Musial	Stan	1944	SLN	NL	12	94	0.347	24	**38.717**
7	Musial	Stan	1952	SLN	NL	21	91	0.336	32	**37.889**
8	Hornsby	Rogers	1925	SLN	NL	39	143	0.403	29	**37.655**
9	Musial	Stan	1951	SLN	NL	32	108	0.355	31	**37.339**
10	Pujols	Albert	2010	SLN	NL	42	118	0.312	30	**36.842**
11	Pujols	Albert	2006	SLN	NL	49	137	0.331	26	**36.422**
12	Musial	Stan	1948	SLN	NL	39	131	0.376	28	**35.811**
13	Torre	Joe	1971	SLN	NL	24	137	0.363	31	**34.530**
14	Hornsby	Rogers	1924	SLN	NL	25	94	0.424	28	**34.349**
15	Pujols	Albert	2003	SLN	NL	43	124	0.359	23	**33.510**
16	Pujols	Albert	2008	SLN	NL	37	116	0.357	28	**33.388**
17	Musial	Stan	1954	SLN	NL	35	126	0.330	34	**33.159**
18	Burkett	Jesse	1901	SLN	NL	10	75	0.376	33	**32.782**
19	Pujols	Albert	2005	SLN	NL	41	117	0.330	25	**32.329**
20	O'Neill	Tip	1887	SL4	AA	14	123	0.435	29	**31.789**

Note: Previous Names of Franchise:

St. Louis Brown Stockings (SL4) AA 1882; St. Louis Browns (SL4) AA 1883-1898; St. Louis Perfectos 1899

Top Position Players By Franchise (1871-2010)

Rank	Name	First	Year	Team	Lg	HR	RBI	AVG	Age	PEVA-B
	St. Paul Apostles									
1	Brown	Jim	1884	SPU	UA	0		0.313	24	**0.724**
2	Tilley	John	1884	SPU	UA	0		0.154	NA	**0.260**
3	Hengle	Moxie	1884	SPU	UA	0		0.152	27	**0.258**
4	Ganzel	Charlie	1884	SPU	UA	0		0.217	22	**0.200**
5	Dealy	Pat	1884	SPU	UA	0		0.133	23	**0.200**

Top Position Players By Franchise (1871-2010)

Rank	Name	First	Year	Team	Lg	HR	RBI	AVG	Age	PEVA-B
	Syracuse Stars									
1	Farrell	Jack	1879	SR1	NL	1	21	0.303	22	**4.629**

Rank	Name	First	Year	Team	Lg	HR	RBI	AVG	Age	PEVA-B
2	Purcell	Blondie	1879	SR1	NL	0	25	0.260	25	**4.088**
3	Dorgan	Mike	1879	SR1	NL	1	17	0.267	26	**2.974**
4	Richmond	John	1879	SR1	NL	1	23	0.213	25	**2.060**
5	Mansell	Mike	1879	SR1	NL	1	13	0.215	21	**1.971**

Top Position Players By Franchise (1871-2010)

Rank	Name	First	Year	Team	Lg	HR	RBI	AVG	Age	PEVA-B
	Syracuse Stars (American Association)									
1	Childs	Cupid	1890	SR2	AA	2	89	0.345	23	**24.650**
2	Simon	Hank	1890	SR2	AA	2	23	0.301	28	**9.569**
3	McQuery	Mox	1890	SR2	AA	2	55	0.308	29	**7.634**
4	Wright	Rasty	1890	SR2	AA	0	27	0.305	27	**7.468**
5	Ely	Bones	1890	SR2	AA	0	64	0.262	27	**6.237**

Top Position Players By Franchise (1871-2010)

Rank	Name	First	Year	Team	Lg	HR	RBI	AVG	Age	PEVA-B
	Tampa Bay Devil Rays									
1	Pena	Carlos	2007	TBA	AL	46	121	0.282	29	**23.368**
2	Longoria	Evan	2010	TBA	AL	22	104	0.294	25	**22.846**
3	Longoria	Evan	2009	TBA	AL	33	113	0.281	24	**19.499**
4	Crawford	Carl	2010	TBA	AL	19	90	0.307	29	**18.841**
5	Zobrist	Ben	2009	TBA	AL	27	91	0.297	28	**16.721**
6	Huff	Aubrey	2003	TBA	AL	34	107	0.311	27	**14.625**
7	Crawford	Carl	2005	TBA	AL	15	81	0.301	24	**12.677**
8	McGriff	Fred	1999	TBA	AL	32	104	0.310	36	**12.146**
9	Huff	Aubrey	2004	TBA	AL	29	104	0.297	28	**11.992**
10	Baldelli	Rocco	2003	TBA	AL	11	78	0.289	22	**11.878**
11	Winn	Randy	2002	TBA	AL	14	75	0.298	28	**11.748**
12	Lugo	Julio	2005	TBA	AL	6	57	0.295	30	**11.736**
13	Crawford	Carl	2009	TBA	AL	15	68	0.305	28	**11.622**
14	Upton	B.J.	2007	TBA	AL	24	82	0.300	23	**11.351**
15	Crawford	Carl	2004	TBA	AL	11	55	0.296	23	**11.209**
16	Pena	Carlos	2009	TBA	AL	39	100	0.227	31	**11.191**
17	Upton	B.J.	2008	TBA	AL	9	67	0.273	24	**11.160**
18	Bartlett	Jason	2009	TBA	AL	14	66	0.320	30	**11.149**
19	Crawford	Carl	2006	TBA	AL	18	77	0.305	25	**10.846**
20	Pena	Carlos	2008	TBA	AL	31	102	0.247	30	**10.440**

Top Position Players By Franchise (1871-2010)

Rank	Name	First	Year	Team	Lg	HR	RBI	AVG	Age	PEVA-B
	Texas Rangers									
1	Howard	Frank	1969	WS2	AL	48	111	0.296	33	**30.491**
2	Howard	Frank	1970	WS2	AL	44	126	0.283	34	**29.338**
3	Howard	Frank	1968	WS2	AL	44	106	0.274	32	**28.584**
4	Rodriguez	Alex	2001	TEX	AL	52	135	0.318	26	**27.632**
5	Rodriguez	Alex	2002	TEX	AL	57	142	0.300	27	**25.524**
6	Palmeiro	Rafael	1991	TEX	AL	26	88	0.322	27	**25.081**
7	Burroughs	Jeff	1974	TEX	AL	25	118	0.301	23	**24.708**
8	Rodriguez	Alex	2003	TEX	AL	47	118	0.298	28	**23.712**
9	Sierra	Ruben	1991	TEX	AL	25	116	0.307	26	**23.040**
10	Teixeira	Mark	2005	TEX	AL	43	144	0.301	25	**22.742**
11	Hamilton	Josh	2010	TEX	AL	32	100	0.359	29	**22.732**
12	Palmeiro	Rafael	1999	TEX	AL	47	148	0.324	35	**22.557**
13	Sierra	Ruben	1989	TEX	AL	29	119	0.306	24	**21.583**
14	Bell	Buddy	1984	TEX	AL	11	83	0.315	33	**20.732**
15	Palmeiro	Rafael	1993	TEX	AL	37	105	0.295	29	**20.393**

16	Gonzalez	Juan	1993	TEX	AL	46	118	0.310	24	**19.823**
17	Hamilton	Josh	2008	TEX	AL	32	130	0.304	27	**19.261**
18	Rodriguez	Ivan	1999	TEX	AL	35	113	0.332	28	**19.153**
19	Harrah	Toby	1975	TEX	AL	20	93	0.293	27	**18.954**
20	Franco	Julio	1991	TEX	AL	15	78	0.341	33	**18.845**

Note: Previous Names of Franchise: Washington Senators (WS2) 1961-1971

Top Position Players By Franchise (1871-2010)

Rank	Name	First	Year	Team	Lg	HR	RBI	AVG	Age	PEVA-B
	Toledo Blue Stockings									
1	Barkley	Sam	1884	TL1	AA	1	0	0.306	26	**11.577**
2	Welch	Curt	1884	TL1	AA	0	0	0.224	22	**5.719**
3	Miller	Joe	1884	TL1	AA	1	0	0.239	23	**5.237**
4	Mullane	Tony	1884	TL1	AA	3	0	0.276	25	**4.851**
5	Poorman	Tom	1884	TL1	AA	0	0	0.233	27	**4.162**

Top Position Players By Franchise (1871-2010)

Rank	Name	First	Year	Team	Lg	HR	RBI	AVG	Age	PEVA-B
	Toledo Maumees									
1	Swartwood	Ed	1890	TL2	AA	3	64	0.327	31	**17.677**
2	Werden	Perry	1890	TL2	AA	6	72	0.295	25	**13.212**
3	Nicholson	Parson	1890	TL2	AA	4	72	0.268	27	**6.869**
4	Scheibeck	Frank	1890	TL2	AA	1	49	0.241	25	**6.814**
5	Alvord	Billy	1890	TL2	AA	2	52	0.273	27	**5.814**

Top Position Players By Franchise (1871-2010)

Rank	Name	First	Year	Team	Lg	HR	RBI	AVG	Age	PEVA-B
	Toronto Blue Jays									
1	Delgado	Carlos	2000	TOR	AL	41	137	0.344	28	**34.143**
2	Bautista	Jose	2010	TOR	AL	54	124	0.260	30	**29.447**
3	Olerud	John	1993	TOR	AL	24	107	0.363	25	**28.968**
4	Barfield	Jesse	1986	TOR	AL	40	108	0.289	27	**28.916**
5	McGriff	Fred	1989	TOR	AL	36	92	0.269	26	**25.384**
6	Delgado	Carlos	2003	TOR	AL	42	145	0.302	31	**24.444**
7	Bell	George	1987	TOR	AL	47	134	0.308	28	**24.361**
8	Moseby	Lloyd	1984	TOR	AL	18	92	0.280	25	**22.132**
9	Green	Shawn	1999	TOR	AL	42	123	0.309	27	**21.898**
10	McGriff	Fred	1990	TOR	AL	35	88	0.300	27	**21.863**
11	McGriff	Fred	1988	TOR	AL	34	82	0.282	25	**21.613**
12	Bell	George	1986	TOR	AL	31	108	0.309	27	**21.052**
13	Molitor	Paul	1993	TOR	AL	22	111	0.332	37	**19.716**
14	Gruber	Kelly	1990	TOR	AL	31	118	0.274	28	**19.660**
15	Barfield	Jesse	1985	TOR	AL	27	84	0.289	26	**19.220**
16	Lind	Adam	2009	TOR	AL	35	114	0.305	26	**18.508**
17	Moseby	Lloyd	1983	TOR	AL	18	81	0.315	24	**18.437**
18	Wells	Vernon	2003	TOR	AL	33	117	0.317	25	**17.993**
19	Upshaw	Willie	1983	TOR	AL	27	104	0.306	26	**17.849**
20	Carter	Joe	1991	TOR	AL	33	108	0.273	31	**17.061**

Top Position Players By Franchise (1871-2010)

Rank	Name	First	Year	Team	Lg	HR	RBI	AVG	Age	PEVA-B
	Troy Trojans									
1	Connor	Roger	1880	TRN	NL	3	47	0.332	23	**24.092**
2	Connor	Roger	1882	TRN	NL	4	42	0.330	25	**22.844**
3	Ewing	Buck	1882	TRN	NL	2	29	0.271	23	**11.039**
4	Ferguson	Bob	1881	TRN	NL	1	35	0.283	36	**9.032**

Rank	Name	First	Year	Team	Lg	HR	RBI	AVG	Age	PEVA-B
5	Gillespie	Pete	1880	TRN	NL	2	24	0.243	29	**8.975**

Top Position Players By Franchise (1871-2010)

Rank	Name	First	Year	Team	Lg	HR	RBI	AVG	Age	PEVA-B
	Troy Haymakers									
1	Force	Davy	1872	TRO	NA	0	16	0.408	23	**18.801**
2	King	Steve	1871	TRO	NA	0	34	0.396	29	**17.901**
3	Pike	Lip	1871	TRO	NA	4	39	0.377	26	**16.526**
4	Allison	Doug	1872	TRO	NA	0	20	0.304	26	**7.853**
5	York	Tom	1871	TRO	NA	2	23	0.255	21	**7.002**

Top Position Players By Franchise (1871-2010)

Rank	Name	First	Year	Team	Lg	HR	RBI	AVG	Age	PEVA-B
	Washington Nationals									
1	Carter	Gary	1982	MON	NL	29	97	0.293	28	**33.521**
2	Carter	Gary	1984	MON	NL	27	106	0.294	30	**31.258**
3	Singleton	Ken	1973	MON	NL	23	103	0.302	26	**28.768**
4	Raines	Tim	1984	MON	NL	8	60	0.309	25	**26.337**
5	Oliver	Al	1982	MON	NL	22	109	0.331	36	**25.696**
6	Dawson	Andre	1983	MON	NL	32	113	0.299	29	**25.631**
7	Staub	Rusty	1969	MON	NL	29	79	0.302	25	**24.146**
8	Staub	Rusty	1971	MON	NL	19	97	0.311	27	**23.695**
9	Staub	Rusty	1970	MON	NL	30	94	0.274	26	**22.827**
10	Guerrero	Vladimir	2000	MON	NL	44	123	0.345	24	**22.091**
11	Dawson	Andre	1981	MON	NL	24	64	0.302	27	**22.054**
12	Raines	Tim	1983	MON	NL	11	71	0.298	24	**21.966**
13	Guerrero	Vladimir	2002	MON	NL	39	111	0.336	26	**21.831**
14	Raines	Tim	1987	MON	NL	18	68	0.330	28	**21.635**
15	Zimmerman	Ryan	2009	WAS	NL	33	106	0.292	25	**21.206**
16	Raines	Tim	1985	MON	NL	11	41	0.320	26	**20.439**
17	Raines	Tim	1986	MON	NL	9	62	0.334	27	**19.897**
18	Carter	Gary	1977	MON	NL	31	84	0.284	23	**19.370**
19	Carter	Gary	1980	MON	NL	29	101	0.264	26	**18.884**
20	Soriano	Alfonso	2006	WAS	NL	46	95	0.277	30	**18.543**

Note: Previous Names of Franchise: Montreal Expos (MON) 1969-2004

Top Position Players By Franchise (1871-2010)

Rank	Name	First	Year	Team	Lg	HR	RBI	AVG	Age	PEVA-B
	Wilmington Quicksteps									
1	Burns	Oyster	1884	WIL	UA	0		0.143	20	**2.444**
2	Lynch	Tom	1884	WIL	UA	0		0.276	24	**1.832**
3	Casey	Dennis	1884	WIL	UA	0		0.250	26	**1.198**
4	Benners	Ike	1884	WIL	UA	0		0.045	28	**1.111**
5	Bastian	Charlie	1884	WIL	UA	2		0.200	24	**0.646**

Top Position Players By Franchise (1871-2010)

Rank	Name	First	Year	Team	Lg	HR	RBI	AVG	Age	PEVA-B
	Worcester Ruby Legs									
1	Stovey	Harry	1880	WOR	NL	6	28	0.265	24	**15.558**
2	Stovey	Harry	1882	WOR	NL	5	26	0.289	26	**15.306**
3	Irwin	Arthur	1880	WOR	NL	1	35	0.259	22	**11.326**
4	Dickerson	Buttercup	1881	WOR	NL	1	31	0.316	23	**10.689**
5	Hotaling	Pete	1881	WOR	NL	1	35	0.309	25	**9.080**

Top Position Players By Franchise (1871-2010)

Rank	Name	First	Year	Team	Lg	HR	RBI	AVG	Age	PEVA-B

	Washington Olympics									
1	Force	Davy	1871	WS3	NA	0	29	0.278	22	**13.071**
2	Waterman	Fred	1871	WS3	NA	0	17	0.316	26	**13.059**
3	Mills	Everett	1871	WS3	NA	1	24	0.274	26	**7.910**
4	Allison	Doug	1871	WS3	NA	2	27	0.331	25	**7.859**
5	Hall	George	1871	WS3	NA	2	17	0.294	22	**7.458**

Top Position Players By Franchise (1871-2010)

Rank	Name	First	Year	Team	Lg	HR	RBI	AVG	Age	PEVA-B
	Washington Nationals (National Association)									
1	Hines	Paul	1872	WS4	NA	0	5	0.245	17	**0.200**
2	Bielaski	Oscar	1872	WS4	NA	0	3	0.196	25	**0.200**
3	Coughlin	Dennis	1872	WS4	NA	0	7	0.324	NA	**0.200**
4	Studley	Seem	1872	WS4	NA	0	2	0.095	NA	**0.200**
5	Hollingshead	Holly	1872	WS4	NA	0	6	0.341	19	**0.200**

Top Position Players By Franchise (1871-2010)

Rank	Name	First	Year	Team	Lg	HR	RBI	AVG	Age	PEVA-B
	Washington Blue Legs									
1	Hines	Paul	1873	WS5	NA	1	29	0.331	18	**3.798**
2	Beals	Tommy	1873	WS5	NA	0	24	0.272	23	**2.452**
3	Bielaski	Oscar	1873	WS5	NA	0	23	0.283	26	**2.134**
4	White	Warren	1873	WS5	NA	0	21	0.269	29	**1.811**
5	Glenn	John	1873	WS5	NA	1	21	0.265	23	**1.739**

Top Position Players By Franchise (1871-2010)

Rank	Name	First	Year	Team	Lg	HR	RBI	AVG	Age	PEVA-B
	Washington Nationals (National Association)									
1	Allison	Art	1875	WS6	NA	0	3	0.214	26	**3.744**
2	Hollingshead	Holly	1875	WS6	NA	0	5	0.247	22	**0.719**
3	Stearns	Bill	1875	WS6	NA	0	7	0.256	22	**0.461**
4	Dailey	John	1875	WS6	NA	0	13	0.182	22	**0.445**
5	Ressler	Larry	1875	WS6	NA	0	5	0.194	27	**0.254**

Top Position Players By Franchise (1871-2010)

Rank	Name	First	Year	Team	Lg	HR	RBI	AVG	Age	PEVA-B
	Washington Nationals (American Association)									
1	Fennelly	Frank	1884	WS7	AA	2	0	0.292	24	**16.134**
2	Olin	Frank	1884	WS7	AA	0	0	0.386	24	**2.457**
3	Humphries	John	1884	WS7	AA	0	0	0.176	23	**1.958**
4	Hawkes	Thorny	1884	WS7	AA	0	0	0.278	32	**1.209**
5	Gladman	Buck	1884	WS7	AA	1	0	0.156	21	**0.999**

Top Position Players By Franchise (1871-2010)

Rank	Name	First	Year	Team	Lg	HR	RBI	AVG	Age	PEVA-B
	Washington Nationals (National League)									
1	Hoy	Dummy	1888	WS8	NL	2	29	0.274	26	**15.180**
2	Hines	Paul	1886	WS8	NL	9	56	0.312	31	**12.931**
3	Wilmot	Walt	1889	WS8	NL	9	57	0.289	26	**12.119**
4	Hines	Paul	1887	WS8	NL	10	72	0.308	32	**9.751**
5	Hoy	Dummy	1889	WS8	NL	0	39	0.274	27	**9.367**

Top Position Players By Franchise (1871-2010)

Rank	Name	First	Year	Team	Lg	HR	RBI	AVG	Age	PEVA-B
	Washington Statesmen/Senators									
1	Joyce	Bill	1896	WSN	NL	8	51	0.313	31	**26.720**
2	Freeman	Buck	1899	WSN	NL	25	122	0.318	28	**20.576**

3	Joyce	Bill	1895	WSN	NL	17	95	0.312	30	**19.182**
4	Selbach	Kip	1897	WSN	NL	5	59	0.313	25	**17.820**
5	McGuire	Deacon	1895	WSN	NL	10	97	0.336	32	**17.188**
6	Anderson	John	1898	WSN	NL	9	71	0.305	25	**16.096**
7	Hoy	Dummy	1892	WSN	NL	3	75	0.280	30	**15.479**
8	Joyce	Bill	1894	WSN	NL	17	89	0.355	29	**15.302**
9	Selbach	Kip	1896	WSN	NL	5	100	0.304	24	**15.214**
10	Selbach	Kip	1898	WSN	NL	3	60	0.303	26	**14.725**
11	Selbach	Kip	1895	WSN	NL	6	55	0.322	23	**13.602**
12	DeMontrevill	Gene	1896	WSN	NL	8	77	0.343	23	**12.897**
13	McGuire	Deacon	1891	WS9	AA	3	66	0.303	28	**11.914**
14	DeMontrevill	Gene	1897	WSN	NL	3	93	0.341	24	**11.581**
15	McGann	Dan	1899	WSN	NL	5	58	0.343	28	**11.364**
16	Abbey	Charlie	1894	WSN	NL	7	101	0.314	28	**11.185**

Note: Previous Names of Franchise: Washington Statesmen (WS9) 1899; Washington Senators (WSN) 1892-1899

Top Position Players By Franchise (1871-2010)

Rank	Name	First	Year	Team	Lg	HR	RBI	AVG	Age	PEVA-B
	Washington Nationals (Union League)									
1	Moore	Harry	1884	WSU	UA	1		0.336	NA	**14.018**
2	McCormick	Jerry	1884	WSU	UA	0		0.217	NA	**5.672**
3	Baker	Phil	1884	WSU	UA	1		0.288	28	**5.563**
4	Evers	Tom	1884	WSU	UA	0		0.232	32	**3.744**
5	Wise	Bill	1884	WSU	UA	2		0.233	23	**2.817**

Note1: For players with more than one team during the year, actual stats reflect stint with team only, PEVA values reflect Total Year.
Note2: Franchises with less than 5 players above 10.000 PEVA are listed with the five highest players.
Note3: Age = Player age at end of year.

All Franchises (Current and Past) - Top Pitching Seasons

Stat Geek Baseball Franchise Players Best Years

Top 20 Pitchers By Franchise (1871-2010)

Rank	Name	First	Year	Team	Lg	W	L	SV	IP	ERA	Age	PEVA-P
	Altoona Mountain City											
1	Murphy	John	1884	ALT	UA	5	6	0	111.7	3.87	NA	**1.076**
2	Brown	Jim	1884	ALT	UA	1	9	0	74.0	5.35	24	**0.944**
3	Leary	Jack	1884	ALT	UA	0	3	0	24.0	5.25	27	**0.680**
4	Smith	Germany	1884	ALT	UA	0	0	0	1.0	9.00	21	**0.200**
5	Connors	Joe	1884	ALT	UA	0	1	0	9.0	7.00	NA	**0.200**

Rank	Name	First	Year	Team	Lg	W	L	SV	IP	ERA	Age	PEVA-P
	Los Angeles Angels of Anaheim											
1	Chance	Dean	1964	LAA	AL	20	9	4	278.3	1.65	23	**29.474**
2	Tanana	Frank	1976	CAL	AL	19	10	0	288.3	2.43	23	**28.912**
3	Lackey	John	2007	LAA	AL	19	9	0	224.0	3.01	29	**28.594**
4	Langston	Mark	1993	CAL	AL	16	11	0	256.3	3.20	33	**23.167**
5	Langston	Mark	1991	CAL	AL	19	8	0	246.3	3.00	31	**23.149**
6	Wright	Clyde	1970	CAL	AL	22	12	0	260.7	2.83	29	**22.880**
7	Santana	Ervin	2008	LAA	AL	16	7	0	219.0	3.49	25	**22.414**
8	Ryan	Nolan	1974	CAL	AL	22	16	0	332.7	2.89	27	**21.613**
9	Witt	Mike	1986	CAL	AL	18	10	0	269.0	2.84	26	**21.604**
10	Weaver	Jered	2010	LAA	AL	13	12	0	224.3	3.01	28	**21.199**
11	Ryan	Nolan	1977	CAL	AL	19	16	0	299.0	2.77	30	**20.766**
12	Abbott	Jim	1991	CAL	AL	18	11	0	243.0	2.89	24	**20.695**

Rank	Name	First	Year	Team	Lg	W	L	SV	IP	ERA	Age	PEVA-P
13	Blyleven	Bert	1989	CAL	AL	17	5	0	241.0	2.73	38	**20.160**
14	Tanana	Frank	1977	CAL	AL	15	9	0	241.3	2.54	24	**19.687**
15	Finley	Chuck	1993	CAL	AL	16	14	0	251.3	3.15	31	**19.355**
16	Escobar	Kelvim	2007	LAA	AL	18	7	0	195.7	3.40	31	**18.934**
17	Colon	Bartolo	2005	LAA	AL	21	8	0	222.7	3.48	32	**18.290**
18	Lackey	John	2006	LAA	AL	13	11	0	217.7	3.56	28	**17.660**
19	Ryan	Nolan	1973	CAL	AL	21	16	1	326.0	2.87	26	**17.473**
20	Harvey	Bryan	1991	CAL	AL	2	4	46	78.7	1.60	28	**17.346**

Note: Previous Names of Franchise: Los Angeles Angels (LAA), California Angels (CAL), Anaheim Angels (ANA)

Top Pitchers By Franchise (1871-2010)

Rank	Name	First	Year	Team	Lg	W	L	SV	IP	ERA	Age	PEVA-P
	Arizona Diamondbacks											
1	Johnson	Randy	2001	ARI	NL	21	6	0	249.7	2.49	38	**52.009**
2	Johnson	Randy	2004	ARI	NL	16	14	0	245.7	2.60	41	**44.961**
3	Johnson	Randy	2002	ARI	NL	24	5	0	260.0	2.32	39	**43.254**
4	Schilling	Curt	2001	ARI	NL	22	6	0	256.7	2.98	35	**38.613**
5	Schilling	Curt	2002	ARI	NL	23	7	0	259.3	3.23	36	**37.153**
6	Webb	Brandon	2006	ARI	NL	16	8	0	235.0	3.10	27	**36.142**
7	Webb	Brandon	2007	ARI	NL	18	10	0	236.3	3.01	28	**35.177**
8	Johnson	Randy	1999	ARI	NL	17	9	0	271.7	2.48	36	**34.970**
9	Webb	Brandon	2008	ARI	NL	22	7	0	226.7	3.30	29	**31.968**
10	Johnson	Randy	2000	ARI	NL	19	7	0	248.7	2.64	37	**31.544**
11	Haren	Danny	2009	ARI	NL	14	10	0	229.3	3.14	29	**29.954**
12	Haren	Danny	2008	ARI	NL	16	8	0	216.0	3.33	28	**29.540**
13	Webb	Brandon	2005	ARI	NL	14	12	0	229.0	3.54	26	**14.087**
14	Valverde	Jose	2007	ARI	NL	1	4	47	64.3	2.66	28	**13.931**
15	Webb	Brandon	2003	ARI	NL	10	9	0	180.7	2.84	24	**13.770**
16	Kim	Byung-Hyu	2002	ARI	NL	8	3	36	84.0	2.04	23	**13.742**
17	Schilling	Curt	2003	ARI	NL	8	9	0	168.0	2.95	37	**13.437**
18	Daal	Omar	1999	ARI	NL	16	9	0	214.7	3.65	27	**13.208**
19	Benes	Andy	1998	ARI	NL	14	13	0	231.3	3.97	31	**12.773**
20	Davis	Doug	2007	ARI	NL	13	12	0	192.7	4.25	32	**11.480**

Top Pitchers By Franchise (1871-2010)

Rank	Name	First	Year	Team	Lg	W	L	SV	IP	ERA	Age	PEVA-P
	Atlanta Braves											
1	Maddux	Greg	1994	ATL	NL	16	6	0	202.0	1.56	28	**57.974**
2	Maddux	Greg	1995	ATL	NL	19	2	0	209.7	1.63	29	**52.118**
3	Maddux	Greg	1993	ATL	NL	20	10	0	267.0	2.36	27	**50.920**
4	Nichols	Kid	1897	BSN	NL	31	11	3	368.0	2.64	28	**47.498**
5	Maddux	Greg	1997	ATL	NL	19	4	0	232.7	2.20	31	**47.132**
6	Maddux	Greg	1998	ATL	NL	18	9	0	251.0	2.22	32	**46.792**
7	Bond	Tommy	1879	BSN	NL	43	19	0	555.3	1.96	23	**41.857**
8	Smoltz	John	1996	ATL	NL	24	8	0	253.7	2.94	29	**41.196**
9	Spahn	Warren	1947	BSN	NL	21	10	3	289.7	2.33	26	**39.610**
10	Nichols	Kid	1898	BSN	NL	31	12	4	388.0	2.13	29	**38.155**
11	Willis	Vic	1899	BSN	NL	27	8	2	342.7	2.50	23	**37.022**
12	Maddux	Greg	2001	ATL	NL	17	11	0	233.0	3.05	35	**35.904**
13	Maddux	Greg	1996	ATL	NL	15	11	0	245.0	2.72	30	**35.136**
14	Glavine	Tom	1991	ATL	NL	20	11	0	246.7	2.55	25	**35.128**
15	Clarkson	John	1889	BSN	NL	49	19	1	620.0	2.73	28	**32.959**
16	Spahn	Warren	1953	ML1	NL	23	7	3	265.7	2.10	32	**32.929**
17	Nichols	Kid	1893	BSN	NL	34	14	1	425.0	3.52	24	**32.804**
18	Bond	Tommy	1877	BSN	NL	40	17	0	521.0	2.11	21	**32.408**
19	Maddux	Greg	2000	ATL	NL	19	9	0	249.3	3.00	34	**32.027**

20	Sain	Johnny	1948	BSN	NL	24	15	1	314.7	2.60	31	**31.497**

Note: Previous Names of Franchise:
Boston Red Caps, Beaneaters, Doves, Braves (BSN) 1876-1952; Milwaukee Braves (ML1) 1953-1965

Top Pitchers By Franchise (1871-2010)

Rank	Name	First	Year	Team	Lg	W	L	SV	IP	ERA	Age	PEVA-P
	Baltimore Orioles											
1	Palmer	Jim	1975	BAL	AL	23	11	1	323.0	2.09	30	**36.388**
2	Shocker	Urban	1922	SLA	AL	24	17	3	348.0	2.97	32	**36.261**
3	Palmer	Jim	1976	BAL	AL	22	13	0	315.0	2.51	31	**32.669**
4	Cuellar	Mike	1969	BAL	AL	23	11	0	290.7	2.38	32	**32.370**
5	Boddicker	Mike	1984	BAL	AL	20	11	0	261.3	2.79	27	**29.576**
6	Palmer	Jim	1970	BAL	AL	20	10	0	305.0	2.71	25	**26.828**
7	Mussina	Mike	1992	BAL	AL	18	5	0	241.0	2.54	24	**25.430**
8	Mussina	Mike	1995	BAL	AL	19	9	0	221.7	3.29	27	**25.181**
9	McNally	Dave	1970	BAL	AL	24	9	0	296.0	3.22	28	**23.420**
10	Palmer	Jim	1977	BAL	AL	20	11	0	319.0	2.91	32	**22.845**
11	Cuellar	Mike	1970	BAL	AL	24	8	0	297.7	3.48	33	**22.342**
12	Flanagan	Mike	1979	BAL	AL	23	9	0	265.7	3.08	28	**21.747**
13	Mussina	Mike	1994	BAL	AL	16	5	0	176.3	3.06	26	**21.206**
14	Shocker	Urban	1921	SLA	AL	27	12	4	326.7	3.55	31	**20.956**
15	Palmer	Jim	1973	BAL	AL	22	9	1	296.3	2.40	28	**20.679**
16	McNally	Dave	1968	BAL	AL	22	10	0	273.0	1.95	26	**20.047**
17	O'Dell	Billy	1958	BAL	AL	14	11	8	221.3	2.97	26	**19.683**
18	Stewart	Lefty	1930	SLA	AL	20	12	0	271.0	3.45	30	**19.662**
19	Gray	Dolly	1929	SLA	AL	18	15	1	305.0	3.72	32	**19.648**
20	Potter	Nels	1945	SLA	AL	15	11	0	255.3	2.47	34	**19.161**

Note: Previous Names of Franchise: Milwaukee Brewers (MLA) 1901; St. Louis Browns (SLA) 1902-1953

Top Pitchers By Franchise (1871-2010)

Rank	Name	First	Year	Team	Lg	W	L	SV	IP	ERA	Age	PEVA-P
	Buffalo Bisons											
1	Galvin	Pud	1884	BFN	NL	46	22	0	636.3	1.99	28	**25.160**
2	Galvin	Pud	1883	BFN	NL	46	29	0	656.3	2.72	27	**23.316**
3	Galvin	Pud	1879	BFN	NL	37	27	0	593.0	2.28	23	**23.153**
4	Galvin	Pud	1881	BFN	NL	28	24	0	474.0	2.37	25	**18.358**
5	Galvin	Pud	1882	BFN	NL	28	23	0	445.3	3.17	26	**6.267**

Top Pitchers By Franchise (1871-2010)

Rank	Name	First	Year	Team	Lg	W	L	SV	IP	ERA	Age	PEVA-P
	Buffalo Bisons (Pacific Coast League)											
1	Cunningham	Bert	1890	BFP	PL	9	15	0	211.0	5.84	25	**2.308**
2	Haddock	George	1890	BFP	PL	9	26	0	290.7	5.76	24	**1.851**
3	Keefe	George	1890	BFP	PL	6	16	0	196.0	6.52	23	**0.966**
4	Twitchell	Larry	1890	BFP	PL	5	7	0	104.3	4.57	26	**0.600**
5	Stafford	General	1890	BFP	PL	3	9	0	98.0	5.14	22	**0.578**

Top Pitchers By Franchise (1871-2010)

Rank	Name	First	Year	Team	Lg	W	L	SV	IP	ERA	Age	PEVA-P
	Baltimore Canaries											
1	Cummings	Candy	1873	BL1	NA	28	14	0	382.0	2.66	25	**11.758**
2	Mathews	Bobby	1872	BL1	NA	25	18	0	405.3	3.15	21	**11.404**
3	Fisher	Cherokee	1872	BL1	NA	10	1	1	110.3	2.53	27	**5.924**
4	Brainard	Asa	1874	BL1	NA	5	22	0	239.0	4.93	33	**0.944**
5	Brainard	Asa	1873	BL1	NA	5	7	0	108.7	4.14	32	**0.909**

Top Pitchers By Franchise (1871-2010)

Rank	Name	First	Year	Team	Lg	W	L	SV	IP	ERA	Age	PEVA-P
	Baltimore Orioles (American Association/National League)											
1	McGinnity	Joe	1899	BLN	NL	28	16	2	366.3	2.68	28	**34.184**
2	Kilroy	Matt	1887	BL2	AA	46	19	0	589.3	3.07	21	**29.854**
3	McJames	Doc	1898	BLN	NL	27	15	0	374.0	2.36	24	**23.828**
4	Kitson	Frank	1899	BLN	NL	22	16	0	327.7	2.77	30	**23.707**
5	McMahon	Sadie	1891	BL3	AA	35	24	1	503.0	2.81	24	**22.650**
6	Corbett	Joe	1897	BLN	NL	24	8	0	313.0	3.11	22	**20.492**
7	Kilroy	Matt	1889	BL2	AA	29	25	0	480.7	2.85	23	**19.075**
8	Hoffer	Bill	1895	BLN	NL	31	6	0	314.0	3.21	25	**17.949**
9	Hoffer	Bill	1896	BLN	NL	25	7	0	309.0	3.38	26	**16.948**
10	Maul	Al	1898	BLN	NL	20	7	0	239.7	2.10	33	**13.619**
11	Kilroy	Matt	1886	BL2	AA	29	34	0	583.0	3.37	20	**13.318**
12	Foreman	Frank	1889	BL2	AA	23	21	0	414.0	3.52	26	**11.443**
13	McMahon	Sadie	1894	BLN	NL	25	8	0	275.7	4.21	27	**10.426**
14	Nops	Jerry	1897	BLN	NL	20	6	0	220.7	2.81	22	**10.388**

Note: Previous Names of Franchise: Baltimore Orioles (BL2) 1882-1889; (BL3) 1890-1891; (BLN) 1892-1899

Top Pitchers By Franchise (1871-2010)

Rank	Name	First	Year	Team	Lg	W	L	SV	IP	ERA	Age	PEVA-P
	Baltimore Marylands											
1	Stratton	Ed	1873	BL4	NA	0	3	0	27.0	8.33	NA	**0.377**
2	Selman	Frank	1873	BL4	NA	0	1	0	9.0	8.00	21	**0.200**
3	French	Bill	1873	BL4	NA	0	1	0	9.0	12.00	NA	**0.200**
4	McDoolan		1873	BL4	NA	0	1	0	9.0	3.00	NA	**0.200**

Top Pitchers By Franchise (1871-2010)

Rank	Name	First	Year	Team	Lg	W	L	SV	IP	ERA	Age	PEVA-P
	Baltimore Terrapins											
1	Quinn	Jack	1914	BLF	FL	26	14	1	342.7	2.60	31	**18.114**
2	Suggs	George	1914	BLF	FL	24	14	4	319.3	2.90	32	**10.789**
3	Quinn	Jack	1915	BLF	FL	9	22	1	273.7	3.45	32	**3.286**
4	Wilhelm	Kaiser	1914	BLF	FL	12	17	5	243.7	4.03	40	**3.270**
5	Suggs	George	1915	BLF	FL	11	17	3	232.7	4.14	33	**2.635**

Top Pitchers By Franchise (1871-2010)

Rank	Name	First	Year	Team	Lg	W	L	SV	IP	ERA	Age	PEVA-P
	Baltimore Monumentals											
1	Sweeney	Bill	1884	BLU	UA	40	21	0	538.0	2.59	NA	**12.261**
2	Atkinson	Al	1884	BLU	UA	3	5	0	69.7	2.33	23	**4.069**
3	Lee	Tom	1884	BLU	UA	5	8	0	122.0	3.39	22	**1.034**
4	Robinson	Yank	1884	BLU	UA	3	3	0	75.0	3.48	25	**0.874**
5	Ryan	John	1884	BLU	UA	3	2	0	51.0	3.35	NA	**0.713**

Top Pitchers By Franchise (1871-2010)

Rank	Name	First	Year	Team	Lg	W	L	SV	IP	ERA	Age	PEVA-P
	Boston Red Sox											
1	Young	Cy	1901	BOS	AL	33	10	0	371.3	1.62	34	**48.628**
2	Martinez	Pedro	1999	BOS	AL	23	4	0	213.3	2.07	28	**45.790**
3	Clemens	Roger	1987	BOS	AL	20	9	0	281.7	2.97	25	**43.210**
4	Clemens	Roger	1991	BOS	AL	18	10	0	271.3	2.62	29	**43.004**
5	Clemens	Roger	1990	BOS	AL	21	6	0	228.3	1.93	28	**41.426**
6	Martinez	Pedro	2000	BOS	AL	18	6	0	217.0	1.74	29	**40.251**
7	Wood	Joe	1912	BOS	AL	34	5	1	344.0	1.91	23	**38.082**

Rank	Name	First	Year	Team	Lg	W	L	SV	IP	ERA	Age	PEVA-P
8	Martinez	Pedro	2002	BOS	AL	20	4	0	199.3	2.26	31	**37.759**
9	Clemens	Roger	1986	BOS	AL	24	4	0	254.0	2.48	24	**35.030**
10	Clemens	Roger	1992	BOS	AL	18	11	0	246.7	2.41	30	**34.707**
11	Schilling	Curt	2004	BOS	AL	21	6	0	226.7	3.26	38	**31.192**
12	Lowe	Derek	2002	BOS	AL	21	8	0	219.7	2.58	29	**31.179**
13	Beckett	Josh	2007	BOS	AL	20	7	0	200.7	3.27	27	**30.976**
14	Clemens	Roger	1988	BOS	AL	18	12	0	264.0	2.93	26	**28.435**
15	Young	Cy	1902	BOS	AL	32	11	0	384.7	2.15	35	**28.273**
16	Sullivan	Frank	1957	BOS	AL	14	11	0	240.7	2.73	27	**28.105**
17	Wood	Joe	1911	BOS	AL	23	17	3	275.7	2.02	22	**28.038**
18	Parnell	Mel	1949	BOS	AL	25	7	2	295.3	2.77	27	**27.333**
19	Martinez	Pedro	2003	BOS	AL	14	4	0	186.7	2.22	32	**27.294**
20	Grove	Lefty	1935	BOS	AL	20	12	1	273.0	2.70	35	**26.486**

Note: Previous Names of Franchise: Boston Americans, Somersets, Pilgrims

Top Pitchers By Franchise (1871-2010)

Rank	Name	First	Year	Team	Lg	W	L	SV	IP	ERA	Age	PEVA-P
	Brooklyn Eckfords											
1	Zettlein	George	1872	BR1	NA	1	8	0	75.3	2.99	28	**6.833**
2	Martin	Phonney	1872	BR1	NA	2	7	0	85.0	4.45	27	**0.686**
3	McDermott	Joe	1872	BR1	NA	0	7	0	63.0	8.29	NA	**0.444**
4	Malone	Martin	1872	BR1	NA	0	3	0	27.0	11.33	NA	**0.377**
5	O'Rourke		1872	BR1	NA	0	1	0	9.0	6.00	NA	**0.200**

Top Pitchers By Franchise (1871-2010)

Rank	Name	First	Year	Team	Lg	W	L	SV	IP	ERA	Age	PEVA-P
	Brooklyn Atlantics											
1	Britt	Jim	1873	BR2	NA	17	36	0	480.7	3.89	17	**7.291**
2	Bond	Tommy	1874	BR2	NA	22	32	0	497.0	3.19	18	**4.309**
3	Britt	Jim	1872	BR2	NA	9	28	0	336.0	5.06	16	**2.874**
4	Cassidy	John	1875	BR2	NA	1	20	0	214.7	3.98	18	**1.076**
5	Clinton	Jim	1875	BR2	NA	1	14	0	123.0	3.22	25	**1.039**

Top Pitchers By Franchise (1871-2010)

Rank	Name	First	Year	Team	Lg	W	L	SV	IP	ERA	Age	PEVA-P
	Brooklyn Gladiators											
1	Daily	Ed	1890	BR4	AA	10	15	0	235.7	4.05	28	**5.199**
2	McCullough	Charlie	1890	BR4	AA	4	21	0	215.7	4.59	24	**1.280**
3	Mattimore	Mike	1890	BR4	AA	6	13	0	178.3	4.54	32	**0.955**
4	Toole	Steve	1890	BR4	AA	2	4	0	53.3	4.05	31	**0.831**
5	Murphy	Bob	1890	BR4	AA	3	9	0	96.0	5.72	24	**0.639**

Top Pitchers By Franchise (1871-2010)

Rank	Name	First	Year	Team	Lg	W	L	SV	IP	ERA	Age	PEVA-P
	Brooklyn Tip-Tops											
1	Seaton	Tom	1914	BRF	FL	25	14	2	302.7	3.03	27	**9.669**
2	Lafitte	Ed	1914	BRF	FL	18	15	2	290.7	2.63	28	**7.529**
3	Seaton	Tom	1915	BRF	FL	12	11	3	189.3	4.56	28	**4.319**
4	Wiltse	Hooks	1915	BRF	FL	3	5	5	59.3	2.28	36	**3.540**
5	Falkenberg	Cy	1915	BRF	FL	3	3	0	48.0	1.50	35	**3.257**

Top Pitchers By Franchise (1871-2010)

Rank	Name	First	Year	Team	Lg	W	L	SV	IP	ERA	Age	PEVA-P
	Brooklyn Ward's Wonders											
1	Weyhing	Gus	1890	BRP	PL	30	16	0	390.0	3.60	24	**7.634**
2	Sowders	John	1890	BRP	PL	19	16	0	309.0	3.82	24	**4.114**

3	Van Haltren	George	1890	BRP	PL	15	10	2	223.0	4.28	24	**1.564**
4	Murphy	Con	1890	BRP	PL	4	10	2	139.0	4.79	27	**0.907**
5	Hemming	George	1890	BRP	PL	8	4	3	123.0	3.80	22	**0.891**

Top Pitchers By Franchise (1871-2010)

Rank	Name	First	Year	Team	Lg	W	L	SV	IP	ERA	Age	PEVA-P
	Boston Red Stockings											
1	Spalding	Al	1872	BS1	NA	38	8	0	404.7	1.98	22	**31.314**
2	Spalding	Al	1874	BS1	NA	52	16	0	616.3	2.35	24	**30.873**
3	Spalding	Al	1871	BS1	NA	19	10	0	257.3	3.36	21	**28.490**
4	Spalding	Al	1875	BS1	NA	55	5	8	575.0	1.52	25	**27.761**
5	Spalding	Al	1873	BS1	NA	41	14	2	497.7	2.46	23	**26.662**

Top Pitchers By Franchise (1871-2010)

Rank	Name	First	Year	Team	Lg	W	L	SV	IP	ERA	Age	PEVA-P
	Boston Reds (Pacific Coast League/American Association)											
1	Haddock	George	1891	BS2	AA	34	11	1	379.7	2.49	25	**17.269**
2	Buffinton	Charlie	1891	BS2	AA	29	9	1	363.7	2.55	30	**16.050**
3	Radbourn	Charley	1890	BSP	PL	27	12	0	343.0	3.31	36	**7.308**
4	Gumbert	Ad	1890	BSP	PL	23	12	0	277.3	3.96	22	**3.317**
5	Daley	Bill	1890	BSP	PL	18	7	2	235.0	3.60	22	**2.526**

Top Pitchers By Franchise (1871-2010)

Rank	Name	First	Year	Team	Lg	W	L	SV	IP	ERA	Age	PEVA-P
	Boston Reds (UA)											
1	Shaw	Dupee	1884	BSU	UA	21	15	0	315.7	1.77	25	**12.916**
2	Burke	James	1884	BSU	UA	19	15	0	322.0	2.85	1884	**3.920**
3	Tenney	Fred	1884	BSU	UA	3	1	0	35.0	2.31	25	**2.110**
4	Bond	Tommy	1884	BSU	UA	13	9	0	189.0	3.00	28	**1.831**
5	McCarthy	Tommy	1884	BSU	UA	0	7	0	56.0	4.82	21	**0.557**

Top Pitchers By Franchise (1871-2010)

Rank	Name	First	Year	Team	Lg	W	L	SV	IP	ERA	Age	PEVA-P
	Buffalo Buffeds & Blues											
1	Ford	Russ	1914	BUF	FL	21	6	6	247.3	1.82	31	**13.668**
2	Bedient	Hugh	1915	BUF	FL	16	18	10	269.3	3.17	26	**8.280**
3	Schulz	Al	1915	BUF	FL	21	14	0	309.7	3.08	26	**7.604**
4	Anderson	Fred	1915	BUF	FL	19	13	0	240.0	2.51	30	**6.632**
5	Krapp	Gene	1914	BUF	FL	16	14	0	252.7	2.49	27	**5.677**

Top Pitchers By Franchise (1871-2010)

Rank	Name	First	Year	Team	Lg	W	L	SV	IP	ERA	Age	PEVA-P
	Chicago White Stockings (National Association)											
1	Zettlein	George	1871	CH1	NA	18	9	0	240.7	2.73	27	**32.500**
2	Zettlein	George	1875	CH2	NA	17	14	0	282.0	1.82	31	**11.035**
3	Zettlein	George	1874	CH2	NA	27	30	0	515.7	3.07	30	**6.230**
4	Golden	Mike	1875	CH2	NA	6	7	0	119.0	2.87	24	**1.716**
5	Devlin	Jim	1875	CH2	NA	7	16	0	224.0	2.89	26	**1.544**

Top Pitchers By Franchise (1871-2010)

Rank	Name	First	Year	Team	Lg	W	L	SV	IP	ERA	Age	PEVA-P
	Chicago White Sox											
1	Walsh	Ed	1908	CHA	AL	40	15	6	464.0	1.42	27	**46.136**
2	Faber	Red	1921	CHA	AL	25	15	1	330.7	2.48	33	**37.013**
3	Walsh	Ed	1911	CHA	AL	27	18	4	368.7	2.22	30	**36.402**
4	Faber	Red	1922	CHA	AL	21	17	2	352.0	2.81	34	**36.363**
5	Cicotte	Eddie	1917	CHA	AL	28	12	4	346.7	1.53	33	**35.909**

Rank	Name	First	Year	Team	Lg	W	L	SV	IP	ERA	Age	PEVA-P
6	Wood	Wilbur	1971	CHA	AL	22	13	1	334.0	1.91	30	**35.043**
7	Walsh	Ed	1910	CHA	AL	18	20	5	369.7	1.27	29	**34.646**
8	Cicotte	Eddie	1919	CHA	AL	29	7	1	306.7	1.82	35	**31.775**
9	Walsh	Ed	1907	CHA	AL	24	18	4	422.3	1.60	26	**28.794**
10	Walsh	Ed	1912	CHA	AL	27	17	10	393.0	2.15	31	**28.368**
11	Horlen	Joe	1967	CHA	AL	19	7	0	258.0	2.06	30	**28.268**
12	Lyons	Ted	1927	CHA	AL	22	14	2	307.7	2.84	27	**27.787**
13	Smith	Frank	1909	CHA	AL	25	17	1	365.0	1.80	30	**27.566**
14	Loaiza	Esteban	2003	CHA	AL	21	9	0	226.3	2.90	32	**27.439**
15	Lee	Thornton	1941	CHA	AL	22	11	1	300.3	2.37	35	**26.776**
16	Wood	Wilbur	1972	CHA	AL	24	17	0	376.7	2.51	31	**25.752**
17	Thomas	Tommy	1927	CHA	AL	19	16	1	307.7	2.98	28	**24.131**
18	Pierce	Billy	1957	CHA	AL	20	12	2	257.0	3.26	30	**23.698**
19	Hoyt	La Marr	1983	CHA	AL	24	10	0	260.7	3.66	28	**23.565**
20	Buehrle	Mark	2001	CHA	AL	16	8	0	221.3	3.29	22	**22.701**

Top Pitchers By Franchise (1871-2010)

Rank	Name	First	Year	Team	Lg	W	L	SV	IP	ERA	Age	PEVA-P
	Chicago Chi-Feds											
1	Hendrix	Claude	1914	CHF	FL	29	10	5	362.0	1.69	25	**27.456**
2	McConnell	George	1915	CHF	FL	25	10	1	303.0	2.20	38	**11.452**
3	Brown	Mordecai	1915	CHF	FL	17	8	4	236.3	2.09	39	**7.083**
4	Prendergast	Mike	1915	CHF	FL	14	12	0	253.7	2.48	27	**5.064**
5	Hendrix	Claude	1915	CHF	FL	16	15	4	285.0	3.00	26	**4.944**

Top Pitchers By Franchise (1871-2010)

Rank	Name	First	Year	Team	Lg	W	L	SV	IP	ERA	Age	PEVA-P
	Chicago Cubs											
1	Maddux	Greg	1992	CHN	NL	20	11	0	268.0	2.18	26	**49.969**
2	Clarkson	John	1885	CHN	NL	53	16	0	623.0	1.85	24	**42.218**
3	Brown	Mordecai	1909	CHN	NL	27	9	7	342.7	1.31	33	**40.455**
4	Alexander	Pete	1920	CHN	NL	27	14	5	363.3	1.91	33	**40.230**
5	Passeau	Claude	1940	CHN	NL	20	13	5	280.7	2.50	31	**35.806**
6	Lee	Bill	1938	CHN	NL	22	9	2	291.0	2.66	29	**34.833**
7	Jenkins	Fergie	1970	CHN	NL	22	16	0	313.0	3.39	28	**34.695**
8	Hutchison	Bill	1890	CHN	NL	42	25	2	603.0	2.70	31	**33.191**
9	Clarkson	John	1887	CHN	NL	38	21	0	523.0	3.08	26	**32.014**
10	Hutchison	Bill	1891	CHN	NL	44	19	1	561.0	2.81	32	**31.786**
11	Brown	Mordecai	1906	CHN	NL	26	6	3	277.3	1.04	30	**30.326**
12	Ellsworth	Dick	1963	CHN	NL	22	10	0	290.7	2.11	23	**30.293**
13	Hands	Bill	1969	CHN	NL	20	14	0	300.0	2.49	29	**29.345**
14	Vaughn	Hippo	1918	CHN	NL	22	10	0	290.3	1.74	30	**28.883**
15	Hutchison	Bill	1892	CHN	NL	36	36	1	622.0	2.76	33	**28.107**
16	Jenkins	Fergie	1971	CHN	NL	24	13	0	325.0	2.77	29	**27.507**
17	Prior	Mark	2003	CHN	NL	18	6	0	211.3	2.43	23	**27.373**
18	Griffith	Clark	1898	CHN	NL	24	10	0	325.7	1.88	29	**27.152**
19	Jenkins	Fergie	1969	CHN	NL	21	15	1	311.3	3.21	27	**26.754**
20	Warneke	Lon	1932	CHN	NL	22	6	0	277.0	2.37	23	**26.554**

Note: Previous Names of Franchise: Chicago White Stockings, Colts, Orphans

Top Pitchers By Franchise (1871-2010)

Rank	Name	First	Year	Team	Lg	W	L	SV	IP	ERA	Age	PEVA-P
	Chicago Pirates											
1	King	Silver	1890	CHP	PL	30	22	0	461.0	2.69	22	**18.015**
2	Baldwin	Mark	1890	CHP	PL	34	24	0	501.0	3.31	27	**13.556**
3	Bartson	Charlie	1890	CHP	PL	8	10	1	188.0	4.26	25	**1.003**

Rank	Name	First	Year	Team	Lg	W	L	SV	IP	ERA	Age	PEVA-P
4	Dwyer	Frank	1890	CHP	PL	3	6	1	69.3	6.23	22	**0.543**

Top Pitchers By Franchise (1871-2010)

Rank	Name	First	Year	Team	Lg	W	L	SV	IP	ERA	Age	PEVA-P
	Chicago/Pittsburgh (Union League)											
1	Daily	Hugh	1884	CHU	UA	27	27	0	484.7	2.43	27	**10.851**
2	Atkinson	Al	1884	CHU	UA	6	10	0	140.0	2.76	23	**4.069**
3	Horan	John	1884	CHU	UA	3	6	0	98.0	3.49	21	**0.984**
4	Cady	Charlie	1884	CHU	UA	3	1	0	35.0	2.83	19	**0.933**
5	Foreman	Frank	1884	CHU	UA	1	2	0	18.0	4.00	21	**0.743**

Top Pitchers By Franchise (1871-2010)

Rank	Name	First	Year	Team	Lg	W	L	SV	IP	ERA	Age	PEVA-P
	Cincinnati Reds											
1	Blackwell	Ewell	1947	CIN	NL	22	8	0	273.0	2.47	25	**46.689**
2	Luque	Dolf	1923	CIN	NL	27	8	2	322.0	1.93	33	**42.654**
3	Donohue	Pete	1925	CIN	NL	21	14	2	301.0	3.08	25	**42.024**
4	Walters	Bucky	1939	CIN	NL	27	11	0	319.0	2.29	30	**41.246**
5	Walters	Bucky	1940	CIN	NL	22	10	0	305.0	2.48	31	**36.538**
6	Luque	Dolf	1925	CIN	NL	16	18	0	291.0	2.63	35	**35.862**
7	Rixey	Eppa	1925	CIN	NL	21	11	1	287.3	2.88	34	**34.428**
8	Derringer	Paul	1939	CIN	NL	25	7	0	301.0	2.93	33	**34.402**
9	Derringer	Paul	1938	CIN	NL	21	14	3	307.0	2.93	32	**34.132**
10	Hahn	Noodles	1899	CIN	NL	23	8	0	309.0	2.68	20	**33.291**
11	Rijo	Jose	1993	CIN	NL	14	9	0	257.3	2.48	28	**32.798**
12	Derringer	Paul	1940	CIN	NL	20	12	0	296.7	3.06	34	**31.160**
13	Seaver	Tom	1977	CIN	NL	14	3	0	165.3	2.34	33	**30.175**
14	Jackson	Danny	1988	CIN	NL	23	8	0	260.7	2.73	26	**29.265**
15	Rijo	Jose	1991	CIN	NL	15	6	0	204.3	2.51	26	**27.931**
16	White	Will	1879	CN1	NL	43	31	0	680.0	1.99	25	**27.082**
17	Soto	Mario	1982	CIN	NL	14	13	0	257.7	2.79	26	**27.077**
18	Arroyo	Bronson	2006	CIN	NL	14	11	0	240.7	3.29	29	**27.027**
19	Purkey	Bob	1962	CIN	NL	23	5	0	288.3	2.81	33	**26.852**
20	White	Will	1883	CN2	AA	43	22	0	577.0	2.09	29	**26.681**

Note: Previous Names of Franchise:

Cincinnati Reds (CN1) 1876-1880; Cincinnati Red Stockings (CN2) AA 1882-1889; Cincinnati Reds/Red Legs 1890-present

Top Pitchers By Franchise (1871-2010)

Rank	Name	First	Year	Team	Lg	W	L	SV	IP	ERA	Age	PEVA-P
	Cleveland Forest Citys											
1	Pratt	Al	1871	CL1	NA	10	17	0	224.7	3.77	23	**6.902**
2	Pratt	Al	1872	CL1	NA	2	9	0	104.7	4.39	24	**0.570**
3	Wolters	Rynie	1872	CL1	NA	3	6	0	76.3	5.19	30	**0.506**
4	Pabor	Charlie	1871	CL1	NA	0	2	0	29.3	6.75	25	**0.377**
5	Pabor	Charlie	1872	CL1	NA	1	1	0	18.0	3.00	26	**0.200**

Top Pitchers By Franchise (1871-2010)

Rank	Name	First	Year	Team	Lg	W	L	SV	IP	ERA	Age	PEVA-P
	Cleveland Blues											
1	McCormick	Jim	1880	CL2	NL	45	28	0	657.7	1.85	24	**21.965**
2	McCormick	Jim	1882	CL2	NL	36	30	0	595.7	2.37	26	**18.813**
3	McCormick	Jim	1881	CL2	NL	26	30	0	526.0	2.45	25	**17.833**
4	McCormick	Jim	1884	CL2	NL	19	22	0	359.0	2.86	28	**15.537**
5	McCormick	Jim	1883	CL2	NL	28	12	1	342.0	1.84	27	**10.611**

Top Pitchers By Franchise (1871-2010)

Rank	Name	First	Year	Team	Lg	W	L	SV	IP	ERA	Age	PEVA-P
	Cleveland Blues & Spiders											
1	Young	Cy	1893	CL4	NL	34	16	1	422.7	3.36	26	**43.542**
2	Young	Cy	1896	CL4	NL	28	15	3	414.3	3.24	29	**38.520**
3	Young	Cy	1895	CL4	NL	35	10	0	369.7	3.26	28	**32.941**
4	Young	Cy	1892	CL4	NL	36	12	0	453.0	1.93	25	**31.345**
5	Young	Cy	1898	CL4	NL	25	13	0	377.7	2.53	31	**26.356**
6	Young	Cy	1894	CL4	NL	26	21	1	408.7	3.94	27	**26.323**
7	Cuppy	Nig	1896	CL4	NL	25	14	1	358.0	3.12	27	**22.116**
8	Young	Cy	1897	CL4	NL	21	19	0	333.7	3.80	30	**18.713**
9	Cuppy	Nig	1895	CL4	NL	26	14	2	353.0	3.54	26	**16.908**
10	Young	Cy	1891	CL4	NL	27	22	2	423.7	2.85	24	**14.082**
11	Powell	Jack	1898	CL4	NL	23	15	0	342.0	3.00	24	**12.785**
12	Cuppy	Nig	1892	CL4	NL	28	13	1	376.0	2.51	23	**11.417**
13	Clarkson	John	1892	CL4	NL	17	10	1	243.3	2.55	31	**10.721**

Note: Previous Names of Franchise: Cleveland Blues (CL3) AA 1887-1888; Cleveland Spiders (CL4) NL 1889-1899

Top Pitchers By Franchise (1871-2010)

Rank	Name	First	Year	Team	Lg	W	L	SV	IP	ERA	Age	PEVA-P
	Columbus Buckeyes											
1	Morris	Ed	1884	CL5	AA	34	13	0	429.7	2.18	22	**15.285**
2	Mountain	Frank	1883	CL5	AA	26	33	0	503.0	3.60	23	**5.599**
3	Mountain	Frank	1884	CL5	AA	23	17	1	360.7	2.45	24	**5.549**
4	Valentine	John	1883	CL5	AA	2	10	0	102.0	3.53	28	**0.859**
5	Dundon	Ed	1883	CL5	AA	3	16	0	166.7	4.48	24	**0.777**

Top Pitchers By Franchise (1871-2010)

Rank	Name	First	Year	Team	Lg	W	L	SV	IP	ERA	Age	PEVA-P
	Columbus Solons											
1	Knell	Phil	1891	CL6	AA	28	27	0	462.0	2.92	26	**14.684**
2	Gastright	Hank	1890	CL6	AA	30	14	0	401.3	2.94	25	**13.178**
3	Baldwin	Mark	1889	CL6	AA	27	34	1	513.7	3.61	26	**10.311**
4	Knauss	Frank	1890	CL6	AA	17	12	2	275.7	2.81	22	**8.110**
5	Chamberlain	Elton	1890	CL6	AA	12	6	0	175.0	2.21	23	**4.450**

Top Pitchers By Franchise (1871-2010)

Rank	Name	First	Year	Team	Lg	W	L	SV	IP	ERA	Age	PEVA-P
	Cleveland Indians											
1	Lee	Cliff	2008	CLE	AL	22	3	0	223.3	2.54	30	**47.417**
2	Feller	Bob	1940	CLE	AL	27	11	4	320.3	2.61	22	**46.643**
3	Sabathia	C.C.	2007	CLE	AL	19	7	0	241.0	3.21	27	**39.488**
4	Feller	Bob	1946	CLE	AL	26	15	4	371.3	2.18	28	**34.939**
5	Perry	Gaylord	1972	CLE	AL	24	16	1	342.7	1.92	34	**33.015**
6	Perry	Gaylord	1974	CLE	AL	21	13	0	322.3	2.51	36	**30.630**
7	Feller	Bob	1947	CLE	AL	20	11	3	299.0	2.68	29	**30.546**
8	Coveleski	Stan	1918	CLE	AL	22	13	1	311.0	1.82	29	**27.680**
9	Garcia	Mike	1954	CLE	AL	19	8	5	258.7	2.64	31	**27.607**
10	McDowell	Sam	1970	CLE	AL	20	12	0	305.0	2.92	28	**27.561**
11	Wynn	Early	1954	CLE	AL	23	11	2	270.7	2.73	34	**27.349**
12	Uhle	George	1926	CLE	AL	27	11	1	318.3	2.83	28	**27.341**
13	Blyleven	Bert	1984	CLE	AL	19	7	0	245.0	2.87	33	**27.257**
14	Carmona	Fausto	2007	CLE	AL	19	8	0	215.0	3.06	24	**26.017**
15	Joss	Addie	1908	CLE	AL	24	11	2	325.0	1.16	28	**25.793**
16	Garcia	Mike	1952	CLE	AL	22	11	4	292.3	2.37	29	**25.421**
17	Feller	Bob	1939	CLE	AL	24	9	1	296.7	2.85	21	**25.193**

Rank	Name	First	Year	Team	Lg	W	L	SV	IP	ERA	Age	PEVA-P
18	Ferrell	Wes	1930	CLE	AL	25	13	3	296.7	3.31	22	**24.202**
19	Lemon	Bob	1952	CLE	AL	22	11	4	309.7	2.50	32	**23.697**
20	Coveleski	Stan	1920	CLE	AL	24	14	2	315.0	2.49	31	**23.627**

Note: Previous Names of Franchise: Cleveland Blues 1901, Broncos 1902, Naps 1903-1914.

Top Pitchers By Franchise (1871-2010)

Rank	Name	First	Year	Team	Lg	W	L	SV	IP	ERA	Age	PEVA-P
	Cleveland Infants											
1	Gruber	Henry	1890	CLP	PL	22	23	0	383.3	4.27	27	**4.143**
2	Bakely	Jersey	1890	CLP	PL	12	25	0	326.3	4.47	26	**2.445**
3	O'Brien	Darby	1890	CLP	PL	8	16	0	206.3	3.40	23	**1.719**
4	McGill	Willie	1890	CLP	PL	11	9	0	183.7	4.12	17	**1.069**
5	Hemming	George	1890	CLP	PL	0	1	0	21.0	6.86	22	**0.891**

Top Pitchers By Franchise (1871-2010)

Rank	Name	First	Year	Team	Lg	W	L	SV	IP	ERA	Age	PEVA-P
	Cincinnati Kelly's Killers											
1	Crane	Ed	1891	CN3	AA	14	14	0	250.0	2.45	29	**7.790**
2	McGill	Willie	1891	CN3	AA	2	5	0	65.0	4.98	18	**5.515**
3	Dwyer	Frank	1891	CN3	AA	13	19	0	289.0	4.52	23	**4.025**
4	Mains	Willard	1891	CN3	AA	12	12	0	204.0	2.69	23	**3.145**
5	Kilroy	Matt	1891	CN3	AA	1	4	0	45.3	2.98	25	**0.973**

Top Pitchers By Franchise (1871-2010)

Rank	Name	First	Year	Team	Lg	W	L	SV	IP	ERA	Age	PEVA-P
	Cincinnati Outlaw Reds											
1	McCormick	Jim	1884	CNU	UA	21	3	0	210.0	1.54	28	**15.537**
2	Bradley	George	1884	CNU	UA	25	15	0	342.0	2.71	32	**6.254**
3	Burns	Dick	1884	CNU	UA	23	15	0	329.7	2.46	21	**5.938**
4	Sylvester	Lou	1884	CNU	UA	0	1	1	32.7	3.58	29	**0.904**

Top Pitchers By Franchise (1871-2010)

Rank	Name	First	Year	Team	Lg	W	L	SV	IP	ERA	Age	PEVA-P
	Colorado Rockies											
1	Jimenez	Ubaldo	2010	COL	NL	19	8	0	221.7	2.88	26	**28.985**
2	Jimenez	Ubaldo	2009	COL	NL	15	12	0	218.0	3.47	25	**20.035**
3	Francis	Jeff	2007	COL	NL	17	9	0	215.3	4.22	26	**15.645**
4	Marquis	Jason	2009	COL	NL	15	13	0	216.0	4.04	31	**14.173**
5	Cook	Aaron	2008	COL	NL	16	9	0	211.3	3.96	29	**13.582**
6	White	Gabe	2000	COL	NL	11	2	5	83.0	2.17	29	**13.270**
7	Francis	Jeff	2006	COL	NL	13	11	0	199.0	4.16	25	**12.808**
8	Astacio	Pedro	1999	COL	NL	17	11	0	232.0	5.04	30	**12.605**
9	Fuentes	Brian	2008	COL	NL	1	5	30	62.7	2.73	33	**12.332**
10	Jennings	Jason	2006	COL	NL	9	13	0	212.0	3.78	28	**12.146**
11	Jimenez	Ubaldo	2008	COL	NL	12	12	0	198.7	3.99	24	**12.017**
12	de la Rosa	Jorge	2009	COL	NL	16	9	0	185.0	4.38	28	**11.140**
13	Ritz	Kevin	1996	COL	NL	17	11	0	213.0	5.28	31	**10.928**
14	Corpas	Manuel	2007	COL	NL	4	2	19	78.0	2.08	25	**10.715**
15	Street	Huston	2009	COL	NL	4	1	35	61.7	3.06	26	**10.678**
16	Cook	Aaron	2006	COL	NL	9	15	0	212.7	4.23	27	**10.214**
17	Freeman	Marvin	1994	COL	NL	10	2	0	112.7	2.80	31	**10.157**

Top Pitchers By Franchise (1871-2010)

Rank	Name	First	Year	Team	Lg	W	L	SV	IP	ERA	Age	PEVA-P
	Detroit Tigers											
1	Newhouser	Hal	1945	DET	AL	25	9	2	313.3	1.81	24	**48.483**

Rank	Name	First	Year	Team	Lg	W	L	SV	IP	ERA	Age	PEVA-P
2	Trout	Dizzy	1944	DET	AL	27	14	0	352.3	2.12	29	**39.350**
3	Newhouser	Hal	1946	DET	AL	26	9	1	292.7	1.94	25	**38.936**
4	Newhouser	Hal	1944	DET	AL	29	9	2	312.3	2.22	23	**32.575**
5	McLain	Denny	1968	DET	AL	31	6	0	336.0	1.96	24	**32.509**
6	Bunning	Jim	1957	DET	AL	20	8	1	267.3	2.69	26	**32.324**
7	McLain	Denny	1969	DET	AL	24	9	0	325.0	2.80	25	**29.720**
8	Verlander	Justin	2009	DET	AL	19	9	0	240.0	3.45	26	**27.965**
9	Newhouser	Hal	1947	DET	AL	17	17	2	285.0	2.87	26	**26.043**
10	Lary	Frank	1961	DET	AL	23	9	0	275.3	3.24	31	**25.029**
11	Bunning	Jim	1961	DET	AL	17	11	1	268.0	3.19	30	**24.083**
12	Fidrych	Mark	1976	DET	AL	19	9	0	250.3	2.34	22	**23.725**
13	Bunning	Jim	1960	DET	AL	11	14	0	252.0	2.79	29	**23.650**
14	Lolich	Mickey	1971	DET	AL	25	14	0	376.0	2.92	31	**23.606**
15	Newsom	Bobo	1940	DET	AL	21	5	0	264.0	2.83	33	**23.291**
16	Newhouser	Hal	1948	DET	AL	21	12	1	272.3	3.01	27	**21.891**
17	Trucks	Virgil	1949	DET	AL	19	11	4	275.0	2.81	32	**21.862**
18	Hernandez	Willie	1984	DET	AL	9	3	32	140.3	1.92	30	**21.765**
19	Morris	Jack	1981	DET	AL	14	7	0	198.0	3.05	26	**21.619**
20	Lary	Frank	1958	DET	AL	16	15	1	260.3	2.90	28	**21.531**

Top Pitchers By Franchise (1871-2010)

Rank	Name	First	Year	Team	Lg	W	L	SV	IP	ERA	Age	PEVA-P
	Detroit Wolverines											
1	Baldwin	Lady	1886	DTN	NL	42	13	0	487.0	2.24	27	**30.125**
2	Derby	George	1881	DTN	NL	29	26	0	494.7	2.20	24	**22.265**
3	Shaw	Dupee	1884	DTN	NL	9	18	0	227.7	3.04	25	**12.916**
4	Conway	Pete	1888	DTN	NL	30	14	0	391.0	2.26	22	**12.719**
5	Getzein	Charlie	1887	DTN	NL	29	13	0	366.7	3.73	23	**7.624**

Top Pitchers By Franchise (1871-2010)

Rank	Name	First	Year	Team	Lg	W	L	SV	IP	ERA	Age	PEVA-P
	Elizabeth Resolutes											
1	Campbell	Hugh	1873	ELI	NA	2	16	0	165.0	2.84	27	**1.710**
2	Fleet	Frank	1873	ELI	NA	0	3	0	24.0	5.62	25	**0.200**
3	Wolters	Rynie	1873	ELI	NA	0	1	0	9.0	0.00	31	**0.200**
4	Lovett	Len	1873	ELI	NA	0	1	0	9.0	7.00	21	**0.200**

Top Pitchers By Franchise (1871-2010)

Rank	Name	First	Year	Team	Lg	W	L	SV	IP	ERA	Age	PEVA-P
	Florida Marlins											
1	Brown	Kevin	1996	FLO	NL	17	11	0	233.0	1.89	31	**40.004**
2	Willis	Dontrelle	2005	FLO	NL	22	10	0	236.3	2.63	23	**30.817**
3	Brown	Kevin	1997	FLO	NL	16	8	0	237.3	2.69	32	**22.694**
4	Johnson	Josh	2009	FLO	NL	15	5	0	209.0	3.23	25	**21.840**
5	Johnson	Josh	2010	FLO	NL	11	8	0	183.7	2.30	26	**21.832**
6	Pavano	Carl	2004	FLO	NL	18	8	0	222.3	3.00	28	**19.631**
7	Nolasco	Ricky	2008	FLO	NL	15	8	0	212.3	3.52	26	**19.070**
8	Harvey	Bryan	1993	FLO	NL	1	5	45	69.0	1.70	30	**16.133**
9	Nen	Robb	1996	FLO	NL	5	1	35	83.0	1.95	27	**15.924**
10	Leiter	Al	1996	FLO	NL	16	12	0	215.3	2.93	31	**15.083**
11	Willis	Dontrelle	2006	FLO	NL	12	12	0	223.3	3.87	24	**14.550**
12	Fernandez	Alex	1997	FLO	NL	17	12	0	220.7	3.59	28	**14.499**
13	Jones	Todd	2005	FLO	NL	1	5	40	73.0	2.10	37	**14.477**
14	Dempster	Ryan	2000	FLO	NL	14	10	0	226.3	3.66	23	**14.130**
15	Benitez	Armando	2004	FLO	NL	2	2	47	69.7	1.29	32	**13.658**
16	Penny	Brad	2001	FLO	NL	10	10	0	205.0	3.69	23	**12.356**
17	Burnett	A.J.	2005	FLO	NL	12	12	0	209.0	3.44	28	**12.274**

18	Sanchez	Anibel	2010	FLO	NL	13	12	0	195.0	3.55	26	**12.197**
19	Burnett	A.J.	2002	FLO	NL	12	9	0	204.3	3.30	25	**11.366**
20	Rapp	Pat	1995	FLO	NL	14	7	0	167.3	3.44	28	**11.235**

Top Pitchers By Franchise (1871-2010)

Rank	Name	First	Year	Team	Lg	W	L	SV	IP	ERA	Age	PEVA-P
	Fort Wayne Kekiongas											
1	Mathews	Bobby	1871	FW1	NA	6	11	0	169.0	5.17	20	**1.663**

Top Pitchers By Franchise (1871-2010)

Rank	Name	First	Year	Team	Lg	W	L	SV	IP	ERA	Age	PEVA-P
	Hartford Dark Blues (National League)											
1	Bond	Tommy	1876	HAR	NL	31	13	0	408.0	1.68	20	**19.434**
2	Larkin	Terry	1877	HAR	NL	29	25	0	501.0	2.14	NA	**18.656**
3	Cummings	Candy	1876	HAR	NL	16	8	0	216.0	1.67	28	**4.764**
4	Ferguson	Bob	1877	HAR	NL	1	1	0	25.0	3.96	32	**0.629**
5	Cassidy	John	1877	HAR	NL	1	1	0	18.0	5.00	20	**0.200**

Top Pitchers By Franchise (1871-2010)

Rank	Name	First	Year	Team	Lg	W	L	SV	IP	ERA	Age	PEVA-P
	Houston Astros											
1	Scott	Mike	1986	HOU	NL	18	10	0	275.3	2.22	31	**44.955**
2	Richard	J.R.	1979	HOU	NL	18	13	0	292.3	2.71	29	**35.177**
3	Dierker	Larry	1969	HOU	NL	20	13	0	305.3	2.33	23	**34.302**
4	Pettitte	Andy	2005	HOU	NL	17	9	0	222.3	2.39	33	**29.658**
5	Oswalt	Roy	2006	HOU	NL	15	8	0	220.7	2.98	29	**29.269**
6	Niekro	Joe	1982	HOU	NL	17	12	0	270.0	2.47	38	**28.453**
7	Clemens	Roger	2005	HOU	NL	13	8	0	211.3	1.87	43	**28.104**
8	Scott	Mike	1987	HOU	NL	16	13	0	247.7	3.23	32	**27.688**
9	Oswalt	Roy	2005	HOU	NL	20	12	0	241.7	2.94	28	**24.970**
10	Kile	Darryl	1997	HOU	NL	19	7	0	255.7	2.57	29	**24.432**
11	Hampton	Mike	1999	HOU	NL	22	4	0	239.0	2.90	27	**23.528**
12	Sutton	Don	1981	HOU	NL	11	9	0	158.7	2.61	36	**23.269**
13	Oswalt	Roy	2004	HOU	NL	20	10	0	237.0	3.49	27	**22.905**
14	Oswalt	Roy	2002	HOU	NL	19	9	0	233.0	3.01	25	**22.588**
15	Clemens	Roger	2004	HOU	NL	18	4	0	214.3	2.98	42	**22.379**
16	Johnson	Randy	1998	HOU	NL	10	1	0	84.3	1.28	35	**22.017**
17	Ryan	Nolan	1981	HOU	NL	11	5	0	149.0	1.69	34	**21.204**
18	Richard	J.R.	1976	HOU	NL	20	15	0	291.0	2.75	26	**20.879**
19	Ryan	Nolan	1987	HOU	NL	8	16	0	211.7	2.76	40	**20.466**
20	Lima	Jose	1999	HOU	NL	21	10	0	246.3	3.58	27	**19.877**

Note: Previous Names of Franchise: Houston Colt 45's

Top Pitchers By Franchise (1871-2010)

Rank	Name	First	Year	Team	Lg	W	L	SV	IP	ERA	Age	PEVA-P
	Hartford Dark Blues (National Association)											
1	Cummings	Candy	1875	HR1	NA	35	12	0	417.0	1.60	27	**19.393**
2	Bond	Tommy	1875	HR1	NA	19	16	0	352.0	1.56	19	**13.404**
3	Fisher	Cherokee	1874	HR1	NA	14	23	0	317.0	3.04	29	**2.732**
4	Stearns	Bill	1874	HR1	NA	2	14	1	164.0	4.50	21	**0.877**
5	Ferguson	Bob	1875	HR1	NA	0	0	0	2.0	18.00	30	**0.200**

Top Pitchers By Franchise (1871-2010)

Rank	Name	First	Year	Team	Lg	W	L	SV	IP	ERA	Age	PEVA-P

	Indianapolis Blues											
1	Nolan	The Only	1878	IN1	NL	13	22	0	347.0	2.57	21	**3.014**
2	McCormick	Jim	1878	IN1	NL	5	8	0	117.0	1.69	22	**2.216**
3	Healey	Tom	1878	IN1	NL	6	4	1	89.0	2.22	25	**0.596**
4	McKelvy	Russ	1878	IN1	NL	0	2	0	25.0	2.16	24	**0.456**

Top Pitchers By Franchise (1871-2010)

Rank	Name	First	Year	Team	Lg	W	L	SV	IP	ERA	Age	PEVA-P
	Indianapolis Hoosiers (American Association)											
1	McKeon	Larry	1884	IN2	AA	18	41	0	512.0	3.50	18	**5.217**
2	Barr	Bob	1884	IN2	AA	3	11	0	132.0	4.98	28	**3.228**
3	Bond	Tommy	1884	IN2	AA	0	5	0	43.0	5.65	28	**1.831**
4	Aydelott	Jake	1884	IN2	AA	5	7	0	106.0	4.92	23	**1.005**
5	McCauley	Al	1884	IN2	AA	2	7	0	76.0	5.09	21	**0.613**

Top Pitchers By Franchise (1871-2010)

Rank	Name	First	Year	Team	Lg	W	L	SV	IP	ERA	Age	PEVA-P
	Indianapolis Hoosiers (National League)											
1	Boyle	Henry	1887	IN3	NL	13	24	0	328.0	3.65	27	**5.295**
2	Boyle	Henry	1889	IN3	NL	21	23	0	378.7	3.92	29	**4.998**
3	Getzein	Charlie	1889	IN3	NL	18	22	1	349.0	4.54	25	**3.074**
4	Healy	John	1887	IN3	NL	12	29	0	341.0	5.17	21	**2.641**
5	Boyle	Henry	1888	IN3	NL	15	22	0	323.0	3.26	28	**2.508**

Top Pitchers By Franchise (1871-2010)

Rank	Name	First	Year	Team	Lg	W	L	SV	IP	ERA	Age	PEVA-P
	Indianapolis Hoosiers & Newark Peppers (Federal League)											
1	Falkenberg	Cy	1914	IND	FL	25	16	3	377.3	2.22	34	**23.241**
2	Moseley	Earl	1915	NEW	FL	15	15	1	268.0	1.91	31	**8.189**
3	Reulbach	Ed	1915	NEW	FL	21	10	1	270.0	2.23	33	**8.067**
4	Moseley	Earl	1914	IND	FL	19	18	1	316.7	3.47	30	**7.377**
5	Kaiserling	George	1915	NEW	FL	15	15	2	261.3	2.24	22	**6.622**

Note: Previous Names of Franchise: Indianapolis Hoosiers (IND) 1914; Newark Peppers (NEW) 1915

Top Pitchers By Franchise (1871-2010)

Rank	Name	First	Year	Team	Lg	W	L	SV	IP	ERA	Age	PEVA-P
	Kansas City Cowboys (American Association)											
1	Conway	Jim	1889	KC2	AA	19	19	0	335.0	3.25	31	**7.431**
2	Porter	Henry	1888	KC2	AA	18	37	0	474.0	4.16	30	**5.473**
3	Swartzel	Park	1889	KC2	AA	19	27	1	410.3	4.32	24	**4.020**
4	Sullivan	Tom	1888	KC2	AA	8	16	0	214.7	3.40	28	**1.319**
5	Hoffman	Frank	1888	KC2	AA	3	9	0	104.0	2.77	NA	**1.120**

Top Pitchers By Franchise (1871-2010)

Rank	Name	First	Year	Team	Lg	W	L	SV	IP	ERA	Age	PEVA-P
	Kansas City Royals											
1	Saberhagen	Bret	1989	KCA	AL	23	6	0	262.3	2.16	25	**53.447**
2	Greinke	Zack	2009	KCA	AL	16	8	0	229.3	2.16	26	**42.305**
3	Appier	Kevin	1993	KCA	AL	18	8	0	238.7	2.56	26	**36.409**
4	Leonard	Dennis	1977	KCA	AL	20	12	1	292.7	3.04	26	**27.380**
5	Saberhagen	Bret	1987	KCA	AL	18	10	0	257.0	3.36	23	**26.693**
6	Black	Bud	1984	KCA	AL	17	12	0	257.0	3.12	27	**25.197**
7	Quisenberry	Dan	1983	KCA	AL	5	3	45	139.0	1.94	30	**24.535**
8	Gubicza	Mark	1988	KCA	AL	20	8	0	269.7	2.70	26	**23.085**
9	Cone	David	1994	KCA	AL	16	5	0	171.7	2.94	31	**22.648**
10	Gura	Larry	1981	KCA	AL	11	8	0	172.3	2.72	34	**20.054**

11	Leonard	Dennis	1981	KCA	AL	13	11	0	201.7	2.99	30	**19.981**
12	Saberhagen	Bret	1985	KCA	AL	20	6	0	235.3	2.87	21	**19.446**
13	Appier	Kevin	1992	KCA	AL	15	8	0	208.3	2.46	25	**19.157**
14	Quisenberry	Dan	1984	KCA	AL	6	3	44	129.3	2.64	31	**18.502**
15	Montgomery	Jeff	1993	KCA	AL	7	5	45	87.3	2.27	31	**18.356**
16	Gubicza	Mark	1989	KCA	AL	15	11	0	255.0	3.04	27	**18.146**
17	Quisenberry	Dan	1982	KCA	AL	9	7	35	136.7	2.57	29	**18.009**
18	Gura	Larry	1980	KCA	AL	18	10	0	283.3	2.95	33	**17.423**
19	Byrd	Paul	2002	KCA	AL	17	11	0	228.3	3.90	32	**17.059**
20	Saberhagen	Bret	1991	KCA	AL	13	8	0	196.3	3.07	27	**16.377**

Top Pitchers By Franchise (1871-2010)

Rank	Name	First	Year	Team	Lg	W	L	SV	IP	ERA	Age	PEVA-P
	Kansas City Packers											
1	Cullop	Nick	1915	KCF	FL	22	11	2	302.3	2.44	28	**9.555**
2	Cullop	Nick	1914	KCF	FL	14	19	1	295.7	2.34	27	**8.157**
3	Packard	Gene	1914	KCF	FL	20	14	5	302.0	2.89	27	**7.764**
4	Packard	Gene	1915	KCF	FL	20	12	2	281.7	2.68	28	**6.620**
5	Johnson	Chief	1915	KCF	FL	17	17	2	281.3	2.75	29	**6.210**

Top Pitchers By Franchise (1871-2010)

Rank	Name	First	Year	Team	Lg	W	L	SV	IP	ERA	Age	PEVA-P
	Kansas City Cowboys (National League)											
1	Wiedman	Stump	1886	KCN	NL	12	36	0	427.7	4.50	25	**3.572**
2	Whitney	Jim	1886	KCN	NL	12	32	0	393.0	4.49	29	**3.297**
3	Conway	Pete	1886	KCN	NL	5	15	0	180.0	5.75	20	**1.622**
4	McKeon	Larry	1886	KCN	NL	0	2	0	21.0	10.71	20	**0.933**
5	King	Silver	1886	KCN	NL	1	3	0	39.0	4.85	18	**0.763**

Top Pitchers By Franchise (1871-2010)

Rank	Name	First	Year	Team	Lg	W	L	SV	IP	ERA	Age	PEVA-P
	Kansas City Cowboys (Union League)											
1	Bakely	Jersey	1884	KCU	UA	2	3	0	33.0	2.45	20	**3.870**
2	Veach	Peek-A-Bo	1884	KCU	UA	3	9	0	104.0	2.42	22	**2.327**
3	Voss	Alex	1884	KCU	UA	0	6	0	53.0	4.25	26	**1.420**
4	Crothers	Doug	1884	KCU	UA	1	2	0	25.0	1.80	25	**1.348**
5	Black	Bob	1884	KCU	UA	4	9	0	123.0	3.22	22	**1.201**

Top Pitchers By Franchise (1871-2010)

Rank	Name	First	Year	Team	Lg	W	L	SV	IP	ERA	Age	PEVA-P
	Keokuk Westerns											
1	Golden	Mike	1875	KEO	NA	1	12	0	112.0	2.81	24	**1.716**

Top Pitchers By Franchise (1871-2010)

Rank	Name	First	Year	Team	Lg	W	L	SV	IP	ERA	Age	PEVA-P
	Los Angeles Dodgers											
1	Koufax	Sandy	1963	LAN	NL	25	5	0	311.0	1.88	28	**50.345**
2	Koufax	Sandy	1965	LAN	NL	26	8	2	335.7	2.04	30	**46.321**
3	Vance	Dazzy	1924	BRO	NL	28	6	0	308.3	2.16	33	**44.534**
4	Wyatt	Whit	1941	BRO	NL	22	10	1	288.3	2.34	34	**42.854**
5	Koufax	Sandy	1966	LAN	NL	27	9	0	323.0	1.73	31	**37.199**
6	Vance	Dazzy	1928	BRO	NL	22	10	2	280.3	2.09	37	**34.912**
7	Singer	Bill	1969	LAN	NL	20	12	1	315.7	2.34	25	**33.351**
8	Drysdale	Don	1962	LAN	NL	25	9	1	314.3	2.83	26	**33.051**
9	Drysdale	Don	1964	LAN	NL	18	16	0	321.3	2.18	28	**32.039**

Rank	Name	First	Year	Team	Lg	W	L	SV	IP	ERA	Age	PEVA-P
10	Hershiser	Orel	1988	LAN	NL	23	8	1	267.0	2.26	30	**31.540**
11	Drysdale	Don	1960	LAN	NL	15	14	2	269.0	2.84	24	**31.247**
12	Vance	Dazzy	1925	BRO	NL	22	9	0	265.3	3.53	34	**31.049**
13	Newcombe	Don	1956	BRO	NL	27	7	0	268.0	3.06	30	**30.309**
14	Branca	Ralph	1947	BRO	NL	21	12	1	280.0	2.67	21	**30.169**
15	Messersmith	Andy	1975	LAN	NL	19	14	1	321.7	2.29	30	**29.822**
16	Brown	Kevin	1999	LAN	NL	18	9	0	252.3	3.00	34	**29.265**
17	Vance	Dazzy	1930	BRO	NL	17	15	0	258.7	2.61	39	**27.614**
18	Valenzuela	Fernando	1981	LAN	NL	13	7	0	192.3	2.48	21	**27.207**
19	Hershiser	Orel	1985	LAN	NL	19	3	0	239.7	2.03	27	**26.576**
20	Valenzuela	Fernando	1982	LAN	NL	19	13	0	285.0	2.87	22	**26.495**

Note: Previous Names of Franchise:
Brooklyn Atlantics (BR3) 1884, Grays (BR3) 1885-1887, Bridegrooms (BR3) 1888-9; Bridegrooms, Grooms, Superbas, Dodgers, Robins (BRO) 1890-1957

Top Pitchers By Franchise (1871-2010)

Rank	Name	First	Year	Team	Lg	W	L	SV	IP	ERA	Age	PEVA-P
	Louisville Grays											
1	Devlin	Jim	1877	LS1	NL	35	25	0	559.0	2.25	28	**41.423**
2	Devlin	Jim	1876	LS1	NL	30	35	0	622.0	1.56	27	**31.066**
3	Pearce	Frank	1876	LS1	NL	0	0	0	4.0	4.50	16	**0.200**
4	Ryan	Johnny	1876	LS1	NL	0	0	0	8.0	5.62	23	**0.200**
5	Clinton	Jim	1876	LS1	NL	0	1	0	9.0	6.00	26	**0.200**

Top Pitchers By Franchise (1871-2010)

Rank	Name	First	Year	Team	Lg	W	L	SV	IP	ERA	Age	PEVA-P
	Louisville Eclipse & Colonels											
1	Hecker	Guy	1884	LS2	AA	52	20	0	670.7	1.80	28	**39.319**
2	Ramsey	Toad	1886	LS2	AA	38	27	0	588.7	2.45	22	**32.963**
3	Stratton	Scott	1890	LS2	AA	34	14	0	431.0	2.36	21	**32.787**
4	Ramsey	Toad	1887	LS2	AA	37	27	0	561.0	3.43	23	**26.103**
5	Phillippe	Deacon	1899	LS3	NL	21	17	1	321.0	3.17	27	**17.432**
6	Cunningham	Bert	1898	LS3	NL	28	15	0	362.0	3.16	33	**15.085**
7	Hecker	Guy	1885	LS2	AA	30	23	0	480.0	2.18	29	**14.512**
8	Ehret	Red	1890	LS2	AA	25	14	2	359.0	2.53	22	**14.253**
9	Mullane	Tony	1882	LS2	AA	30	24	0	460.3	1.88	23	**13.305**

Note: Previous Names of Franchise:
Louisville Eclipse (LS2) AA 1882-1884; Louisville Colonels AA (LS2) 1885-1891; Louisville Colonels (LS3) NL 1892-1899

Top Pitchers By Franchise (1871-2010)

Rank	Name	First	Year	Team	Lg	W	L	SV	IP	ERA	Age	PEVA-P
	Middletown Mansfields											
1	Bentley	Cy	1872	MID	NA	2	15	0	154.0	6.14	22	**0.737**
2	Brainard	Asa	1872	MID	NA	0	2	0	8.0	7.88	31	**0.696**
3	Buttery	Frank	1872	MID	NA	3	2	0	49.0	5.14	21	**0.630**

Top Pitchers By Franchise (1871-2010)

Rank	Name	First	Year	Team	Lg	W	L	SV	IP	ERA	Age	PEVA-P
	Minnesota Twins											
1	Johnson	Walter	1912	WS1	AL	33	12	2	369.0	1.39	25	**52.347**
2	Santana	Johan	2006	MIN	AL	19	6	0	233.7	2.77	27	**46.748**
3	Johnson	Walter	1910	WS1	AL	25	17	1	370.0	1.36	23	**45.752**
4	Johnson	Walter	1913	WS1	AL	36	7	2	346.0	1.14	26	**43.737**
5	Johnson	Walter	1918	WS1	AL	23	13	3	326.0	1.27	31	**43.252**
6	Santana	Johan	2004	MIN	AL	20	6	0	228.0	2.61	25	**39.021**

Rank	Name	First	Year	Team	Lg	W	L	SV	IP	ERA	Age	PEVA-P
7	Johnson	Walter	1914	WS1	AL	28	18	1	371.7	1.72	27	**36.936**
8	Johnson	Walter	1915	WS1	AL	27	13	4	336.7	1.55	28	**36.871**
9	Marberry	Firpo	1929	WS1	AL	19	12	11	250.3	3.06	31	**33.863**
10	Johnson	Walter	1919	WS1	AL	20	14	2	290.3	1.49	32	**32.304**
11	Blyleven	Bert	1973	MIN	AL	20	17	0	325.0	2.52	22	**32.194**
12	Johnson	Walter	1916	WS1	AL	25	20	1	369.7	1.90	29	**31.062**
13	Viola	Frank	1987	MIN	AL	17	10	0	251.7	2.90	27	**30.300**
14	Perry	Jim	1970	MIN	AL	24	12	0	278.7	3.04	35	**30.121**
15	Santana	Johan	2005	MIN	AL	16	7	0	231.7	2.87	26	**29.297**
16	Johnson	Walter	1911	WS1	AL	25	13	1	322.3	1.90	24	**27.861**
17	Viola	Frank	1988	MIN	AL	24	7	0	255.3	2.64	28	**26.964**
18	Johnson	Walter	1924	WS1	AL	23	7	0	277.7	2.72	37	**25.968**
19	Viola	Frank	1984	MIN	AL	18	12	0	257.7	3.21	24	**25.678**
20	Wolff	Roger	1945	WS1	AL	20	10	2	250.0	2.12	34	**25.554**

Note: Previous Names of Franchise: Washington Senators (WS1) 1901-1960

Top Pitchers By Franchise (1871-2010)

Rank	Name	First	Year	Team	Lg	W	L	SV	IP	ERA	Age	PEVA-P
	Milwaukee Grays											
1	Weaver	Sam	1878	ML2	NL	12	31	0	383.0	1.95	23	**12.351**
2	Golden	Mike	1878	ML2	NL	3	13	0	161.0	4.14	27	**0.822**
3	Ellick	Joe	1878	ML2	NL	0	1	0	3.0	3.00	24	**0.200**

Top Pitchers By Franchise (1871-2010)

Rank	Name	First	Year	Team	Lg	W	L	SV	IP	ERA	Age	PEVA-P
	Milwaukee Brewers (American Association)											
1	Davies	George	1891	ML3	AA	7	5	0	102.0	2.65	23	**4.217**
2	Dwyer	Frank	1891	ML3	AA	6	4	0	86.0	2.20	23	**4.025**
3	Killen	Frank	1891	ML3	AA	7	4	0	96.7	1.68	21	**3.860**
4	Mains	Willard	1891	ML3	AA	0	2	0	10.0	10.80	23	**3.145**
5	Hughey	Jim	1891	ML3	AA	1	0	0	15.0	3.00	22	**0.200**

Top Pitchers By Franchise (1871-2010)

Rank	Name	First	Year	Team	Lg	W	L	SV	IP	ERA	Age	PEVA-P
	Milwaukee Brewers											
1	Sabathia	C.C.	2008	MIL	NL	11	2	0	130.7	1.65	28	**39.539**
2	Sheets	Ben	2004	MIL	NL	12	14	0	237.0	2.70	26	**27.681**
3	Caldwell	Mike	1978	ML4	AL	22	9	1	293.3	2.36	29	**24.297**
4	Higuera	Teddy	1988	ML4	AL	16	9	0	227.3	2.45	30	**22.403**
5	Higuera	Teddy	1987	ML4	AL	18	10	0	261.7	3.85	29	**20.419**
6	Fingers	Rollie	1981	ML4	AL	6	3	28	78.0	1.04	35	**20.222**
7	Higuera	Teddy	1986	ML4	AL	20	11	0	248.3	2.79	28	**19.290**
8	Jones	Doug	1997	ML4	AL	6	6	36	80.3	2.02	40	**18.476**
9	Bosio	Chris	1989	ML4	AL	15	10	0	234.7	2.95	26	**16.011**
10	Navarro	Jaime	1992	ML4	AL	17	11	0	246.0	3.33	25	**15.093**
11	Wegman	Bill	1992	ML4	AL	13	14	0	261.7	3.20	30	**14.900**
12	Capuano	Chris	2005	MIL	NL	18	12	0	219.0	3.99	27	**14.873**
13	Wegman	Bill	1991	ML4	AL	15	7	0	193.3	2.84	29	**14.785**
14	Sanders	Ken	1971	ML4	AL	7	12	31	136.3	1.91	30	**14.657**
15	Cordero	Francisco	2007	MIL	NL	0	4	44	63.3	2.98	32	**14.631**
16	Davis	Doug	2004	MIL	NL	12	12	0	207.3	3.39	29	**14.047**
17	Hoffman	Trevor	2009	MIL	NL	3	2	37	54.0	1.83	42	**13.498**
18	Eldred	Cal	1993	ML4	AL	16	16	0	258.0	4.01	26	**13.472**
19	Capuano	Chris	2006	MIL	NL	11	12	0	221.3	4.03	28	**13.396**
20	Bones	Ricky	1994	ML4	AL	10	9	0	170.7	3.43	25	**13.105**

Note: Previous Names of Franchise:
Seattle Pilots (SE1) 1969; Milwaukee Brewers (ML4) AL 1970-1997; Milwaukee Brewers (MIL) NL 1998-present

Top Pitchers By Franchise (1871-2010)

Rank	Name	First	Year	Team	Lg	W	L	SV	IP	ERA	Age	PEVA-P
	Milwaukee Brewers (Union League)											
1	Cushman	Ed	1884	MLU	UA	4	0	0	36.0	1.00	32	**5.306**
2	Porter	Henry	1884	MLU	UA	3	3	0	51.0	3.00	26	**1.144**
3	Baldwin	Lady	1884	MLU	UA	1	1	0	17.0	2.65	25	**0.200**

Top Pitchers By Franchise (1871-2010)

Rank	Name	First	Year	Team	Lg	W	L	SV	IP	ERA	Age	PEVA-P
	New Haven Elm Citys											
1	Nichols	Tricky	1875	NH1	NA	4	29	0	288.0	3.03	25	**1.569**
2	Luff	Henry	1875	NH1	NA	1	6	0	69.7	4.78	19	**0.789**
3	Ryan	Johnny	1875	NH1	NA	1	5	0	59.3	3.34	22	**0.535**
4	Knight	George	1875	NH1	NA	1	0	0	9.0	2.00	20	**0.200**

Top Pitchers By Franchise (1871-2010)

Rank	Name	First	Year	Team	Lg	W	L	SV	IP	ERA	Age	PEVA-P
	New York Mutuals (National Association)											
1	Mathews	Bobby	1874	NY2	NA	42	22	0	578.0	2.30	23	**29.885**
2	Cummings	Candy	1872	NY2	NA	33	20	0	497.0	2.52	24	**26.548**
3	Wolters	Rynie	1871	NY2	NA	16	16	0	283.0	3.43	29	**21.840**
4	Mathews	Bobby	1873	NY2	NA	29	23	0	443.0	2.56	22	**20.558**
5	Mathews	Bobby	1875	NY2	NA	29	38	0	626.7	2.41	24	**12.808**

Top Pitchers By Franchise (1871-2010)

Rank	Name	First	Year	Team	Lg	W	L	SV	IP	ERA	Age	PEVA-P
	New York Mutuals (National League)											
1	Mathews	Bobby	1876	NY3	NL	21	34	0	516.0	2.86	25	**5.545**
2	Booth	Eddie	1876	NY3	NL	0	0	0	5.0	10.80	NA	**0.200**
3	Larkin	Terry	1876	NY3	NL	0	1	0	9.0	3.00	NA	**0.200**

Top Pitchers By Franchise (1871-2010)

Rank	Name	First	Year	Team	Lg	W	L	SV	IP	ERA	Age	PEVA-P
	New York Metropolitans											
1	Keefe	Tim	1883	NY4	AA	41	27	0	619.0	2.41	26	**28.675**
2	Keefe	Tim	1884	NY4	AA	37	17	0	483.0	2.25	27	**13.989**
3	Lynch	Jack	1884	NY4	AA	37	15	0	496.0	2.67	27	**12.508**
4	Lynch	Jack	1886	NY4	AA	20	30	0	432.7	3.95	29	**4.657**
5	Mays	Al	1887	NY4	AA	17	34	0	441.3	4.73	22	**4.609**

Top Pitchers By Franchise (1871-2010)

Rank	Name	First	Year	Team	Lg	W	L	SV	IP	ERA	Age	PEVA-P
	New York Yankees											
1	Chandler	Spud	1943	NYA	AL	20	4	0	253.0	1.64	36	**42.384**
2	Chesbro	Jack	1904	NYA	AL	41	12	0	454.7	1.82	30	**41.233**
3	Guidry	Ron	1978	NYA	AL	25	3	0	273.7	1.74	28	**39.576**
4	Gomez	Lefty	1937	NYA	AL	21	11	0	278.3	2.33	29	**35.814**
5	Hunter	Catfish	1975	NYA	AL	23	14	0	328.0	2.58	29	**34.002**
6	Mussina	Mike	2001	NYA	AL	17	11	0	228.7	3.15	33	**31.025**
7	Mays	Carl	1921	NYA	AL	27	9	7	336.7	3.05	30	**28.720**
8	Peterson	Fritz	1969	NYA	AL	17	16	0	272.0	2.55	27	**27.915**
9	Ford	Whitey	1958	NYA	AL	14	7	1	219.3	2.01	30	**26.457**
10	Gomez	Lefty	1934	NYA	AL	26	5	1	281.7	2.33	26	**26.325**
11	Ford	Russ	1910	NYA	AL	26	6	1	299.7	1.65	27	**25.360**

Rank	Name	First	Year	Team	Lg	W	L	SV	IP	ERA	Age	PEVA-P
12	Mussina	Mike	2008	NYA	AL	20	9	0	200.3	3.37	40	**25.277**
13	Terry	Ralph	1962	NYA	AL	23	12	2	298.7	3.19	26	**25.227**
14	Key	Jimmy	1993	NYA	AL	18	6	0	236.7	3.00	32	**25.024**
15	Sabathia	C.C.	2010	NYA	AL	21	7	0	237.7	3.18	30	**24.810**
16	Pettitte	Andy	1997	NYA	AL	18	7	0	240.3	2.88	25	**24.405**
17	Bonham	Tiny	1942	NYA	AL	21	5	0	226.0	2.27	29	**24.208**
18	Sabathia	C.C.	2009	NYA	AL	19	8	0	230.0	3.37	29	**23.949**
19	Ford	Whitey	1961	NYA	AL	25	4	0	283.0	3.21	33	**23.895**
20	John	Tommy	1979	NYA	AL	21	9	0	276.3	2.96	36	**23.417**

Note: Previous Names of Franchise: Baltimore Orioles (BLA) 1901-1902; New York Highlanders 1903-1912

Top Pitchers By Franchise (1871-2010)

Rank	Name	First	Year	Team	Lg	W	L	SV	IP	ERA	Age	PEVA-P
	New York Mets											
1	Gooden	Dwight	1985	NYN	NL	24	4	0	276.7	1.53	21	**45.645**
2	Gooden	Dwight	1984	NYN	NL	17	9	0	218.0	2.60	20	**37.956**
3	Seaver	Tom	1969	NYN	NL	25	7	0	273.3	2.21	25	**31.954**
4	Seaver	Tom	1971	NYN	NL	20	10	0	286.3	1.76	27	**31.022**
5	Seaver	Tom	1973	NYN	NL	19	10	0	290.0	2.08	29	**30.181**
6	Seaver	Tom	1977	NYN	NL	7	3	0	96.0	3.00	33	**30.175**
7	Seaver	Tom	1975	NYN	NL	22	9	0	280.3	2.38	31	**29.900**
8	Seaver	Tom	1970	NYN	NL	18	12	0	290.7	2.82	26	**29.421**
9	Santana	Johan	2008	NYN	NL	16	7	0	234.3	2.53	29	**28.994**
10	Saberhagen	Bret	1994	NYN	NL	14	4	0	177.3	2.74	30	**28.916**
11	Viola	Frank	1990	NYN	NL	20	12	0	249.7	2.67	30	**25.299**
12	Martinez	Pedro	2005	NYN	NL	15	8	0	217.0	2.82	34	**24.921**
13	Koosman	Jerry	1969	NYN	NL	17	9	0	241.0	2.28	27	**21.934**
14	Leiter	Al	1998	NYN	NL	17	6	0	193.0	2.47	33	**20.800**
15	Cone	David	1988	NYN	NL	20	3	0	231.3	2.22	25	**20.404**
16	Matlack	Jon	1974	NYN	NL	13	15	0	265.3	2.41	24	**19.734**
17	Fernandez	Sid	1992	NYN	NL	14	11	0	214.7	2.73	30	**19.470**
18	Koosman	Jerry	1976	NYN	NL	21	10	0	247.3	2.69	34	**19.282**
19	Cone	David	1992	NYN	NL	13	7	0	196.7	2.88	29	**18.790**
20	Seaver	Tom	1976	NYN	NL	14	11	0	271.0	2.59	32	**18.758**

Top Pitchers By Franchise (1871-2010)

Rank	Name	First	Year	Team	Lg	W	L	SV	IP	ERA	Age	PEVA-P
	New York Giants (Pacific League)											
1	Keefe	Tim	1890	NYP	PL	17	11	0	229.0	3.38	33	**3.709**
2	O'Day	Hank	1890	NYP	PL	22	13	3	329.0	4.21	28	**3.441**
3	Crane	Ed	1890	NYP	PL	16	19	0	330.3	4.63	28	**2.641**
4	Ewing	John	1890	NYP	PL	18	12	2	267.3	4.24	27	**2.579**
5	Hatfield	Gil	1890	NYP	PL	1	1	1	7.7	3.52	35	**0.200**

Top Pitchers By Franchise (1871-2010)

Rank	Name	First	Year	Team	Lg	W	L	SV	IP	ERA	Age	PEVA-P
	Oakland Athletics											
1	Grove	Lefty	1930	PHA	AL	28	5	9	291.0	2.54	30	**52.344**
2	Grove	Lefty	1931	PHA	AL	31	4	5	288.7	2.06	31	**46.335**
3	Grove	Lefty	1929	PHA	AL	20	6	4	275.3	2.81	29	**38.544**
4	Hunter	Catfish	1974	OAK	AL	25	12	0	318.3	2.49	28	**35.949**
5	Blue	Vida	1971	OAK	AL	24	8	0	312.0	1.82	22	**34.160**
6	Coombs	Jack	1910	PHA	AL	31	9	1	353.0	1.30	28	**32.981**
7	Blue	Vida	1976	OAK	AL	18	13	0	298.3	2.35	27	**29.573**
8	Grove	Lefty	1928	PHA	AL	24	8	4	261.7	2.58	28	**28.141**
9	Zito	Barry	2002	OAK	AL	23	5	0	229.3	2.75	24	**27.825**

Rank	Name	First	Year	Team	Lg	W	L	SV	IP	ERA	Age	PEVA-P
10	Rommel	Eddie	1922	PHA	AL	27	13	2	294.0	3.28	25	**27.744**
11	Shantz	Bobby	1952	PHA	AL	24	7	0	279.7	2.48	27	**26.992**
12	Grove	Lefty	1932	PHA	AL	25	10	7	291.7	2.84	32	**26.783**
13	Norris	Mike	1980	OAK	AL	22	9	0	284.3	2.53	25	**26.018**
14	Stewart	Dave	1990	OAK	AL	22	11	0	267.0	2.56	33	**24.907**
15	Mulder	Mark	2001	OAK	AL	21	8	0	229.3	3.45	24	**24.581**
16	McCatty	Steve	1981	OAK	AL	14	7	0	185.7	2.33	27	**24.432**
17	Hudson	Tim	2003	OAK	AL	16	7	0	240.0	2.70	28	**24.250**
18	Earnshaw	George	1929	PHA	AL	24	8	1	254.7	3.29	29	**23.280**
19	Waddell	Rube	1905	PHA	AL	27	10	0	328.7	1.48	29	**23.151**
20	Perry	Scott	1918	PHA	AL	20	19	2	332.3	1.98	27	**22.807**

Note: Previous Names of Franchise: Philadelphia Athletics (PHA) 1901-1954; Kansas City Athletics (KC1) 1955-1967

Top Pitchers By Franchise (1871-2010)

Rank	Name	First	Year	Team	Lg	W	L	SV	IP	ERA	Age	PEVA-P
	Philadelphia Athletics (National Association)											
1	McBride	Dick	1874	PH1	NA	33	22	0	487.0	2.55	29	**17.928**
2	McBride	Dick	1875	PH1	NA	44	14	0	538.0	1.97	30	**16.714**
3	McBride	Dick	1872	PH1	NA	30	14	0	419.0	3.01	27	**16.138**
4	McBride	Dick	1873	PH1	NA	24	19	0	382.7	3.32	28	**9.485**
5	McBride	Dick	1871	PH1	NA	18	5	0	222.0	4.58	26	**7.556**

Top Pitchers By Franchise (1871-2010)

Rank	Name	First	Year	Team	Lg	W	L	SV	IP	ERA	Age	PEVA-P
	Philadelphia Athletics (American Association)											
1	Seward	Ed	1888	PH4	AA	35	19	0	518.7	2.01	21	**23.438**
2	McMahon	Sadie	1890	PH4	AA	29	18	1	410.0	3.34	23	**17.442**
3	Taylor	Billy	1884	PH4	AA	18	12	0	260.0	2.53	29	**16.180**
4	Mathews	Bobby	1885	PH4	AA	30	17	0	422.3	2.43	34	**15.060**
5	Weyhing	Gus	1891	PH4	AA	31	20	0	450.0	3.18	25	**13.895**
6	Weyhing	Gus	1888	PH4	AA	28	18	0	404.0	2.25	22	**11.171**
7	Weyhing	Gus	1889	PH4	AA	30	21	0	449.0	2.95	23	**11.167**
8	Mathews	Bobby	1883	PH4	AA	30	13	0	381.0	2.46	32	**10.018**

Top Pitchers By Franchise (1871-2010)

Rank	Name	First	Year	Team	Lg	W	L	SV	IP	ERA	Age	PEVA-P
	Philadelphia Whites											
1	Zettlein	George	1873	PH2	NA	36	15	0	460.0	2.70	29	**17.432**
2	Zettlein	George	1875	PH2	NA	12	8	0	180.3	2.40	31	**11.035**
3	Fisher	Cherokee	1875	PH2	NA	22	19	0	356.7	1.92	30	**8.534**
4	Cummings	Candy	1874	PH2	NA	28	26	0	482.0	2.88	26	**6.770**
5	Borden	Joe	1875	PH2	NA	2	4	0	66.0	1.64	21	**3.528**

Top Pitchers By Franchise (1871-2010)

Rank	Name	First	Year	Team	Lg	W	L	SV	IP	ERA	Age	PEVA-P
	Philadelphia Centennials											
1	Bechtel	George	1875	PH3	NA	2	12	0	126.0	3.93	27	**1.542**

Top Pitchers By Franchise (1871-2010)

Rank	Name	First	Year	Team	Lg	W	L	SV	IP	ERA	Age	PEVA-P
	Philadelphia Phillies											
1	Alexander	Pete	1915	PHI	NL	31	10	3	376.3	1.22	28	**52.205**
2	Carlton	Steve	1980	PHI	NL	24	9	0	304.0	2.34	36	**48.494**
3	Alexander	Pete	1916	PHI	NL	33	12	3	389.0	1.55	29	**47.561**

Rank	Name	First	Year	Team	Lg	W	L	SV	IP	ERA	Age	PEVA-P
4	Halladay	Roy	2010	PHI	NL	21	10	0	250.7	2.44	33	**43.340**
5	Alexander	Pete	1917	PHI	NL	30	13	0	388.0	1.83	30	**39.958**
6	Carlton	Steve	1972	PHI	NL	27	10	0	346.3	1.97	28	**39.105**
7	Roberts	Robin	1951	PHI	NL	21	15	2	315.0	3.03	25	**38.640**
8	Roberts	Robin	1952	PHI	NL	28	7	2	330.0	2.59	26	**38.501**
9	Roberts	Robin	1953	PHI	NL	23	16	2	346.7	2.75	27	**38.372**
10	Roberts	Robin	1954	PHI	NL	23	15	4	336.7	2.97	28	**35.774**
11	Carlton	Steve	1982	PHI	NL	23	11	0	295.7	3.10	38	**35.010**
12	Roberts	Robin	1950	PHI	NL	20	11	1	304.3	3.02	24	**30.767**
13	Schilling	Curt	1997	PHI	NL	17	11	0	254.3	2.97	31	**30.295**
14	Roberts	Robin	1955	PHI	NL	23	14	3	305.0	3.28	29	**28.350**
15	Bunning	Jim	1967	PHI	NL	17	15	0	302.3	2.29	36	**28.202**
16	Schilling	Curt	1998	PHI	NL	15	14	0	268.7	3.25	32	**28.183**
17	Denny	John	1983	PHI	NL	19	6	0	242.7	2.37	31	**28.177**
18	Carlton	Steve	1977	PHI	NL	23	10	0	283.0	2.64	33	**28.165**
19	Carlton	Steve	1981	PHI	NL	13	4	0	190.0	2.42	37	**27.567**
20	Alexander	Pete	1911	PHI	NL	28	13	3	367.0	2.57	24	**27.469**

Note: Previous Names of Franchise: Philadelphia Quakers 1883-1889; Philadelphia Blue Jays 1943-1944

Top Pitchers By Franchise (1871-2010)

Rank	Name	First	Year	Team	Lg	W	L	SV	IP	ERA	Age	PEVA-P
	Philadelphia Athletics (National League)											
1	Knight	Lon	1876	PHN	NL	10	22	0	282.0	2.62	23	**1.878**
2	Zettlein	George	1876	PHN	NL	4	20	2	234.0	3.88	32	**1.477**
3	Coon	William	1876	PHN	NL	0	0	0	7.0	5.14	21	**0.200**
4	Lafferty	Flip	1876	PHN	NL	0	1	0	9.0	0.00	22	**0.200**
5	Meyerle	Levi	1876	PHN	NL	0	2	0	18.0	5.00	31	**0.200**

Top Pitchers By Franchise (1871-2010)

Rank	Name	First	Year	Team	Lg	W	L	SV	IP	ERA	Age	PEVA-P
	Philadelphia Athletics (Pacific League)											
1	Sanders	Ben	1890	PHP	PL	19	18	1	346.7	3.76	25	**4.233**
2	Knell	Phil	1890	PHP	PL	22	11	0	286.7	3.83	25	**3.221**
3	Buffinton	Charlie	1890	PHP	PL	19	15	1	283.3	3.81	29	**2.990**
4	Cunningham	Bert	1890	PHP	PL	3	9	0	108.7	5.22	25	**2.308**
5	Husted	Bill	1890	PHP	PL	5	10	0	129.0	4.88	24	**0.841**

Top Pitchers By Franchise (1871-2010)

Rank	Name	First	Year	Team	Lg	W	L	SV	IP	ERA	Age	PEVA-P
	Philadelphia Keystones											
1	Bakely	Jersey	1884	PHU	UA	14	25	0	344.7	4.47	20	**3.870**
2	Fisher	J.	1884	PHU	UA	1	7	0	70.7	3.57	NA	**0.919**
3	Weaver	Sam	1884	PHU	UA	5	10	0	136.0	5.76	29	**0.903**
4	Gallagher	Bill	1884	PHU	UA	1	2	0	25.0	3.24	NA	**0.462**
5	McCormick	Jerry	1884	PHU	UA	0	0	0	2.0	9.00	NA	**0.200**

Top Pitchers By Franchise (1871-2010)

Rank	Name	First	Year	Team	Lg	W	L	SV	IP	ERA	Age	PEVA-P
	Pittsburgh Pirates											
1	Friend	Bob	1960	PIT	NL	18	12	1	275.7	3.00	30	**38.546**
2	Morris	Ed	1886	PT1	AA	41	20	1	555.3	2.45	24	**33.289**
3	Blanton	Cy	1935	PIT	NL	18	13	1	254.3	2.58	27	**32.884**
4	Hawley	Pink	1895	PIT	NL	31	22	1	444.3	3.18	23	**31.890**
5	Killen	Frank	1893	PIT	NL	36	14	0	415.0	3.64	23	**30.320**
6	Cooper	Wilbur	1920	PIT	NL	24	15	2	327.0	2.39	28	**29.361**

Rank	Name	First	Year	Team	Lg	W	L	SV	IP	ERA	Age	PEVA-P
7	Law	Vern	1960	PIT	NL	20	9	0	271.7	3.08	30	**28.689**
8	Grimes	Burleigh	1928	PIT	NL	25	14	3	330.7	2.99	35	**28.344**
9	Adams	Babe	1911	PIT	NL	22	12	0	293.3	2.33	29	**28.103**
10	Adams	Babe	1919	PIT	NL	17	10	1	263.3	1.98	37	**26.879**
11	Reuschel	Rick	1987	PIT	NL	8	6	0	177.0	2.75	38	**25.570**
12	Candelaria	John	1977	PIT	NL	20	5	0	230.7	2.34	24	**24.674**
13	Adams	Babe	1920	PIT	NL	17	13	2	263.0	2.16	38	**23.725**
14	Killen	Frank	1896	PIT	NL	30	18	0	432.3	3.41	26	**23.686**
15	Drabek	Doug	1992	PIT	NL	15	11	0	256.7	2.77	30	**23.331**
16	Morris	Ed	1885	PT1	AA	39	24	0	581.0	2.35	23	**23.155**
17	Drabek	Doug	1990	PIT	NL	22	6	0	231.3	2.76	28	**22.627**
18	Friend	Bob	1963	PIT	NL	17	16	0	268.7	2.34	33	**22.183**
19	Kremer	Ray	1927	PIT	NL	19	8	2	226.0	2.47	34	**21.882**
20	Rhoden	Rick	1984	PIT	NL	14	9	0	238.3	2.72	31	**21.773**

Note: Previous Names of Franchise: Pittsburgh Alleghenys (PT1) 1882-1886; Pittsburgh Alleghenys (PIT) 1887-1890

Top Pitchers By Franchise (1871-2010)

Rank	Name	First	Year	Team	Lg	W	L	SV	IP	ERA	Age	PEVA-P
	Providence Grays											
1	Radbourn	Charley	1883	PRO	NL	48	25	1	632.3	2.05	29	**38.272**
2	Radbourn	Charley	1884	PRO	NL	59	12	1	678.7	1.38	30	**36.790**
3	Sweeney	Charlie	1884	PRO	NL	17	8	1	221.0	1.55	21	**27.423**
4	Ward	John	1879	PRO	NL	47	19	1	587.0	2.15	19	**24.362**
5	Ward	John	1880	PRO	NL	39	24	1	595.0	1.74	20	**19.056**
6	Ward	John	1878	PRO	NL	22	13	0	334.0	1.51	18	**14.937**
7	Radbourn	Charley	1882	PRO	NL	33	20	0	474.0	2.09	28	**13.913**

Top Pitchers By Franchise (1871-2010)

Rank	Name	First	Year	Team	Lg	W	L	SV	IP	ERA	Age	PEVA-P
	Pittsburgh Rebels											
1	Allen	Frank	1915	PTF	FL	23	13	0	283.3	2.51	27	**9.639**
2	Knetzer	Elmer	1915	PTF	FL	18	14	3	279.0	2.58	30	**6.974**
3	Knetzer	Elmer	1914	PTF	FL	20	12	1	272.0	2.88	29	**6.059**
4	Rogge	Clint	1915	PTF	FL	17	11	0	254.3	2.55	26	**5.847**
5	Camnitz	Howie	1914	PTF	FL	14	19	1	262.0	3.23	33	**4.541**

Top Pitchers By Franchise (1871-2010)

Rank	Name	First	Year	Team	Lg	W	L	SV	IP	ERA	Age	PEVA-P
	Pittsburgh Burghers											
1	Staley	Harry	1890	PTP	PL	21	25	0	387.7	3.23	24	**10.092**
2	Maul	Al	1890	PTP	PL	16	12	0	246.7	3.79	25	**2.071**
3	Galvin	Pud	1890	PTP	PL	12	13	0	217.0	4.35	34	**1.564**
4	Morris	Ed	1890	PTP	PL	8	7	0	144.3	4.86	28	**0.778**
5	Tener	John	1890	PTP	PL	3	11	0	117.0	7.31	27	**0.631**

Top Pitchers By Franchise (1871-2010)

Rank	Name	First	Year	Team	Lg	W	L	SV	IP	ERA	Age	PEVA-P
	Rockford Forest Citys											
1	Fisher	Cherokee	1871	RC1	NA	4	16	0	213.0	4.35	26	**2.834**
2	Mack	Denny	1871	RC1	NA	0	1	0	13.0	3.46	20	**0.200**

Top Pitchers By Franchise (1871-2010)

Rank	Name	First	Year	Team	Lg	W	L	SV	IP	ERA	Age	PEVA-P
	Rochester Broncos											
1	Barr	Bob	1890	RC2	AA	28	24	0	493.3	3.25	34	**10.690**
2	Calihan	Will	1890	RC2	AA	18	15	0	296.3	3.28	21	**4.396**

Rank	Name	First	Year	Team	Lg	W	L	SV	IP	ERA	Age	PEVA-P
3	Titcomb	Cannonbal	1890	RC2	AA	10	9	0	168.7	3.74	24	**1.121**
4	Miller	Bob	1890	RC2	AA	3	7	1	92.3	4.29	28	**0.942**
5	Fitzgerald	John	1890	RC2	AA	3	8	0	78.0	4.04	NA	**0.825**

Top Pitchers By Franchise (1871-2010)

Rank	Name	First	Year	Team	Lg	W	L	SV	IP	ERA	Age	PEVA-P
	Richmond Virginians											
1	Meegan	Pete	1884	RIC	AA	7	12	0	179.0	4.32	21	**1.107**
2	Dugan	Ed	1884	RIC	AA	5	14	0	166.3	4.49	20	**0.933**
3	Firth	Ted	1884	RIC	AA	0	1	0	9.0	8.00	29	**0.200**
4	Curry	Wes	1884	RIC	AA	0	2	0	16.0	5.06	24	**0.200**

Top Pitchers By Franchise (1871-2010)

Rank	Name	First	Year	Team	Lg	W	L	SV	IP	ERA	Age	PEVA-P
	San Diego Padres											
1	Brown	Kevin	1998	SDN	NL	18	7	0	257.0	2.38	33	**42.712**
2	Peavy	Jake	2007	SDN	NL	19	6	0	223.3	2.54	26	**41.135**
3	Jones	Randy	1976	SDN	NL	22	14	0	315.3	2.74	26	**31.744**
4	Jones	Randy	1975	SDN	NL	20	12	0	285.0	2.24	25	**27.108**
5	Whitson	Ed	1989	SDN	NL	16	11	0	227.0	2.66	34	**19.683**
6	Benes	Andy	1991	SDN	NL	15	11	0	223.0	3.03	24	**19.087**
7	Whitson	Ed	1990	SDN	NL	14	9	0	228.7	2.60	35	**18.998**
8	Hoffman	Trevor	1998	SDN	NL	4	2	53	73.0	1.48	31	**18.234**
9	Hurst	Bruce	1989	SDN	NL	15	11	0	244.7	2.69	31	**18.221**
10	Perry	Gaylord	1978	SDN	NL	21	6	0	260.7	2.73	40	**16.718**
11	Peavy	Jake	2005	SDN	NL	13	7	0	203.0	2.88	24	**16.518**
12	Hoffman	Trevor	1996	SDN	NL	9	5	42	88.0	2.25	29	**16.509**
13	Hoffman	Trevor	1999	SDN	NL	2	3	40	67.3	2.14	32	**15.899**
14	Peavy	Jake	2004	SDN	NL	15	6	0	166.3	2.27	23	**15.750**
15	Latos	Matt	2010	SDN	NL	14	10	0	184.7	2.92	23	**15.624**
16	Ashby	Andy	1998	SDN	NL	17	9	0	226.7	3.34	31	**15.490**
17	Roberts	Dave	1971	SDN	NL	14	17	0	269.7	2.10	27	**14.958**
18	Hurst	Bruce	1991	SDN	NL	15	8	0	221.7	3.29	33	**14.829**
19	Young	Chris	2007	SDN	NL	9	8	0	173.0	3.12	28	**14.773**
20	Ashby	Andy	1995	SDN	NL	12	10	0	192.7	2.94	28	**14.730**

Top Pitchers By Franchise (1871-2010)

Rank	Name	First	Year	Team	Lg	W	L	SV	IP	ERA	Age	PEVA-P
	Seattle Mariners											
1	Johnson	Randy	1995	SEA	AL	18	2	0	214.3	2.48	32	**34.064**
2	Hernandez	Felix	2009	SEA	AL	19	5	0	238.7	2.49	23	**33.947**
3	Hernandez	Felix	2010	SEA	AL	13	12	0	249.7	2.27	24	**28.666**
4	Johnson	Randy	1993	SEA	AL	19	8	1	255.3	3.24	30	**26.278**
5	Johnson	Randy	1997	SEA	AL	20	4	0	213.0	2.28	34	**25.435**
6	Garcia	Freddy	2001	SEA	AL	18	6	0	238.7	3.05	26	**25.267**
7	Johnson	Randy	1998	SEA	AL	9	10	0	160.0	4.33	35	**22.017**
8	Moyer	Jamie	2001	SEA	AL	20	6	0	209.7	3.43	39	**20.937**
9	Putz	J.J.	2006	SEA	AL	4	1	36	78.3	2.30	29	**19.452**
10	Moyer	Jamie	2003	SEA	AL	21	7	0	215.0	3.27	41	**18.423**
11	Langston	Mark	1989	SEA	AL	4	5	0	73.3	3.56	29	**18.250**
12	Hanson	Erik	1990	SEA	AL	18	9	0	236.0	3.24	25	**18.227**
13	Putz	J.J.	2007	SEA	AL	6	1	40	71.7	1.38	30	**17.859**
14	Langston	Mark	1987	SEA	AL	19	13	0	272.0	3.84	27	**17.850**
15	Moyer	Jamie	1998	SEA	AL	15	9	0	234.3	3.53	36	**17.212**
16	Fassero	Jeff	1997	SEA	AL	16	9	0	234.3	3.61	34	**16.405**

Rank	Name	First	Year	Team	Lg	W	L	SV	IP	ERA	Age	PEVA-P
17	Moyer	Jamie	2002	SEA	AL	13	8	0	230.7	3.32	40	**15.827**
18	Johnson	Randy	1994	SEA	AL	13	6	0	172.0	3.19	31	**15.782**
19	Langston	Mark	1988	SEA	AL	15	11	0	261.3	3.34	28	**13.939**
20	Sele	Aaron	2001	SEA	AL	15	5	0	215.0	3.60	31	**13.898**

Top Pitchers By Franchise (1871-2010)

Rank	Name	First	Year	Team	Lg	W	L	SV	IP	ERA	Age	PEVA-P
	San Francisco Giants											
1	Rusie	Amos	1894	NY1	NL	36	13	1	444.0	2.78	23	**55.357**
2	Hubbell	Carl	1933	NY1	NL	23	12	5	308.7	1.66	30	**52.001**
3	Hubbell	Carl	1936	NY1	NL	26	6	3	304.0	2.31	33	**48.808**
4	Hubbell	Carl	1934	NY1	NL	21	12	8	313.0	2.30	31	**43.860**
5	Marichal	Juan	1969	SFN	NL	21	11	0	299.7	2.10	32	**42.465**
6	Rusie	Amos	1893	NY1	NL	33	21	1	482.0	3.23	22	**37.905**
7	Lincecum	Tim	2009	SFN	NL	15	7	0	225.3	2.48	25	**37.546**
8	Maglie	Sal	1951	NY1	NL	23	6	4	298.0	2.93	34	**35.603**
9	Lincecum	Tim	2008	SFN	NL	18	5	0	227.0	2.62	24	**34.883**
10	Mathewson	Christy	1909	NY1	NL	25	6	2	275.3	1.14	29	**34.761**
11	McGinnity	Joe	1904	NY1	NL	35	8	5	408.0	1.61	33	**33.860**
12	Marichal	Juan	1966	SFN	NL	25	6	0	307.3	2.23	29	**33.441**
13	Melton	Cliff	1937	NY1	NL	20	9	7	248.0	2.61	25	**32.906**
14	Mathewson	Christy	1905	NY1	NL	31	9	2	338.7	1.28	25	**32.837**
15	Marichal	Juan	1965	SFN	NL	22	13	1	295.3	2.13	28	**32.690**
16	Meekin	Jouett	1894	NY1	NL	33	9	2	409.0	3.70	27	**32.597**
17	Mathewson	Christy	1911	NY1	NL	26	13	3	307.0	1.99	31	**32.300**
18	Mathewson	Christy	1908	NY1	NL	37	11	5	390.7	1.43	28	**32.234**
19	Jansen	Larry	1951	NY1	NL	23	11	0	278.7	3.04	31	**32.060**
20	Rusie	Amos	1897	NY1	NL	28	10	0	322.3	2.54	26	**31.574**

Note: Previous Names of Franchise: New York Gothams (NY1) 1883-1884; New York Giants (NY1) 1885-1957

Top Pitchers By Franchise (1871-2010)

Rank	Name	First	Year	Team	Lg	W	L	SV	IP	ERA	Age	PEVA-P
	St. Louis Red Stockings											
1	Morgan	Pidgey	1875	SL1	NA	1	3	0	42.0	3.43	22	**1.981**
2	Blong	Joe	1875	SL1	NA	3	12	0	129.0	3.35	22	**1.374**

Top Pitchers By Franchise (1871-2010)

Rank	Name	First	Year	Team	Lg	W	L	SV	IP	ERA	Age	PEVA-P
	St. Louis Brown Stockings (National Association)											
1	Bradley	George	1875	SL2	NA	33	26	0	535.7	2.05	23	**11.903**
2	Galvin	Pud	1875	SL2	NA	4	2	1	62.0	2.18	19	**2.934**
3	Fleet	Frank	1875	SL2	NA	2	1	0	27.0	2.33	27	**0.657**
4	Pearce	Dickey	1875	SL2	NA	0	0	0	5.3	3.38	39	**0.200**

Top Pitchers By Franchise (1871-2010)

Rank	Name	First	Year	Team	Lg	W	L	SV	IP	ERA	Age	PEVA-P
	St. Louis Brown Stockings (National League)											
1	Bradley	George	1876	SL3	NL	45	19	0	573.0	1.23	24	**36.677**
2	Nichols	Tricky	1877	SL3	NL	18	23	0	350.0	2.60	27	**4.735**
3	Blong	Joe	1877	SL3	NL	10	9	0	187.3	2.74	24	**1.578**
4	Blong	Joe	1876	SL3	NL	0	0	0	4.0	0.00	23	**0.200**
5	Battin	Joe	1877	SL3	NL	0	0	0	3.7	4.91	26	**0.200**

Top Pitchers By Franchise (1871-2010)

Rank	Name	First	Year	Team	Lg	W	L	SV	IP	ERA	Age	PEVA-P
	St. Louis Maroons											

Rank	Name	First	Year	Team	Lg	W	L	SV	IP	ERA	Age	PEVA-P
1	Sweeney	Charlie	1884	SLU	UA	24	7	0	271.0	1.83	21	**27.423**
2	Taylor	Billy	1884	SLU	UA	25	4	4	263.0	1.68	29	**16.180**
3	Boyle	Henry	1884	SLU	UA	15	3	1	150.0	1.74	24	**6.366**
4	Healy	John	1886	SL5	NL	17	23	0	353.7	2.88	20	**4.808**
5	Boyle	Henry	1886	SL5	NL	9	15	0	210.0	1.76	26	**4.806**

Top Pitchers By Franchise (1871-2010)

Rank	Name	First	Year	Team	Lg	W	L	SV	IP	ERA	Age	PEVA-P
	St. Louis Terriers											
1	Davenport	Dave	1915	SLF	FL	22	18	1	392.7	2.20	25	**21.910**
2	Plank	Eddie	1915	SLF	FL	21	11	3	268.3	2.08	40	**16.422**
3	Crandall	Doc	1915	SLF	FL	21	15	1	312.7	2.59	28	**8.983**
4	Groom	Bob	1914	SLF	FL	13	20	1	280.7	3.24	30	**4.791**
5	Watson	Doc	1914	SLF	FL	3	4	0	56.0	1.93	28	**4.762**

Top Pitchers By Franchise (1871-2010)

Rank	Name	First	Year	Team	Lg	W	L	SV	IP	ERA	Age	PEVA-P
	St. Louis Cardinals											
1	King	Silver	1888	SL4	AA	45	21	0	585.7	1.64	20	**56.728**
2	Cooper	Mort	1942	SLN	NL	22	7	0	278.7	1.78	29	**47.763**
3	Young	Cy	1899	SLN	NL	26	16	1	369.3	2.58	32	**40.907**
4	Tudor	John	1985	SLN	NL	21	8	0	275.0	1.93	31	**38.832**
5	Brecheen	Harry	1948	SLN	NL	20	7	1	233.3	2.24	34	**37.896**
6	Gibson	Bob	1968	SLN	NL	22	9	0	304.7	1.12	33	**36.485**
7	Wainwright	Adam	2010	SLN	NL	20	11	0	230.3	2.42	29	**35.709**
8	Dean	Dizzy	1934	SLN	NL	30	7	7	311.7	2.66	24	**35.666**
9	Carpenter	Chris	2009	SLN	NL	17	4	0	192.7	2.24	34	**35.125**
10	Andujar	Joaquin	1982	SLN	NL	15	10	0	265.7	2.47	30	**34.352**
11	Tewksbury	Bob	1992	SLN	NL	16	5	0	233.0	2.16	32	**32.658**
12	Haines	Jesse	1927	SLN	NL	24	10	1	300.7	2.72	34	**32.652**
13	Gibson	Bob	1969	SLN	NL	20	13	0	314.0	2.18	34	**32.616**
14	Dean	Dizzy	1935	SLN	NL	28	12	5	325.3	3.04	25	**32.595**
15	Carpenter	Chris	2005	SLN	NL	21	5	0	241.7	2.83	30	**32.538**
16	Dean	Dizzy	1936	SLN	NL	24	13	11	315.0	3.17	26	**30.183**
17	Carpenter	Chris	2006	SLN	NL	15	8	0	221.7	3.09	31	**29.268**
18	Cooper	Mort	1943	SLN	NL	21	8	3	274.0	2.30	30	**28.795**
19	Wainwright	Adam	2009	SLN	NL	19	8	0	233.0	2.63	28	**28.015**
20	Gibson	Bob	1970	SLN	NL	23	7	0	294.0	3.12	35	**27.659**

Note: Previous Names of Franchise:

St. Louis Brown Stockings (SL4) AA 1882; St. Louis Browns (SL4) AA 1883-1898; St. Louis Perfectos 1899

Top Pitchers By Franchise (1871-2010)

Rank	Name	First	Year	Team	Lg	W	L	SV	IP	ERA	Age	PEVA-P
	St. Paul Apostles											
1	Brown	Jim	1884	SPU	UA	1	4	0	36.0	3.75	24	**0.944**
2	Galvin	Lou	1884	SPU	UA	0	2	0	25.0	2.88	22	**0.666**
3	O'Brien	Billy	1884	SPU	UA	1	0	0	10.0	1.80	24	**0.200**

Top Pitchers By Franchise (1871-2010)

Rank	Name	First	Year	Team	Lg	W	L	SV	IP	ERA	Age	PEVA-P
	Syracuse Stars											
1	McCormick	Harry	1879	SR1	NL	18	33	0	457.3	2.99	24	**3.295**
2	Purcell	Blondie	1879	SR1	NL	4	15	0	179.7	3.76	25	**0.775**
3	Dorgan	Mike	1879	SR1	NL	0	0	0	12.0	2.25	26	**0.200**

Top Pitchers By Franchise (1871-2010)

Rank	Name	First	Year	Team	Lg	W	L	SV	IP	ERA	Age	PEVA-P
	Syracuse Stars (American Association)											
1	Casey	Dan	1890	SR2	AA	19	22	0	360.7	4.14	28	**3.522**
2	Keefe	John	1890	SR2	AA	17	24	0	352.3	4.32	23	**3.209**
3	McCullough	Charlie	1890	SR2	AA	1	2	0	26.0	7.27	24	**1.280**
4	Mars	Ed	1890	SR2	AA	9	5	0	121.3	4.67	24	**0.971**
5	Lincoln	Ezra	1890	SR2	AA	0	3	0	20.0	10.35	22	**0.871**

Top Pitchers By Franchise (1871-2010)

Rank	Name	First	Year	Team	Lg	W	L	SV	IP	ERA	Age	PEVA-P
	Tampa Bay Devil Rays											
1	Price	David	2010	TBA	AL	19	8	0	208.7	2.72	25	**19.172**
2	Shields	James	2008	TBA	AL	14	8	0	215.0	3.56	27	**18.346**
3	Shields	James	2007	TBA	AL	12	8	0	215.0	3.85	26	**16.713**
4	Kazmir	Scott	2007	TBA	AL	13	9	0	206.7	3.48	23	**16.131**
5	Soriano	Rafael	2010	TBA	AL	3	2	45	62.3	1.73	31	**14.495**
6	Arrojo	Rolando	1998	TBA	AL	14	12	0	202.0	3.56	30	**12.370**
7	Shields	James	2009	TBA	AL	11	12	0	219.7	4.14	28	**10.877**
8	Garza	Matt	2008	TBA	AL	11	9	0	184.7	3.70	25	**10.615**
9	Garza	Matt	2010	TBA	AL	15	10	1	204.7	3.91	27	**10.151**

Top Pitchers By Franchise (1871-2010)

Rank	Name	First	Year	Team	Lg	W	L	SV	IP	ERA	Age	PEVA-P
	Texas Rangers											
1	Jenkins	Fergie	1974	TEX	AL	25	12	0	328.3	2.82	32	**34.395**
2	Hough	Charlie	1987	TEX	AL	18	13	0	285.3	3.79	39	**19.948**
3	Brown	Kevin	1992	TEX	AL	21	11	0	265.7	3.32	27	**19.711**
4	Kern	Jim	1979	TEX	AL	13	5	29	143.0	1.57	30	**18.109**
5	Matlack	Jon	1978	TEX	AL	15	13	1	270.0	2.27	28	**17.730**
6	Rogers	Kenny	1995	TEX	AL	17	7	0	208.0	3.38	31	**17.288**
7	Perry	Gaylord	1975	TEX	AL	12	8	0	184.0	3.03	37	**16.820**
8	Ryan	Nolan	1989	TEX	AL	16	10	0	239.3	3.20	42	**16.581**
9	Blyleven	Bert	1977	TEX	AL	14	12	0	234.7	2.72	26	**16.187**
10	Tanana	Frank	1984	TEX	AL	15	15	0	246.3	3.25	31	**15.957**
11	Hill	Ken	1996	TEX	AL	16	10	0	250.7	3.63	31	**15.910**
12	Wilson	C.J.	2010	TEX	AL	15	8	0	204.0	3.35	30	**15.168**
13	Russell	Jeff	1989	TEX	AL	6	4	38	72.7	1.98	28	**14.931**
14	Bosman	Dick	1969	WS2	AL	14	5	1	193.0	2.19	25	**14.855**
15	Ryan	Nolan	1991	TEX	AL	12	6	0	173.0	2.91	44	**14.595**
16	Hough	Charlie	1984	TEX	AL	16	14	0	266.0	3.76	36	**14.414**
17	Helling	Rick	2000	TEX	AL	16	13	0	217.0	4.48	30	**13.951**
18	Helling	Rick	1998	TEX	AL	20	7	0	216.3	4.41	28	**13.633**
19	Brown	Kevin	1993	TEX	AL	15	12	0	233.0	3.59	28	**13.519**
20	Sele	Aaron	1998	TEX	AL	19	11	0	212.7	4.23	28	**13.483**

Note: Previous Names of Franchise: Washington Senators (WS2) 1961-1971

Top Pitchers By Franchise (1871-2010)

Rank	Name	First	Year	Team	Lg	W	L	SV	IP	ERA	Age	PEVA-P
	Toledo Blue Stockings											
1	Mullane	Tony	1884	TL1	AA	36	26	0	567.0	2.52	25	**15.319**
2	O'Day	Hank	1884	TL1	AA	9	28	1	326.7	3.75	22	**2.252**
3	Kent	Ed	1884	TL1	AA	0	1	0	9.0	6.00	25	**0.200**
4	Morton	Charlie	1884	TL1	AA	0	1	0	23.3	3.09	30	**0.200**
5	Brown	Ed	1884	TL1	AA	0	1	0	9.0	9.00	1884	**0.200**

Top Pitchers By Franchise (1871-2010)

Rank	Name	First	Year	Team	Lg	W	L	SV	IP	ERA	Age	PEVA-P
	Toledo Maumees											
1	Healy	John	1890	TL2	AA	22	21	0	389.0	2.89	24	**14.100**
2	Smith	Fred	1890	TL2	AA	19	13	0	286.0	3.27	27	**5.225**
3	Cushman	Ed	1890	TL2	AA	17	21	1	315.7	4.19	38	**3.263**
4	Sprague	Charlie	1890	TL2	AA	9	5	0	122.7	3.89	26	**1.399**
5	O'Neil	Ed	1890	TL2	AA	0	2	0	16.0	7.88	31	**0.747**

Top Pitchers By Franchise (1871-2010)

Rank	Name	First	Year	Team	Lg	W	L	SV	IP	ERA	Age	PEVA-P
	Toronto Blue Jays											
1	Clemens	Roger	1997	TOR	AL	21	7	0	264.0	2.05	35	**47.079**
2	Halladay	Roy	2008	TOR	AL	20	11	0	246.0	2.78	31	**46.107**
3	Halladay	Roy	2003	TOR	AL	22	7	0	266.0	3.25	26	**36.858**
4	Key	Jimmy	1987	TOR	AL	17	8	0	261.0	2.76	26	**36.007**
5	Clemens	Roger	1998	TOR	AL	20	6	0	234.7	2.65	36	**33.173**
6	Halladay	Roy	2009	TOR	AL	17	10	0	239.0	2.79	32	**31.554**
7	Halladay	Roy	2006	TOR	AL	16	5	0	220.0	3.19	29	**30.499**
8	Stieb	Dave	1984	TOR	AL	16	8	0	267.0	2.83	27	**29.332**
9	Alexander	Doyle	1984	TOR	AL	17	6	0	261.7	3.13	34	**26.207**
10	Halladay	Roy	2002	TOR	AL	19	7	0	239.3	2.93	25	**23.939**
11	Stieb	Dave	1982	TOR	AL	17	14	0	288.3	3.25	25	**22.711**
12	Stieb	Dave	1983	TOR	AL	17	12	0	278.0	3.04	26	**22.592**
13	Hentgen	Pat	1996	TOR	AL	20	10	0	265.7	3.22	28	**22.520**
14	Candiotti	Tom	1991	TOR	AL	6	7	0	129.7	2.98	34	**22.447**
15	Wells	David	2000	TOR	AL	20	8	0	229.7	4.11	37	**21.659**
16	Cone	David	1992	TOR	AL	4	3	0	53.0	2.55	29	**18.790**
17	Cone	David	1995	TOR	AL	9	6	0	130.3	3.38	32	**18.701**
18	Key	Jimmy	1991	TOR	AL	16	12	0	209.3	3.05	30	**18.636**
20	Stieb	Dave	1981	TOR	AL	11	10	0	183.7	3.19	24	**18.245**

Top Pitchers By Franchise (1871-2010)

Rank	Name	First	Year	Team	Lg	W	L	SV	IP	ERA	Age	PEVA-P
	Troy Trojans											
1	Welch	Mickey	1880	TRN	NL	34	30	0	574.0	2.54	21	**9.344**
2	Welch	Mickey	1881	TRN	NL	21	18	0	368.0	2.67	22	**5.545**
3	Goldsmith	Fred	1879	TRN	NL	2	4	0	63.0	1.57	23	**4.900**
4	Keefe	Tim	1882	TRN	NL	17	26	0	375.0	2.50	25	**4.868**
5	Keefe	Tim	1881	TRN	NL	18	27	0	402.0	3.25	24	**4.088**

Top Pitchers By Franchise (1871-2010)

Rank	Name	First	Year	Team	Lg	W	L	SV	IP	ERA	Age	PEVA-P
	Troy Haymakers											
1	Zettlein	George	1872	TRO	NA	14	8	0	187.7	2.54	28	**6.833**
2	McMullin	John	1871	TRO	NA	12	15	0	249.0	5.53	23	**6.373**
3	Martin	Phonney	1872	TRO	NA	1	2	0	37.3	5.79	27	**0.686**
4	Flowers	Dickie	1871	TRO	NA	0	0	0	1.0	0.00	21	**0.200**

Top Pitchers By Franchise (1871-2010)

Rank	Name	First	Year	Team	Lg	W	L	SV	IP	ERA	Age	PEVA-P
	Washington Nationals											
1	Martinez	Pedro	1997	MON	NL	17	8	0	241.3	1.90	26	**38.574**
2	Rogers	Steve	1982	MON	NL	19	8	0	277.0	2.40	33	**36.926**
3	Martinez	Dennis	1992	MON	NL	16	11	0	226.3	2.47	37	**24.770**
4	Vazquez	Javier	2001	MON	NL	16	11	0	223.7	3.42	25	**24.621**
5	Martinez	Dennis	1991	MON	NL	14	11	0	222.0	2.39	36	**23.324**

Rank	Name	First	Year	Team	Lg	W	L	SV	IP	ERA	Age	PEVA-P
6	Rogers	Steve	1977	MON	NL	17	16	0	301.7	3.10	28	**21.900**
7	Vazquez	Javier	2003	MON	NL	13	12	0	230.7	3.24	27	**21.294**
8	Wetteland	John	1993	MON	NL	9	3	43	85.3	1.37	27	**20.171**
9	Hernandez	Livan	2003	MON	NL	15	10	0	233.3	3.20	28	**19.856**
10	Colon	Bartolo	2002	MON	NL	10	4	0	117.0	3.31	29	**19.341**
11	Fassero	Jeff	1996	MON	NL	15	11	0	231.7	3.30	33	**18.992**
12	Langston	Mark	1989	MON	NL	12	9	0	176.7	2.39	29	**18.250**
13	Rogers	Steve	1980	MON	NL	16	11	0	281.0	2.98	31	**17.538**
14	Martinez	Dennis	1988	MON	NL	15	13	0	235.3	2.72	33	**17.293**
15	Smith	Bryn	1985	MON	NL	18	5	0	222.3	2.91	30	**16.655**
16	Hill	Ken	1992	MON	NL	16	9	0	218.0	2.68	27	**16.391**
17	Perez	Pascual	1988	MON	NL	12	8	0	188.0	2.44	31	**15.887**
18	Rogers	Steve	1983	MON	NL	17	12	0	273.0	3.23	34	**15.775**
19	Hill	Ken	1994	MON	NL	16	5	0	154.7	3.32	29	**15.672**
20	Martinez	Dennis	1990	MON	NL	10	11	0	226.0	2.95	35	**15.470**

Note: Previous Names of Franchise: Montreal Expos (MON) 1969-2004

Top Pitchers By Franchise (1871-2010)

Rank	Name	First	Year	Team	Lg	W	L	SV	IP	ERA	Age	PEVA-P
	Wilmington Quicksteps											
1	Bakely	Jersey	1884	WIL	UA	0	2	0	17.0	4.24	20	**3.870**
2	Tenney	Fred	1884	WIL	UA	0	1	0	8.0	1.12	25	**2.110**
3	Murphy	John	1884	WIL	UA	0	6	0	48.0	3.00	NA	**1.076**
4	Nolan	The Only	1884	WIL	UA	1	4	0	40.0	2.92	27	**0.999**
5	McElroy	Jim	1884	WIL	UA	0	1	0	5.0	10.80	22	**0.896**

Top Pitchers By Franchise (1871-2010)

Rank	Name	First	Year	Team	Lg	W	L	SV	IP	ERA	Age	PEVA-P
	Worcester Ruby Legs											
1	Richmond	Lee	1880	WOR	NL	32	32	3	590.7	2.15	23	**13.353**
2	Richmond	Lee	1881	WOR	NL	25	26	0	462.3	3.39	24	**7.387**
3	Richmond	Lee	1882	WOR	NL	14	33	0	411.0	3.74	25	**3.698**
4	Corey	Fred	1880	WOR	NL	8	9	2	148.3	2.43	25	**1.403**
5	Mountain	Frank	1882	WOR	NL	0	5	0	42.0	3.00	22	**1.132**
	Mountain	Frank	1882	WOR	NL	2	11	0	102.0	3.97	22	**1.132**

Note: Frank Mountain statistics reflect two stints with Worcester in 1882

Top Pitchers By Franchise (1871-2010)

Rank	Name	First	Year	Team	Lg	W	L	SV	IP	ERA	Age	PEVA-P
	Washington Olympics											
1	Brainard	Asa	1871	WS3	NA	12	15	0	264.0	4.50	30	**7.299**
2	Brainard	Asa	1872	WS3	NA	2	7	0	79.0	6.38	31	**0.696**
3	Stearns	Bill	1871	WS3	NA	2	0	0	18.0	2.50	18	**0.200**

Top Pitchers By Franchise (1871-2010)

Rank	Name	First	Year	Team	Lg	W	L	SV	IP	ERA	Age	PEVA-P
	Washington Nationals (National Association)											
1	Stearns	Bill	1872	WS4	NA	0	11	0	99.0	6.91	19	**0.491**

Top Pitchers By Franchise (1871-2010)

Rank	Name	First	Year	Team	Lg	W	L	SV	IP	ERA	Age	PEVA-P
	Washington Blue Legs											
1	Stearns	Bill	1873	WS5	NA	7	25	0	283.0	4.55	20	**2.029**
2	Greason	John	1873	WS5	NA	1	6	0	63.0	5.43	22	**0.482**

Top Pitchers By Franchise (1871-2010)

Rank	Name	First	Year	Team	Lg	W	L	SV	IP	ERA	Age	PEVA-P
	Washington Nationals (National Association)											
1	Stearns	Bill	1875	WS6	NA	1	14	0	141.0	5.36	22	**1.014**
2	Parks	Bill	1875	WS6	NA	4	8	0	106.7	4.05	26	**0.671**
3	Witherow	Charles	1875	WS6	NA	0	1	0	1.0	18.00	23	**0.200**
4	Mason	Charlie	1875	WS6	NA	0	0	0	2.0	4.50	22	**0.200**

Top Pitchers By Franchise (1871-2010)

Rank	Name	First	Year	Team	Lg	W	L	SV	IP	ERA	Age	PEVA-P
	Washington Nationals (American Association)											
1	Barr	Bob	1884	WS7	AA	9	23	0	281.0	3.46	28	**3.228**
2	Hamill	John	1884	WS7	AA	2	17	0	156.7	4.48	24	**0.855**
3	Trumbull	Ed	1884	WS7	AA	1	9	0	84.0	4.71	24	**0.622**
4	Smith	Edgar (EE)	1884	WS7	AA	0	2	0	22.0	4.91	22	**0.200**

Top Pitchers By Franchise (1871-2010)

Rank	Name	First	Year	Team	Lg	W	L	SV	IP	ERA	Age	PEVA-P
	Washington Nationals (National League)											
1	Whitney	Jim	1887	WS8	NL	24	21	0	404.7	3.22	30	**15.607**
2	O'Day	Hank	1888	WS8	NL	16	29	0	403.0	3.10	26	**3.935**
3	Shaw	Dupee	1886	WS8	NL	13	31	0	385.7	3.34	27	**3.651**
4	Whitney	Jim	1888	WS8	NL	18	21	0	325.0	3.05	31	**3.090**
5	Gilmore	Frank	1886	WS8	NL	4	4	0	75.0	2.52	22	**2.525**

Top Pitchers By Franchise (1871-2010)

Rank	Name	First	Year	Team	Lg	W	L	SV	IP	ERA	Age	PEVA-P
	Washington Statesmen/Senators											
1	Mercer	Win	1897	WSN	NL	20	20	3	333.0	3.24	23	**18.174**
2	Mercer	Win	1896	WSN	NL	25	18	0	366.3	4.13	22	**12.345**
3	McJames	Doc	1897	WSN	NL	15	23	2	323.7	3.61	23	**11.355**
4	Mercer	Win	1894	WSN	NL	17	23	3	336.3	3.85	20	**9.525**
5	Killen	Frank	1892	WSN	NL	29	26	0	459.7	3.31	22	**7.801**

Note: Previous Names of Franchise: Washington Statesmen (WS9) 1899; Washington Senators (WSN) 1892-1899

Top Pitchers By Franchise (1871-2010)

Rank	Name	First	Year	Team	Lg	W	L	SV	IP	ERA	Age	PEVA-P
	Washington Nationals (Union League)											
1	Daily	Hugh	1884	WSU	UA	1	1	0	16.0	2.25	27	**10.851**
2	Wise	Bill	1884	WSU	UA	23	18	0	364.3	3.04	23	**4.409**
3	Geggus	Charlie	1884	WSU	UA	10	9	0	177.3	2.54	22	**3.031**
4	Voss	Alex	1884	WSU	UA	5	14	0	186.3	3.57	26	**1.420**
5	Powell	Abner	1884	WSU	UA	6	12	0	134.0	3.43	24	**0.990**

Note1: For players with more than one team during the year, actual stats reflect stint with team only, PEVA values reflect Total Year.

Note2: Franchises with less than 5 players above 10.000 PEVA are listed with the five highest players.

Note3: Age = Player age at end of year.

Note: 2010 PEVA = Reflects preliminary final PEVA numbers. Marginal adjustment may be made with final stats and park factors.

Chapter 4 - BEST CAREERS EVER, POSITION PLAYERS

The list contains the best of all eras, from the days just past the turn of the century, to the years of World War II, the great 1950s, as well as today. It holds the key players of teams from coast to coast, and everywhere in between. And the players still plying their trade today, yes, they are moving up the list with lightning speed, particularly a man named Albert and Rodriguez, although not yet quite into the Top Ten. But for now as we start to discuss that exclusive of exclusive clubs, the men on that list are truly amazing, and names are within the lexicon of any baseball fan with no surprises, except for the debate of who should place where.

The rankings of the list below reflect Regular Season values only and only include PEVA-B (Batting) Ratings, so you won't see Babe's pitching numbers included. The PEVA-B per year number is purely a division of the total rating divided by the number of years played, but the stat that might be even more appropriate to compare is the Per EQ Year stat. This stat shows the average rating for a player for his average full season. It's a pretty good indicator of how a player would rank, pushing longevity to the side for a while. That's not perfect either, however, as some players with shorter careers may seem better than they are, because they truncated the downturn seasons of their late 30s while others played longer, with limited stats, into their 40s.

But let's stop the explanation of the explanation now and get to the rankings of the Position Players who had the best Batting Careers ever in the history of baseball, and the man at the top will come as no surprise again. It is Babe Ruth. And after Babe, you'll see a list of the players from 2010 who are moving up the list. Look out folks, there a Pujols on the rise.

1. Babe Ruth (1914-1935)

When you dominate the game during an entire decade to the point where almost every season ranks as one of the best seasons ever, it's not hard to imagine that when you cobble a career together, it ranks as the best in the long history of the game. Babe Ruth was, and still is, an icon that's hard for the public to imagine today. Yes, he was the Tiger Woods of baseball, if Tiger laps Nicklaus for the majors record by a mile. There was nobody in the game prior to Babe Ruth who was so dominant and nobody who has come along in later times who accomplished as much. There are some who prefer the grace and power of Mays or the dogged determination and consistency of the magnificent Hank, and there's logic in that passion. However, when the passion subsides and the newsreel footage rolls, it is Babe Ruth who accumulated statistics that almost defy logic. In a career of 154 game seasons, he hit 714 Home Runs, one per every 11.6 At Bats. All with a glass of beer and no PED allegations.

Where Ruth Ranks		
Category		**Rank**
HR	714	3
RBI	2213	2
AVE	0.342	10
OBP	0.474	2
SLG	0.690	1
HITS	2873	39
RUNS	2174	5

It wasn't until he was 24 years old that Ruth even became a full-time position player, dividing time before then, while with the Boston Red Sox, as one of the best pitchers in the game. And yes, he could really pitch, winning 94 games with an ERA of 2.28. If you add his pitching rating to the Batting PEVA, it would rise by nearly 50 points. But it would be with that big stick where Ruth would amaze the crowds of folks who filled the seats at Yankee Stadium. His Slugging Percentage is the highest in baseball history, at 0.690 over 56 points higher than the man in second place, Ted Williams, at 0.634.

He had 17 seasons with more than 20 home runs and hit 0.342 for his entire career. There was no more dominant player in the game during his career than the man that built a stadium and rekindled a game after scandal. And he did it with a style that marveled the people of the times. Babe Ruth ranks as the Best Ever Career Batter in baseball history, with nobody really close.

Babe Ruth Stats					**641.541 PEVA RATING**		
Year	**Team**	**Lg**	**HR**	**RBI**	**AVE**	**Age**	**PEVA-B**
1914	BOS	AL	**0**	**2**	**0.200**	19	0.200
1915	BOS	AL	**4**	**21**	**0.315**	20	2.467
1916	BOS	AL	**3**	**15**	**0.272**	21	1.693

1917	BOS	AL	**2**	**12**	**0.325**	22	2.020
1918	BOS	AL	**11**	**66**	**0.300**	23	22.063
1919	BOS	AL	**29**	**114**	**0.322**	24	45.484
1920	NYA	AL	**54**	**137**	**0.376**	25	55.754
1921	NYA	AL	**59**	**171**	**0.378**	26	54.876
1922	NYA	AL	**35**	**99**	**0.315**	27	15.781
1923	NYA	AL	**41**	**131**	**0.393**	28	58.931
1924	NYA	AL	**46**	**121**	**0.378**	29	51.324
1925	NYA	AL	**25**	**66**	**0.290**	30	7.608
1926	NYA	AL	**47**	**146**	**0.372**	31	51.603
1927	NYA	AL	**60**	**164**	**0.356**	32	51.850
1928	NYA	AL	**54**	**142**	**0.323**	33	40.061
1929	NYA	AL	**46**	**154**	**0.345**	34	31.530
1930	NYA	AL	**49**	**153**	**0.359**	35	39.019
1931	NYA	AL	**46**	**163**	**0.373**	36	41.968
1932	NYA	AL	**41**	**137**	**0.341**	37	29.556
1933	NYA	AL	**34**	**103**	**0.301**	38	23.392
1934	NYA	AL	**22**	**84**	**0.288**	39	13.487
1935	BSN	NL	**6**	**12**	**0.181**	40	0.872
		Total	**714**	**2213**	**0.342**		

2. Barry Bonds (1986-2007)

It's kind of a shame really. But there's no good way to start this explanation of the second best player in baseball history, from a statistical standpoint, than to put it out front. PED allegations are a part of his story, whether we like it or not. Barry Bonds played in the steroid era. We'll leave it at that. How many, if any of his exploits are enhanced by playing in that time, we don't know. But he played in the era, enough said. There's no doubt that Barry Bonds was a great player. He had speed, power, defensive ability in his early years with Pittsburgh, and was one of the best players of his time. When Bonds came to the plate, managers shuddered. And they walked him just to avoid the inevitable, "Why did you pitch to him?" questions if the next pitch was deposited in the seats or even the bay. Bonds was a complete player from the start of his career, and by the end of it, had stolen 514 bases, #32 All-Time, to go along with those Home Runs and RBIs.

Where Bonds Ranks		
Category		**Rank**
HR	762	1
RBI	1996	3
AVE	0.298	230
OBP	0.444	6
SLG	0.607	6
HITS	2935	31
RUNS	2227	3

To get an idea of just how great a player Bonds was in his early years, just take a gander at the last three years he played for the Pirates. 33, 116, 0.301 in 1990 followed by 25, 116, 0.292 in 1991 then 34, 103, 0.311. Bonds was well on his way to a stellar career then. Now just where it would rank if he remained on the path begun in Pittsburgh, it's only speculation. It's doubtful that he becomes a Top Ten player in baseball history, but he'd still rank pretty high. But according to the stats of his entire career, Bonds ranks as the 2nd best player. Discount the controversial side of the equation as much as you'd like. It's the best we can do at this moment in time. Bonds dominated the era that he played in, at a level that surpassed the others in the same boat. That's what the numbers say. And no, we don't like it.

Barry Bonds Stats					**606.700 PEVA RATING**		
Year	**Team**	**Lg**	**HR**	**RBI**	**AVE**	**Age**	**PEVA-B**
1986	PIT	NL	**16**	**48**	**0.223**	22	5.543
1987	PIT	NL	**25**	**59**	**0.261**	23	12.642
1988	PIT	NL	**24**	**58**	**0.283**	24	15.439
1989	PIT	NL	**19**	**58**	**0.248**	25	14.892
1990	PIT	NL	**33**	**114**	**0.301**	26	36.095
1991	PIT	NL	**25**	**116**	**0.292**	27	28.310
1992	PIT	NL	**34**	**103**	**0.311**	28	35.418

Year	Team	Lg	HR	RBI	AVE	Age	PEVA-B
1993	SFN	NL	**46**	**123**	**0.336**	29	43.404
1994	SFN	NL	**37**	**81**	**0.312**	30	27.257
1995	SFN	NL	**33**	**104**	**0.294**	31	30.088
1996	SFN	NL	**42**	**129**	**0.308**	32	33.905
1997	SFN	NL	**40**	**101**	**0.291**	33	31.253
1998	SFN	NL	**37**	**122**	**0.303**	34	27.324
1999	SFN	NL	**34**	**83**	**0.262**	35	11.750
2000	SFN	NL	**49**	**106**	**0.306**	36	32.041
2001	SFN	NL	**73**	**137**	**0.328**	37	55.207
2002	SFN	NL	**46**	**110**	**0.370**	38	54.848
2003	SFN	NL	**45**	**90**	**0.341**	39	31.974
2004	SFN	NL	**45**	**101**	**0.362**	40	48.632
2005	SFN	NL	**5**	**10**	**0.286**	41	1.555
2006	SFN	NL	**26**	**77**	**0.270**	42	15.305
2007	SFN	NL	**28**	**66**	**0.276**	43	13.819
		Total	**762**	**1996**	**0.298**		

3. Ty Cobb (1905-1928)

They say he was one of the toughest, grittiest baseball players in the history of the game, and he was certainly one of the best. Look at those career ranks, the best Batting Average ever, and the 2nd most Hits and Runs Scored. Also number 6 in Runs Batted In. A phenom from the time he was 19, this Detroit Tiger had more than one in his tank, he had a pride. And longevity was his game as well, batting 0.357 and knocking in 93 runs in 1927 at the age of 41.

Where Cobb Ranks		
Category		**Rank**
HR	117	594
RBI	1937	6
AVE	0.366	1
OBP	0.433	9
SLG	0.512	75
HITS	4189	2
RUNS	2246	2

But then, by that time, he was getting used to the heights of the batting average stratosphere. When you swing it at a clip of 0.366 for your career, you should think of nothing less. Not one season between the ages of 20 and 42 when he hit less than 0.300. Three seasons over 0.400. PEVA numbers for individual seasons that peaked above 40.000 for three seasons between 1911 and 1917, including his 1911 season when he hit an astounding 0.420.

There hasn't been a lot of chatter about Ty Cobb lately, really not since the Pete Rose chase of his hits record a generation of baseball ago. Ruth, Mays, Williams, and Aaron get more publicity in the debates about best players ever. Maybe it's because he was dissed by Joe Jackson in the movie Field of Dreams. But whatever the reason, on the merits of the debate, Ty Cobb has got to be mentioned way up in the ladder of success. Ty Cobb is the 3rd Best Batter Ever in our list of the Best of the Best.

Ty Cobb Stats					**569.252 PEVA RATING**		
Year	**Team**	**Lg**	**HR**	**RBI**	**AVE**	**Age**	**PEVA-B**
1905	DET	AL	**1**	**15**	**0.240**	19	0.625
1906	DET	AL	**1**	**34**	**0.316**	20	6.582
1907	DET	AL	**5**	**119**	**0.350**	21	31.905
1908	DET	AL	**4**	**108**	**0.324**	22	25.112
1909	DET	AL	**9**	**107**	**0.377**	23	39.724
1910	DET	AL	**8**	**91**	**0.383**	24	32.427
1911	DET	AL	**8**	**127**	**0.420**	25	42.464
1912	DET	AL	**7**	**83**	**0.409**	26	36.188
1913	DET	AL	**4**	**67**	**0.390**	27	23.325
1914	DET	AL	**2**	**57**	**0.368**	28	13.516
1915	DET	AL	**3**	**99**	**0.369**	29	43.671
1916	DET	AL	**5**	**68**	**0.371**	30	34.488
1917	DET	AL	**6**	**102**	**0.383**	31	48.911

1918	DET	AL	**3**	**64**	**0.382**	32	37.701
1919	DET	AL	**1**	**70**	**0.384**	33	25.384
1920	DET	AL	**2**	**63**	**0.334**	34	9.602
1921	DET	AL	**12**	**101**	**0.389**	35	21.150
1922	DET	AL	**4**	**99**	**0.401**	36	24.441
1923	DET	AL	**6**	**88**	**0.340**	37	15.619
1924	DET	AL	**4**	**78**	**0.338**	38	20.008
1925	DET	AL	**12**	**102**	**0.378**	39	17.033
1926	DET	AL	**4**	**62**	**0.339**	40	3.646
1927	PHA	AL	**5**	**93**	**0.357**	41	12.198
1928	PHA	AL	**1**	**40**	**0.323**	42	3.534
		Total	**117**	**1937**	**0.366**		

4. Hank Aaron (1954-1976)

Who was better, Aaron or Mays? Now that's a debate that's been going round and round since they both took their first swings in the early 1950s. Yes, Mays was the more flashy player, patrolling the outfields of the Polo Grounds and Candlestick Park like a graceful football player with a fantastic glove. But Aaron was the more consistent, never deviating from his steady man, stat producing season batter from Milwaukee to Atlanta as the Braves traversed the nation. So it becomes a little ironic that the two men who were so compared to each other during their career should sit right next to each other in the Best Ever rankings, with this time Hank taking a higher spot at #4. And they were only 15 PEVA points apart.

Where Aaron Ranks		
Category		**Rank**
HR	755	2
RBI	2297	1
AVE	0.305	146
OBP	0.374	222
SLG	0.555	26
HITS	3771	3
RUNS	2174	4

When Hank Aaron hit the Home Run that bested the record of Ruth, it was national news, and news that even surpassed the game itself. For most fans of baseball, Hank Aaron not only has the record for most Runs Batted In All-Time, actually the better record if you think about it, but he still has the Home Run Record. And whether that record ever gets an asterick next to it post steroid ball days or not, the point is likely moot to most. Aaron slugged his way into the record books on pure guts and determination and it was a pleasure to watch both.

Hank was not the most selective hitter, which might account for the reason his ranking isn't a notch or two higher. He was productive and durable. In the end, that's what a team wants, someone to be counted on,. day after day after day, to help win games. That's certainly what Aaron was all about from Milwaukee to Atlanta to Cooperstown.

Hank Aaron Stats					**535.808 PEVA RATING**		
Year	**Team**	**Lg**	**HR**	**RBI**	**AVE**	**Age**	**PEVA-B**
1954	ML1	NL	**13**	**69**	**0.280**	20	6.510
1955	ML1	NL	**27**	**106**	**0.314**	21	19.849
1956	ML1	NL	**26**	**92**	**0.328**	22	19.663
1957	ML1	NL	**44**	**132**	**0.322**	23	29.093
1958	ML1	NL	**30**	**95**	**0.326**	24	28.666
1959	ML1	NL	**39**	**123**	**0.355**	25	41.920
1960	ML1	NL	**40**	**126**	**0.292**	26	33.522
1961	ML1	NL	**34**	**120**	**0.327**	27	27.731
1962	ML1	NL	**45**	**128**	**0.323**	28	37.264
1963	ML1	NL	**44**	**130**	**0.319**	29	40.268
1964	ML1	NL	**24**	**95**	**0.328**	30	24.270
1965	ML1	NL	**32**	**89**	**0.318**	31	28.502
1966	ATL	NL	**44**	**127**	**0.279**	32	27.049
1967	ATL	NL	**39**	**109**	**0.307**	33	32.133
1968	ATL	NL	**29**	**86**	**0.287**	34	27.112

1969	ATL	NL	**44**	**97**	**0.300**	35	25.597
1970	ATL	NL	**38**	**118**	**0.298**	36	19.248
1971	ATL	NL	**47**	**118**	**0.327**	37	27.080
1972	ATL	NL	**34**	**77**	**0.265**	38	12.901
1973	ATL	NL	**40**	**96**	**0.301**	39	15.849
1974	ATL	NL	**20**	**69**	**0.268**	40	5.635
1975	ML4	AL	**12**	**60**	**0.234**	41	4.436
1976	ML4	AL	**10**	**35**	**0.229**	42	1.510
		Total	**755**	**2297**	**0.305**		

5. Willie Mays (1951-1973)

It's an indelible impression that over the shoulder catch, one of the most played baseball highlights in baseball history. Boy, just imagine if there were a Sports Center highlights show every night in those days. It's hard to believe Mays struggled when he first came to the majors, because once Willie got his footing on the turf of the Polo Grounds, there was no stopping him. By the age of 23, Mays was knocking out 40 home run seasons and batting 0.345. Beside the comparisons made with Aaron, the streets of New York were having. their own debates of spectacular outfielders all within the same city. Who was better? Mays of the Giants. Snider of the Dodgers. Mantle of the Yankees. Well, for a career, it was Willie, sitting at #5 on the All-Time Best Careers ever by a position player.

Where Mays Ranks		
Category		**Rank**
HR	660	4
RBI	1903	9
AVE	0.302	187
OBP	0.384	141
SLG	0.558	21
HITS	3283	10
RUNS	2062	7

Mays had 660 home runs in his career, good enough for 4th place on the career list, and he stole 338 bases. This was the ultimate 5 tool baseball player. He could field, hit, hit for power, run, and throw. Say Hey Willie Mays from the streets of New York City to the San Francisco Bay. Can you imagine waking up after the 1957 season a fan of a player of this caliber and finding out he was now going to play in San Francisco, along with the rest of his team when the migration of teams to the west took both the Giants, Dodgers, and Athletics out of their northeasten homes. But oh those new fans in California were in for a treat. Mays' career waned during the final years as he came back east for a few seasons, and he got some criticism for that. But, boy, that's about the only thing bad you could say about the marvelous player called Mays who made being a baseball fan for two decades from 1951 forward a fantastic journey.

Willie Mays Stats					**520.998 PEVA RATING**		
Year	**Team**	**Lg**	**HR**	**RBI**	**AVE**	**Age**	**PEVA-B**
1951	NY1	NL	**20**	**68**	**0.274**	20	8.512
1952	NY1	NL	**4**	**23**	**0.236**	21	1.484
1953							
1954	NY1	NL	**41**	**110**	**0.345**	23	33.994
1955	NY1	NL	**51**	**127**	**0.319**	24	39.404
1956	NY1	NL	**36**	**84**	**0.296**	25	19.264
1957	NY1	NL	**35**	**97**	**0.333**	26	29.482
1958	SFN	NL	**29**	**96**	**0.347**	27	36.606
1959	SFN	NL	**34**	**104**	**0.313**	28	29.575
1960	SFN	NL	**29**	**103**	**0.319**	29	38.898
1961	SFN	NL	**40**	**123**	**0.308**	30	27.605
1962	SFN	NL	**49**	**141**	**0.304**	31	40.947
1963	SFN	NL	**38**	**103**	**0.314**	32	34.477
1964	SFN	NL	**47**	**111**	**0.296**	33	36.258
1965	SFN	NL	**52**	**112**	**0.317**	34	43.109
1966	SFN	NL	**37**	**103**	**0.288**	35	24.388
1967	SFN	NL	**22**	**70**	**0.263**	36	11.108
1968	SFN	NL	**23**	**79**	**0.289**	37	22.259

1969	SFN	NL	**13**	**58**	**0.283**	38	7.176
1970	SFN	NL	**28**	**83**	**0.291**	39	15.837
1971	SFN	NL	**18**	**61**	**0.271**	40	14.789
1972	SFN	NL	**0**	**3**	**0.184**	41	4.898
1972	NYN	NL	**8**	**19**	**0.267**	41	
1973	NYN	NL	**6**	**25**	**0.211**	42	0.927
		Total	**660**	**1903**	**0.302**		

6. Ted Williams (1939-1960)

Just imagine how great the numbers would be if World War II hadn't intervened and taken three of Ted's most productive seasons away from him. He was the best player in baseball before the war and the best after, too. Even without those years, Williams ranks high in the counting stats, but just imagine where they'd be if they looked the same as the Batting Average, ranking #7 All-Time, the On Base Percentage, at the top of the list, and Slugging Percentage at #2.

Where Williams Ranks		
Category		**Rank**
HR	521	18
RBI	1839	13
AVE	0.344	7
OBP	0.482	1
SLG	0.634	2
HITS	2654	69
RUNS	1798	16

If you just do a little quick math, during the three seasons before his first missed season in 1943, Ted averaged 32 Home Runs, 123 RBI, and a PEVA Rating of 44. For the three seasons after returning, Williams averaged 32 HR, 121 RBI, and PEVA of 37. Add those averages in to the already gaudy numbers and that raises Ted Williams to 617 HR, 2205 RBI, and a Career PEVA Rating of 614.5. He'd be #6 on the Home Run list then, #3 on the RBI list, and #2 on the Best Ever list. But we know, that's only projection, and there really is no need to do that anyway, because Ted Williams was one of the Top Ten position players in the history of baseball with or without them, and it doesn't really matter whether that's at #2 or #6, it's still remarkable. The Splendid Splinter, the splendid player, and maybe the best pure hitter the game has ever seen.

Ted Williams Stats					**493.074 PEVA RATING**		
Year	**Team**	**Lg**	**HR**	**RBI**	**AVE**	**Age**	**PEVA-B**
1939	BOS	AL	**31**	**145**	**0.327**	21	28.548
1940	BOS	AL	**23**	**113**	**0.344**	22	29.157
1941	BOS	AL	**37**	**120**	**0.406**	23	51.730
1942	BOS	AL	**36**	**137**	**0.356**	24	52.075
1946	BOS	AL	**38**	**123**	**0.342**	28	44.043
1947	BOS	AL	**32**	**114**	**0.343**	29	36.436
1948	BOS	AL	**25**	**127**	**0.369**	30	30.567
1949	BOS	AL	**43**	**159**	**0.343**	31	41.778
1950	BOS	AL	**28**	**97**	**0.317**	32	12.326
1951	BOS	AL	**30**	**126**	**0.318**	33	30.939
1952	BOS	AL	**1**	**3**	**0.400**	34	0.200
1953	BOS	AL	**13**	**34**	**0.407**	35	5.650
1954	BOS	AL	**29**	**89**	**0.345**	36	20.173
1955	BOS	AL	**28**	**83**	**0.356**	37	19.146
1956	BOS	AL	**24**	**82**	**0.345**	38	15.207
1957	BOS	AL	**38**	**87**	**0.388**	39	32.674
1958	BOS	AL	**26**	**85**	**0.328**	40	17.101
1959	BOS	AL	**10**	**43**	**0.254**	41	3.478
1960	BOS	AL	**29**	**72**	**0.316**	42	21.846
		Total	**521**	**1839**	**0.344**		

7. Stan Musial (1941-1963)

Stan the Man. And boy, what a baseball man. Musial began his career in 1941 and played until he was in his 43rd year. Along that path, he was the consummate hitter. Musial could hit for average, had patience at the plate, and could slug the ball out of the park. There aren't a whole lot of 0.300/0.400/0.500 players in AVE/OBP/SLG in baseball history; even three of the batters ranked above him could not accomplish that.

Where Musial Ranks		
Category		**Rank**
HR	475	29
RBI	1951	5
AVE	0.331	28
OBP	0.417	23
SLG	0.559	20
HITS	3630	4
RUNS	1949	8

Musial also missed time for service in World War II, losing the entire season of 1945 due to Naval service, and like Williams, although not nearly as much, would have numbers that rise to higher levels if that hadn't been so. But Musial became even better once the war years were over, adding power to his game when he slugged 39 home runs in 1948. Musial is the best player to have ever donned the redbird suits of Cardinal nation, but there won't be a whole lot of surprise if someday in the future, pending the Cardinals being able to retain his service, that Albert Pujols won't challenge that position. But for now, and for all those seasons of the 40s and 50s, it was Stan the Man who was the toast of St. Louis. And what a baseball man.

Stan Musial Stats					**481.184 PEVA RATING**		
Year	**Team**	**Lg**	**HR**	**RBI**	**AVE**	**Age**	**PEVA-B**
1941	SLN	NL	**1**	**7**	**0.426**	21	0.200
1942	SLN	NL	**10**	**72**	**0.315**	22	13.326
1943	SLN	NL	**13**	**81**	**0.357**	23	41.278
1944	SLN	NL	**12**	**94**	**0.347**	24	38.717
1946	SLN	NL	**16**	**103**	**0.365**	26	29.109
1947	SLN	NL	**19**	**95**	**0.312**	27	15.709
1948	SLN	NL	**39**	**131**	**0.376**	28	35.811
1949	SLN	NL	**36**	**123**	**0.338**	29	30.709
1950	SLN	NL	**28**	**109**	**0.346**	30	25.751
1951	SLN	NL	**32**	**108**	**0.355**	31	37.339
1952	SLN	NL	**21**	**91**	**0.336**	32	37.889
1953	SLN	NL	**30**	**113**	**0.337**	33	40.502
1954	SLN	NL	**35**	**126**	**0.330**	34	33.159
1955	SLN	NL	**33**	**108**	**0.319**	35	23.404
1956	SLN	NL	**27**	**109**	**0.310**	36	16.380
1957	SLN	NL	**29**	**102**	**0.351**	37	18.848
1958	SLN	NL	**17**	**62**	**0.337**	38	12.769
1959	SLN	NL	**14**	**44**	**0.255**	39	3.702
1960	SLN	NL	**17**	**63**	**0.275**	40	5.546
1961	SLN	NL	**15**	**70**	**0.288**	41	6.170
1962	SLN	NL	**19**	**82**	**0.330**	42	11.477
1963	SLN	NL	**12**	**58**	**0.255**	43	3.387
		Total	**475**	**1951**	**0.331**		

8. Tris Speaker (1907-1928)

This was a bit of a surprise to us, not judging by the stats, but by how much Speaker is recalled in discussion around baseball. I guess it's a bit of the old era ballplayer syndrome, too far back, lack of video footage, not someone many people saw play in person. However, that doesn't take away his fantastic accomplishments. Speaker has the 5th highest career batting average at 0.345 as well as the 5th most hits in history. Although he did not hit a plethora of home runs, he was productive, knocking in over 1,500 runs during his tenure with Boston, Cleveland, Washington, and the Philadelphia Athletics.

Where Speaker Ranks		
Category		**Rank**
HR	117	594
RBI	1529	45
AVE	0.345	5
OBP	0.428	11
SLG	0.500	98
HITS	3514	5
RUNS	1882	10

Often compared with Cobb because they played during the same era, Speaker was the 2nd best player at the end of the dead ball age, and deserves the 8th spot on the Best Ever List of Career Batters. He had six seasons when his PEVA Rating was above 30.000, beginning with the 1912 season when he hit 0.383. Speaker played at a high level until he was 38 years old with only two seasons at the end of his career when he didn't reach the 10.000 PEVA Rating level of an above average starting player.

Tris Speaker Stats					**479.576 PEVA RATING**		
Year	**Team**	**Lg**	**HR**	**RBI**	**AVE**	**Age**	**PEVA-B**
1907	BOS	AL	**0**	**1**	**0.158**	19	0.200
1908	BOS	AL	**0**	**9**	**0.224**	20	0.574
1909	BOS	AL	**7**	**77**	**0.309**	21	21.221
1910	BOS	AL	**7**	**65**	**0.340**	22	21.964
1911	BOS	AL	**8**	**70**	**0.334**	23	19.171
1912	BOS	AL	**10**	**90**	**0.383**	24	43.198
1913	BOS	AL	**3**	**71**	**0.363**	25	30.292
1914	BOS	AL	**4**	**90**	**0.338**	26	37.327
1915	BOS	AL	**0**	**69**	**0.322**	27	26.160
1916	CLE	AL	**2**	**79**	**0.386**	28	42.388
1917	CLE	AL	**2**	**60**	**0.352**	29	25.522
1918	CLE	AL	**0**	**61**	**0.318**	30	26.230
1919	CLE	AL	**2**	**63**	**0.296**	31	18.309
1920	CLE	AL	**8**	**107**	**0.388**	32	35.002
1921	CLE	AL	**3**	**75**	**0.362**	33	15.221
1922	CLE	AL	**11**	**71**	**0.378**	34	20.068
1923	CLE	AL	**17**	**130**	**0.380**	35	35.232
1924	CLE	AL	**9**	**65**	**0.344**	36	15.563
1925	CLE	AL	**12**	**87**	**0.389**	37	17.730
1926	CLE	AL	**7**	**86**	**0.304**	38	17.283
1927	WS1	AL	**2**	**73**	**0.327**	39	9.573
1928	PHA	AL	**3**	**30**	**0.267**	40	1.349
		Total	**117**	**1529**	**0.345**		

9. Lou Gehrig (1923-1939)

It must have been a daunting sight for the manager of the opposing team to not only have to face the best hitter in baseball history, but then know that #9 was also in the same lineup. Lou Gehrig complimented the exploits of Babe Ruth in a tandem that would never be duplicated. Career OBP ranked #5 of All-Time. Career Slugging Percentage at #3. The 4th most Runs Batted In in history as well.

Where Gehrig Ranks		
Category		**Rank**
HR	493	27
RBI	1995	4
AVE	0.340	14
OBP	0.447	5
SLG	0.632	3
HITS	2721	54
RUNS	1888	9

Now it certainly helped both men to have the other surrounding them in the batting order, but with those rate stats, Gehrig and Ruth could have batted for today's worst team and still knocked in 100 runs. The PEVA system rates players against the players of their own era and measures how dominant they were against the maximum and average player of the day. Gehrig and Ruth had themselves to be compared to; each did it well. Well enough to be the only teammates present within the Best Ever Top Ten.

Lou Gehrig Stats					479.522 PEVA RATING		
Year	Team	Lg	HR	RBI	AVE	Age	PEVA-B
1923	NYA	AL	**1**	**9**	**0.423**	20	0.200
1924	NYA	AL	**0**	**5**	**0.500**	21	0.200
1925	NYA	AL	**20**	**68**	**0.295**	22	6.536
1926	NYA	AL	**16**	**112**	**0.313**	23	23.484
1927	NYA	AL	**47**	**175**	**0.373**	24	50.953
1928	NYA	AL	**27**	**142**	**0.374**	25	34.257
1929	NYA	AL	**35**	**126**	**0.300**	26	28.894
1930	NYA	AL	**41**	**174**	**0.379**	27	39.839
1931	NYA	AL	**46**	**184**	**0.341**	28	40.131
1932	NYA	AL	**34**	**151**	**0.349**	29	34.089
1933	NYA	AL	**32**	**139**	**0.334**	30	36.863
1934	NYA	AL	**49**	**165**	**0.363**	31	50.136
1935	NYA	AL	**30**	**119**	**0.329**	32	31.877
1936	NYA	AL	**49**	**152**	**0.354**	33	45.995
1937	NYA	AL	**37**	**159**	**0.351**	34	37.760
1938	NYA	AL	**29**	**114**	**0.295**	35	18.108
1939	NYA	AL	**0**	**1**	**0.143**	36	0.200
		Total	**493**	**1995**	**0.340**		

10. Mickey Mantle (1951-1968)

There were Mantle fans. There were Mays fans. There were Snider fans, too. But even though each of those fan bases for the Yanks, Giants, and Dodgers were in constant battle for supremacy within the outfields of the three New York teams at the time, there was a begrudging respect for the ability of the other two men. Mantle played with reckless abandon, and his body did not hold up as well as Mays. Yes, injuries caused more than a few seasons for Mantle when his ability couldn't overcome them. By the time Mantle was 33, his career was over as a premier player. But before then, when making those comparisons, the majority in the city likely would have chosen Mantle as the best player in the game.

Where Mantle Ranks		
Category		Rank
HR	536	16
RBI	1509	50
AVE	0.298	230
OBP	0.421	19
SLG	0.557	22
HITS	2415	110
RUNS	1677	27

Mantle was only 21 years old when he burst onto the scene of his sophomore season and took the league by storm, batting 0.311 with 87 RBI's and a PEVA Rating of 31.043. By the time he was 25, he had hit 52 homers, knocked in 130, and batted 0.353, all grading out to the tune of 47.236. For the six seasons thereafter, Mantle did not go below 23 in PEVA Ratings; he hit at least 30 Home Runs in each year, too. Mickey Mantle comes in at #10 in the Best Ever list of position players, because, at his peak, he was the best player in baseball. The fact that he does not rank higher comes because his peak just didn't last quite long enough, allowing his compatriots of the time, both Mays and Aaron to surpass him. Not far past, but past.

Mickey Mantle Stats					455.611 PEVA RATING		
Year	Team	Lg	HR	RBI	AVE	Age	PEVA-B
1951	NYA	AL	**13**	**65**	**0.267**	20	4.497
1952	NYA	AL	**23**	**87**	**0.311**	21	31.043
1953	NYA	AL	**21**	**92**	**0.295**	22	17.160
1954	NYA	AL	**27**	**102**	**0.300**	23	24.795
1955	NYA	AL	**37**	**99**	**0.306**	24	36.784
1956	NYA	AL	**52**	**130**	**0.353**	25	47.236
1957	NYA	AL	**34**	**94**	**0.365**	26	40.637
1958	NYA	AL	**42**	**97**	**0.304**	27	36.436
1959	NYA	AL	**31**	**75**	**0.285**	28	23.336
1960	NYA	AL	**40**	**94**	**0.275**	29	42.394

1961	NYA	AL	**54**	**128**	**0.317**	30	42.398
1962	NYA	AL	**30**	**89**	**0.321**	31	29.086
1963	NYA	AL	**15**	**35**	**0.314**	32	8.024
1964	NYA	AL	**35**	**111**	**0.303**	33	28.313
1965	NYA	AL	**19**	**46**	**0.255**	34	8.096
1966	NYA	AL	**23**	**56**	**0.288**	35	11.410
1967	NYA	AL	**22**	**55**	**0.245**	36	12.369
1968	NYA	AL	**18**	**54**	**0.237**	37	11.597
		Total	**536**	**1509**	**0.298**		

Who Jumped Up the Best Careers Ever List in 2010

It may be a marginal move, but it is important. Alex Rodriguez has now jumped four spots on the best ever position player list to land at #23 after his 2010 season. When you've passed the Yaz you know you've accomplished something significant. Yes, we know that Alex would be lower on the list pending PED calculations, but that's a debate for another day. Even though Alex is at the highest point on the current player list, that likely won't last long, at least not with the pace that Albert Pujols is playing. Jumping fifteen spots to #33, the Cardinal first baseman is clipping off players at a grand pace. If Pujols has another year like this past season, he'll pass Yaz next year. Further down the list you see some younger movers and shakers, who are passing players at a rate of over 100 notches per season. Miguel Cabrera moved up one hundred and twenty-five spots to #220 and David Wright became the biggest mover on the countdown by clipping past one hundred and thirty-two of the best baseball men on the planet to land at #368. Now that's some fine driving on the corners, don't you think.

Moving Up the Career List in 2010

2009 Rank	Name		2010 Rank	Up
27	Alex	Rodriguez	23	4
42	Ken	Griffey	44	-2
43	Manny	Ramirez	42	1
48	Albert	Pujols	33	15
59	Chipper	Jones	55	4
66	Jim	Thome	63	3
94	Derek	Jeter	82	12
112	Jason	Giambi	112	0
116	Bobby	Abreu	97	19
137	Vladimir	Guerrero	114	23
157	Ivan	Rodriguez	145	12
171	Todd	Helton	168	3
188	Lance	Berkman	176	12
221	Miguel	Tejada	193	28
224	Scott	Rolen	187	37
225	Andruw	Jones	214	11
260	Carlos	Beltran	251	9
261	Johnny	Damon	234	27
268	Magglio	Ordonez	257	11
272	Jorge	Posada	253	19

2009 Rank	Name		2010 Rank	Up
293	Omar	Vizquel	285	8
304	Jason	Kendall	291	13
314	David	Ortiz	271	43
315	Ichiro	Suzuki	278	37
345	Miguel	Cabrera	220	125
368	Derrek	Lee	325	43
390	Carlos	Lee	338	52
423	Edgar	Renteria	415	8
440	Adrian	Beltre	321	119
445	Mike	Cameror	445	0
448	Mark	Teixeira	340	108
466	Troy	Glaus	441	25
496	Adam	Dunn	393	103
500	David	Wright	368	132
501	Jimmy	Rollins	491	10
525	Mike	Lowell	526	-1
529	Michae	Young	448	81
547	Torii	Hunter	457	90
549	Joe	Mauer	434	115

Joining the Over 100 PEVA Career List in 2010

Name		2010 Rank	2010 Team
Paul	Konerko	425	Chicago White Sox
Matt	Holliday	489	St. Louis Cardinals
Ryan	Howard	496	Philadelphia Phillies

Vernon	Wells	512	Toronto Blue Jays
Raul	Ibanez	525	Philadelphia Phillies
Aramis	Ramirez	534	Chicago Cubs
J.D.	Drew	535	Boston Red Sox
Alfonso	Soriano	538	Chicago Cubs
Jason	Bay	542	New York Mets
Orlando	Cabrera	558	Cincinnati Reds

Best Position Players of All-Time 1871-2010 (Regular Season)

			HOF	PEVA-B		PEVA-B	CAREER				PEVA-B	EQ
Rank	**First**	**Last**	**Cat.**	**TOTAL**	**YRS.**	**PER YR.**	**Hits**	**HR**	**RBI**	**Ave.**	**EQ YR.**	**YRS**
1	Babe	Ruth	HOFP	**641.541**	**22**	**29.161**	2873	714	2217	0.342	**31.665**	20.26
2	Barry	Bonds		**606.700**	**22**	**27.577**	2935	762	1996	0.298	**28.987**	20.93
3	Ty	Cobb	HOFP	**569.252**	**24**	**23.719**	4189	117	1937	0.366	**25.300**	22.50
4	Hank	Aaron	HOFP	**535.808**	**23**	**23.296**	3771	755	2297	0.305	**23.708**	22.60
5	Willie	Mays	HOFP	**520.998**	**22**	**23.682**	3283	660	1903	0.302	**25.353**	20.55
6	Ted	Williams	HOFP	**493.074**	**19**	**25.951**	2654	521	1839	0.344	**29.141**	16.92
7	Stan	Musial	HOFP	**481.184**	**22**	**21.872**	3630	475	1951	0.331	**22.837**	21.07
8	Tris	Speaker	HOFP	**479.576**	**22**	**21.799**	3514	117	1529	0.345	**24.136**	19.87
9	Lou	Gehrig	HOFP	**479.522**	**17**	**28.207**	2721	493	1995	0.340	**33.627**	14.26
10	Mickey	Mantle	HOFP	**455.611**	**18**	**25.312**	2415	536	1509	0.298	**26.474**	17.21
11	Honus	Wagner	HOFP	**451.472**	**21**	**21.499**	3415	101	1732	0.327	**22.372**	20.18
12	Cap	Anson	HOFP	**432.886**	**27**	**16.033**	3418	97	2076	0.333	**16.171**	26.77
13	Frank	Robinson	HOFP	**411.403**	**21**	**19.591**	2943	586	1812	0.294	**21.011**	19.58
14	Mike	Schmidt	HOFP	**383.574**	**18**	**21.310**	2234	548	1595	0.267	**23.460**	16.35
15	Mel	Ott	HOFP	**369.518**	**22**	**16.796**	2876	511	1860	0.304	**19.156**	19.29
16	Rogers	Hornsby	HOFP	**361.117**	**23**	**15.701**	2930	301	1584	0.358	**21.886**	16.50
17	Eddie	Mathews	HOFP	**360.187**	**17**	**21.187**	2315	512	1453	0.271	**22.165**	16.25
18	Jimmie	Foxx	HOFP	**359.936**	**20**	**17.997**	2646	534	1922	0.325	**21.735**	16.56
19	Ed	Delahanty	HOFP	**351.660**	**16**	**21.979**	2596	101	1464	0.346	**23.955**	14.68
20	Eddie	Collins	HOFP	**349.132**	**25**	**13.965**	3315	47	1300	0.333	**17.301**	20.18
21	Pete	Rose		**349.021**	**24**	**14.543**	4256	160	1314	0.303	**14.808**	23.57
22	Dan	Brouthers	HOFP	**348.085**	**19**	**18.320**	2296	106	1296	0.342	**22.501**	15.47
23	Alex	Rodriguez		**346.418**	**17**	**20.378**	2672	613	1831	0.303	**22.149**	15.64
24	Carl	Yastrzemsk	HOFP	**339.140**	**23**	**14.745**	3419	452	1844	0.285	**14.855**	22.83
25	Nap	Lajoie	HOFP	**337.934**	**21**	**16.092**	3242	83	1599	0.338	**17.555**	19.25
26	Roger	Connor	HOFP	**336.978**	**18**	**18.721**	2467	138	1322	0.317	**19.558**	17.23
27	Frank	Thomas		**333.924**	**19**	**17.575**	2468	521	1704	0.301	**20.689**	16.14
28	Rickey	Henderson	HOFP	**331.490**	**25**	**13.260**	3055	297	1115	0.279	**14.106**	23.50
29	Jim	O'Rourke	HOFP	**329.815**	**23**	**14.340**	2643	62	1203	0.311	**15.298**	21.56
30	Dave	Winfield	HOFP	**327.768**	**22**	**14.899**	3110	465	1833	0.283	**15.766**	20.79
31	Eddie	Murray	HOFP	**324.361**	**21**	**15.446**	3255	504	1917	0.287	**15.861**	20.45
32	Cal	Ripken	HOFP	**320.188**	**21**	**15.247**	3184	431	1695	0.276	**16.286**	19.66
33	Albert	Pujols		**308.345**	**10**	**30.835**	1900	408	1230	0.331	**30.835**	10.00
34	George	Brett	HOFP	**301.897**	**21**	**14.376**	3154	317	1595	0.305	**15.125**	19.96
35	Reggie	Jackson	HOFP	**301.318**	**21**	**14.348**	2584	563	1702	0.262	**14.931**	20.18
36	Joe	DiMaggio	HOFP	**297.630**	**13**	**22.895**	2214	361	1537	0.325	**23.509**	12.66
37	Al	Kaline	HOFP	**297.259**	**22**	**13.512**	3007	399	1583	0.297	**14.416**	20.62
38	Sam	Crawford	HOFP	**295.985**	**19**	**15.578**	2961	97	1525	0.309	**16.846**	17.57
39	Jeff	Bagwell		**293.606**	**15**	**19.574**	2314	449	1529	0.297	**20.503**	14.32
40	Wade	Boggs	HOFP	**293.035**	**18**	**16.280**	3010	118	1014	0.328	**16.812**	17.43
41	Robin	Yount	HOFP	**291.976**	**20**	**14.599**	3142	251	1406	0.285	**14.687**	19.88
42	Manny	Ramirez		**288.532**	**18**	**16.030**	2573	555	1830	0.313	**17.073**	16.90
42	Joe	Morgan	HOFP	**287.830**	**22**	**13.083**	2517	268	1133	0.271	**15.007**	19.18

44	Ken	Griffey		**286.056**	**22**	**13.003**	2781	630	1836	0.284	**14.677**	19.49
44	Jesse	Burkett	HOFP	**276.152**	**16**	**17.259**	2850	75	952	0.338	**17.897**	15.43
45	Billy	Williams	HOFP	**275.510**	**18**	**15.306**	2711	426	1475	0.290	**16.996**	16.21
46	Deacon	White		**274.720**	**20**	**13.736**	2066	23	977	0.312	**14.652**	18.75
47	Paul	Hines		**274.652**	**20**	**13.733**	2134	57	855	0.302	**14.617**	18.79
49	Harmon	Killebrew	HOFP	**270.652**	**22**	**12.302**	2086	573	1584	0.256	**15.772**	17.16
50	Gary	Sheffield		**267.943**	**22**	**12.179**	2689	509	1676	0.292	**14.231**	19.13

Best Position Players of All-Time 1871-2010

Rank	First	Last	HOF Cat.	PEVA-B TOTAL	YRS.	PEVA-B PER YR.	CAREER Hits	HR	RBI	Ave.	PEVA-B EQ YR.	EQ YRS
51	Ernie	Banks	HOFP	**264.788**	**19**	**13.936**	2583	512	1636	0.274	**15.585**	16.99
52	Roberto	Clemente	HOFP	**264.635**	**18**	**14.702**	3000	240	1305	0.317	**14.960**	17.69
53	Dwight	Evans		**264.273**	**20**	**13.214**	2446	385	1384	0.272	**14.231**	18.57
54	Mark	McGwire		**261.187**	**16**	**16.324**	1626	583	1414	0.263	**19.506**	13.39
55	Chipper	Jones		**261.173**	**17**	**15.363**	2490	436	1491	0.306	**16.667**	15.67
56	Willie	McCovey	HOFP	**261.038**	**22**	**11.865**	2211	521	1555	0.270	**13.476**	19.37
57	Rafael	Palmeiro		**260.423**	**20**	**13.021**	3020	569	1835	0.288	**13.874**	18.77
58	Ron	Santo		**259.712**	**15**	**17.314**	2254	342	1331	0.277	**17.584**	14.77
59	Johnny	Bench	HOFP	**256.662**	**17**	**15.098**	2048	389	1376	0.267	**16.432**	15.62
60	Mike	Piazza		**255.854**	**16**	**15.991**	2127	427	1335	0.308	**17.842**	14.34
61	Fred	Clarke	HOFP	**255.011**	**21**	**12.143**	2672	67	1015	0.312	**14.605**	17.46
62	Duke	Snider	HOFP	**253.177**	**18**	**14.065**	2116	407	1333	0.295	**16.188**	15.64
63	Jim	Thome		**252.394**	**20**	**12.620**	2216	589	1624	0.278	**14.581**	17.31
64	Yogi	Berra	HOFP	**251.624**	**19**	**13.243**	2150	358	1430	0.285	**15.756**	15.97
65	Harry	Heilmann	HOFP	**249.516**	**17**	**14.677**	2660	183	1539	0.342	**16.077**	15.52
66	George	Davis	HOFP	**245.441**	**20**	**12.272**	2660	73	1437	0.295	**13.464**	18.23
67	Paul	Waner	HOFP	**242.638**	**20**	**12.132**	3152	113	1309	0.333	**13.391**	18.12
68	Paul	Molitor	HOFP	**242.403**	**21**	**11.543**	3319	234	1307	0.306	**12.243**	19.80
69	Sherry	Magee		**242.236**	**16**	**15.140**	2169	83	1176	0.291	**15.832**	15.30
70	Willie	Stargell	HOFP	**241.847**	**21**	**11.517**	2232	475	1540	0.282	**13.429**	18.01
71	Ted	Simmons		**239.551**	**21**	**11.407**	2472	248	1389	0.285	**13.619**	17.59
72	Tony	Gwynn	HOFP	**238.944**	**20**	**11.947**	3141	135	1138	0.338	**13.334**	17.92
73	Bob	Johnson		**237.767**	**13**	**18.290**	2051	288	1283	0.296	**18.290**	13.00
74	Al	Simmons	HOFP	**237.604**	**20**	**11.880**	2927	307	1827	0.334	**14.400**	16.50
75	Dick	Allen		**237.460**	**15**	**15.831**	1848	351	1119	0.292	**18.566**	12.79
76	Harry	Stovey		**236.464**	**14**	**16.890**	1771	122	908	0.289	**17.400**	13.59
77	Joe	Jackson		**235.944**	**13**	**18.150**	1772	54	785	0.356	**24.968**	9.45
78	King	Kelly	HOFP	**235.100**	**16**	**14.694**	1813	69	950	0.308	**15.982**	14.71
79	Rusty	Staub		**234.389**	**23**	**10.191**	2716	292	1466	0.279	**11.301**	20.74
80	Gary	Carter	HOFP	**233.612**	**19**	**12.295**	2092	324	1225	0.262	**14.073**	16.60
81	Charlie	Gehringer	HOFP	**233.202**	**19**	**12.274**	2839	184	1427	0.320	**14.254**	16.36
82	Derek	Jeter		**233.157**	**16**	**14.572**	2926	234	1135	0.314	**15.400**	15.14
83	Brooks	Robinson	HOFP	**232.881**	**23**	**10.125**	2848	268	1357	0.267	**12.167**	19.14
84	Sammy	Sosa		**232.726**	**18**	**12.929**	2408	609	1667	0.273	**13.869**	16.78
85	Billy	Hamilton	HOFP	**231.802**	**14**	**16.557**	2158	40	736	0.344	**17.735**	13.07
86	Andre	Dawson	HOFP	**230.234**	**21**	**10.964**	2774	438	1591	0.279	**12.073**	19.07
87	Dale	Murphy		**228.943**	**18**	**12.719**	2111	398	1266	0.265	**15.767**	14.52
88	Craig	Biggio		**226.358**	**20**	**11.318**	3060	291	1175	0.281	**11.710**	19.33
89	Tim	Raines		**226.257**	**23**	**9.837**	2605	170	980	0.294	**12.029**	18.81
90	Joe	Kelley	HOFP	**226.052**	**17**	**13.297**	2220	65	1194	0.317	**15.040**	15.03
91	Edgar	Martinez		**225.436**	**18**	**12.524**	2247	309	1261	0.312	**15.130**	14.90
92	Jimmy	Wynn		**224.887**	**15**	**14.992**	1665	291	964	0.250	**16.560**	13.58
93	Fred	McGriff		**224.159**	**19**	**11.798**	2490	493	1550	0.284	**13.319**	16.83
94	Jim	Rice	HOFP	**224.084**	**16**	**14.005**	2452	382	1451	0.298	**15.380**	14.57
95	Goose	Goslin	HOFP	**220.227**	**18**	**12.235**	2735	248	1609	0.316	**13.790**	15.97
96	Willie	Keeler	HOFP	**220.111**	**19**	**11.585**	2932	33	810	0.341	**13.638**	16.14

97	Bobby	Abreu		**218.502**	**15**	**14.567**	2257	276	1265	0.296	**16.055**	13.61
98	Bill	Dahlen		**216.119**	**21**	**10.291**	2457	84	1233	0.272	**11.746**	18.40
99	Tony	Perez	HOFP	**215.659**	**23**	**9.376**	2732	379	1652	0.279	**10.859**	19.86
100	Ken	Singleton		**215.557**	**15**	**14.370**	2029	246	1065	0.282	**14.959**	14.41

Best Position Players of All-Time 1871-2010

			HOF	PEVA-B		PEVA-B	CAREER				PEVA-B	EQ
Rank	**First**	**Last**	**Cat.**	**TOTAL**	**YRS.**	**PER YR.**	**Hits**	**HR**	**RBI**	**Ave.**	**EQ YR.**	**YRS**
101	Hugh	Duffy	HOFP	**215.369**	**17**	**12.669**	2282	106	1302	0.324	**16.291**	13.22
102	Hank	Greenberg	HOFP	**215.163**	**13**	**16.551**	1628	331	1276	0.313	**21.624**	9.95
103	Carlton	Fisk	HOFP	**214.953**	**24**	**8.956**	2356	376	1330	0.269	**11.337**	18.96
104	Joe	Torre		**214.765**	**18**	**11.931**	2342	252	1185	0.297	**13.758**	15.61
105	Darrell	Evans		**214.577**	**21**	**10.218**	2223	414	1354	0.248	**11.414**	18.80
106	Charley	Jones		**213.948**	**12**	**17.829**	1114	56	552	0.298	**20.853**	10.26
107	Zack	Wheat	HOFP	**213.678**	**19**	**11.246**	2884	132	1248	0.317	**12.086**	17.68
108	Jake	Beckley	HOFP	**213.234**	**20**	**10.662**	2930	86	1575	0.308	**11.421**	18.67
109	George	Van Haltren		**210.974**	**17**	**12.410**	2532	69	1014	0.316	**13.161**	16.03
110	Ross	Barnes		**210.856**	**9**	**23.428**	859	6	350	0.359	**24.748**	8.52
111	Luke	Appling	HOFP	**210.671**	**20**	**10.534**	2749	45	1116	0.310	**12.284**	17.15
112	Jason	Giambi		**210.366**	**16**	**13.148**	1914	415	1365	0.281	**14.609**	14.40
113	Orlando	Cepeda	HOFP	**209.487**	**17**	**12.323**	2351	379	1365	0.297	**14.568**	14.38
114	Vladimir	Guerrero		**208.999**	**15**	**13.933**	2427	436	1433	0.320	**15.413**	13.56
115	Ralph	Kiner	HOFP	**208.610**	**10**	**20.861**	1451	369	1015	0.279	**20.924**	9.97
116	Rod	Carew	HOFP	**208.083**	**19**	**10.952**	3053	92	1015	0.328	**11.465**	18.15
117	Elmer	Flick	HOFP	**207.666**	**13**	**15.974**	1752	48	756	0.313	**19.175**	10.83
118	Richie	Ashburn	HOFP	**207.541**	**15**	**13.836**	2574	29	586	0.308	**13.938**	14.89
119	Albert	Belle		**206.268**	**12**	**17.189**	1726	381	1239	0.295	**19.496**	10.58
120	Sam	Thompson	HOFP	**206.207**	**15**	**13.747**	1979	127	1299	0.331	**17.776**	11.60
121	Jimmy	Sheckard		**205.718**	**17**	**12.101**	2084	56	813	0.274	**13.128**	15.67
122	Rocky	Colavito		**204.838**	**14**	**14.631**	1730	374	1159	0.266	**16.335**	12.54
123	Frank	Baker	HOFP	**204.500**	**13**	**15.731**	1838	96	987	0.307	**18.081**	11.31
124	George	Gore		**204.244**	**14**	**14.589**	1612	46	618	0.301	**15.322**	13.33
125	Kirby	Puckett	HOFP	**204.074**	**12**	**17.006**	2304	207	1085	0.318	**17.006**	12.00
126	Carlos	Delgado		**203.096**	**17**	**11.947**	2038	473	1512	0.280	**14.647**	14.07
127	Minnie	Minoso		**202.978**	**17**	**11.940**	1963	186	1023	0.298	**16.148**	12.57
128	Bill	Dickey	HOFP	**202.680**	**17**	**11.922**	1969	202	1209	0.313	**13.920**	14.56
129	Jimmy	Ryan		**202.408**	**18**	**11.245**	2502	118	1093	0.306	**12.062**	16.78
130	Dave	Parker		**202.394**	**19**	**10.652**	2712	339	1493	0.290	**11.719**	17.27
131	Bernie	Williams		**201.884**	**16**	**12.618**	2336	287	1257	0.297	**13.169**	15.33
132	Vada	Pinson		**201.635**	**18**	**11.202**	2757	256	1170	0.286	**11.854**	17.01
133	Keith	Hernandez		**199.518**	**17**	**11.736**	2182	162	1071	0.296	**13.875**	14.38
134	Bobby	Veach		**199.426**	**14**	**14.245**	2063	64	1166	0.310	**15.568**	12.81
135	Bobby	Bonds		**199.147**	**14**	**14.225**	1886	332	1024	0.268	**15.571**	12.79
136	Joe	Cronin	HOFP	**196.868**	**20**	**9.843**	2285	170	1424	0.301	**12.943**	15.21
137	Luis	Gonzalez		**196.581**	**19**	**10.346**	2591	354	1439	0.283	**10.921**	18.00
138	Al	Oliver		**195.039**	**18**	**10.836**	2743	219	1326	0.303	**11.616**	16.79
139	Will	Clark		**194.626**	**15**	**12.975**	2176	284	1205	0.303	**13.018**	14.95
140	Max	Carey	HOFP	**194.330**	**20**	**9.717**	2665	70	800	0.285	**10.961**	17.73
141	Jack	Clark		**193.774**	**18**	**10.765**	1826	340	1180	0.267	**13.419**	14.44
142	Don	Mattingly		**193.054**	**14**	**13.790**	2153	222	1099	0.307	**15.261**	12.65
143	George	Sisler	HOFP	**192.404**	**15**	**12.827**	2812	102	1175	0.340	**13.115**	14.67
144	Frank	Howard		**192.157**	**16**	**12.010**	1774	382	1119	0.273	**14.202**	13.53
145	Ivan	Rodriguez		**191.791**	**20**	**9.590**	2817	309	1313	0.298	**10.005**	19.17
146	George	Foster		**191.583**	**18**	**10.643**	1925	348	1239	0.274	**13.332**	14.37
147	Johnny	Mize	HOFP	**191.547**	**15**	**12.770**	2011	359	1337	0.312	**13.701**	13.98
148	Arky	Vaughan	HOFP	**191.281**	**14**	**13.663**	2103	96	926	0.318	**14.714**	13.00
149	Harry	Hooper	HOFP	**191.097**	**17**	**11.241**	2466	75	817	0.281	**11.457**	16.68

150	Fred	Lynn		**190.089**	**17**	**11.182**	1960	306	1111	0.283	**12.319**	15.43

Best Position Players of All-Time 1871-2010

Rank	First	Last	HOF Cat.	PEVA-B TOTAL	YRS.	PEVA-B PER YR.	CAREER Hits	HR	RBI	Ave.	PEVA-B EQ YR.	EQ YRS
151	Sal	Bando		**189.399**	**16**	**11.837**	1790	242	1039	0.254	**14.040**	13.49
152	Alan	Trammell		**189.326**	**20**	**9.466**	2365	185	1003	0.285	**10.988**	17.23
153	Brian	Downing		**189.089**	**20**	**9.454**	2099	275	1073	0.267	**10.842**	17.44
154	Jim	Edmonds		**188.605**	**16**	**11.788**	1881	382	1176	0.284	**13.143**	14.35
155	Harold	Baines		**188.522**	**22**	**8.569**	2866	384	1628	0.289	**9.042**	20.85
156	Reggie	Smith		**188.010**	**17**	**11.059**	2020	314	1092	0.287	**12.851**	14.63
157	Earl	Averill	HOFP	**187.958**	**13**	**14.458**	2019	238	1164	0.318	**16.259**	11.56
158	Buddy	Bell		**187.015**	**18**	**10.390**	2514	201	1106	0.279	**11.001**	17.00
159	Ron	Cey		**186.886**	**17**	**10.993**	1868	316	1139	0.261	**13.096**	14.27
160	Joe	Medwick	HOFP	**186.522**	**17**	**10.972**	2471	205	1383	0.324	**13.172**	14.16
161	Eddie	Yost		**185.714**	**18**	**10.317**	1863	139	683	0.254	**12.250**	15.16
162	Brian	Giles		**185.683**	**15**	**12.379**	1897	287	1078	0.291	**14.437**	12.89
163	Brett	Butler		**185.209**	**17**	**10.895**	2375	54	578	0.290	**12.058**	15.36
164	Graig	Nettles		**185.124**	**22**	**8.415**	2225	390	1314	0.248	**9.597**	19.29
165	Jose	Canseco		**184.776**	**17**	**10.869**	1877	462	1407	0.266	**12.805**	14.43
166	Roberto	Alomar		**184.493**	**17**	**10.853**	2724	210	1134	0.300	**11.256**	16.39
167	Pete	Browning		**184.248**	**13**	**14.173**	1646	46	659	0.341	**16.262**	11.33
168	Todd	Helton		**183.922**	**14**	**13.137**	2236	333	1239	0.324	**14.170**	12.98
169	Jose	Cruz		**183.794**	**19**	**9.673**	2251	165	1077	0.284	**10.875**	16.90
170	Larry	Doby	HOFP	**183.611**	**13**	**14.124**	1515	253	970	0.283	**16.191**	11.34
171	Sam	Rice	HOFP	**182.580**	**20**	**9.129**	2987	34	1078	0.322	**10.721**	17.03
172	Bob	Elliott		**180.955**	**15**	**12.064**	2061	170	1195	0.289	**12.843**	14.09
173	Steve	Garvey		**180.570**	**19**	**9.504**	2599	272	1308	0.294	**11.436**	15.79
174	John	Olerud		**180.482**	**17**	**10.617**	2239	255	1230	0.295	**11.518**	15.67
175	Enos	Slaughter	HOFP	**180.421**	**19**	**9.496**	2383	169	1304	0.300	**10.188**	17.71
176	Lance	Berkman		**180.132**	**12**	**15.011**	1675	327	1099	0.296	**16.126**	11.17
177	Mike	Tiernan		**180.026**	**13**	**13.848**	1834	106	851	0.311	**14.660**	12.28
178	Larry	Walker		**179.515**	**17**	**10.560**	2160	383	1311	0.313	**11.912**	15.07
179	Lou	Brock	HOFP	**179.216**	**19**	**9.432**	3023	149	900	0.293	**10.091**	17.76
180	Bobby	Wallace	HOFP	**179.185**	**25**	**7.167**	2309	34	1121	0.268	**9.299**	19.27
181	Jack	Glasscock		**178.962**	**17**	**10.527**	2040	27	825	0.290	**11.013**	16.25
182	Ken	Boyer		**178.433**	**15**	**11.896**	2143	282	1141	0.287	**12.855**	13.88
183	Chili	Davis		**177.938**	**19**	**9.365**	2380	350	1372	0.274	**10.268**	17.33
184	Bobby	Murcer		**177.478**	**17**	**10.440**	1862	252	1043	0.277	**13.354**	13.29
185	Jeff	Kent		**177.365**	**17**	**10.433**	2461	377	1518	0.290	**10.570**	16.78
186	Ryne	Sandberg	HOFP	**177.295**	**16**	**11.081**	2386	282	1061	0.285	**11.955**	14.83
187	Scott	Rolen		**176.523**	**15**	**11.768**	1944	304	1212	0.284	**12.970**	13.61
188	Joe	Carter		**175.685**	**16**	**10.980**	2184	396	1445	0.259	**11.927**	14.73
189	Bobby	Bonilla		**175.138**	**16**	**10.946**	2010	287	1173	0.279	**11.668**	15.01
190	Norm	Cash		**174.984**	**17**	**10.293**	1820	377	1103	0.271	**11.642**	15.03
191	Mike	Griffin		**174.674**	**12**	**14.556**	1753	42	719	0.296	**14.556**	12.00
192	Cesar	Cedeno		**173.719**	**17**	**10.219**	2087	199	976	0.285	**11.535**	15.06
193	Miguel	Tejada		**173.278**	**14**	**12.377**	2285	300	1256	0.287	**13.258**	13.07
194	Mickey	Cochrane	HOFP	**173.444**	**13**	**13.342**	1652	119	832	0.320	**14.914**	11.63
195	Edd	Roush	HOFP	**173.402**	**18**	**9.633**	2376	68	981	0.323	**11.378**	15.24
196	George	Burns		**170.571**	**15**	**11.371**	2077	41	611	0.287	**13.233**	12.89
197	Willie	Davis		**170.488**	**18**	**9.472**	2561	182	1053	0.279	**10.301**	16.55
198	Toby	Harrah		**170.485**	**17**	**10.029**	1954	195	918	0.264	**10.964**	15.55
199	Lou	Whitaker		**170.425**	**19**	**8.970**	2369	244	1084	0.276	**9.585**	17.78
200	Cal	McVey		**170.048**	**9**	**18.894**	869	11	448	0.346	**18.894**	9.00

Best Position Players of All-Time 1871-2010

Rank	First	Last	HOF Cat.	PEVA-B TOTAL	YRS.	PEVA-B PER YR.	CAREER Hits	HR	RBI	Ave.	PEVA-B EQ YR.	EQ YRS
201	Greg	Luzinski		169.911	15	11.327	1795	307	1128	0.276	13.212	12.86
202	Chuck	Klein	HOFP	169.818	17	9.989	2076	300	1201	0.320	12.914	13.15
203	Gil	Hodges		169.363	18	9.409	1921	370	1274	0.273	11.656	14.53
204	Steve	Finley		169.123	19	8.901	2548	304	1167	0.271	10.617	15.93
205	Mickey	Vernon		168.791	20	8.440	2495	172	1311	0.286	9.774	17.27
206	Lou	Boudreau	HOFP	168.772	15	11.251	1779	68	789	0.295	14.339	11.77
207	Ed	Konetchy		168.590	15	11.239	2150	74	992	0.281	11.414	14.77
208	Fielder	Jones		167.891	15	11.193	1920	21	631	0.285	12.925	12.99
209	Paul	O'Neill		167.694	17	9.864	2105	281	1269	0.288	11.400	14.71
210	Don	Baylor		167.552	19	8.819	2135	338	1276	0.260	10.204	16.42
211	Jimmy	Collins	HOFP	167.223	14	11.944	1999	65	983	0.294	12.746	13.12
212	Tom	York		167.187	15	11.146	1095	15	502	0.274	11.874	14.08
213	Boog	Powell		167.004	17	9.824	1776	339	1187	0.266	11.089	15.06
214	Andruw	Jones		166.956	15	11.130	1840	407	1222	0.256	12.432	13.43
215	Pee Wee	Reese	HOFP	166.690	16	10.418	2170	126	885	0.269	11.017	15.13
216	Bid	McPhee	HOFP	166.066	18	9.226	2250	53	1067	0.271	9.361	17.74
217	Barry	Larkin		165.994	19	8.737	2340	198	960	0.295	10.159	16.34
218	Roy	Thomas		165.828	13	12.756	1537	7	299	0.290	15.048	11.02
219	Bobby	Grich		165.495	17	9.735	1833	224	864	0.266	11.421	14.49
220	Miguel	Cabrera		164.926	8	20.616	1400	247	879	0.313	21.391	7.71
221	Robin	Ventura		164.910	16	10.307	1885	294	1182	0.267	11.444	14.41
222	Roy	White		164.619	15	10.975	1803	160	758	0.271	12.538	13.13
223	Joe	Start		164.545	16	10.284	1418	15	544	0.299	10.762	15.29
224	Tommy	Leach		164.542	19	8.660	2143	63	810	0.269	10.039	16.39
225	Lave	Cross		164.121	21	7.815	2645	47	1371	0.292	8.978	18.28
226	Del	Ennis		163.943	14	11.710	2063	288	1284	0.284	12.448	13.17
227	Bill	Nicholson		163.306	16	10.207	1484	235	948	0.268	13.419	12.17
228	Dummy	Hoy		163.266	14	11.662	2044	40	726	0.287	11.865	13.76
229	Roy	Campanella	HOFP	162.658	10	16.266	1161	242	856	0.276	17.050	9.54
230	Amos	Otis		162.237	17	9.543	2020	193	1007	0.277	11.143	14.56
231	Ozzie	Smith	HOFP	162.052	19	8.529	2460	28	793	0.262	8.963	18.08
232	Harlond	Clift		161.666	12	13.472	1558	178	829	0.272	14.591	11.08
233	Kiki	Cuyler	HOFP	161.663	18	8.981	2299	128	1065	0.321	11.506	14.05
234	Johnny	Damon		160.796	16	10.050	2571	215	1047	0.287	10.401	15.46
235	Ezra	Sutton		160.677	18	8.927	1574	25	671	0.294	9.645	16.66
236	Hardy	Richardson		160.021	14	11.430	1688	70	822	0.299	12.570	12.73
237	Gary	Gaetti		159.980	20	7.999	2280	360	1341	0.255	8.932	17.91
238	Chet	Lemon		159.807	16	9.988	1875	215	884	0.273	10.805	14.79
239	Julio	Franco		159.393	23	6.930	2586	173	1194	0.298	8.579	18.58
240	Stan	Hack		159.139	16	9.946	2193	57	642	0.301	11.270	14.12
241	Joe	Sewell	HOFP	159.048	14	11.361	2226	49	1055	0.312	12.123	13.12
242	Vern	Stephens		158.792	15	10.586	1859	247	1174	0.286	12.643	12.56
243	Lance	Parrish		158.480	19	8.341	1782	324	1070	0.252	10.244	15.47
244	Kip	Selbach		157.810	13	12.139	1803	44	779	0.293	12.615	12.51
245	Cecil	Cooper		157.313	17	9.254	2192	241	1125	0.298	11.441	13.75
246	Bill	Terry	HOFP	157.226	14	11.230	2193	154	1078	0.341	12.909	12.18
247	Bobby	Doerr	HOFP	157.073	14	11.219	2042	223	1247	0.288	11.713	13.41
248	Ed	McKean		156.981	13	12.075	2083	66	1124	0.302	12.439	12.62
249	Dixie	Walker		156.831	18	8.713	2064	105	1023	0.306	11.226	13.97
250	Mark	Grace		156.827	16	9.802	2445	173	1146	0.303	10.092	15.54

Best Position Players of All-Time 1871-2010

Rank	First	Last	HOF Cat.	PEVA-B TOTAL	YRS.	PEVA-B PER YR.	CAREER Hits	HR	RBI	Ave.	PEVA-B EQ YR.	EQ YRS

251	Carlos	Beltran		**156.861**	**13**	**12.066**	1761	280	1062	0.282	**14.031**	11.18
252	Luis	Aparicio	HOFP	**156.787**	**18**	**8.710**	2677	83	791	0.262	**8.720**	17.98
253	Jorge	Posada		**156.632**	**16**	**9.790**	1583	261	1021	0.275	**12.314**	12.72
254	Tom	Brown		**156.353**	**17**	**9.197**	1951	64	736	0.265	**10.016**	15.61
255	Darryl	Strawberry		**156.264**	**17**	**9.192**	1401	335	1000	0.259	**13.243**	11.80
256	Heinie	Manush	HOFP	**156.227**	**17**	**9.190**	2524	110	1183	0.330	**10.535**	14.83
257	Magglio	Ordonez		**156.022**	**14**	**11.144**	2072	289	1204	0.312	**12.980**	12.02
258	Tony	Oliva		**155.848**	**15**	**10.390**	1917	220	947	0.304	**13.241**	11.77
259	George	Wright	HOF-PI	**155.730**	**12**	**12.978**	867	11	330	0.302	**16.088**	9.68
260	Nellie	Fox	HOFP	**155.463**	**19**	**8.182**	2663	35	790	0.288	**9.710**	16.01
261	Gabby	Hartnett	HOFP	**155.361**	**20**	**7.768**	1912	236	1179	0.297	**9.543**	16.28
262	Al	Rosen		**154.897**	**10**	**15.490**	1063	192	717	0.285	**21.219**	7.30
263	Shawn	Green		**154.634**	**15**	**10.309**	2003	328	1070	0.283	**11.732**	13.18
264	Juan	Gonzalez		**154.558**	**17**	**9.092**	1936	434	1404	0.295	**12.103**	12.77
265	Kenny	Lofton		**152.858**	**17**	**8.992**	2428	130	781	0.299	**9.705**	15.75
266	Tim	Wallach		**152.754**	**17**	**8.986**	2085	260	1125	0.257	**9.779**	15.62
267	Hughie	Jennings	HOFP	**152.306**	**17**	**8.959**	1527	18	840	0.311	**14.561**	10.46
268	Pedro	Guerrero		**152.252**	**15**	**10.150**	1618	215	898	0.300	**13.521**	11.26
269	Johnny	Callison		**152.023**	**16**	**9.501**	1757	226	840	0.264	**11.345**	13.40
270	Henry	Larkin		**151.769**	**10**	**15.177**	1429	53	836	0.303	**15.377**	9.87
271	David	Ortiz		**150.947**	**14**	**10.782**	1598	349	1170	0.281	**14.213**	10.62
272	Donie	Bush		**150.871**	**16**	**9.429**	1804	9	436	0.250	**11.093**	13.60
273	Dusty	Baker		**150.737**	**19**	**7.934**	1981	242	1013	0.278	**10.226**	14.74
274	Tim	Salmon		**150.732**	**14**	**10.767**	1674	299	1016	0.282	**12.416**	12.14
275	Matt	Williams		**149.951**	**17**	**8.821**	1878	378	1218	0.268	**10.772**	13.92
276	Hack	Wilson	HOFP	**149.587**	**12**	**12.466**	1461	244	1063	0.307	**14.884**	10.05
277	Moises	Alou		**149.281**	**17**	**8.781**	2134	332	1287	0.303	**10.239**	14.58
278	Ichiro	Suzuki		**148.997**	**10**	**14.900**	2244	90	558	0.331	**14.900**	10.00
279	Augie	Galan		**148.693**	**16**	**9.293**	1706	100	830	0.287	**11.473**	12.96
280	Rabbit	Maranville	HOFP	**148.551**	**23**	**6.459**	2605	28	884	0.258	**7.872**	18.87
281	Charlie	Keller		**146.592**	**13**	**11.276**	1085	189	760	0.286	**16.453**	8.91
282	Buck	Ewing	HOFP	**146.265**	**18**	**8.126**	1625	71	883	0.303	**10.538**	13.88
283	Jackie	Robinson	HOFP	**146.209**	**10**	**14.621**	1518	137	734	0.311	**14.783**	9.89
284	Ruben	Sierra		**145.729**	**20**	**7.286**	2152	306	1322	0.268	**9.241**	15.77
285	Omar	Vizquel		**145.546**	**22**	**6.616**	2799	80	936	0.273	**7.177**	20.28
286	Jim	Bottomley	HOFP	**145.525**	**16**	**9.095**	2313	219	1422	0.310	**10.036**	14.50
287	Heinie	Groh		**145.428**	**16**	**9.089**	1774	26	566	0.292	**11.606**	12.53
288	Bill	Freehan		**145.340**	**15**	**9.689**	1591	200	758	0.262	**10.986**	13.23
289	Rudy	York		**145.001**	**13**	**11.154**	1621	277	1152	0.275	**13.146**	11.03
290	Ben	Chapman		**144.921**	**15**	**9.661**	1958	90	977	0.302	**11.474**	12.63
291	Jason	Kendall		**144.916**	**15**	**9.661**	2195	75	744	0.288	**9.899**	14.64
292	Abner	Dalrymple		**144.723**	**12**	**12.060**	1202	43	407	0.288	**13.627**	10.62
293	George	Hendrick		**144.471**	**18**	**8.026**	1980	267	1111	0.278	**9.387**	15.39
294	Roy	Sievers		**143.956**	**17**	**8.468**	1703	318	1147	0.267	**10.409**	13.83
295	Ned	Williamson		**143.763**	**13**	**11.059**	1159	64	667	0.255	**11.823**	12.16
296	Ellis	Burks		**143.485**	**18**	**7.971**	2107	352	1206	0.291	**9.348**	15.35
297	Carney	Lansford		**143.016**	**15**	**9.534**	2074	151	874	0.290	**10.562**	13.54
298	Orator	Shaffer		**142.757**	**13**	**10.981**	1000	11	317	0.282	**13.583**	10.51
299	Lee	May		**142.473**	**18**	**7.915**	2031	354	1244	0.267	**9.799**	14.54
300	Garret	Anderson		**141.967**	**16**	**8.873**	2501	285	1353	0.295	**9.683**	14.87

Best Position Players of All-Time 1871-2010

Rank	First	Last	HOF Cat.	PEVA-B TOTAL	YRS.	PEVA-B PER YR.	CAREER Hits	HR	RBI	Ave.	PEVA-B EQ YR.	EQ YRS
301	Frankie	Frisch	HOFP	**141.796**	**19**	**7.463**	2880	105	1244	0.316	**8.196**	17.30
302	Pie	Traynor	HOFP	**141.423**	**17**	**8.319**	2416	58	1273	0.320	**10.300**	13.73
303	Cupid	Childs		**141.383**	**13**	**10.876**	1720	20	743	0.306	**12.146**	11.64

304	Andres	Galarraga		**140.938**	**19**	**7.418**	2333	399	1425	0.288	**8.526**	16.53
305	John	Ward	HOFP	**140.850**	**17**	**8.285**	2104	26	867	0.275	**8.285**	17.00
306	Herman	Long		**140.840**	**16**	**8.803**	2127	91	1055	0.277	**9.478**	14.86
307	Gary	Matthews		**140.782**	**16**	**8.799**	2011	234	978	0.281	**9.597**	14.67
308	Ginger	Beaumont		**140.093**	**12**	**11.674**	1759	39	617	0.311	**12.486**	11.22
309	Clyde	Milan		**139.763**	**16**	**8.735**	2100	17	617	0.285	**9.599**	14.56
310	Cy	Williams		**139.687**	**19**	**7.352**	1981	251	1005	0.292	**9.018**	15.49
311	Cy	Seymour		**139.379**	**16**	**8.711**	1723	52	799	0.303	**10.294**	13.54
312	Ken	Caminiti		**138.421**	**15**	**9.228**	1710	239	983	0.272	**10.739**	12.89
313	Roger	Maris		**138.127**	**12**	**11.511**	1325	275	851	0.260	**12.684**	10.89
314	Denny	Lyons		**137.809**	**13**	**10.601**	1333	62	755	0.310	**14.296**	9.64
315	Bob	Watson		**137.412**	**19**	**7.232**	1826	184	989	0.295	**10.149**	13.54
316	Tony	Phillips		**137.398**	**18**	**7.633**	2023	160	819	0.266	**8.713**	15.77
317	Andy	Pafko		**136.513**	**17**	**8.030**	1796	213	976	0.285	**9.800**	13.93
318	Larry	Gardner		**136.459**	**17**	**8.027**	1931	27	934	0.289	**9.817**	13.90
319	Tip	O'Neill		**136.175**	**10**	**13.617**	1386	52	757	0.326	**14.170**	9.61
320	Bobby	Thomson		**135.795**	**15**	**9.053**	1705	264	1026	0.270	**10.651**	12.75
321	Adrian	Beltre		**135.769**	**13**	**10.444**	1889	278	1008	0.275	**10.792**	12.58
322	Bob	Allison		**135.670**	**13**	**10.436**	1281	256	796	0.255	**12.668**	10.71
323	Ted	Kluszewski		**135.572**	**15**	**9.038**	1766	279	1028	0.298	**10.567**	12.83
324	Jay	Bell		**135.350**	**18**	**7.519**	1963	195	860	0.265	**9.373**	14.44
325	Derrek	Lee		**135.122**	**14**	**9.652**	1843	312	1019	0.282	**11.103**	12.17
326	George	Kell	HOFP	**134.828**	**15**	**8.989**	2054	78	870	0.306	**9.965**	13.53
327	Doc	Cramer		**134.475**	**20**	**6.724**	2705	37	842	0.296	**8.458**	15.90
328	Thurman	Munson		**134.322**	**11**	**12.211**	1558	113	701	0.292	**13.379**	10.04
329	Gavvy	Cravath		**134.238**	**11**	**12.203**	1134	119	719	0.287	**14.655**	9.16
330	Kent	Hrbek		**134.078**	**14**	**9.577**	1749	293	1086	0.282	**10.290**	13.03
331	Jim	Fregosi		**133.839**	**18**	**7.435**	1726	151	706	0.265	**10.025**	13.35
332	George	Scott		**133.682**	**14**	**9.549**	1992	271	1051	0.268	**9.659**	13.84
333	Babe	Herman		**133.641**	**13**	**10.280**	1818	181	997	0.324	**11.662**	11.46
334	Felipe	Alou		**133.359**	**17**	**7.845**	2101	206	852	0.286	**8.728**	15.28
335	Tony	Fernandez		**133.085**	**17**	**7.829**	2276	94	844	0.288	**8.614**	15.45
336	Brady	Anderson		**132.779**	**15**	**8.852**	1661	210	761	0.256	**9.850**	13.48
337	Sid	Gordon		**132.470**	**13**	**10.190**	1415	202	805	0.283	**12.232**	10.83
338	Carlos	Lee		**132.374**	**12**	**11.031**	1967	331	1192	0.287	**11.068**	11.96
339	Mo	Vaughn		**132.140**	**12**	**11.012**	1620	328	1064	0.293	**12.281**	10.76
340	Mark	Teixeira		**131.960**	**8**	**16.495**	1321	275	906	0.286	**16.495**	8.00
341	George	Bell		**131.710**	**12**	**10.976**	1702	265	1002	0.278	**12.083**	10.90
342	Willie	Randolph		**131.448**	**18**	**7.303**	2210	54	687	0.276	**7.801**	16.85
343	Ken	Griffey		**131.432**	**19**	**6.917**	2143	152	859	0.296	**8.138**	16.15
344	Earle	Combs	HOFP	**131.405**	**12**	**10.950**	1866	58	632	0.325	**12.420**	10.58
345	Red	Schoendier	HOFP	**131.155**	**19**	**6.903**	2449	84	773	0.289	**8.187**	16.02
346	Del	Pratt		**131.095**	**13**	**10.084**	1996	43	970	0.292	**10.194**	12.86
347	Billy	Herman	HOFP	**131.084**	**15**	**8.739**	2345	47	839	0.304	**9.812**	13.36
348	Rick	Monday		**131.076**	**19**	**6.899**	1619	241	775	0.264	**8.435**	15.54
349	Hal	McRae		**131.020**	**19**	**6.896**	2091	191	1097	0.290	**8.377**	15.64
350	Larry	Doyle		**130.652**	**14**	**9.332**	1887	74	793	0.290	**9.883**	13.22

Best Position Players of All-Time 1871-2010

Rank	First	Last	HOF Cat.	PEVA-B TOTAL	YRS.	PEVA-B PER YR.	CAREER Hits	HR	RBI	Ave.	PEVA-B EQ YR.	EQ YRS
351	Gene	Woodling		**130.338**	**17**	**7.667**	1585	147	830	0.284	**9.115**	14.30
352	Wally	Joyner		**130.181**	**16**	**8.136**	2060	204	1106	0.289	**8.667**	15.02
353	Gene	Tenace		**130.075**	**15**	**8.672**	1060	201	674	0.241	**11.470**	11.34
354	Ken	Keltner		**130.026**	**13**	**10.002**	1570	163	852	0.276	**12.095**	10.75
355	Dave	Concepcion		**130.003**	**19**	**6.842**	2326	101	950	0.267	**7.287**	17.84
356	John	Titus		**129.977**	**11**	**11.816**	1401	38	561	0.282	**12.895**	10.08

Rank	First	Last	HOF Cat.	PEVA-B TOTAL	YRS.	PEVA-B PER YR.	CAREER Hits	HR	RBI	Ave.	PEVA-B EQ YR.	EQ YRS
357	Tommy	Corcoran		**129.562**	**18**	**7.198**	2252	34	1135	0.256	**7.707**	16.81
358	Jackie	Jensen		**129.098**	**11**	**11.736**	1463	199	929	0.279	**13.106**	9.85
359	Jake	Daubert		**129.093**	**15**	**8.606**	2326	56	722	0.303	**8.652**	14.92
360	Darrell	Porter		**129.013**	**17**	**7.589**	1369	188	826	0.247	**9.275**	13.91
361	Frank	Schulte		**128.353**	**15**	**8.557**	1766	92	792	0.270	**9.410**	13.64
362	George	Wood		**128.190**	**13**	**9.861**	1467	68	601	0.273	**10.313**	12.43
363	Bill	Madlock		**128.170**	**15**	**8.545**	2008	163	860	0.305	**9.274**	13.82
364	John	Reilly		**128.015**	**10**	**12.802**	1352	69	740	0.289	**12.802**	10.00
365	Alvin	Dark		**127.973**	**14**	**9.141**	2089	126	757	0.289	**9.829**	13.02
366	Harry	Davis		**127.900**	**22**	**5.814**	1841	75	951	0.277	**9.432**	13.56
367	Joe	Gordon	HOFP	**127.893**	**11**	**11.627**	1530	253	975	0.268	**11.680**	10.95
368	David	Wright		**127.862**	**7**	**18.266**	1149	169	664	0.305	**19.461**	6.57
369	Joe	Kuhel		**127.834**	**18**	**7.102**	2212	131	1049	0.277	**8.344**	15.32
370	Hal	Trosky		**127.735**	**11**	**11.612**	1561	228	1012	0.302	**13.306**	9.60
371	Tommy	Holmes		**127.603**	**11**	**11.600**	1507	88	581	0.302	**13.604**	9.38
372	Ron	Fairly		**127.559**	**21**	**6.074**	1913	215	1044	0.266	**7.059**	18.07
373	Marquis	Grissom		**127.400**	**17**	**7.494**	2251	227	967	0.272	**8.332**	15.29
374	Andy	Van Slyke		**127.396**	**13**	**9.800**	1562	164	792	0.274	**10.391**	12.26
375	Doug	DeCinces		**127.103**	**15**	**8.474**	1505	237	879	0.259	**10.275**	12.37
376	Dom	DiMaggio		**127.048**	**11**	**11.550**	1680	87	618	0.298	**12.718**	9.99
377	Ernie	Lombardi	HOFP	**126.894**	**17**	**7.464**	1792	190	990	0.306	**8.267**	15.35
378	Fred	Dunlap		**126.889**	**12**	**10.574**	1159	41	366	0.292	**13.081**	9.70
379	Tony	Lazzeri	HOFP	**126.796**	**14**	**9.057**	1840	178	1191	0.292	**9.992**	12.69
380	Benny	Kauff		**126.714**	**8**	**15.839**	961	49	454	0.311	**20.307**	6.24
381	Ken	Williams		**126.511**	**14**	**9.036**	1552	196	913	0.319	**11.714**	10.80
382	Fred	Tenney		**126.468**	**17**	**7.439**	2231	22	688	0.294	**8.165**	15.49
383	Stuffy	McInnis		**126.365**	**19**	**6.651**	2405	20	1063	0.307	**8.216**	15.38
384	Travis	Fryman		**126.328**	**13**	**9.718**	1776	223	1022	0.274	**10.510**	12.02
385	Ray	Lankford		**126.135**	**14**	**9.010**	1561	238	874	0.272	**9.901**	12.74
386	Bill	Buckner		**125.901**	**22**	**5.723**	2715	174	1208	0.289	**6.831**	18.43
387	Wally	Moses		**125.889**	**17**	**7.405**	2138	89	679	0.291	**8.201**	15.35
388	Carl	Furillo		**125.828**	**15**	**8.389**	1910	192	1058	0.299	**9.383**	13.41
389	Jeff	Heath		**125.783**	**14**	**8.985**	1447	194	887	0.293	**11.647**	10.80
390	John	Mayberry		**125.747**	**15**	**8.383**	1379	255	879	0.253	**10.878**	11.56
391	Vic	Wertz		**125.655**	**17**	**7.391**	1692	266	1178	0.277	**8.781**	14.31
392	Pinky	Higgins		**125.489**	**14**	**8.964**	1941	140	1075	0.292	**9.789**	12.82
393	Adam	Dunn		**125.480**	**10**	**12.548**	1246	354	880	0.250	**13.181**	9.52
394	Phil	Cavarretta		**125.119**	**22**	**5.687**	1977	95	920	0.293	**7.864**	15.91
395	Chick	Stahl		**124.306**	**10**	**12.431**	1546	36	622	0.305	**12.749**	9.75
396	Todd	Zeile		**124.224**	**16**	**7.764**	2004	253	1110	0.265	**8.254**	15.05
397	Bert	Campaneris		**123.969**	**19**	**6.525**	2249	79	646	0.259	**7.331**	16.91
398	Danny	Tartabull		**123.347**	**14**	**8.810**	1366	262	925	0.273	**11.474**	10.75
399	Oyster	Burns		**123.319**	**11**	**11.211**	1389	65	832	0.300	**12.419**	9.93
400	Elmer	Smith		**123.216**	**14**	**8.801**	1454	37	663	0.310	**9.842**	12.52

Best Position Players of All-Time 1871-2010

Rank	First	Last	HOF Cat.	PEVA-B TOTAL	YRS.	PEVA-B PER YR.	CAREER Hits	HR	RBI	Ave.	PEVA-B EQ YR.	EQ YRS
401	Wally	Berger		**123.076**	**11**	**11.189**	1550	242	898	0.300	**12.636**	9.74
402	Joe	Adcock		**122.934**	**17**	**7.231**	1832	336	1122	0.277	**8.004**	15.36
403	Joe	Judge		**122.894**	**20**	**6.145**	2352	71	1034	0.298	**7.412**	16.58
404	Buddy	Myer		**122.894**	**17**	**7.229**	2131	38	850	0.303	**8.636**	14.23
405	Dave	Bancroft	HOFP	**122.761**	**16**	**7.673**	2004	32	591	0.279	**8.455**	14.52
406	Billy	Nash		**122.428**	**15**	**8.162**	1606	60	977	0.275	**9.686**	12.64
407	B.J.	Surhoff		**122.381**	**19**	**6.441**	2326	188	1153	0.282	**7.224**	16.94
408	Roger	Peckinpaugh		**122.294**	**17**	**7.194**	1876	48	740	0.259	**8.388**	14.58
409	Dode	Paskert		**122.199**	**15**	**8.147**	1613	42	577	0.268	**9.422**	12.97

410	Willie	Horton		**122.172**	**18**	**6.787**	1993	325	1163	0.273	**8.038**	15.20
411	Kirk	Gibson		**122.078**	**17**	**7.181**	1553	255	870	0.268	**9.298**	13.13
412	Gus	Bell		**122.017**	**15**	**8.134**	1823	206	942	0.281	**9.630**	12.67
413	Arlie	Latham		**122.015**	**17**	**7.177**	1833	27	563	0.269	**9.038**	13.50
414	Dick	Groat		**121.479**	**14**	**8.677**	2138	39	707	0.286	**9.196**	13.21
415	Edgar	Renteria		**121.457**	**15**	**8.097**	2252	135	887	0.287	**8.365**	14.52
416	Curt	Flood		**121.336**	**15**	**8.089**	1861	85	636	0.293	**9.970**	12.17
417	Dolph	Camilli		**121.287**	**12**	**10.107**	1482	239	950	0.277	**11.518**	10.53
418	Terry	Pendleton		**121.272**	**15**	**8.085**	1897	140	946	0.270	**9.043**	13.41
419	Bill	Bradley		**120.886**	**14**	**8.635**	1471	33	552	0.271	**11.080**	10.91
420	Nomar	Garciaparra		**120.880**	**14**	**8.634**	1747	229	936	0.313	**11.745**	10.57
421	Harvey	Kuenn		**120.870**	**15**	**8.058**	2092	87	671	0.303	**9.013**	13.41
422	Jeff	Burroughs		**120.741**	**16**	**7.546**	1443	240	882	0.261	**9.605**	12.57
423	Hal	Chase		**120.683**	**15**	**8.046**	2158	57	941	0.291	**8.238**	14.65
424	Jack	Fournier		**120.401**	**15**	**8.027**	1631	136	859	0.313	**10.308**	11.68
425	Paul	Konerko		**120.359**	**14**	**8.597**	1861	365	1156	0.280	**9.507**	12.66
426	Maury	Wills		**120.338**	**14**	**8.596**	2134	20	458	0.281	**9.041**	13.31
427	David	Justice		**120.261**	**14**	**8.590**	1571	305	1017	0.279	**9.793**	12.28
428	Wally	Schang		**120.177**	**19**	**6.325**	1506	59	710	0.284	**7.718**	15.57
429	Chuck	Knoblauch		**119.682**	**12**	**9.973**	1839	98	615	0.289	**10.344**	11.57
430	Roger	Bresnahan	HOFP	**119.654**	**17**	**7.038**	1252	26	530	0.279	**9.696**	12.34
431	Joe	Tinker	HOFP	**119.627**	**15**	**7.975**	1687	31	782	0.262	**9.042**	13.23
432	Dave	Kingman		**119.486**	**16**	**7.468**	1575	442	1210	0.236	**8.128**	14.70
433	Tino	Martinez		**119.460**	**16**	**7.466**	1925	339	1271	0.271	**8.290**	14.41
434	Joe	Mauer		**119.391**	**7**	**17.056**	1011	81	472	0.327	**19.164**	6.23
435	Cecil	Fielder		**118.951**	**13**	**9.150**	1313	319	1008	0.255	**11.211**	10.61
436	Wally	Pipp		**118.828**	**15**	**7.922**	1941	90	998	0.281	**8.868**	13.40
437	Mickey	Tettleton		**118.622**	**14**	**8.473**	1132	245	732	0.241	**10.667**	11.12
438	Tommy	Davis		**118.537**	**18**	**6.585**	2121	153	1052	0.294	**8.158**	14.53
439	Greg	Vaughn		**118.489**	**15**	**7.899**	1475	355	1072	0.242	**9.073**	13.06
440	Andre	Thornton		**118.476**	**14**	**8.463**	1342	253	895	0.254	**9.964**	11.89
441	Troy	Glaus		**118.405**	**13**	**9.108**	1375	320	950	0.254	**11.087**	10.68
442	Devon	White		**118.194**	**17**	**6.953**	1934	208	846	0.263	**8.208**	14.40
443	Rico	Carty		**118.157**	**15**	**7.877**	1677	204	890	0.299	**9.483**	12.46
444	Ron	Gant		**117.848**	**16**	**7.365**	1651	321	1008	0.256	**8.723**	13.51
445	Mike	Cameron		**117.212**	**16**	**7.326**	1652	269	941	0.250	**8.760**	13.38
446	Jimmy	Williams		**117.083**	**11**	**10.644**	1507	49	796	0.275	**10.722**	10.92
447	Jay	Buhner		**117.017**	**15**	**7.801**	1273	310	965	0.254	**10.619**	11.02
448	Michael	Young		**116.915**	**11**	**10.629**	1848	158	811	0.300	**11.822**	9.89
449	Lip	Pike		**116.787**	**10**	**11.679**	637	20	332	0.321	**14.653**	7.97
450	Willie	Jones		**116.648**	**15**	**7.777**	1502	190	812	0.258	**9.680**	12.05

Best Position Players of All-Time 1871-2010

			HOF	PEVA-B		PEVA-B	CAREER				PEVA-B	EQ
Rank	**First**	**Last**	**Cat.**	**TOTAL**	**YRS.**	**PER YR.**	**Hits**	**HR**	**RBI**	**Ave.**	**EQ YR.**	**YRS**
451	John	Morrill		**116.594**	**15**	**7.773**	1275	43	643	0.260	**8.675**	13.44
452	Heinie	Zimmerman		**116.315**	**13**	**8.947**	1566	58	796	0.295	**10.770**	10.80
453	Jermaine	Dye		**116.116**	**14**	**8.294**	1779	325	1072	0.274	**9.517**	12.44
454	Ben	Oglivie		**116.102**	**16**	**7.256**	1615	235	901	0.273	**8.549**	13.58
455	Bill	White		**115.924**	**13**	**8.917**	1706	202	870	0.286	**10.063**	11.52
456	Eric	Davis		**115.804**	**17**	**6.812**	1430	282	934	0.269	**8.984**	12.89
457	Torii	Hunter		**115.766**	**14**	**8.269**	1667	258	969	0.275	**9.903**	11.69
458	Tom	Brunansky		**115.438**	**14**	**8.246**	1543	271	919	0.245	**9.220**	12.52
459	Kevin	McReynolds		**115.206**	**12**	**9.600**	1439	211	807	0.265	**10.767**	10.70
460	Bill	Joyce		**115.097**	**8**	**14.387**	970	70	607	0.294	**15.065**	7.64
461	Patsy	Donovan		**115.066**	**17**	**6.769**	2253	16	736	0.301	**7.844**	14.67
462	Jesse	Barfield		**114.802**	**12**	**9.567**	1219	241	716	0.256	**11.322**	10.14

463	Gorman	Thomas		**114.636**	**13**	**8.818**	1051	268	782	0.225	**10.876**	10.54
464	Elston	Howard		**114.562**	**14**	**8.183**	1471	167	762	0.274	**8.985**	12.75
465	Mike	Hargrove		**114.409**	**12**	**9.534**	1614	80	686	0.290	**9.630**	11.88
466	John	Anderson		**114.302**	**14**	**8.164**	1841	49	976	0.290	**8.847**	12.92
467	Willie	Wilson		**114.276**	**19**	**6.015**	2207	41	585	0.285	**7.120**	16.05
468	Jose	Cardenal		**114.219**	**18**	**6.346**	1913	138	775	0.275	**7.937**	14.39
469	Topsy	Hartsel		**113.898**	**14**	**8.136**	1336	31	341	0.276	**11.123**	10.24
470	Tommy	Henrich		**113.881**	**11**	**10.353**	1297	183	795	0.282	**11.621**	9.80
471	Dave	Orr		**113.756**	**8**	**14.219**	1125	37	627	0.342	**16.368**	6.95
472	Lloyd	Moseby		**113.446**	**12**	**9.454**	1494	169	737	0.257	**9.831**	11.54
473	Willie	McGee		**112.994**	**18**	**6.277**	2254	79	856	0.295	**6.815**	16.58
474	Sherm	Lollar		**112.715**	**18**	**6.262**	1415	155	808	0.264	**8.097**	13.92
475	Jimmy	Wolf		**112.472**	**11**	**10.225**	1440	18	593	0.290	**11.214**	10.03
476	Frank	McCormick		**112.425**	**13**	**8.648**	1711	128	951	0.299	**10.448**	10.76
477	Baby Doll	Jacobson		**112.422**	**11**	**10.220**	1714	83	821	0.311	**10.820**	10.39
478	Eddie	Joost		**112.405**	**17**	**6.612**	1339	134	601	0.239	**9.624**	11.68
479	Willie	Kamm		**112.279**	**13**	**8.637**	1643	29	826	0.281	**9.349**	12.01
480	Hank	Sauer		**112.180**	**15**	**7.479**	1278	288	876	0.266	**10.694**	10.49
481	Phil	Rizzuto	HOFP	**111.989**	**13**	**8.615**	1588	38	563	0.273	**9.364**	11.96
482	Larry	Parrish		**111.965**	**15**	**7.464**	1789	256	992	0.263	**8.020**	13.96
483	Kevin	Mitchell		**111.893**	**13**	**8.607**	1173	234	760	0.284	**11.717**	9.55
484	Charlie	Bennett		**111.839**	**15**	**7.456**	978	55	533	0.256	**9.510**	11.76
485	Fred	Pfeffer		**111.734**	**16**	**6.983**	1671	94	1019	0.255	**7.759**	14.40
486	Richie	Hebner		**111.657**	**18**	**6.203**	1694	203	890	0.276	**7.390**	15.11
487	Dwayne	Murphy		**111.600**	**12**	**9.300**	1069	166	609	0.246	**10.995**	10.15
488	Chris	Chambliss		**111.205**	**17**	**6.541**	2109	185	972	0.279	**7.156**	15.54
489	Matt	Holliday		**111.044**	**7**	**15.863**	1216	180	695	0.317	**15.863**	7.00
490	Ross	Youngs	HOFP	**110.998**	**10**	**11.100**	1491	42	592	0.322	**12.444**	8.92
491	Jimmy	Rollins		**110.925**	**11**	**10.084**	1714	154	662	0.272	**11.216**	9.89
492	George	Burns		**110.840**	**16**	**6.927**	2018	72	951	0.307	**7.789**	14.23
493	Stan	Spence		**110.548**	**9**	**12.283**	1090	95	575	0.282	**13.648**	8.10
494	Miller	Huggins	HOFM	**110.533**	**13**	**8.503**	1474	9	318	0.265	**9.595**	11.52
495	Ned	Hanlon		**110.182**	**13**	**8.476**	1317	30	517	0.260	**9.113**	12.09
496	Ryan	Howard		**109.947**	**7**	**15.650**	902	253	748	0.279	**18.630**	5.88
497	Rick	Ferrell	HOFP	**109.810**	**18**	**6.101**	1692	28	734	0.281	**7.177**	15.30
498	Jim	Gilliam		**109.763**	**14**	**7.840**	1889	65	558	0.265	**8.089**	13.57
499	Harry	Steinfeldt		**109.603**	**14**	**7.829**	1576	27	762	0.267	**8.712**	12.58
500	Richie	Zisk		**109.541**	**13**	**8.426**	1477	207	792	0.287	**10.152**	10.79

Best Position Players of All-Time 1871-2010

Rank	First	Last	HOF Cat.	PEVA-B TOTAL	YRS.	PEVA-B PER YR.	CAREER Hits	HR	RBI	Ave.	PEVA-B EQ YR.	EQ YRS
501	Pete	Runnels		**109.470**	**14**	**7.819**	1854	49	630	0.291	**8.526**	12.84
502	Dick	Bartell		**109.427**	**18**	**6.079**	2165	79	710	0.284	**7.242**	15.11
503	Vinny	Castilla		**109.374**	**16**	**6.836**	1884	320	1105	0.276	**8.388**	13.04
504	Raul	Mondesi		**109.331**	**13**	**8.410**	1589	271	860	0.273	**10.114**	10.81
505	Frank	Chance	HOFP	**108.846**	**17**	**6.403**	1273	20	596	0.296	**10.182**	10.69
506	Mike	Donlin		**108.775**	**12**	**9.065**	1282	51	543	0.333	**13.074**	8.32
507	Roy	Cullenbine		**108.557**	**10**	**10.856**	1072	110	599	0.276	**12.435**	8.73
508	Jim	Sundberg		**108.073**	**16**	**6.755**	1493	95	624	0.248	**7.469**	14.47
509	Bret	Boone		**108.071**	**14**	**7.719**	1775	252	1021	0.266	**8.570**	12.61
510	Jack	Tobin		**108.007**	**13**	**8.308**	1906	64	581	0.309	**9.069**	11.91
511	Bob	Bailey		**107.922**	**17**	**6.348**	1564	189	773	0.257	**7.698**	14.02
512	Vernon	Wells		**107.684**	**12**	**8.974**	1529	223	813	0.280	**11.456**	9.40
513	Reggie	Sanders		**107.604**	**17**	**6.330**	1643	303	972	0.266	**7.725**	13.93
514	Buck	Freeman		**107.429**	**11**	**9.766**	1235	82	713	0.293	**12.506**	8.59
515	Duffy	Lewis		**107.398**	**11**	**9.763**	1518	38	793	0.284	**10.581**	10.15

516	Lloyd	Waner	HOFP	107.291	18	5.961	2459	27	598	0.316	7.124	15.06
517	Frank	Thomas		107.278	16	6.705	1671	286	962	0.266	8.421	12.74
518	Buddy	Lewis		107.224	11	9.748	1563	71	607	0.297	11.287	9.50
519	Sam	West		106.717	16	6.670	1838	75	838	0.299	7.946	13.43
520	John	McGraw	HOFM	106.609	16	6.663	1309	13	462	0.334	11.082	9.62
521	George	Stone		106.369	7	15.196	984	23	268	0.301	17.877	5.95
522	Rico	Petrocelli		106.231	13	8.172	1352	210	773	0.251	9.393	11.31
523	Ryan	Klesko		106.153	16	6.635	1564	278	987	0.279	8.006	13.26
524	Tommy	Tucker		106.088	13	8.161	1882	42	932	0.290	8.186	12.96
525	Raul	Ibanez		105.633	15	7.042	1660	232	970	0.284	8.990	11.75
526	Mike	Lowell		105.498	13	8.115	1619	223	952	0.279	9.279	11.37
527	Javy	Lopez		105.482	15	7.032	1527	260	864	0.287	8.776	12.02
528	Bob	Boone		105.468	19	5.551	1838	105	826	0.254	6.153	17.14
529	Dante	Bichette		105.445	14	7.532	1906	274	1141	0.299	8.601	12.26
530	Steve	Brodie		104.939	12	8.745	1726	25	900	0.303	9.262	11.33
531	Tilly	Walker		104.891	13	8.069	1423	118	679	0.281	9.840	10.66
532	Ray	Durham		104.684	14	7.477	2054	192	875	0.277	7.515	13.93
533	Alvin	Davis		104.590	9	11.621	1189	160	683	0.280	12.556	8.33
534	Aramis	Ramirez		104.491	13	8.038	1615	289	1029	0.282	9.491	11.01
535	J.D.	Drew		104.392	13	8.030	1382	238	773	0.281	9.321	11.20
536	Jimmie	Dykes		104.326	22	4.742	2256	108	1071	0.280	5.864	17.79
537	Howard	Johnson		104.313	14	7.451	1229	228	760	0.249	9.039	11.54
538	Alfonso	Soriano		104.229	12	8.686	1634	314	839	0.277	10.169	10.25
539	Walker	Cooper		104.205	18	5.789	1341	173	812	0.285	7.829	13.31
540	Art	Devlin		103.969	10	10.397	1185	10	505	0.269	11.049	9.41
541	Curt	Welch		103.802	10	10.380	1152	16	503	0.263	11.598	8.95
542	Jason	Bay		103.732	8	12.967	1017	191	657	0.278	14.347	7.23
543	Eric	Karros		103.652	14	7.404	1724	284	1027	0.268	8.379	12.37
544	Davy	Force		103.607	15	6.907	1060	1	373	0.249	7.155	14.48
545	Burt	Shotton		103.306	14	7.379	1338	9	290	0.271	10.238	10.09
546	Larry	Bowa		103.182	16	6.449	2191	15	525	0.260	6.572	15.70
547	Jeromy	Burnitz		102.981	14	7.356	1447	315	981	0.253	8.596	11.98
548	Tony	Gonzalez		102.882	12	8.573	1485	103	615	0.286	8.771	11.73
549	Amos	Strunk		102.810	17	6.048	1418	15	530	0.284	8.434	12.19
550	Dan	McGann		102.776	12	8.565	1482	42	727	0.284	9.234	11.13

Best Position Players of All-Time 1871-2010

			HOF	PEVA-B		PEVA-B	CAREER				PEVA-B	EQ
Rank	First	Last	Cat.	TOTAL	YRS.	PER YR.	Hits	HR	RBI	Ave.	EQ YR.	YRS
551	Billy	Shindle		102.604	13	7.893	1561	31	758	0.269	9.104	11.27
552	Tony	Pena		102.410	18	5.689	1687	107	708	0.260	6.836	14.98
553	Johnny	Evers	HOFP	102.257	18	5.681	1659	12	538	0.270	7.591	13.47
554	Ray	Schalk	HOFP	102.242	18	5.680	1345	11	594	0.253	7.534	13.57
555	Benito	Santiago		102.224	20	5.111	1830	217	920	0.263	6.612	15.46
556	Danny	Murphy		101.645	16	6.353	1563	44	702	0.289	9.165	11.09
557	Tommy	Harper		101.585	15	6.772	1609	146	567	0.257	7.760	13.09
558	Orlando	Cabrera		101.273	14	7.234	1948	118	803	0.274	8.018	12.63
559	Sam	Chapman		101.112	11	9.192	1329	180	773	0.266	10.121	9.99
560	Bill	Bruton		100.894	12	8.408	1651	94	545	0.273	8.850	11.40
561	Bob	Meusel		100.870	11	9.170	1693	156	1068	0.309	9.314	10.83
562	Bob	Ferguson		100.837	14	7.203	920	1	356	0.265	8.048	12.53
563	Leo	Cardenas		100.684	16	6.293	1725	118	689	0.257	7.622	13.21
564	Joe	Hornung		100.565	12	8.380	1230	31	564	0.257	8.892	11.31
565	Earl	Torgeson		100.467	15	6.698	1318	149	740	0.265	7.788	12.90
566	Jason	Thompson		100.311	11	9.119	1253	208	782	0.261	9.786	10.25
567	Joe	Vosmik		100.295	13	7.715	1682	65	874	0.307	9.709	10.33
568	Charlie	Jamieson		100.173	18	5.565	1990	18	552	0.303	7.222	13.87

569	Mike	Greenwell		**100.166**	**12**	**8.347**	1400	130	726	0.303	**10.622**	9.43
570	Tommy	McCarthy	HOFP	**100.079**	**13**	**7.698**	1496	44	735	0.292	**9.688**	10.33
571	Von	Hayes		**100.034**	**12**	**8.336**	1402	143	696	0.267	**9.280**	10.78

Note: PEVA-B represents Batting PEVA for all regular seasons. It does not include Pitching PEVA values.

Note: PEVA Per EQ Year represents Average PEVA for a Full Season. PEVA per Year equals Average PEVA for Total Seasons.

Note: EQ Year is defined as the baseballevaluation.com statistic that reflects the Experience Equivalant of a player's experience for any or all seasons. It is meant to approximate Major League Service Time data for players prior to its inception, as well as players within the MLST era.

Note: 2010 PEVA = Reflects preliminary final PEVA numbers. Marginal adjustment may be made with final stats and park factors.

Chapter 5 - BEST CAREERS EVER, PITCHERS

When you think of the best pitchers of the game, it's not too hard to think of a great Top Ten, from the names emblazoned with the Best Pitcher of the Year Award, yes, that man Cy Young, to the foursome of great pitchers who hurled during the steroid era and got less credit than they deserved. But pitching stats when compared to each other can get even trickier over the generations of baseball than hitting stats do, and that's why it's important to go beneath the stats themselves and see how they compared against each other for the era in which the pitcher pitched. And that's just what the PEVA Pitching Rating does, measuring the dominance of a pitcher for a season in a number of categories, those of durability (Innings Pitched and Games Played and Games Started) to those of production (Wins and Saves) to Rate Stats (ERA, WHIP, Strike Out to Walk Ratio, and HR allowed per 9 Innings Pitched), then adding their numbers together to get a Career PEVA Rating.

The list below contains players who pitched in every era, with the exception, for the most part, of those who pitched in the first two decades of major league ball. They show up a bit further down the list than the Top Ten. It includes pitchers who threw the ball last year. And in the end, the top spot goes to a pitcher most consider in the bottom half of the Top Ten in their lists, but for us, and the statistics we measure, he pops to the top. I guess that's a bit controversial, but it shouldn't be when you dig further and further into the reasons why. So we'll start with an explanation of why, and know others will have a different opinion.

1. Greg Maddux (1986-2008)

For twenty straight seasons, Greg Maddux won at least fifteen games. He came to the park, usually one with a hitter's pedigree, pitched, gave up few hits, walks, and runs, then won. He led his team to the playoffs most seasons. He struck people out, #10 on the All-Time list, that is likely a shock to some less sabermetric fans of the game. And at the end of the day, that consistency, year after year after year, makes Greg Maddux the number one pitcher in baseball history and the next first ballot pitching entrant into Cooperstown's Hall of Fame. We're going to admit something here that we likely should not. When the PEVA calculations were first run, done by season first and then tototaled, we were surprised by this result. But when we dug deeper into the question, the more we found reasons for Maddux stepping into the spotlight at the pinnacle. Maddux won 355 games for the #8 spot on the career list in a time when pitchers pitched in five man rotations and were removed for closers in tight games. No, he didn't rack up the innings like the men of Cy Young's or Walter Johnson's time, that will likely never happen again due to the evolution of the relief pitcher, but Maddux did what he did in the game he was given.

Where Maddux Ranks		
Category		**Rank**
W	355	8
SO	3371	10
ERA	3.16	230
WHIP9	10.29	53
SO/W	3.37	19
IP	5008.3	13

His ERA + adjusted for park was #31 on the list, and suffered an increase at the end of his career; it's really the only blemish on his record. The other rate stats hold up really well. We shouldn't forget that most of the pitchers who rank higher in ERA or WHIP9 pitched in eras when people did not get lots of hits or score many runs.

Maddux won twenty games only twice, but would have had two more seasons in that league had strike seasons of 1994 and 1995, his best two seasons, not been shortened. In fact, we believe that he would rank higher on other lists if those seasons were given their full credit within baseball history for being among the Top Ten Seasons of All-Time. But we'll stop defending it now, and let you ruminate over the stats below, the string of seasons from 1992 to 1998 that boggle your mind and boggled the hitters, too. And we're perfectly fine with the disagreement some might have, or a personal ranking for Maddux below this statistical one. That's what makes the baseball world go round after all.

Greg Maddux Stats					**594.209 PEVA RATING**			
Year	**Team**	**Lg**	**W**	**SV**	**IP**	**ERA**	**Age**	**PEVA-P**
1986	CHN	NL	2	0	31.0	5.52	20	0.492

1987	CHN	NL	6	0	155.7	5.61	21	2.470
1988	CHN	NL	18	0	249.0	3.18	22	15.180
1989	CHN	NL	19	0	238.3	2.95	23	18.870
1990	CHN	NL	15	0	237.0	3.46	24	13.693
1991	CHN	NL	15	0	263.0	3.35	25	22.395
1992	CHN	NL	20	0	268.0	2.18	26	49.969
1993	ATL	NL	20	0	267.0	2.36	27	50.920
1994	ATL	NL	16	0	202.0	1.56	28	57.974
1995	ATL	NL	19	0	209.7	1.63	29	52.118
1996	ATL	NL	15	0	245.0	2.72	30	35.136
1997	ATL	NL	19	0	232.7	2.20	31	47.132
1998	ATL	NL	18	0	251.0	2.22	32	46.792
1999	ATL	NL	19	0	219.3	3.57	33	18.052
2000	ATL	NL	19	0	249.3	3.00	34	32.027
2001	ATL	NL	17	0	233.0	3.05	35	35.904
2002	ATL	NL	16	0	199.3	2.62	36	18.307
2003	ATL	NL	16	0	218.3	3.96	37	14.031
2004	CHN	NL	16	0	212.7	4.02	38	14.915
2005	CHN	NL	13	0	225.0	4.24	39	11.578
2006	CHN	NL	9	0	136.3	4.69	40	15.846
2006	LAN	NL	6	0	73.7	3.30	40	
2007	SDN	NL	14	0	198.0	4.14	41	13.315
2008	SDN	NL	6	0	153.3	3.99	42	7.092
2008	LAN	NL	2	0	40.7	5.09	42	
		Total	355	0	5008.3	3.16		

2. Cy Young (1890-1911)

Yes, Cy Young would have won a whole lot of his own awards. He was the best pitcher at the turn of the century and many think the best pitcher in the game. And it's hard to quibble, although some of his exploits need to be put into some historic context. Cy Young won 511 games and pitched over 400 innings five times. It's that durability, as well as skill, that pushes Young toward the top. His ERA of 2.63 when adjusted for the age and park ranks #17 on the career list. He wasn't as precise as Maddux and didn't strike out as many batters as hurlers whose careers would start after the turn of the century, but those were the sole items where some could criticize him.

Where Cy Young Ranks		
Category		**Rank**
W	511	1
SO	2803	20
ERA	2.63	58
WHIP9	10.17	37
SO/W	2.30	134
IP	7354.7	1

Only one season in a span of 19 years from 1891 to 1909 did Cy Young fail to win 15 games, including over 30 wins in four separate seasons. There were six seasons with an ERA south of the pitcher's Mendoza line of 2.00. What all that adds up to is the 2nd Best Pitching Career in baseball history within the baseballevaluation universe, plus an annual award given to the best pitchers in the American and National Leagues. It's good to give out props where they're deserved, as the awards remind us of Cy Young each fall.

Geez, wonder what he would think of pitch counts and two hundred inning seasons considered as overuse. Perhaps just progress, perhaps not.

Cy Young Stats						**504.145 PEVA RATING**		
Year	**Team**	**Lg**	**W**	**SV**	**IP**	**ERA**	**Age**	**PEVA-P**
1890	CL4	NL	9	0	147.7	3.47	23	2.182
1891	CL4	NL	27	2	423.7	2.85	24	14.082
1892	CL4	NL	36	0	453.0	1.93	25	31.345
1893	CL4	NL	34	1	422.7	3.36	26	43.542

1894	CL4	NL	26	1	408.7	3.94	27	26.323
1895	CL4	NL	35	0	369.7	3.26	28	32.941
1896	CL4	NL	28	3	414.3	3.24	29	38.520
1897	CL4	NL	21	0	333.7	3.80	30	18.713
1898	CL4	NL	25	0	377.7	2.53	31	26.356
1899	SLN	NL	26	1	369.3	2.58	32	40.907
1900	SLN	NL	19	0	321.3	3.00	33	21.483
1901	BOS	AL	33	0	371.3	1.62	34	48.628
1902	BOS	AL	32	0	384.7	2.15	35	28.273
1903	BOS	AL	28	2	341.7	2.08	36	26.153
1904	BOS	AL	26	1	380.0	1.97	37	21.996
1905	BOS	AL	18	0	320.7	1.82	38	22.243
1906	BOS	AL	13	2	287.7	3.19	39	6.429
1907	BOS	AL	21	2	343.3	1.99	40	18.401
1908	BOS	AL	21	2	299.0	1.26	41	19.816
1909	CLE	AL	19	0	295.0	2.26	42	10.301
1910	CLE	AL	7	0	163.3	2.53	43	4.048
1911	CLE	AL	3	0	46.3	3.88	44	1.462
1911	BSN	NL	4	0	80.0	3.71	44	
		Total	511	17	7354.7	2.63		

3. Roger Clemens (1984-2007)

As Clemens pitched throughout his career, it was pretty much acknowledged that you were witnessing one of the games best pitchers, right up there with Young and Walter Johnson. Most thought of him as a notch above Maddux, particularly because he could dominate a game with the strikeout in a way Maddux could not, plus he was considered a better big game pitcher. And he certainly was up there in class. Now, we once again don't know if or how the PED allegations of recent years affect those thoughts; they certainly don't add to the luster. For some reason, we don't think Clemens will suffer the demotions as much as Bonds might when the full blush of the era is in better view, but that's only a perception way on the outside of the actual facts. But with the career that we know about and the seasons and stats that are in clear view, Clemens ranks at the #3 Pitcher of All-Time for his career.

Where Clemens Ranks		
Category		**Rank**
W	354	9
SO	4672	3
ERA	3.13	212
WHIP9	10.55	87
SO/W	2.96	38
IP	4916.7	16

Let's answer the question right up front. Why is Clemens ranked below his contemporary on the mound, Greg Maddux? It's really pretty easy; he was less durable. They pitched during the same years with Clemens beginning his career only two years earlier and stopping one year before, but Maddux pitched more innings despite a career that was solely within the National League where the manager had to pinch hit for the pitcher in close games. Now, Maddux did not get pinch hit for as much as some; he was a fine hitter, but if you look down the list of seasons, you'll see why Clemens fell short in that regard, logging five seasons below 200 innings from the age of 24-40. Maddux had only one. Clemens had four seasons, 1993, 1995, 1999, and 2002, when his PEVA Pitching rating fell below 10.000; Maddux had none. So while Clemens may have been the pitcher you'd select for that one game; Maddux took the ball more often and won the game he took it in. There's not a whole lot to choose between the two in rate stats; the 10.55 vs. 10.34 for Clemens in WHIP9 probably better considering the league differentials with the same acknowledgment for the ERA (in Roger's favor anyway), and Roger's prowess did not wane in his final years quite as much, although there might be a reason why.

Roger Clemens Stats					487.448 PEVA RATING			
Year	**Team**	**Lg**	**W**	**SV**	**IP**	**ERA**	**Age**	**PEVA-P**
1984	BOS	AL	9	0	133.3	4.32	22	2.586
1985	BOS	AL	7	0	98.3	3.29	23	2.682
1986	BOS	AL	24	0	254.0	2.48	24	35.030

1987	BOS	AL	20	0	281.7	2.97	25	43.210
1988	BOS	AL	18	0	264.0	2.93	26	28.435
1989	BOS	AL	17	0	253.3	3.13	27	17.353
1990	BOS	AL	21	0	228.3	1.93	28	41.426
1991	BOS	AL	18	0	271.3	2.62	29	43.004
1992	BOS	AL	18	0	246.7	2.41	30	34.707
1993	BOS	AL	11	0	191.7	4.46	31	6.679
1994	BOS	AL	9	0	170.7	2.85	32	15.705
1995	BOS	AL	10	0	140.0	4.18	33	4.859
1996	BOS	AL	10	0	242.7	3.63	34	11.918
1997	TOR	AL	21	0	264.0	2.05	35	47.079
1998	TOR	AL	20	0	234.7	2.65	36	33.173
1999	NYA	AL	14	0	187.7	4.60	37	6.966
2000	NYA	AL	13	0	204.3	3.70	38	12.249
2001	NYA	AL	20	0	220.3	3.51	39	18.346
2002	NYA	AL	13	0	180.0	4.35	40	5.994
2003	NYA	AL	17	0	211.7	3.91	41	12.288
2004	HOU	NL	18	0	214.3	2.98	42	22.379
2005	HOU	NL	13	0	211.3	1.87	43	28.104
2006	HOU	NL	7	0	113.3	2.30	44	10.811
2007	NYA	AL	6	0	99.0	4.18	45	2.466
		Total	354	0	4916.7	3.12		

4. Walter Johnson (1907-1927)

The Big Train is coming, coming round the bend, and everyone knew when and how. Walter Johnson ranks as the 4th best pitcher in baseball history and gets a whole lot of kudos from baseball historians as deserving of a spot several notches higher. And we can understand the sentiment. Johnson was really the first pitcher in history to dominate the game with the strikeout. He was the pitching version of Babe Ruth. Dominant on the mound in every way, particularly through the age of 32. But Johnson waned near the end of his career, with only three above 10.000 PEVA seasons after that age. But up until that time, there was no doubt to his achievements and dominance. Ranking #2 in All-Time Wins, #3 in Innings Pitched, and #4 in ERA plus adjusted show what many people knew each time they watched him start a game. The Big Train was coming, coming fast, and going to give his team a great chance to win.

Where W. Johnson Ranks		
Category		**Rank**
W	417	2
SO	3509	9
ERA	2.17	11
WHIP9	9.55	8
SO/W	2.57	68
IP	5914.7	3

Johnson pitched for the old Washington Senators, today's Minnesota Twins, and didn't have the benefit of pitching for a club that won with the tradition of the Yankees or some other teams at the time, so his 0.599 winning percentage is the 2nd lowest of the Top Ten on this list. Washington only had 7 winning seasons in the 21 years he pitched for the Senators, winning the pennant in 1924 and 1925 when Johnson was 37 and 38 years old. It's likely that if he'd been on some better teams, his win total would be even higher. But it was certainly high enough to place Johnson in the Top Ten, particularly when combined with those eleven seasons of an ERA under 2.00. Oh, my.

Walter Johnson Stats					**479.881 PEVA RATING**			
Year	**Team**	**Lg**	**W**	**SV**	**IP**	**ERA**	**Age**	**PEVA-P**
1907	WS1	AL	5	0	110.3	1.88	20	2.169
1908	WS1	AL	14	1	256.3	1.65	21	9.455
1909	WS1	AL	13	1	296.3	2.22	22	9.771
1910	WS1	AL	25	1	370.0	1.36	23	45.752
1911	WS1	AL	25	1	322.3	1.90	24	27.861
1912	WS1	AL	33	2	369.0	1.39	25	52.347
1913	WS1	AL	36	2	346.0	1.14	26	43.737
1914	WS1	AL	28	1	371.7	1.72	27	36.936

1915	WS1	AL	27	4	336.7	1.55	28	36.871
1916	WS1	AL	25	1	369.7	1.90	29	31.062
1917	WS1	AL	23	3	326.0	2.21	30	18.739
1918	WS1	AL	23	3	326.0	1.27	31	43.252
1919	WS1	AL	20	2	290.3	1.49	32	32.304
1920	WS1	AL	8	3	143.7	3.13	33	3.207
1921	WS1	AL	17	1	264.0	3.51	34	9.385
1922	WS1	AL	15	4	280.0	2.99	35	12.127
1923	WS1	AL	17	4	261.3	3.48	36	9.689
1924	WS1	AL	23	0	277.7	2.72	37	25.968
1925	WS1	AL	20	0	229.0	3.07	38	18.743
1926	WS1	AL	15	0	260.7	3.63	39	9.206
1927	WS1	AL	5	0	107.7	5.10	40	1.299
		Total	417	34	5914.7	2.17		

5. Randy Johnson (1988-2009)

Standing 6'10 inches tall, Johnson looked, at times, like he was pitching from all angles, angles that propelled the ball so quickly toward the plate that the batter had no chance to react. Often regarded as the last pitcher who is ever going to win 300 games when he accomplished the feat last season (we don't thnk that's going to be true, by the way), Johnson was a compelling figure on the mound throughout his stays with Montreal, Seattle, the Yanks, and Arizona. What Johnson lacked compared to both Clemens and Maddux, his two mates who are ranked above him, was consistency in some seasons and a rather late start to his dominance It wasn't until Johnson pitched for Seattle in 1993 in the year he turned 30 years old that Randy had a stellar, All-Star caliber season. But once he hit that stride, for the next dozen seasons, with exceptions for 1996 and 2003, Johnson was likely the man most hitters did not want to face. Part delivery. Part intimidation. Part 6'10" of nasty.

Where Randy Ranks		
Category		**Rank**
W	303	22
SO	4875	2
ERA	3.29	289
WHIP9	10.54	85
SO/W	3.25	24
IP	4135.3	38

Johnson struck more men out than everyone but Nolan Ryan and his Strikeout to Walk ratio was amazing for a pitcher with that many K's. And while he did lack the quick career start or dogged consistency of Maddux in the era, when Johnson dominated, he dominated, leading his team in Arizona to a World Series victory in 2001, and winning the Cy Young Award for four consecutive years from 1999 to 2002.

Randy Johnson Stats					**408.708 PEVA RATING**			
Year	**Team**	**Lg**	**W**	**SV**	**IP**	**ERA**	**Age**	**PEVA-P**
1988	MON	NL	3	0	26.0	2.42	25	2.125
1989	MON	NL	0	0	29.7	6.67	26	3.263
1989	SEA	AL	7	0	131.0	4.40	26	
1990	SEA	AL	14	0	219.7	3.65	27	9.605
1991	SEA	AL	13	0	201.3	3.98	28	8.074
1992	SEA	AL	12	0	210.3	3.77	29	7.487
1993	SEA	AL	19	1	255.3	3.24	30	26.278
1994	SEA	AL	13	0	172.0	3.19	31	15.782
1995	SEA	AL	18	0	214.3	2.48	32	34.064
1996	SEA	AL	5	1	61.3	3.67	33	1.959
1997	SEA	AL	20	0	213.0	2.28	34	25.435
1998	SEA	AL	9	0	160.0	4.33	35	22.017
1998	HOU	NL	10	0	84.3	1.28	35	
1999	ARI	NL	17	0	271.7	2.48	36	34.970
2000	ARI	NL	19	0	248.7	2.64	37	31.544

2001	ARI	NL	21	0	249.7	2.49	38	52.009
2002	ARI	NL	24	0	260.0	2.32	39	43.254
2003	ARI	NL	6	0	114.0	4.26	40	2.746
2004	ARI	NL	16	0	245.7	2.60	41	44.961
2005	NYA	AL	17	0	225.7	3.79	42	17.377
2006	NYA	AL	17	0	205.0	5.00	43	10.209
2007	ARI	NL	4	0	56.7	3.81	44	3.001
2008	ARI	NL	11	0	184.0	3.91	45	10.713
2009	SFN	NL	8	0	96.0	4.88	46	1.834
		Total	295	2	4039.3	3.26		

6. Grover Cleveland Alexander (1911-1930)

If the most recent era of baseball fans saw the foursome of Maddux, Clemens, Johnson, and Martinez stand out on the pitching mound, it was the trio of Young, Johnson, and Alexander who dominated the first part of the 20th century on the rubber. Grover Cleveland, you can call him Pete, Alexander started with a splash, winning 28 games in his rookie season in Baker Bowl for the Philadelphia Phillies. He would win more games than that in three more seasons, including 33 in 1916. Unlike Walter Johnson, Alexander won more with control and not the strikeout, ranking only #52 on the All-Time list in that category. But no matter the method, he would win 373 games, 3rd most All-Time for the Phils, Cubs, and Cardinals over a twenty year career at a winning percentage clip of 0.642.

Where Alexander Ranks		
Category		**Rank**
W	373	3
SO	2198	52
ERA	2.56	49
WHIP9	10.09	34
SO/W	2.31	132
IP	5190.0	10

Even though not considered the most prolific SO pitcher, Alexander won the pitchers Triple Crown, leading the league in wins, strikeouts, and ERA in four seasons, 1915-17 and in the 1920 season for the Chicago Cubs. Like some of the World War II era players like Ted Williams, there were some in the World War I time that lost baseball time to military service. Alexander lost most of the 1918 season, serving in France. And yes, he was named after the President, although that name was probably better known for the baseball player than the politician after both of their tenures were done.

Grover Cleveland Alexander Stats						**386.297 PEVA RATING**		
Year	**Team**	**Lg**	**W**	**SV**	**IP**	**ERA**	**Age**	**PEVA-P**
1911	PHI	NL	28	3	367.0	2.57	24	27.469
1912	PHI	NL	19	3	310.3	2.81	25	10.315
1913	PHI	NL	22	2	306.3	2.79	26	12.045
1914	PHI	NL	27	1	355.0	2.38	27	17.359
1915	PHI	NL	31	3	376.3	1.22	28	52.205
1916	PHI	NL	33	3	389.0	1.55	29	47.561
1917	PHI	NL	30	0	388.0	1.83	30	39.958
1918	CHN	NL	2	0	26.0	1.73	31	4.178
1919	CHN	NL	16	1	235.0	1.72	32	20.354
1920	CHN	NL	27	5	363.3	1.91	33	40.230
1921	CHN	NL	15	1	252.0	3.39	34	10.352
1922	CHN	NL	16	1	245.7	3.63	35	10.770
1923	CHN	NL	22	2	305.0	3.19	36	21.352
1924	CHN	NL	12	0	169.3	3.03	37	6.062
1925	CHN	NL	15	0	236.0	3.39	38	15.070
1926	CHN	NL	3	0	52.0	3.46	39	9.795
1926	SLN	NL	9	2	148.3	2.91	39	
1927	SLN	NL	21	3	268.0	2.52	40	26.529
1928	SLN	NL	16	2	243.7	3.36	41	10.666
1929	SLN	NL	9	0	132.0	3.89	42	3.826
1930	PHI	NL	0	0	21.7	9.14	43	0.200

Total	373	32	5190.0	2.56

7. Warren Spahn (1942-1965)

For fans of the Braves, they have one long and fine history, and the pitcher who sits 2nd on their All-Time list of greats also sits in the Top Ten of All Pitchers at the #7 spot on the list. Spahn pitched for the Braves in their Boston and Milwaukee days. For younger fans, that's not talked about much, the Braves history as the National League entrant in Beantown until 1952, but it was during his days in the land of Paul Revere that Spahn began his mastery. Before he was done, he would win 363 games, the most by a left-handed pitcher in the history of baseball. It was not until Spahn was 26 that he had his breakout year, winning 21 games in 1947 over 289.7 innings with an ERA of just 2.33.

Where Spahn Ranks		
Category		**Rank**
W	363	6
SO	2583	25
ERA	3.09	192
WHIP9	10.75	123
SO/W	1.80	375
IP	5190.0	10

Spahn lost time to service in World War II like a lot of players, although it is a bit unknown how much that affected his career. He had been in the minor leagues prior to service and not yet an established major league player, beyond a cup of coffee in 1942. It is likely, however, that Spahn would have been a full-time player prior to 1946 had the war not intervened, adding to his stats. Spahn was a remarkably durable player once his career took off, pitching over 240 innings every year from the age of 26 to 44. At the age of 42, Spahn went 23-7 with a 2.60 ERA.

Warren Spahn Stats					370.496 PEVA RATING			
Year	**Team**	**Lg**	**W**	**SV**	**IP**	**ERA**	**Age**	**PEVA-P**
1942	BSN	NL	0	0	15.7	5.74	21	0.200
1946	BSN	NL	8	1	125.7	2.94	25	4.616
1947	BSN	NL	21	3	289.7	2.33	26	39.610
1948	BSN	NL	15	1	257.0	3.71	27	11.707
1949	BSN	NL	21	0	302.3	3.07	28	24.764
1950	BSN	NL	21	1	293.0	3.16	29	25.606
1951	BSN	NL	22	0	310.7	2.98	30	25.179
1952	BSN	NL	14	3	290.0	2.98	31	14.102
1953	ML1	NL	23	3	265.7	2.10	32	32.929
1954	ML1	NL	21	3	283.3	3.14	33	16.161
1955	ML1	NL	17	1	245.7	3.26	34	11.620
1956	ML1	NL	20	3	281.3	2.78	35	21.443
1957	ML1	NL	21	3	271.0	2.69	36	25.579
1958	ML1	NL	22	1	290.0	3.07	37	21.895
1959	ML1	NL	21	0	292.0	2.96	38	21.369
1960	ML1	NL	21	2	267.7	3.50	39	15.729
1961	ML1	NL	21	0	262.7	3.02	40	20.112
1962	ML1	NL	18	0	269.3	3.04	41	16.832
1963	ML1	NL	23	0	259.7	2.60	42	15.877
1964	ML1	NL	6	4	173.7	5.29	43	2.254
1965	NYN	NL	4	0	126.0	4.36	44	2.912
1965	SFN	NL	3	0	71.7	3.39	44	
		Total	363	29	5243.7	3.09		

8. Lefty Grove (1925-1941)

The best stats for Lefty Grove are the ones not listed below. He won 68% of his games for #8 on the Career List and his adjusted ERA plus ranks as #3. On those two stats alone, it's understood why Lefty Grove comes into the Top Ten Career list for pitchers at #8. Lefty Grove reached the Major Leagues a bit late in the game, not reaching the Philadelphia Athletics until the year he turned 25 because his minor league team, which was not affiliated with a Major League club, had refused offers for Grove's service until the price was right. By 1924, Grove was sold to the Philadelphia Athletics for $100,500, the most ever at the time.

Where Grove Ranks		
Category		Rank
W	300	23
SO	2266	46
ERA	3.06	180
WHIP9	11.43	366
SO/W	1.91	304
IP	5190.0	10

Once Grove reached the major leagues, it didn't take him long to make an impact. In his second year, he won his first ERA title and won 20 games, starting a streak of twenty win seasons that would last seven years. Grove was particularly stingy giving up runs, winning eight more ERA titles in his career after the first in 1926. And geez, he even went to the pen now and then and saved fifty-five games.

Lefty Grove Stats						342.625 PEVA RATING		
Year	Team	Lg	W	SV	IP	ERA	Age	PEVA-P
1925	PHA	AL	10	1	197.0	4.75	25	2.784
1926	PHA	AL	13	6	258.0	2.51	26	20.438
1927	PHA	AL	20	9	262.3	3.19	27	21.317
1928	PHA	AL	24	4	261.7	2.58	28	28.141
1929	PHA	AL	20	4	275.3	2.81	29	38.544
1930	PHA	AL	28	9	291.0	2.54	30	52.344
1931	PHA	AL	31	5	288.7	2.06	31	46.335
1932	PHA	AL	25	7	291.7	2.84	32	26.783
1933	PHA	AL	24	6	275.3	3.20	33	14.801
1934	BOS	AL	8	0	109.3	6.50	34	1.117
1935	BOS	AL	20	1	273.0	2.70	35	26.486
1936	BOS	AL	17	2	253.3	2.81	36	19.044
1937	BOS	AL	17	0	262.0	3.02	37	18.819
1938	BOS	AL	14	1	163.7	3.08	38	8.803
1939	BOS	AL	15	0	191.0	2.54	39	12.387
1940	BOS	AL	7	0	153.3	3.99	40	2.757
1941	BOS	AL	7	0	134.0	4.37	41	1.727
		Total	300	55	3940.7	3.06		

9. Tom Seaver (1967-1986)

He was known as Tom Terrific and that's just what he was, whether he was part of that great Red Machine with Cincinnati or as an Amazing Met. Tom was a strikeout pitcher, tough, stout, and strong, which led to 311 wins over his twenty year career plus a Rookie of the Year Award and three Cy Youngs. Three of his seasons were 20 win years, although it certainly seemed like there had been more. When he was elected ot the Hall of Fame, Seaver had the highest percentage of writer's voting for his candidacy than ever before, 98.8% wanted Tom Terrific in Cooperstown.

Where Seaver Ranks		
Category		Rank
W	311	18
SO	3640	6
ERA	2.86	123
WHIP9	10.09	32
SO/W	2.62	61
IP	4782.7	18

Like most power pitchers, his seasons began to dwindle in effectiveness once he reached his mid-30s, except for the season of 1981 when he rekindled the magic again. Not that there weren't good seasons after that. There were. But the ERA began to climb out of the 2 plus ballpark from his earlier years. That's the way of the pitching world in an era of less medicinal means. All hail, Tom Terrific, for leading the Mets from the wilderness to the World Series podium in 1969 and for all the great years of his career.

Tom Seaver Stats						336.016 PEVA RATING		
Year	Team	Lg	W	SV	IP	ERA	Age	PEVA-P
1967	NYN	NL	16	0	251.0	2.76	23	12.354
1968	NYN	NL	16	1	277.7	2.20	24	15.914

1969	NYN	NL	25	0	273.3	2.21	25	31.954
1970	NYN	NL	18	0	290.7	2.82	26	29.421
1971	NYN	NL	20	0	286.3	1.76	27	31.022
1972	NYN	NL	21	0	262.0	2.92	28	9.252
1973	NYN	NL	19	0	290.0	2.08	29	30.181
1974	NYN	NL	11	0	236.0	3.20	30	7.862
1975	NYN	NL	22	0	280.3	2.38	31	29.900
1976	NYN	NL	14	0	271.0	2.59	32	18.758
1977	NYN	NL	7	0	96.0	3.00	33	30.175
1977	CIN	NL	14	0	165.3	2.34	33	
1978	CIN	NL	16	0	259.7	2.88	34	11.720
1979	CIN	NL	16	0	215.0	3.14	35	14.233
1980	CIN	NL	10	0	168.0	3.64	36	4.598
1981	CIN	NL	14	0	166.3	2.54	37	22.737
1982	CIN	NL	5	0	111.3	5.50	38	1.208
1983	NYN	NL	9	0	231.0	3.55	39	7.726
1984	CHA	AL	15	0	236.7	3.95	40	12.063
1985	CHA	AL	16	0	238.7	3.17	41	11.385
1986	CHA	AL	2	0	72.0	4.38	42	3.553
1986	BOS	AL	5	0	104.3	3.80	42	
		Total	311	1	4782.7	2.86		

10. Pedro Martinez (1992-2009)

He has the least amount of wins of any pitcher in the Top Fifteen pitchers on the list and his career in later years has become marred by injury and incomplete campaigns, but beyond that, and way beyond that, too, is a career of First Ballot Hall of Fame stuff and dominance of his days on the pitching mound. Pedro did not give up hits or walks or runs. His adjusted ERA plus ranks #2 on the All-Time Career list, he is #6 in WHIP and #3 in Strikeout to Walk Ratio. He won 68.7% of his games, #6 All-Time. So there's no doubt that Pedro Martinez belongs in the list of Top Ten pitchers of All-Time, even with a lower total of wins or games or even innings pitched. Because in the games he pitched, he won, plus was a pretty interesting interview to listen to, too.

Where Martinez Ranks		
Category		**Rank**
W	219	76
SO	3154	13
ERA	2.93	147
WHIP9	9.49	6
SO/W	4.15	3
IP	2827.3	159

Pedro really did not blast onto the scene, which is something I think most of us who've seen him pitch forget. Yes, he was a fine pitcher for that Montreal club people were starting to forget outside of Hab land, and the season before the first year there had been good in a limited role. However, it really wasn't until Martinez had his final year in Montreal, then his days in Fenway Park that he became known as the pitcher of stellar seasons and dominant stuff. Just look at the nine years from 1997 to 2005. Those are the years that tell the story of Pedro Martinez. He pitched with power that belied his lean 5'11" frame, style, guile, and moxy. He beguiled the crowds, infuriated opponents, and won baseball games.

Pedro Martinez Stats					328.493 PEVA RATING			
Year	**Team**	**Lg**	**W**	**SV**	**IP**	**ERA**	**Age**	**PEVA-P**
1992	LAN	NL	0	0	8.0	2.25	21	0.200
1993	LAN	NL	10	2	107.0	2.61	22	7.247
1994	MON	NL	11	1	144.7	3.42	23	13.718
1995	MON	NL	14	0	194.7	3.51	24	14.606
1996	MON	NL	13	0	216.7	3.70	25	13.342
1997	MON	NL	17	0	241.3	1.90	26	38.574
1998	BOS	AL	19	0	233.7	2.89	27	26.412
1999	BOS	AL	23	0	213.3	2.07	28	45.790
2000	BOS	AL	18	0	217.0	1.74	29	40.251

2001	BOS	AL	7	0	116.7	2.39	30	12.083
2002	BOS	AL	20	0	199.3	2.26	31	37.759
2003	BOS	AL	14	0	186.7	2.22	32	27.294
2004	BOS	AL	16	0	217.0	3.90	33	15.926
2005	NYN	NL	15	0	217.0	2.82	34	24.921
2006	NYN	NL	9	0	132.7	4.48	35	4.857
2007	NYN	NL	3	0	28.0	2.57	36	2.037
2008	NYN	NL	5	0	109.0	5.61	37	1.640
2009	PHI	NL	5	0	44.7	3.63	38	1.835
		Total	214	3	2782.7	2.91		

Note: Rate stats rankings based on players with at least 1,000 IP.

Who Jumped Up the Best Pitcher Careers Ever List in 2010

He's going to be in the Hall of Fame, perhaps right now, but definitely with one more great year. So it comes as no surprise that Roy Halladay has moved up ten spots to the #36 position after his Cy Young caliber first year with the Philadelphia Phillies. It comes as no surprise at all that he now sits in the company of men like Jim Palmer and Fergie Jenkins. For Mariano Rivera, all he has accomplished is to be the best relief pitcher in baseball history, now moving up to the #37 spot in the overall pitcher rankings, right behind the quasi-starter/reliever Eckersley. He's far surpassed the other relievers on the list, such as Fingers and Gossage, and should be compared to all pitchers and not just as a specialist. For those further down the list, there is significant moves up the countdown due to great seasons. Cliff Lee may be moving from team to team to team to team, but that's not stopping him from moving up eighty-four spots this year alone to the #154 point on the list. Now it's likely he'll be packing for another city come free agent time for a boatload of cash, but it's more than likely that before he's done pitching, he'll move well into the top one hundred of all-time.

Moving Up the Career List in 2010

2009 Rank	Name		2010 Rank	Up
36	Roy	Halladay	26	10
46	Mariano	Rivera	37	9
50	Johan	Santana	43	7
51	Andy	Pettitte	49	2
53	Jamie	Moyer	52	1
79	Roy	Oswalt	55	24
81	Trevor	Hoffman	83	-2
82	C.C.	Sabathia	53	29
89	Derek	Lowe	77	12
93	Javier	Vazquez	88	5
109	Tim	Hudson	79	30
121	Mark	Buehrle	108	13
131	Chris	Carpenter	99	32
135	Billy	Wagner	114	21
136	Barry	Zito	124	12
146	Kevin	Millwood	137	9
147	Livan	Hernandez	127	20
166	Tim	Wakefield	165	1
186	Mike	Hampton	189	-3
193	Danny	Haren	153	40
204	Carlos	Zambrano	179	25
206	Jake	Peavy	196	10
238	Cliff	Lee	154	84

Joining the Over 100 PEVA Career List in 2010

Name		2010 Rank	2010 Team
John	Lackey	236	Boston Red Sox
Freddie	Garcia	239	Chicago White Sox
Ted	Lilly	249	Chicago Cubs/LA Dodgers

Best Pitchers of All-Time 1871-2010 (Regular Season)

Rank	First	Last	HOF Cat.	PEVA-P TOTAL	YRS.	PEVA-P PER YR.	CAREER Wins	Loss	SV	ERA	PEVA-P EQ YR.	EQ YRS
1	Greg	Maddux		**594.209**	**23**	**25.835**	355	227	0	3.16	**26.827**	22.15
2	Cy	Young	HOFP	**504.145**	**22**	**22.916**	511	316	17	2.63	**23.137**	21.79
3	Roger	Clemens		**487.448**	**24**	**20.310**	354	184	0	3.12	**21.473**	22.70
4	Walter	Johnson	HOFP	**479.881**	**21**	**22.851**	417	279	34	2.17	**23.628**	20.31
5	Randy	Johnson		**408.708**	**22**	**18.578**	303	166	2	3.29	**21.622**	19.16
6	Pete	Alexander	HOFP	**386.297**	**20**	**19.315**	373	208	32	2.56	**21.179**	18.24
7	Warren	Spahn	HOFP	**370.496**	**21**	**17.643**	363	245	29	3.09	**18.618**	19.90
8	Lefty	Grove	HOFP	**342.625**	**17**	**20.154**	300	141	55	3.06	**20.715**	16.54
9	Tom	Seaver	HOFP	**336.016**	**20**	**16.801**	311	205	1	2.86	**17.074**	19.68
10	Pedro	Martinez		**328.493**	**18**	**18.250**	219	100	3	2.93	**22.634**	14.71
11	Steve	Carlton	HOFP	**328.364**	**24**	**13.682**	329	244	2	3.22	**15.482**	21.21
12	Tom	Glavine		**325.047**	**22**	**14.775**	305	203	0	3.54	**15.703**	20.70
13	Kid	Nichols	HOFP	**315.802**	**15**	**21.053**	361	208	17	2.95	**22.429**	14.08
14	Robin	Roberts	HOFP	**307.347**	**19**	**16.176**	286	245	25	3.41	**16.749**	18.35
15	Carl	Hubbell	HOFP	**301.955**	**16**	**18.872**	253	154	33	2.98	**19.865**	15.20
16	John	Smoltz		**301.456**	**21**	**14.355**	213	155	154	3.33	**16.625**	18.79
17	Christy	Mathewson	HOFP	**290.943**	**17**	**17.114**	373	188	28	2.13	**18.531**	15.70
18	Gaylord	Perry	HOFP	**286.287**	**22**	**13.013**	314	265	11	3.11	**13.698**	20.90
19	Curt	Schilling		**282.056**	**20**	**14.103**	216	146	22	3.46	**16.592**	17.00
20	Mike	Mussina		**276.995**	**18**	**15.389**	270	153	0	3.68	**15.846**	17.48
21	Bert	Blyleven		**271.050**	**22**	**12.320**	287	250	0	3.31	**13.075**	20.73
22	Nolan	Ryan	HOFP	**270.518**	**27**	**10.019**	324	292	3	3.19	**11.033**	24.52
23	Phil	Niekro	HOFP	**269.835**	**24**	**11.243**	318	274	29	3.35	**12.009**	22.47
24	Don	Sutton	HOFP	**266.168**	**23**	**11.573**	324	256	5	3.26	**11.809**	22.54
25	Kevin	Brown		**263.325**	**19**	**13.859**	211	144	0	3.28	**16.934**	15.55
26	Roy	Halladay		**257.438**	**13**	**19.803**	169	87	1	3.32	**23.925**	10.76
27	Fergie	Jenkins	HOFP	**250.554**	**19**	**13.187**	284	226	7	3.34	**13.843**	18.10
28	Bob	Feller	HOFP	**246.330**	**18**	**13.685**	266	162	21	3.25	**15.760**	15.63
29	Jim	Palmer	HOFP	**240.163**	**19**	**12.640**	268	152	4	2.86	**14.653**	16.39
30	Jim	Bunning	HOFP	**232.747**	**17**	**13.691**	224	184	16	3.27	**15.192**	15.32
31	Juan	Marichal	HOFP	**231.714**	**16**	**14.482**	243	142	2	2.89	**16.658**	13.91
32	Don	Drysdale	HOFP	**228.992**	**14**	**16.357**	209	166	6	2.95	**17.588**	13.02
33	Bob	Gibson	HOFP	**225.768**	**17**	**13.280**	251	174	6	2.91	**14.353**	15.73
34	Early	Wynn	HOFP	**220.535**	**23**	**9.588**	300	244	15	3.54	**10.945**	20.15
35	Hal	Newhouser	HOFP	**218.723**	**17**	**12.866**	207	150	26	3.06	**16.545**	13.22
36	Dennis	Eckersley	HOFP	**217.107**	**24**	**9.046**	197	171	390	3.50	**9.046**	24.00
37	Mariano	Rivera		**215.110**	**16**	**13.444**	74	55	559	2.23	**13.977**	15.39
38	Ed	Walsh	HOFP	**213.766**	**14**	**15.269**	195	126	34	1.82	**21.880**	9.77
39	Dazzy	Vance	HOFP	**212.748**	**16**	**13.297**	197	140	11	3.24	**16.569**	12.84
40	Dennis	Martinez		**209.456**	**23**	**9.107**	245	193	8	3.70	**10.416**	20.11
41	Amos	Rusie	HOFP	**208.983**	**10**	**20.898**	245	174	5	3.07	**22.840**	9.15
42	Sandy	Koufax	HOFP	**206.787**	**12**	**17.232**	165	87	9	2.76	**19.999**	10.34
43	Johan	Santana		**205.233**	**11**	**18.658**	133	69	1	3.10	**21.468**	9.56
44	Bret	Saberhagen		**204.005**	**16**	**12.750**	167	117	1	3.34	**15.455**	13.20
45	Orel	Hershiser		**203.300**	**18**	**11.294**	204	150	5	3.48	**13.357**	15.22
46	Paul	Derringer		**202.808**	**15**	**13.521**	223	212	29	3.46	**13.521**	15.00

47	Eppa	Rixey	HOFP	**202.000**	**21**	**9.619**	266	251	14	3.15	**10.228**	19.75
48	Whitey	Ford	HOFP	**200.553**	**16**	**12.535**	236	106	10	2.75	**14.064**	14.26
49	Andy	Pettitte		**198.187**	**16**	**12.387**	240	138	0	3.88	**13.090**	15.14
50	Jack	Morris		**194.913**	**18**	**10.829**	254	186	0	3.90	**11.554**	16.87

Best Pitchers of All-Time 1871-2010

			HOF	PEVA-P		PEVA-P	CAREER				PEVA-P	EQ
Rank	**First**	**Last**	**Cat.**	**TOTAL**	**YRS.**	**PER YR.**	**Wins**	**Loss**	**SV**	**ERA**	**EQ YR.**	**YRS**
51	John	Clarkson	HOFP	**194.134**	**12**	**16.178**	328	178	5	2.81	**17.210**	11.28
52	Jamie	Moyer		**191.352**	**24**	**7.973**	267	204	0	4.24	**9.018**	21.22
53	C.C.	Sabathia		**190.081**	**10**	**19.008**	157	88	0	3.57	**19.008**	10.00
54	Dwight	Gooden		**189.771**	**16**	**11.861**	194	112	3	3.51	**13.653**	13.90
55	Roy	Oswalt		**187.825**	**10**	**18.783**	150	83	0	3.18	**19.445**	9.66
56	Rick	Reuschel		**187.186**	**19**	**9.852**	214	191	5	3.37	**11.598**	16.14
57	Burleigh	Grimes	HOFP	**185.950**	**19**	**9.787**	270	212	18	3.53	**10.811**	17.20
58	David	Cone		**185.152**	**17**	**10.891**	194	126	1	3.46	**13.057**	14.18
59	Jim	Kaat		**184.824**	**25**	**7.393**	283	237	18	3.45	**8.277**	22.33
60	Bucky	Walters		**182.990**	**16**	**11.437**	198	160	4	3.30	**12.542**	14.59
61	Tommy	John		**182.842**	**26**	**7.032**	288	231	4	3.34	**7.953**	22.99
62	Ted	Lyons	HOFP	**178.627**	**21**	**8.506**	260	230	23	3.67	**9.347**	19.11
63	Frank	Viola		**178.435**	**15**	**11.896**	176	150	0	3.73	**14.483**	12.32
64	Frank	Tanana		**177.300**	**21**	**8.443**	240	236	1	3.66	**9.005**	19.69
65	Red	Faber	HOFP	**177.153**	**20**	**8.858**	254	213	28	3.15	**9.334**	18.98
66	Bob	Friend		**176.919**	**16**	**11.057**	197	230	11	3.58	**11.233**	15.75
67	Joe	McGinnity	HOFP	**175.944**	**10**	**17.594**	246	142	24	2.66	**17.594**	10.00
68	Stan	Coveleski	HOFP	**175.862**	**14**	**12.562**	215	142	21	2.89	**15.147**	11.61
69	Billy	Pierce		**175.777**	**18**	**9.765**	211	169	32	3.27	**10.965**	16.03
70	Dave	Stieb		**174.001**	**16**	**10.875**	176	137	3	3.44	**13.102**	13.28
71	David	Wells		**173.846**	**21**	**8.278**	239	157	13	4.13	**9.121**	19.06
72	Jimmy	Key		**173.444**	**15**	**11.563**	186	117	10	3.51	**12.867**	13.48
73	Tim	Keefe	HOFP	**173.362**	**14**	**12.383**	342	225	2	2.62	**12.445**	13.93
74	Eddie	Plank	HOFP	**173.294**	**17**	**10.194**	326	194	23	2.35	**10.297**	16.83
75	Catfish	Hunter	HOFP	**171.445**	**15**	**11.430**	224	166	1	3.26	**12.194**	14.06
76	Vida	Blue		**171.111**	**17**	**10.065**	209	161	2	3.27	**11.601**	14.75
77	Derek	Lowe		**170.787**	**14**	**12.199**	157	129	85	3.85	**12.717**	13.43
78	Dutch	Leonard		**170.527**	**20**	**8.526**	191	181	44	3.25	**9.886**	17.25
79	Tim	Hudson		**169.847**	**12**	**14.154**	165	87	0	3.42	**15.483**	10.97
80	Hoyt	Wilhelm	HOFP	**167.483**	**21**	**7.975**	143	122	227	2.52	**8.723**	19.20
81	Mordecai	Brown	HOFP	**167.406**	**14**	**11.958**	239	130	49	2.06	**13.007**	12.87
82	Al	Spalding	HOFPE	**167.275**	**7**	**23.896**	253	65	11	2.14	**23.896**	7.00
83	Trevor	Hoffman		**166.370**	**18**	**9.243**	61	75	601	2.87	**9.690**	17.17
84	Bob	Lemon	HOFP	**165.840**	**13**	**12.757**	207	128	22	3.23	**14.114**	11.75
85	Wilbur	Cooper		**164.114**	**15**	**10.941**	216	178	14	2.89	**12.330**	13.31
86	Kevin	Appier		**163.015**	**16**	**10.188**	169	137	0	3.74	**12.588**	12.95
87	Dizzy	Dean	HOFP	**162.809**	**12**	**13.567**	150	83	30	3.02	**21.593**	7.54
88	Javier	Vazquez		**162.311**	**13**	**12.485**	152	149	0	4.26	**12.563**	12.92
89	Chuck	Finley		**162.129**	**17**	**9.537**	200	173	0	3.85	**10.178**	15.93
90	Fernando	Valenzuela		**162.020**	**17**	**9.531**	173	153	2	3.54	**11.809**	13.72
91	Kenny	Rogers		**161.713**	**20**	**8.086**	219	156	28	4.27	**8.498**	19.03
92	Dolf	Luque		**159.395**	**20**	**7.970**	194	179	28	3.24	**10.425**	15.29
93	Vic	Willis	HOFP	**159.260**	**13**	**12.251**	249	205	11	2.63	**12.251**	13.00
94	Steve	Rogers		**159.201**	**13**	**12.246**	158	152	2	3.17	**13.201**	12.06
95	Babe	Adams		**158.436**	**19**	**8.339**	194	140	15	2.76	**11.341**	13.97
96	Red	Ruffing	HOFP	**157.733**	**22**	**7.170**	273	225	16	3.80	**8.110**	19.45
97	Bob	Welch		**157.692**	**17**	**9.276**	211	146	8	3.47	**10.063**	15.67
98	Lon	Warneke		**157.469**	**15**	**10.498**	192	121	13	3.18	**12.907**	12.20
99	Chris	Carpenter		**154.715**	**13**	**11.901**	133	83	0	3.80	**15.456**	10.01

100	Larry	Jackson		**153.706**	**14**	**10.979**	194	183	20	3.40	**10.979**	14.00

Best Pitchers of All-Time 1871-2010

			HOF	PEVA-P		PEVA-P	CAREER				PEVA-P	EQ
Rank	**First**	**Last**	**Cat.**	**TOTAL**	**YRS.**	**PER YR.**	**Wins**	**Loss**	**SV**	**ERA**	**EQ YR.**	**YRS**
101	Mark	Langston		**152.005**	**16**	**9.500**	179	158	0	3.97	**10.983**	13.84
102	Jerry	Reuss		**151.558**	**22**	**6.889**	220	191	11	3.64	**8.397**	18.05
103	Jerry	Koosman		**151.117**	**19**	**7.954**	222	209	17	3.36	**8.494**	17.79
104	Charlie	Hough		**150.721**	**25**	**6.029**	216	216	61	3.75	**7.116**	21.18
105	Waite	Hoyt	HOFP	**150.378**	**21**	**7.161**	237	182	52	3.59	**8.368**	17.97
106	Carl	Mays		**149.606**	**15**	**9.974**	207	126	31	2.92	**12.075**	12.39
107	Claude	Passeau		**148.781**	**13**	**11.445**	162	150	21	3.32	**13.225**	11.25
108	Mark	Buehrle		**148.328**	**11**	**13.484**	148	110	0	3.85	**14.033**	10.57
109	Joe	Niekro		**147.104**	**22**	**6.687**	221	204	16	3.59	**7.977**	18.44
110	Luis	Tiant		**147.094**	**19**	**7.742**	229	172	15	3.30	**9.041**	16.27
111	Pud	Galvin	HOFP	**146.925**	**15**	**9.795**	364	310	2	2.86	**9.981**	14.72
112	Doug	Drabek		**146.466**	**13**	**11.267**	155	134	0	3.73	**11.745**	12.47
113	Larry	French		**146.144**	**14**	**10.439**	197	171	17	3.44	**10.660**	13.71
114	Billy	Wagner		**146.050**	**16**	**9.128**	47	40	422	2.31	**10.779**	13.55
115	Bobo	Newsom		**145.446**	**20**	**7.272**	211	222	21	3.98	**9.550**	15.23
116	Ron	Guidry		**144.658**	**14**	**10.333**	170	91	4	3.29	**12.667**	11.42
117	Charley	Root		**144.582**	**17**	**8.505**	201	160	40	3.59	**9.334**	15.49
118	Wilbur	Wood		**143.431**	**17**	**8.437**	164	156	57	3.24	**11.365**	12.62
119	Brandon	Webb		**142.694**	**7**	**20.385**	87	62	0	3.27	**23.766**	6.02
120	Tommy	Bond		**141.227**	**10**	**14.123**	234	163	0	2.31	**16.733**	8.44
121	Eddie	Cicotte		**140.807**	**14**	**10.058**	208	149	25	2.38	**10.740**	13.11
122	Jack	Quinn		**140.371**	**23**	**6.103**	247	218	57	3.29	**7.133**	19.68
123	Don	Newcombe		**140.056**	**10**	**14.006**	149	90	7	3.56	**14.291**	9.80
124	Barry	Zito		**139.208**	**11**	**12.655**	142	120	0	3.86	**13.170**	10.57
125	Mike	Scott		**138.503**	**13**	**10.654**	124	108	3	3.54	**13.408**	10.33
126	Freddie	Fitzsimmons		**138.188**	**19**	**7.273**	217	146	13	3.51	**8.985**	15.38
127	Livan	Hernandez		**138.183**	**15**	**9.212**	166	163	0	4.39	**10.146**	13.62
128	Rich	Gossage	HOFP	**138.013**	**22**	**6.273**	124	107	310	3.01	**7.067**	19.53
129	Mickey	Lolich		**137.305**	**16**	**8.582**	217	191	11	3.44	**9.246**	14.85
130	Charley	Radbourn	HOFP	**136.077**	**11**	**12.371**	309	195	2	2.67	**12.371**	11.00
131	Mike	Cuellar		**135.827**	**15**	**9.055**	185	130	11	3.14	**11.414**	11.90
132	Urban	Shocker		**135.739**	**13**	**10.441**	187	117	25	3.17	**12.141**	11.18
133	Jose	Rijo		**135.731**	**14**	**9.695**	116	91	3	3.24	**12.989**	10.45
134	Lefty	Gomez	HOFP	**135.569**	**14**	**9.683**	189	102	9	3.34	**12.191**	11.12
135	Clark	Griffith	HOF-PI	**133.483**	**20**	**6.674**	237	146	6	3.31	**10.284**	12.98
136	Lew	Burdette		**133.345**	**18**	**7.408**	203	144	31	3.66	**9.053**	14.73
137	Kevin	Millwood		**132.088**	**14**	**9.435**	159	137	0	4.11	**10.216**	12.93
138	John	Candelaria		**131.995**	**19**	**6.947**	177	122	29	3.33	**8.083**	16.33
139	Bill	Lee		**130.593**	**14**	**9.328**	169	157	13	3.54	**9.992**	13.07
140	Rollie	Fingers	HOFP	**130.458**	**17**	**7.674**	114	118	341	2.90	**8.384**	15.56
141	Hippo	Vaughn		**130.329**	**13**	**10.025**	178	137	5	2.49	**11.924**	10.93
142	Brad	Radke		**130.309**	**12**	**10.859**	148	139	0	4.22	**11.109**	11.73
143	Mort	Cooper		**129.805**	**11**	**11.800**	128	75	14	2.97	**15.545**	8.35
144	Wes	Ferrell		**128.949**	**15**	**8.597**	193	128	13	4.04	**12.447**	10.36
145	Dizzy	Trout		**128.842**	**15**	**8.589**	170	161	35	3.23	**9.608**	13.41
146	Jim	Perry		**128.689**	**17**	**7.570**	215	174	10	3.45	**7.983**	16.12
147	Rube	Waddell	HOFP	**128.642**	**13**	**9.896**	193	143	5	2.16	**11.900**	10.81
148	Andy	Benes		**127.349**	**14**	**9.096**	155	139	1	3.97	**10.051**	12.67
149	Al	Leiter		**125.172**	**19**	**6.588**	162	132	2	3.80	**9.397**	13.32
150	Doyle	Alexander		**124.644**	**19**	**6.560**	194	174	3	3.76	**7.406**	16.83

Best Pitchers of All-Time 1871-2010

Rank	First	Last	HOF Cat.	PEVA-P TOTAL	YRS.	PEVA-P PER YR.	CAREER Wins	Loss	SV	ERA	PEVA-P EQ YR.	EQ YRS
151	Herb	Pennock	HOFP	**124.016**	**22**	**5.637**	240	162	32	3.60	**6.944**	17.86
152	Mike	Garcia		**123.707**	**14**	**8.836**	142	97	23	3.27	**11.884**	10.41
153	Danny	Haren		**123.336**	**8**	**15.417**	91	74	0	3.66	**18.326**	6.73
154	Cliff	Lee		**123.282**	**9**	**13.698**	102	61	0	3.85	**17.637**	6.99
155	Claude	Osteen		**122.918**	**18**	**6.829**	196	195	1	3.30	**8.448**	14.55
156	Curt	Simmons		**121.668**	**20**	**6.083**	193	183	5	3.54	**7.074**	17.20
157	Bobby	Mathews		**121.473**	**15**	**8.098**	297	248	3	2.89	**8.430**	14.41
158	John	Burkett		**121.463**	**15**	**8.098**	166	136	1	4.31	**9.195**	13.21
159	Dave	Stewart		**121.446**	**16**	**7.590**	168	129	19	3.95	**8.440**	14.39
160	Addie	Joss	HOFP	**121.395**	**9**	**13.488**	160	97	5	1.89	**14.002**	8.67
161	Camilo	Pascual		**121.302**	**18**	**6.739**	174	170	10	3.63	**7.954**	15.25
162	Tom	Candiotti		**120.634**	**16**	**7.540**	151	164	0	3.73	**8.767**	13.76
163	Jack	Chesbro	HOFP	**120.549**	**11**	**10.959**	198	132	5	2.68	**11.692**	10.31
164	Jason	Schmidt		**119.909**	**14**	**8.565**	130	96	0	3.96	**11.703**	10.38
165	Tim	Wakefield		**119.633**	**18**	**6.646**	193	172	22	4.38	**7.087**	16.88
166	Larry	Jansen		**118.922**	**9**	**13.214**	122	89	10	3.58	**15.856**	7.50
167	Silver	King		**118.857**	**10**	**11.886**	203	154	6	3.18	**12.808**	9.28
168	Frank	Lary		**118.731**	**12**	**9.894**	128	116	11	3.49	**12.753**	9.31
169	Tommy	Bridges		**117.328**	**16**	**7.333**	194	138	10	3.57	**8.743**	13.42
170	Virgil	Trucks		**117.186**	**17**	**6.893**	177	135	30	3.39	**8.382**	13.98
171	Jack	Powell		**116.822**	**16**	**7.301**	245	254	15	2.97	**7.394**	15.80
172	Danny	Darwin		**116.591**	**21**	**5.552**	171	182	32	3.84	**6.417**	18.17
173	Jesse	Haines	HOFP	**116.440**	**19**	**6.128**	210	158	10	3.64	**7.421**	15.69
174	Mel	Harder		**116.071**	**20**	**5.804**	223	186	23	3.80	**6.988**	16.61
175	Andy	Messersmith		**115.459**	**12**	**9.622**	130	99	15	2.86	**11.878**	9.72
176	Tim	Belcher		**114.914**	**14**	**8.208**	146	140	5	4.16	**9.404**	12.22
177	Bartolo	Colon		**114.410**	**13**	**8.801**	153	103	0	4.10	**11.370**	10.17
178	George	Uhle		**114.370**	**17**	**6.728**	200	166	25	3.99	**8.003**	14.29
179	Carlos	Zambrano		**114.288**	**10**	**11.429**	116	74	0	3.50	**13.305**	8.59
180	Bill	Gullickson		**114.091**	**14**	**8.149**	162	136	0	3.93	**8.879**	12.85
181	Deacon	Phillippe		**114.075**	**13**	**8.775**	189	109	12	2.59	**10.612**	10.75
182	Mike	Moore		**113.959**	**14**	**8.140**	161	176	2	4.39	**8.385**	13.59
183	Sam	Jones		**113.860**	**22**	**5.175**	229	217	31	3.84	**5.903**	19.29
184	Lee	Smith		**113.816**	**18**	**6.323**	71	92	478	3.03	**6.743**	16.88
185	Hal	Schumacher		**113.603**	**13**	**8.739**	158	121	7	3.36	**9.930**	11.44
186	Johnny	Antonelli		**112.813**	**12**	**9.401**	126	110	21	3.34	**11.989**	9.41
187	Noodles	Hahn		**112.516**	**8**	**14.064**	130	94	0	2.55	**16.669**	6.75
188	Doug	Jones		**112.347**	**16**	**7.022**	69	79	303	3.30	**7.996**	14.05
189	Mike	Hampton		**112.163**	**16**	**7.010**	148	115	1	4.06	**8.588**	13.06
190	Firpo	Marberry		**111.823**	**14**	**7.987**	148	88	101	3.63	**9.690**	11.54
191	Pete	Donohue		**111.741**	**12**	**9.312**	134	118	12	3.87	**11.496**	9.72
192	Bill	Hutchison		**111.227**	**9**	**12.359**	183	163	4	3.59	**14.950**	7.44
193	Dave	McNally		**111.184**	**14**	**7.942**	184	119	2	3.24	**9.265**	12.00
194	Harry	Brecheen		**111.170**	**12**	**9.264**	133	92	18	2.92	**10.910**	10.19
195	Tom	Gordon		**110.831**	**21**	**5.278**	138	126	158	3.96	**6.297**	17.64
196	Jake	Peavy		**110.503**	**9**	**12.278**	102	74	0	3.36	**14.005**	7.89
197	Bob	Knepper		**110.426**	**15**	**7.362**	146	155	1	3.68	**8.222**	13.43
198	Johnny	Sain		**110.416**	**11**	**10.038**	139	116	51	3.49	**10.658**	10.36
199	Will	White		**110.286**	**10**	**11.029**	229	166	0	2.28	**14.067**	7.84
200	Milt	Pappas		**110.110**	**17**	**6.477**	209	164	4	3.40	**6.951**	15.84

Best Pitchers of All-Time 1871-2010

Rank	First	Last	HOF Cat.	PEVA-P TOTAL	YRS.	PEVA-P PER YR.	CAREER Wins	Loss	SV	ERA	PEVA-P EQ YR.	EQ YRS
201	Burt	Hooton		**109.906**	**15**	**7.327**	151	136	7	3.38	**8.075**	13.61
202	Sam	McDowell		**109.730**	**15**	**7.315**	141	134	14	3.17	**9.634**	11.39
203	Tony	Mullane		**109.564**	**13**	**8.428**	284	220	15	3.05	**8.751**	12.52
204	Darryl	Kile		**109.549**	**12**	**9.129**	133	119	0	4.12	**9.852**	11.12
205	Mario	Soto		**109.458**	**12**	**9.121**	100	92	4	3.47	**13.077**	8.37
206	Ed	Lopat		**109.148**	**12**	**9.096**	166	112	3	3.21	**9.234**	11.82
207	Bruce	Sutter	HOFP	**108.819**	**12**	**9.068**	68	71	300	2.83	**9.804**	11.10
208	Steve	Trachsel		**108.447**	**16**	**6.778**	143	159	0	4.39	**7.992**	13.57
209	Jim	McCormick		**108.323**	**10**	**10.832**	265	214	1	2.43	**10.832**	10.00
210	Jeff	Fassero		**107.708**	**16**	**6.732**	121	124	25	4.11	**7.282**	14.79
211	Eddie	Rommel		**107.495**	**13**	**8.269**	171	119	29	3.54	**9.211**	11.67
212	Dean	Chance		**107.115**	**11**	**9.738**	128	115	23	2.92	**11.555**	9.27
213	John	Franco		**107.107**	**21**	**5.100**	90	87	424	2.89	**5.434**	19.71
214	Chief	Bender	HOFP	**107.025**	**16**	**6.689**	212	127	34	2.46	**7.437**	14.39
215	Dan	Quisenberry		**106.916**	**12**	**8.910**	56	46	244	2.76	**9.974**	10.72
216	Jesse	Tannehill		**106.764**	**15**	**7.118**	197	116	7	2.79	**9.133**	11.69
217	Jack	McDowell		**106.731**	**12**	**8.894**	127	87	0	3.85	**11.885**	8.98
218	Lee	Meadows		**106.712**	**15**	**7.114**	188	180	7	3.37	**8.140**	13.11
219	John	Tudor		**106.658**	**12**	**8.888**	117	72	1	3.12	**11.772**	9.06
220	Earl	Whitehill		**106.106**	**17**	**6.242**	218	185	11	4.36	**6.754**	15.71
221	Mickey	Welch	HOFP	**106.022**	**13**	**8.156**	307	210	4	2.71	**8.813**	12.03
222	Mel	Stottlemyre		**105.938**	**11**	**9.631**	164	139	1	2.97	**10.305**	10.28
223	Allie	Reynolds		**105.891**	**13**	**8.145**	182	107	49	3.30	**8.854**	11.96
224	Scott	Erickson		**105.635**	**15**	**7.042**	142	136	0	4.59	**8.803**	12.00
225	John	Smiley		**105.553**	**12**	**8.796**	126	103	4	3.80	**9.801**	10.77
226	Rick	Sutcliffe		**105.324**	**18**	**5.851**	171	139	6	4.08	**7.825**	13.46
227	Whit	Wyatt		**104.522**	**16**	**6.533**	106	95	13	3.79	**10.798**	9.68
228	Rube	Marquard	HOFP	**104.495**	**18**	**5.805**	201	177	19	3.08	**7.123**	14.67
229	Pat	Hentgen		**104.469**	**14**	**7.462**	131	112	1	4.32	**9.818**	10.64
230	Mike	Morgan		**104.252**	**22**	**4.739**	141	186	8	4.23	**6.338**	16.45
231	Dennis	Leonard		**104.156**	**12**	**8.680**	144	106	1	3.70	**11.128**	9.36
232	Joe	Nathan		**104.010**	**10**	**10.401**	46	22	247	2.75	**12.707**	8.21
233	Scott	Sanderson		**103.846**	**19**	**5.466**	163	143	5	3.84	**7.084**	14.66
234	Vern	Law		**103.746**	**16**	**6.484**	162	147	13	3.77	**7.512**	13.81
235	Kevin	Tapani		**103.341**	**13**	**7.949**	143	125	0	4.35	**8.916**	11.59
236	John	Lackey		**103.104**	**9**	**11.456**	116	82	0	3.89	**11.892**	8.67
237	Rick	Rhoden		**103.081**	**16**	**6.443**	151	125	1	3.59	**7.875**	13.09
238	Denny	Neagle		**102.744**	**13**	**7.903**	124	92	3	4.24	**9.298**	11.05
239	Freddie	Garcia		**102.062**	**12**	**8.505**	133	87	0	4.13	**10.710**	9.53
240	Curt	Davis		**101.725**	**13**	**7.825**	158	131	33	3.42	**8.665**	11.74
241	Bruce	Hurst		**101.575**	**15**	**6.772**	145	113	0	3.92	**8.756**	11.60
242	Charlie	Leibrandt		**101.498**	**14**	**7.250**	140	119	2	3.71	**8.565**	11.85
243	Ray	Kremer		**101.477**	**10**	**10.148**	143	85	10	3.76	**11.924**	8.51
244	Red	Lucas		**101.428**	**15**	**6.762**	157	135	7	3.72	**7.918**	12.81
245	Ken	Holtzman		**101.115**	**15**	**6.741**	174	150	3	3.49	**8.181**	12.36
246	Joaquin	Andujar		**101.015**	**13**	**7.770**	127	118	9	3.58	**9.208**	10.97
247	Mike	Boddicker		**101.012**	**14**	**7.215**	134	116	3	3.80	**9.913**	10.19
248	Lindy	McDaniel		**100.726**	**21**	**4.796**	141	119	172	3.45	**5.448**	18.49
249	Ted	Lilly		**100.677**	**12**	**8.390**	113	96	0	4.18	**10.642**	9.46
250	Greg	Swindell		**100.536**	**17**	**5.914**	123	122	7	3.86	**6.640**	15.14
251	Preacher	Roe		**100.022**	**12**	**8.335**	127	84	10	3.43	**10.227**	9.78

Note: PEVA-P represents Batting PEVA for all regular seasons. It does not include Batting PEVA values.

Note: PEVA Per EQ Year represents Average PEVA for a Full Season. PEVA per Year equals Average PEVA for Total Seasons.
Note: EQ Year is defined as the baseballevaluation.com statistic that reflects the Experience Equivalant of a player's experience for any or all seasons. It is meant to approximate Major League Service Time data for players prior to its inception, as well as players within the MLST era.
Note: 2010 PEVA = Reflects preliminary final PEVA numbers. Marginal adjustment may be made with final stats and park factors.

Chapter 6 - BEST CAREERS EVER - TEAMS/FRANCHISES

PEVA Player Rating Boxscore		
32.000	Fantastic	MVP/CY Young Candidate
20.000	Great	All-League
15.000	Very Good	All-Star Caliber
10.000	Good	Plus Starter
3.500	Average	Bench Player

Stat Geek Baseball Best Players (All Franchises Current and Past)

Top 20 Position Players By Team/Franchise (1871-2010) - Regular Season

What Changed On the Best Careers List for Each Team in 2010

Well, of course, it's a whole lot easier to jump around the top of one of the newer franchises, whether that be the Tampa Bay Rays or the Arizona Diamondbacks, to name two. And that's where much of the movement begins, with Carlos Pena moving up to #2 on the Rays list and Evan Longoria, after only three seasons, rising four spots to #3. But to us, it's the movement within the lists of the historic franchises that gets our attention the most. David Ortiz may only be a #11 now for the Red Sox, but it's the names above him that spark baseball history; next year he may pass Manny Ramirez and Jimmy Foxx who sit just above him. Paul Konerko jumped five spots on the Chicago White Sox list this year and it may be possible for him to pass folks like Harold Baines, Nellie Fox, Minnie Minoso, and Eddie Collins in the next couple seasons. Konerko is one of those players that flies under the radar in the national spotlight, and perhaps even in White Sox lore, but when you look at the numbers, he's making a much better case for himself and these historic best lists. In Phladelphia, the current stars of the Phillies are just moving into the Top 20 with Chase Utley actually one spot below that at #21. How far will they climb over the next five years will be interesting to see? Rollins, Howard, and Utley are already the best at their positions in franchise history. For someone like Howard now signed to remain in Philadelphia, it's certainly possible that he'll be pushing into the Top 5, maybe even to #3, although it's unlikely he can climb into the top two spots with Schmidt and Delahanty so far ahead of him.

Changes to the Top 20 Best Batting Careers Team Lists

Name	Team	Rank 2009	Rank 2010
Chris Young	Arizona Diamondbacks	10	4
Mark Reynolds	Arizona Diamondbacks	7	6
Stephen Drew	Arizona Diamondbacks	8	5
Chris Snyder	Arizona Diamondbacks	15	11
Justin Upton	Arizona Diamondbacks	NA	13
Conor Jackson	Arizona Diamondbacks	13	14
David Ortiz	Boston Red Sox	12	11
Paul Konerko	Chicago White Sox	12	7
Troy Tulowitski	Colorado Rockies	9	8
Brad Hawpe	Colorado Rockies	8	9
Carlos Gonzalez	Colorado Rockies	NA	14
Clint Barmes	Colorado Rockies	17	15
Dan Uggla	Florida Marlins	8	5
Cody Ross	Florida Marlins	15	13
Jorge Cantu	Florida Marlins	NA	18
David DeJesus	Kansas City Royals	15	14
Prince Fielder	Milwaukee Brewers	6	5
Ryan Braun	Milwaukee Brewers	16	9
Joe Mauer	Minnesota Twins	16	13
Alex Rodriguez	New York Yankees	13	11

Carlos Beltran	New York Mets	6	5
Jose Reyes	New York Mets	14	10
Jimmy Rollins	Philadelphia Phillies	18	17
Ryan Howard	Philadelphia Phillies	19	18
Carlos Pena	Tampa Bay Rays	3	2
Evan Longoria	Tampa Bay Rays	7	3
Ben Zobrist	Tampa Bay Rays	11	8
Jason Bartlett	Tampa Bay Rays	15	11
Dioner Navarro	Tampa Bay Rays	14	14
Vernon Wells	Toronto Blue Jays	5	3
Ryan Zimmerman	Washington Nationals	12	6

On the pitching side of the stadium, we have a new number one pitcher in franchise history, albeit a short one, in Colorado with Ubaldo Jimenez taking over the top spot from Aaron Cook. And Josh Johnson in Miami and Chris Carpenter in St. Louis move into the #3 spots for their teams. In Chicago, on the White Sox side, the every season performance of Mark Buehrle has now taken him up the ladder of pitching success to the #5 spot for one of the long standing franchises in the game. And in Philadelphia, there is one debut that will make headlines, and has, for the season of 2010. After only one year, Roy Halladay jumps into the Top 20 at #18. Just goes to show how great a season Halladay has had, plus the pitching difficulty of that franchise over the years. And in San Francisco, Mr. Lincecum and now Mr. Cain are trying to garner some accolades amongst the historic pitching greats of the New York and San Francisco wings of the Giant franchise. Be interesting to see just how far either of them can go over the next couple years.

Joining the Top 20 Best Pitching Seasons for Their Team List

Name	**Team**	**Rank 2009**	**Rank 2010**
Jered Weaver	Los Angeles Angels of Anaheim	16	8
Ervin Santana	Los Angeles Angels of Anaheim	13	11
Joe Saunders	Los Angeles Angels of Anaheim	NA	20
Tim Hudson	Atlanta Braves	20	12
Jonathan Papelbon	Boston Red Sox	15	14
Jon Lester	Boston Red Sox	NA	17
Mark Buehrle	Chicago White Sox	6	5
Carlos Zambrano	Chicago Cubs	10	9
Bronson Arroyo	Cincinnati Reds	NA	15
Aaron Harang	Cincinnati Reds	19	20
Jake Westbrook	Cleveland Indians	NA	20
Ubaldo Jimenez	Colorado Rockies	6	1
Aaron Cook	Colorado Rockies	1	2
Jorge de la Rosa	Colorado Rockies	NA	16
Huston Street	Colorado Rockies	NA	19
Justin Verlander	Detroit Tigers	12	11
Josh Johnson	Florida Marlins	4	3
Ricky Nolasco	Florida Marlins	8	5
Anibal Sanchez	Florida Marlins	NA	13
Wandy Rodriguez	Houston Astros	NA	17
Zach Greinke	Kansas City Royals	9	6
Joakim Soria	Kansas City Royals	18	11
David Bush	Milwaukee Brewers	16	13
Johan Santana	New York Mets	17	13
Roy Halladay	Philadelphia Phillies	NA	18
Chris Young	San Diego Padres	16	17
Heath Bell	San Diego Padres	NA	11
Tim Lincecum	San Francisco Giants	14	11

Matt Cain	San Francisco Giants	NA	15
Adam Wainwright	St. Louis Cardinals	NA	8
Chris Carpenter	St. Louis Cardinals	5	3
David Price	Tampa Bay Rays	NA	4
Grant Balfour	Tampa Bay Rays	16	8
Jeff Niemann	Tampa Bay Rays	NA	7
Dan Wheeler	Tampa Bay Rays	11	9
Rafael Soriano	Tampa Bay Rays	NA	12
Andy Sonnanstine	Tampa Bay Rays	12	15
C.J. Wilson	Texas Rangers	NA	16
Scott Downs	Toronto Blue Jays	NA	18
Livan Hernandez	Washington Nationals	7	6

Altoona Mountain City — **Top Career Batters**

Rank	Name	First	LYear	LTeam	Lg	PEVA-B	YRS	PER YR	HR	RBI	H	AVE
1	Smith	Germany	1884	ALT	UA	**1.415**	1	**1.415**	0	0	34	0.315
2	Moore	Jerrie	1884	ALT	UA	**0.684**	1	**0.684**	1	0	25	0.313
3	Brown	Jim	1884	ALT	UA	**0.596**	1	**0.596**	1	0	22	0.250
4	Dougherty	Charlie	1884	ALT	UA	**0.587**	1	**0.587**	0	0	22	0.259
5	Harris	Frank	1884	ALT	UA	**0.559**	1	**0.559**	0	0	25	0.263

Note: Altoona Mountain City franchise folded after 25 games of their only season.

Los Angeles Angels of Anaheim (ANA, CAL, LAA) — **Top Career Batters**

Rank	Name	First	LYear	LTeam	Lg	PEVA-B	YRS	PER YR	HR	RBI	H	AVE
1	Downing	Brian	1990	CAL	AL	**157.437**	13	**12.111**	222	846	1588	0.271
2	Salmon	Tim	2006	LAA	AL	**150.732**	14	**10.767**	299	1016	1674	0.282
3	Anderson	Garret	2008	LAA	AL	**137.630**	15	**9.175**	272	1292	2368	0.296
4	Fregosi	Jim	1971	CAL	AL	**123.541**	11	**11.231**	115	546	1408	0.268
5	Grich	Bobby	1986	CAL	AL	**95.385**	10	**9.539**	154	557	1103	0.269
6	Guerrero	Vladimir	2009	LAA	AL	**92.851**	6	**15.475**	173	616	1034	0.319
7	Erstad	Darin	2006	LAA	AL	**77.974**	11	**7.089**	114	625	1505	0.286
8	Davis	Chili	1996	CAL	AL	**74.258**	7	**10.608**	156	618	973	0.279
9	Joyner	Wally	2001	ANA	AL	**72.753**	7	**10.393**	117	532	961	0.286
10	Baylor	Don	1982	CAL	AL	**71.347**	6	**11.891**	141	523	813	0.262
11	DeCinces	Doug	1987	CAL	AL	**71.314**	6	**11.886**	130	481	765	0.265
12	Glaus	Troy	2004	ANA	AL	**64.989**	7	**9.284**	182	515	748	0.253
13	Edmonds	Jim	1999	ANA	AL	**60.298**	7	**8.614**	121	408	768	0.290
14	Figgins	Chone	2009	LAA	AL	**55.609**	8	**6.951**	31	341	1045	0.291
15	Carew	Rod	1985	CAL	AL	**51.759**	7	**7.394**	18	282	968	0.314
16	Lynn	Fred	1984	CAL	AL	**51.617**	4	**12.904**	71	270	456	0.271
17	Pearson	Albie	1966	CAL	AL	**48.955**	6	**8.159**	24	167	618	0.275
18	Jackson	Reggie	1986	CAL	AL	**47.312**	5	**9.462**	123	374	557	0.239
19	DiSarcina	Gary	2000	ANA	AL	**42.152**	12	**3.513**	28	355	966	0.258
20	Boone	Bob	1988	CAL	AL	**40.915**	7	**5.845**	39	318	742	0.245

Arizona Diamondbacks — **Top Career Batters**

Rank	Name	First	LYear	LTeam	Lg	PEVA-B	YRS	PER YR	HR	RBI	H	AVE
1	Gonzalez	Luis	2006	ARI	NL	**118.928**	8	**14.866**	224	774	1337	0.298
2	Finley	Steve	2004	ARI	NL	**61.202**	6	**10.200**	153	479	847	0.278
3	Bell	Jay	2002	ARI	NL	**36.706**	5	**7.341**	91	304	573	0.263
4	Young	Chris	2009	ARI	NL	**36.162**	5	**7.232**	98	296	549	0.241
5	Drew	Stephen	2010	ARI	NL	**35.855**	5	**7.171**	65	276	669	0.272
6	Reynolds	Mark	2010	ARI	NL	**35.366**	4	**8.834**	121	346	480	0.242
7	Williams	Matt	2003	ARI	NL	**34.312**	6	**5.719**	99	381	629	0.278

8	Womack	Tony	2003	ARI	NL	**26.177**	5	**5.235**	21	200	677	0.269
9	Tracy	Chad	2009	ARI	NL	**26.100**	6	**4.350**	78	318	654	0.280
10	Counsell	Craig	2006	ARI	NL	**23.861**	6	**3.977**	24	193	611	0.266
11	Snyder	Chris	2010	ARI	NL	**21.568**	7	**3.081**	62	240	400	0.229
12	Byrnes	Eric	2009	ARI	NL	**20.545**	4	**5.136**	61	216	426	0.261
13	Upton	Justin	2010	ARI	NL	**19.900**	4	**4.975**	60	208	413	0.272
14	Jackson	Conor	2010	ARI	NL	**19.212**	6	**3.202**	46	247	492	0.277
15	Miller	Damian	2002	ARI	NL	**18.873**	5	**3.775**	48	194	394	0.269
16	Hudson	Orlando	2008	ARI	NL	**18.610**	3	**6.203**	33	171	442	0.294
17	Green	Shawn	2006	ARI	NL	**16.093**	2	**8.047**	33	124	284	0.285
18	Bautista	Danny	2004	ARI	NL	**14.781**	5	**2.956**	33	197	432	0.296
19	Glaus	Troy	2005	ARI	NL	**13.277**	1	**13.277**	37	97	139	0.258
20	Cintron	Alex	2005	ARI	NL	**13.169**	5	**2.634**	25	152	398	0.279

Atlanta Braves (ATL, BSN, ML1) — **Top Career Batters**

Rank	Name	First	LYear	LTeam	Lg	PEVA-B	YRS	PER YR	HR	RBI	H	AVE
1	Aaron	Hank	1974	ATL	NL	**529.862**	21	**25.232**	733	2202	3600	0.310
2	Mathews	Eddie	1966	ATL	NL	**353.724**	15	**23.582**	493	1388	2201	0.273
3	Jones	Chipper	2010	ATL	NL	**261.173**	17	**15.363**	436	1491	2490	0.306
4	Murphy	Dale	1990	ATL	NL	**215.974**	15	**14.398**	371	1143	1901	0.268
5	Jones	Andruw	2007	ATL	NL	**160.694**	12	**13.391**	368	1117	1683	0.263
6	Duffy	Hugh	1900	BSN	NL	**146.888**	9	**16.321**	69	927	1544	0.332
7	Holmes	Tommy	1951	BSN	NL	**127.403**	10	**12.740**	88	580	1503	0.303
8	Long	Herman	1902	BSN	NL	**125.822**	13	**9.679**	88	964	1900	0.280
9	Sutton	Ezra	1888	BSN	NL	**116.138**	12	**9.678**	20	487	1161	0.287
10	Morrill	John	1888	BSN	NL	**116.103**	13	**8.931**	41	625	1247	0.262
11	Berger	Wally	1937	BSN	NL	**111.484**	8	**13.936**	199	746	1263	0.304
12	Tenney	Fred	1911	BSN	NL	**109.435**	15	**7.296**	17	609	1994	0.300
13	Torre	Joe	1968	ATL	NL	**107.109**	9	**11.901**	142	552	1087	0.294
14	Maranville	Rabbit	1935	BSN	NL	**103.680**	15	**6.912**	23	558	1696	0.252
15	Nash	Billy	1895	BSN	NL	**103.383**	10	**10.338**	51	809	1283	0.281
16	Crandall	Del	1963	ML1	NL	**93.077**	13	**7.160**	170	628	1176	0.257
17	Logan	Johnny	1961	ML1	NL	**92.375**	11	**8.398**	92	521	1329	0.270
18	Adcock	Joe	1962	ML1	NL	**91.345**	10	**9.135**	239	760	1206	0.285
19	O'Rourke	Jim	1880	BSN	NL	**86.924**	4	**21.731**	9	140	369	0.309
20	Elliott	Bob	1951	BSN	NL	**83.856**	5	**16.771**	101	466	763	0.295

Baltimore Orioles (BAL, MLA, SLA) — **Top Career Batters**

Rank	Name	First	LYear	LTeam	Lg	PEVA-B	YRS	PER YR	HR	RBI	H	AVE
1	Ripken	Cal	2001	BAL	AL	**320.188**	21	**15.247**	431	1695	3184	0.276
2	Murray	Eddie	1996	BAL	AL	**239.660**	13	**18.435**	343	1224	2080	0.294
3	Robinson	Brooks	1977	BAL	AL	**232.881**	23	**10.125**	268	1357	2848	0.267
4	Sisler	George	1927	SLA	AL	**175.399**	12	**14.617**	93	959	2295	0.344
5	Singleton	Ken	1984	BAL	AL	**159.850**	10	**15.985**	182	766	1455	0.284
6	Clift	Harlond	1943	SLA	AL	**155.325**	10	**15.532**	170	769	1463	0.277
7	Powell	Boog	1974	BAL	AL	**151.843**	14	**10.846**	303	1063	1574	0.266
8	Robinson	Frank	1971	BAL	AL	**135.352**	6	**22.559**	179	545	882	0.300
9	Anderson	Brady	2001	BAL	AL	**131.590**	14	**9.399**	209	744	1614	0.257
10	Williams	Ken	1927	SLA	AL	**116.736**	10	**11.674**	185	808	1308	0.326
11	Wallace	Bobby	1916	SLA	AL	**106.262**	15	**7.084**	8	607	1424	0.258
12	Stone	George	1910	SLA	AL	**106.169**	6	**17.695**	23	268	986	0.301
13	Jacobson	Baby Doll	1926	SLA	AL	**103.894**	10	**10.389**	76	706	1508	0.317
14	Blair	Paul	1976	BAL	AL	**97.931**	13	**7.533**	126	567	1426	0.254
15	Palmeiro	Rafael	2005	BAL	AL	**90.214**	7	**12.888**	223	701	1071	0.284
16	Shotton	Burt	1917	SLA	AL	**87.173**	8	**10.897**	6	228	1070	0.274

17	Mora	Melvin	2009	BAL	AL	**85.180**	10	**8.518**	158	662	1323	0.280
18	Stephens	Vern	1955	BAL	AL	**84.426**	10	**8.443**	121	591	1100	0.292
19	Bumbry	Al	1984	BAL	AL	**82.275**	13	**6.329**	53	392	1403	0.283
20	Tobin	Jack	1925	SLA	AL	**79.114**	9	**8.790**	48	438	1399	0.318

Buffalo Bisons **Top Career Batters**

Rank	Name	First	LYear	LTeam	Lg	PEVA-B	YRS	PER YR	HR	RBI	H	AVE
1	Brouthers	Dan	1885	BFN	NL	**126.636**	5	**25.327**	38	343	650	0.351
2	Richardson	Hardy	1885	BFN	NL	**90.063**	7	**12.866**	17	324	772	0.292
3	O'Rourke	Jim	1884	BFN	NL	**67.848**	4	**16.962**	8	168	514	0.317
4	White	Deacon	1885	BFN	NL	**56.284**	5	**11.257**	6	264	573	0.301
5	Rowe	Jack	1885	BFN	NL	**53.651**	7	**7.664**	10	279	610	0.289
6	Force	Davy	1885	BFN	NL	**24.398**	7	**3.485**	1	154	454	0.207
7	Foley	Curry	1883	BFN	NL	**20.207**	3	**6.736**	4	80	230	0.278
8	Crowley	Bill	1885	BFN	NL	**18.240**	3	**6.080**	1	86	253	0.264
9	Hornung	Joe	1880	BFN	NL	**16.862**	2	**8.431**	1	80	176	0.266
10	Lillie	Jim	1885	BFN	NL	**15.588**	3	**5.196**	6	112	259	0.235
11	Purcell	Blondie	1882	BFN	NL	**13.142**	2	**6.571**	2	57	138	0.280
12	Shaffer	Orator	1883	BFN	NL	**12.566**	1	**12.566**	0	41	117	0.292

Buffalo Bisons (Pacific Coast League) **Top Career Batters**

Rank	Name	First	LYear	LTeam	Lg	PEVA-B	YRS	PER YR	HR	RBI	H	AVE
1	Hoy	Dummy	1890	BFP	PL	**15.291**	1	**15.291**	1	53	147	0.298
2	Wise	Sam	1890	BFP	PL	**12.640**	1	**12.640**	6	102	148	0.293
3	Mack	Connie	1890	BFP	PL	**10.921**	1	**10.921**	0	53	134	0.266
4	Beecher	Ed	1890	BFP	PL	**9.205**	1	**9.205**	3	90	159	0.297
5	Rowe	Jack	1890	BFP	PL	**8.439**	1	**8.439**	2	76	126	0.250

Baltimore Canaries **Top Career Batters**

Rank	Name	First	LYear	LTeam	Lg	PEVA-B	YRS	PER YR	HR	RBI	H	AVE
1	Pike	Lip	1873	BL1	NA	**28.774**	2	**14.387**	10	110	174	0.303
2	York	Tom	1873	BL1	NA	**23.360**	2	**11.680**	3	90	150	0.286
3	Hall	George	1873	BL1	NA	**18.938**	2	**9.469**	1	67	142	0.340
4	Mills	Everett	1873	BL1	NA	**18.843**	2	**9.422**	0	91	166	0.314
5	Force	Davy	1873	BL1	NA	**18.415**	2	**9.207**	0	44	127	0.386
6	Radcliff	John	1873	BL1	NA	**18.184**	2	**9.092**	1	77	156	0.288
7	Carey	Tom	1873	BL1	NA	**13.845**	2	**6.923**	3	82	154	0.316
8	Higham	Dick	1872	BL1	NA	**12.788**	1	**12.788**	2	38	84	0.343

Baltimore Orioles (American Association/National League) (BL2, BL3, BLN) **Top Career Batters**

Rank	Name	First	LYear	LTeam	Lg	PEVA-B	YRS	PER YR	HR	RBI	H	AVE
1	Kelley	Joe	1898	BLN	NL	**151.911**	7	**21.702**	40	653	1069	0.351
2	Jennings	Hughie	1899	BLN	NL	**127.498**	7	**18.214**	12	529	929	0.359
3	Keeler	Willie	1898	BLN	NL	**108.922**	5	**21.784**	14	372	1097	0.388
4	McGraw	John	1899	BLN	NL	**86.517**	9	**9.613**	10	392	1063	0.336
5	Brodie	Steve	1899	BLN	NL	**51.992**	6	**8.665**	10	459	776	0.331
6	Tucker	Tommy	1889	BL2	AA	**44.242**	3	**14.747**	17	244	489	0.311
7	Griffin	Mike	1889	BL2	AA	**44.007**	3	**14.669**	7	188	447	0.279
8	Shindle	Billy	1893	BLN	NL	**40.560**	4	**10.140**	8	242	577	0.260
9	Burns	Oyster	1888	BL2	AA	**37.883**	4	**9.471**	24	201	398	0.300
10	Van Haltren	George	1892	BLN	NL	**36.236**	2	**18.118**	16	140	348	0.310
11	Sommer	Joe	1890	BL3	AA	**36.221**	7	**5.174**	7	255	670	0.241
12	Robinson	Wilbert	1899	BLN	NL	**34.867**	10	**3.487**	10	456	836	0.295

Rank	Name	First	LYear	LTeam	Lg	PEVA-B	YRS	PER YR	HR	RBI	H	AVE
13	Reitz	Heinie	1897	BLN	NL	**31.908**	5	**6.382**	9	400	618	0.291
14	Stenzel	Jake	1898	BLN	NL	**22.731**	2	**11.365**	4	138	224	0.332
15	Doyle	Jack	1897	BLN	NL	**21.305**	2	**10.653**	3	188	328	0.346
16	Clinton	Jim	1886	BL2	AA	**18.925**	3	**6.308**	4	6	258	0.281
17	Welch	Curt	1892	BLN	NL	**17.721**	3	**5.907**	4	82	203	0.248
18	Purcell	Blondie	1888	BL2	AA	**15.896**	3	**5.299**	6	143	257	0.243
19	DeMontreville	Gene	1899	BLN	NL	**15.641**	2	**7.820**	1	122	253	0.314
20	Holmes	Ducky	1899	BLN	NL	**15.606**	2	**7.803**	5	130	303	0.305

Baltimore Marylands — **Top Career Batters**

Rank	Name	First	LYear	LTeam	Lg	PEVA-B	YRS	PER YR	HR	RBI	H	AVE
1	French	Bill	1873	BL4	NA	**0.200**	1	**0.200**	0	1	4	0.222
2	Hooper	Mike	1873	BL4	NA	**0.200**	1	**0.200**	0	2	3	0.214
3	Lennon	Bill	1873	BL4	NA	**0.200**	1	**0.200**	0	2	4	0.211
4	Say	Lou	1873	BL4	NA	**0.200**	1	**0.200**	0	2	2	0.167
5	Smith	Bill	1873	BL4	NA	**0.200**	1	**0.200**	0	1	4	0.174

Baltimore Terrapins — **Top Career Batters**

Rank	Name	First	LYear	LTeam	Lg	PEVA-B	YRS	PER YR	HR	RBI	H	AVE
1	Duncan	Vern	1915	BLF	FL	**20.247**	2	**10.124**	4	96	302	0.278
2	Walsh	Jimmy	1915	BLF	FL	**14.932**	2	**7.466**	19	125	253	0.305
3	Doolan	Mickey	1915	BLF	FL	**12.543**	2	**6.271**	3	74	194	0.218
4	Meyer	Benny	1915	BLF	FL	**12.041**	2	**6.021**	5	45	181	0.292
5	Swacina	Harry	1915	BLF	FL	**11.880**	2	**5.940**	1	128	247	0.269
6	Jacklitsch	Fred	1915	BLF	FL	**11.714**	2	**5.857**	4	61	125	0.265

Baltimore Monumentals — **Top Career Batters**

Rank	Name	First	LYear	LTeam	Lg	PEVA-B	YRS	PER YR	HR	RBI	H	AVE
1	Seery	Emmett	1884	BLU	UA	**12.494**	1	**12.494**	2	0	144	0.311
2	Robinson	Yank	1884	BLU	UA	**7.959**	1	**7.959**	2	0	111	0.267
3	Fusselback	Eddie	1884	BLU	UA	**3.994**	1	**3.994**	1	0	86	0.284
4	Say	Lou	1884	BLU	UA	**3.749**	1	**3.749**	2	0	81	0.239
5	Phelan	Dick	1884	BLU	UA	**3.499**	1	**3.499**	3	0	99	0.246

Boston Red Sox — **Top Career Batters**

Rank	Name	First	LYear	LTeam	Lg	PEVA-B	YRS	PER YR	HR	RBI	H	AVE
1	Williams	Ted	1960	BOS	AL	**493.074**	19	**25.951**	521	1839	2654	0.344
2	Yastrzemski	Carl	1983	BOS	AL	**339.140**	23	**14.745**	452	1844	3419	0.285
3	Evans	Dwight	1990	BOS	AL	**260.131**	19	**13.691**	379	1346	2373	0.272
4	Boggs	Wade	1992	BOS	AL	**238.122**	11	**21.647**	85	687	2098	0.338
5	Rice	Jim	1989	BOS	AL	**224.084**	16	**14.005**	382	1451	2452	0.298
6	Speaker	Tris	1915	BOS	AL	**200.107**	9	**22.234**	39	542	1327	0.337
7	Doerr	Bobby	1951	BOS	AL	**157.073**	14	**11.219**	223	1247	2042	0.288
8	Hooper	Harry	1920	BOS	AL	**142.344**	12	**11.862**	30	497	1707	0.272
9	Foxx	Jimmie	1942	BOS	AL	**140.084**	7	**20.012**	222	788	1051	0.320
10	Ramirez	Manny	2008	BOS	AL	**139.989**	8	**17.499**	274	868	1232	0.312
11	Ortiz	David	2010	BOS	AL	**136.585**	8	**17.073**	291	932	1205	0.286
12	DiMaggio	Dom	1953	BOS	AL	**127.048**	11	**11.550**	87	618	1680	0.298
13	Vaughn	Mo	1998	BOS	AL	**106.452**	8	**13.306**	230	752	1165	0.304
14	Petrocelli	Rico	1976	BOS	AL	**106.231**	13	**8.172**	210	773	1352	0.251
15	Garciaparra	Nomar	2004	BOS	AL	**106.002**	9	**11.778**	178	690	1281	0.323
16	Lynn	Fred	1980	BOS	AL	**103.569**	7	**14.796**	124	521	944	0.308
17	Cronin	Joe	1945	BOS	AL	**102.213**	11	**9.292**	119	737	1168	0.300

Rank	Name	First	LYear	LTeam	Lg	PEVA-B	YRS	PER YR	HR	RBI	H	AVE
18	Greenwell	Mike	1996	BOS	AL	**100.166**	12	**8.347**	130	726	1400	0.303
19	Jensen	Jackie	1961	BOS	AL	**100.047**	7	**14.292**	170	733	1089	0.282
20	Fisk	Carlton	1980	BOS	AL	**98.875**	11	**8.989**	162	568	1097	0.284

Brooklyn Eckfords — **Top Career Batters**

Rank	Name	First	LYear	LTeam	Lg	PEVA-B	YRS	PER YR	HR	RBI	H	AVE
1	Allison	Doug	1872	BR1	NA	**3.198**	1	**3.198**	0	5	27	0.342
2	Martin	Phonney	1872	BR1	NA	**1.355**	1	**1.355**	0	9	12	0.154
3	Gedney	Count	1872	BR1	NA	**1.298**	1	**1.298**	0	7	13	0.183
4	Wood	Jimmy	1872	BR1	NA	**1.277**	1	**1.277**	0	0	6	0.200
5	Snyder	Jim	1872	BR1	NA	**1.135**	1	**1.135**	0	11	28	0.262

Brooklyn Atlantics — **Top Career Batters**

Rank	Name	First	LYear	LTeam	Lg	PEVA-B	YRS	PER YR	HR	RBI	H	AVE
1	Pearce	Dickey	1874	BR2	NA	**12.454**	2	**6.227**	1	51	147	0.284
2	Pabor	Charlie	1875	BR2	NA	**11.655**	2	**5.828**	0	53	118	0.310
3	Ferguson	Bob	1874	BR2	NA	**10.748**	3	**3.583**	0	62	169	0.265
4	Barlow	Tom	1875	BR2	NA	**8.599**	3	**2.866**	1	24	127	0.285
5	Dehlman	Herman	1874	BR2	NA	**8.068**	3	**2.689**	0	49	137	0.227

Brooklyn Gladiators — **Top Career Batters**

Rank	Name	First	LYear	LTeam	Lg	PEVA-B	YRS	PER YR	HR	RBI	H	AVE
1	Simon	Hank	1890	BR4	AA	**6.747**	1	**6.747**	0	38	96	0.257
2	Daily	Ed	1890	BR4	AA	**4.045**	1	**4.045**	1	39	94	0.239
3	Peltz	John	1890	BR4	AA	**3.983**	1	**3.983**	1	33	87	0.227
4	O'Brien	Billy	1890	BR4	AA	**3.832**	1	**3.832**	4	67	108	0.278
5	Gerhardt	Joe	1890	BR4	AA	**3.565**	1	**3.565**	2	40	75	0.203

Brooklyn Tip-Tops — **Top Career Batters**

Rank	Name	First	LYear	LTeam	Lg	PEVA-B	YRS	PER YR	HR	RBI	H	AVE
1	Evans	Steve	1915	BRF	FL	**35.118**	2	**17.559**	15	126	243	0.333
2	Kauff	Benny	1915	BRF	FL	**32.472**	1	**32.472**	12	83	165	0.342
3	Cooper	Claude	1915	BRF	FL	**19.456**	2	**9.728**	4	88	251	0.271
4	Anderson	George	1915	BRF	FL	**13.206**	2	**6.603**	5	63	250	0.286
5	Shaw	Al	1914	BRF	FL	**11.926**	1	**11.926**	5	49	122	0.324
6	Hofman	Solly	1914	BRF	FL	**11.721**	1	**11.721**	5	83	148	0.287
7	Wisterzil	Tex	1915	BRF	FL	**11.303**	2	**5.652**	0	87	170	0.266

Brooklyn Ward's Wonders — **Top Career Batters**

Rank	Name	First	LYear	LTeam	Lg	PEVA-B	YRS	PER YR	HR	RBI	H	AVE
1	Ward	John	1890	BRP	PL	**15.619**	1	**15.619**	4	60	188	0.335
2	Orr	Dave	1890	BRP	PL	**15.509**	1	**15.509**	6	124	172	0.371
3	Bierbauer	Lou	1890	BRP	PL	**14.454**	1	**14.454**	7	99	180	0.306
4	Joyce	Bill	1890	BRP	PL	**13.176**	1	**13.176**	1	78	123	0.252
5	Van Haltren	George	1890	BRP	PL	**6.382**	1	**6.382**	5	54	126	0.335

Boston Red Stockings — **Top Career Batters**

Rank	Name	First	LYear	LTeam	Lg	PEVA-B	YRS	PER YR	HR	RBI	H	AVE
1	Barnes	Ross	1875	BS1	NA	**151.486**	5	**30.297**	4	239	530	0.390
2	McVey	Cal	1875	BS1	NA	**101.404**	4	**25.351**	6	242	403	0.359
3	Wright	George	1875	BS1	NA	**101.045**	5	**20.209**	9	198	484	0.350

Rank	Name	First	Year	Team	Lg	PEVA-B	YRS	PER YR	HR	RBI	H	AVE
4	White	Deacon	1875	BS1	NA	**75.311**	3	**25.104**	4	178	363	0.351
5	Spalding	Al	1875	BS1	NA	**55.859**	5	**11.172**	2	248	455	0.323
6	Leonard	Andy	1875	BS1	NA	**54.782**	4	**13.695**	3	229	414	0.324
7	O'Rourke	Jim	1875	BS1	NA	**47.885**	3	**15.962**	12	181	308	0.318
8	Schafer	Harry	1875	BS1	NA	**38.416**	5	**7.683**	4	173	337	0.277
9	Wright	Harry	1875	BS1	NA	**23.237**	5	**4.647**	4	111	222	0.274
10	Gould	Charlie	1872	BS1	NA	**11.796**	2	**5.898**	2	65	97	0.268

Boston Reds (Pacific Coast League/American Association) (BS2, BSP) **Top Career Batters**

Rank	Name	First	Year	Team	Lg	PEVA-B	YRS	PER YR	HR	RBI	H	AVE
1	Brouthers	Dan	1891	BS2	AA	**53.797**	2	**26.899**	6	206	322	0.340
2	Brown	Tom	1891	BS2	AA	**40.871**	2	**20.435**	9	133	338	0.299
3	Duffy	Hugh	1891	BS2	AA	**23.866**	1	**23.866**	9	110	180	0.336
4	Richardson	Hardy	1891	BS2	AA	**23.522**	2	**11.761**	20	198	252	0.303
5	Farrell	Duke	1891	BS2	AA	**20.311**	1	**20.311**	12	110	143	0.302
6	Stovey	Harry	1890	BSP	PL	**16.612**	1	**16.612**	12	84	144	0.299
7	Radford	Paul	1891	BS2	AA	**12.663**	1	**12.663**	0	65	118	0.259
8	Nash	Billy	1890	BSP	PL	**11.930**	1	**11.930**	5	90	130	0.266

Boston Reds (UA) **Top Career Batters**

Rank	Name	First	LYear	LTeam	Lg	PEVA-B	YRS	PER YR	HR	RBI	H	AVE
1	Crane	Ed	1884	BSU	UA	**10.742**	1	**10.742**	12	0	122	0.285
2	O'Brien	Tom	1884	BSU	UA	**6.013**	1	**6.013**	4	0	118	0.263
3	Irwin	John	1884	BSU	UA	**5.791**	1	**5.791**	1	0	101	0.234
4	Hackett	Walter	1884	BSU	UA	**5.602**	1	**5.602**	1	0	101	0.243
5	Slattery	Mike	1884	BSU	UA	**4.100**	1	**4.100**	0	0	86	0.208

Buffalo Buffeds & Blues

Rank	Name	First	LYear	LTeam	Lg	PEVA-B	YRS	PER YR	HR	RBI	H	AVE
1	Louden	Baldy	1915	BUF	FL	**20.009**	2	**10.005**	10	111	267	0.297
2	Chase	Hal	1915	BUF	FL	**19.588**	2	**9.794**	20	137	266	0.310
3	Hanford	Charlie	1914	BUF	FL	**17.564**	1	**17.564**	12	90	174	0.291
4	McDonald	Tex	1915	BUF	FL	**8.299**	2	**4.150**	9	71	142	0.283
5	Blair	Walter	1915	BUF	FL	**7.696**	2	**3.848**	2	53	157	0.235

Chicago White Stockings (National Association) (CH1, CH2) **Top Career Batters**

Rank	Name	First	LYear	LTeam	Lg	PEVA-B	YRS	PER YR	HR	RBI	H	AVE
1	Hines	Paul	1875	CH2	NA	**17.390**	2	**8.695**	0	70	181	0.313
2	Wood	Jimmy	1871	CH1	NA	**14.911**	1	**14.911**	1	29	51	0.378
3	Meyerle	Levi	1874	CH2	NA	**14.600**	1	**14.600**	1	47	100	0.394
4	Peters	John	1875	CH2	NA	**11.009**	2	**5.505**	1	59	154	0.287
5	Devlin	Jim	1875	CH2	NA	**9.465**	2	**4.733**	0	67	150	0.288

Chicago White Sox **Top Career Batters**

Rank	Name	First	LYear	LTeam	Lg	PEVA-B	YRS	PER YR	HR	RBI	H	AVE
1	Thomas	Frank	2005	CHA	AL	**309.515**	16	**19.345**	448	1465	2136	0.307
2	Appling	Luke	1950	CHA	AL	**210.671**	20	**10.534**	45	1116	2749	0.310
3	Collins	Eddie	1926	CHA	AL	**187.849**	12	**15.654**	31	804	2007	0.331
4	Minoso	Minnie	1980	CHA	AL	**162.367**	12	**13.531**	135	808	1523	0.304
5	Fox	Nellie	1963	CHA	AL	**148.794**	14	**10.628**	35	740	2470	0.291
6	Baines	Harold	2001	CHA	AL	**124.121**	14	**8.866**	221	981	1773	0.288

Rank	Name	First	LYear	LTeam	Lg	PEVA-B	YRS	PER YR	HR	RBI	H	AVE
7	Konerko	Paul	2010	CHA	AL	**119.553**	12	**9.963**	358	1127	1813	0.282
8	Fisk	Carlton	1993	CHA	AL	**116.078**	13	**8.929**	214	762	1259	0.257
9	Jones	Fielder	1908	CHA	AL	**115.066**	8	**14.383**	10	375	1151	0.269
10	Ventura	Robin	1998	CHA	AL	**113.229**	10	**11.323**	171	741	1244	0.274
11	Jackson	Joe	1920	CHA	AL	**111.528**	6	**18.588**	30	426	829	0.340
12	Schalk	Ray	1928	CHA	AL	**102.042**	17	**6.002**	11	594	1345	0.254
13	Aparicio	Luis	1970	CHA	AL	**98.124**	10	**9.812**	43	464	1576	0.269
14	Lollar	Sherm	1963	CHA	AL	**93.775**	12	**7.815**	124	631	1122	0.265
15	Ordonez	Magglio	2004	CHA	AL	**89.760**	8	**11.220**	187	703	1167	0.307
16	Kamm	Willie	1931	CHA	AL	**81.731**	9	**9.081**	25	587	1136	0.279
17	Falk	Bibb	1928	CHA	AL	**79.810**	9	**8.868**	50	627	1219	0.315
18	Felsch	Happy	1920	CHA	AL	**77.735**	6	**12.956**	38	446	825	0.293
19	Weaver	Buck	1920	CHA	AL	**77.322**	9	**8.591**	21	421	1308	0.272
20	Robinson	Floyd	1966	CHA	AL	**74.771**	7	**10.682**	65	400	875	0.287

Chicago Chi-Feds — **Top Career Batters**

Rank	Name	First	LYear	LTeam	Lg	PEVA-B	YRS	PER YR	HR	RBI	H	AVE
1	Zwilling	Dutch	1915	CHF	FL	**47.895**	2	**23.947**	29	189	342	0.300
2	Wilson	Art	1915	CHF	FL	**38.359**	2	**19.180**	17	95	210	0.296
3	Flack	Max	1915	CHF	FL	**22.698**	2	**11.349**	5	84	288	0.281
4	Wickland	Al	1915	CHF	FL	**19.932**	2	**9.966**	7	73	169	0.272
5	Mann	Les	1915	CHF	FL	**14.224**	1	**14.224**	4	58	144	0.306
6	Beck	Fred	1915	CHF	FL	**13.168**	2	**6.584**	16	115	238	0.256
7	Zeider	Rollie	1915	CHF	FL	**11.358**	2	**5.679**	1	70	236	0.249
8	Fischer	William	1915	CHF	FL	**10.030**	1	**10.030**	4	50	96	0.329

Chicago Cubs — **Top Career Batters**

Rank	Name	First	LYear	LTeam	Lg	PEVA-B	YRS	PER YR	HR	RBI	H	AVE
1	Anson	Cap	1897	CHN	NL	**374.385**	22	**17.017**	97	1879	2995	0.329
2	Banks	Ernie	1971	CHN	NL	**264.788**	19	**13.936**	512	1636	2583	0.274
3	Williams	Billy	1974	CHN	NL	**262.803**	16	**16.425**	392	1353	2510	0.296
4	Santo	Ron	1973	CHN	NL	**257.426**	14	**18.388**	337	1290	2171	0.279
5	Sosa	Sammy	2004	CHN	NL	**214.891**	13	**16.530**	545	1414	1985	0.284
6	Sandberg	Ryne	1997	CHN	NL	**177.095**	15	**11.806**	282	1061	2385	0.285
7	Ryan	Jimmy	1900	CHN	NL	**165.018**	15	**11.001**	99	914	2073	0.307
8	Hack	Stan	1947	CHN	NL	**159.139**	16	**9.946**	57	642	2193	0.301
9	Gore	George	1886	CHN	NL	**157.398**	8	**19.675**	24	380	933	0.315
10	Nicholson	Bill	1948	CHN	NL	**153.688**	10	**15.369**	205	833	1323	0.272
11	Hartnett	Gabby	1940	CHN	NL	**152.886**	19	**8.047**	231	1153	1867	0.297
12	Grace	Mark	2000	CHN	NL	**147.685**	13	**11.360**	148	1004	2201	0.308
13	Kelly	King	1886	CHN	NL	**142.521**	7	**20.360**	33	480	899	0.316
14	Williamson	Ned	1889	CHN	NL	**136.810**	11	**12.437**	61	622	1050	0.260
15	Wilson	Hack	1931	CHN	NL	**126.233**	6	**21.039**	190	769	1017	0.322
16	Cavarretta	Phil	1953	CHN	NL	**123.104**	20	**6.155**	92	896	1927	0.292
17	Dalrymple	Abner	1886	CHN	NL	**118.974**	8	**14.872**	40	325	938	0.295
18	Schulte	Frank	1916	CHN	NL	**115.881**	13	**8.914**	91	712	1590	0.272
19	Dahlen	Bill	1898	CHN	NL	**109.804**	8	**13.725**	57	560	1166	0.299
20	Chance	Frank	1912	CHN	NL	**108.446**	15	**7.230**	20	590	1268	0.297

Chicago Pirates — **Top Career Batters**

Rank	Name	First	LYear	LTeam	Lg	PEVA-B	YRS	PER YR	HR	RBI	H	AVE
1	Duffy	Hugh	1890	CHP	PL	**23.069**	1	**23.069**	7	82	191	0.320
2	Ryan	Jimmy	1890	CHP	PL	**16.850**	1	**16.850**	6	89	165	0.340
3	O'Neill	Tip	1890	CHP	PL	**12.577**	1	**12.577**	3	75	174	0.302

Rank	Name	First	LYear	LTeam	Lg	PEVA-B	YRS	PER YR	HR	RBI	H	AVE
4	Farrell	Duke	1890	CHP	PL	**9.023**	1	**9.023**	2	84	131	0.290
5	Pfeffer	Fred	1890	CHP	PL	**7.185**	1	**7.185**	5	80	128	0.257

Chicago/Pittsburgh (Union League) **Top Career Batters**

Rank	Name	First	LYear	LTeam	Lg	PEVA-B	YRS	PER YR	HR	RBI	H	AVE
1	Schoeneck	Jumbo	1884	CHU	UA	**5.256**	1	**5.256**	2	0	116	0.317
2	Ellick	Joe	1884	CHU	UA	**4.290**	1	**4.290**	0	0	93	0.236
3	Gross	Emil	1884	CHU	UA	**3.217**	1	**3.217**	4	0	34	0.358
4	Krieg	Bill	1884	CHU	UA	**3.184**	1	**3.184**	0	0	69	0.247
5	Householder	Charlie	1884	CHU	UA	**2.194**	1	**2.194**	1	0	74	0.239

Cincinnati Reds (CN1, CN2, CIN) **Top Career Batters**

Rank	Name	First	LYear	LTeam	Lg	PEVA-B	YRS	PER YR	HR	RBI	H	AVE
1	Rose	Pete	1986	CIN	NL	**294.862**	19	**15.519**	152	1036	3358	0.307
2	Bench	Johnny	1983	CIN	NL	**256.662**	17	**15.098**	389	1376	2048	0.267
3	Robinson	Frank	1965	CIN	NL	**237.493**	10	**23.749**	324	1009	1673	0.303
4	Perez	Tony	1986	CIN	NL	**176.730**	16	**11.046**	287	1192	1934	0.283
5	Morgan	Joe	1979	CIN	NL	**176.447**	8	**22.056**	152	612	1155	0.288
6	McPhee	Bid	1899	CIN	NL	**166.066**	18	**9.226**	53	1067	2250	0.271
7	Larkin	Barry	2004	CIN	NL	**165.994**	19	**8.737**	198	960	2340	0.295
8	Pinson	Vada	1968	CIN	NL	**159.185**	11	**14.471**	186	814	1881	0.297
9	Jones	Charley	1887	CN2	AA	**152.726**	9	**16.970**	39	407	835	0.301
10	Foster	George	1981	CIN	NL	**151.263**	11	**13.751**	244	861	1276	0.286
11	Roush	Edd	1931	CIN	NL	**139.644**	12	**11.637**	47	763	1784	0.331
12	Concepcion	Dave	1988	CIN	NL	**130.003**	19	**6.842**	101	950	2326	0.267
13	Reilly	John	1891	CIN	NL	**128.015**	10	**12.802**	69	740	1352	0.289
14	Kluszewski	Ted	1957	CIN	NL	**126.670**	11	**11.515**	251	886	1499	0.302
15	Groh	Heinie	1921	CIN	NL	**122.139**	9	**13.571**	17	408	1323	0.298
16	McCormick	Frank	1945	CIN	NL	**101.295**	10	**10.130**	110	800	1439	0.301
17	Bell	Gus	1961	CIN	NL	**96.043**	9	**10.671**	160	711	1343	0.288
18	Griffey	Ken	1990	CIN	NL	**91.377**	12	**7.615**	71	466	1275	0.303
19	Holliday	Bug	1898	CIN	NL	**90.136**	10	**9.014**	65	617	1134	0.311
20	Davis	Eric	1996	CIN	NL	**89.831**	9	**9.981**	203	615	886	0.271

Cleveland Forest Citys **Top Career Batters**

Rank	Name	First	LYear	LTeam	Lg	PEVA-B	YRS	PER YR	HR	RBI	H	AVE
1	White	Deacon	1872	CL1	NA	**14.827**	2	**7.413**	1	43	84	0.329
2	Sutton	Ezra	1872	CL1	NA	**12.885**	2	**6.443**	3	33	75	0.319
3	Allison	Art	1872	CL1	NA	**5.186**	2	**2.593**	0	27	63	0.281
4	Hastings	Scott	1872	CL1	NA	**5.016**	1	**5.016**	0	16	45	0.391
5	Pratt	Al	1872	CL1	NA	**4.000**	2	**2.000**	0	32	52	0.267

Cleveland Blues **Top Career Batters**

Rank	Name	First	LYear	LTeam	Lg	PEVA-B	YRS	PER YR	HR	RBI	H	AVE
1	Glasscock	Jack	1884	CL2	NL	**61.663**	6	**10.277**	5	203	510	0.258
2	Dunlap	Fred	1883	CL2	NL	**60.683**	4	**15.171**	11	119	448	0.302
3	Phillips	Bill	1884	CL2	NL	**44.308**	6	**7.385**	11	242	590	0.264
4	Shaffer	Orator	1882	CL2	NL	**29.336**	3	**9.779**	4	83	245	0.246
5	Hotaling	Pete	1884	CL2	NL	**23.115**	3	**7.705**	3	98	285	0.248
6	Muldoon	Mike	1884	CL2	NL	**21.802**	3	**7.267**	8	112	271	0.238
7	McCormick	Jim	1884	CL2	NL	**12.089**	6	**2.015**	2	123	356	0.239
8	York	Tom	1883	CL2	NL	**11.469**	1	**11.469**	2	46	99	0.260

Cleveland Blues & Spiders (CL3, CL4) — **Top Career Batters**

Rank	Name	First	LYear	LTeam	Lg	PEVA-B	YRS	PER YR	HR	RBI	H	AVE
1	McKean	Ed	1898	CL4	NL	**155.581**	12	**12.965**	63	1084	2011	0.304
2	Burkett	Jesse	1898	CL4	NL	**138.866**	8	**17.358**	33	512	1453	0.356
3	Childs	Cupid	1898	CL4	NL	**103.775**	8	**12.972**	17	541	1238	0.318
4	Zimmer	Chief	1899	CL4	NL	**51.049**	13	**3.927**	23	503	945	0.272
5	McAleer	Jimmy	1898	CL4	NL	**44.872**	9	**4.986**	11	427	913	0.252
6	Tebeau	Patsy	1898	CL4	NL	**42.527**	9	**4.725**	21	625	1076	0.282
7	Davis	George	1892	CL4	NL	**35.366**	3	**11.789**	14	244	448	0.265
8	Wallace	Bobby	1898	CL4	NL	**33.255**	5	**6.651**	8	239	391	0.286
9	Virtue	Jake	1894	CL4	NL	**29.040**	5	**5.808**	7	256	483	0.274
10	O'Connor	Jack	1898	CL4	NL	**26.275**	7	**3.754**	11	410	765	0.277
11	Stricker	Cub	1889	CL4	NL	**22.024**	3	**7.341**	4	133	398	0.250
12	McGarr	Chippy	1896	CL4	NL	**17.091**	4	**4.273**	5	214	454	0.276
13	Hotaling	Pete	1888	CL3	AA	**16.755**	2	**8.378**	3	149	252	0.278
14	Blake	Harry	1898	CL4	NL	**14.672**	5	**2.934**	6	212	403	0.254
15	Ewing	Buck	1894	CL4	NL	**12.875**	2	**6.438**	8	161	225	0.316
16	Gilks	Bob	1890	CL4	NL	**10.530**	4	**2.633**	1	135	303	0.229

Columbus Buckeyes — **Top Career Batters**

Rank	Name	First	LYear	LTeam	Lg	PEVA-B	YRS	PER YR	HR	RBI	H	AVE
1	Brown	Tom	1884	CL5	AA	**24.575**	2	**12.287**	10	64	238	0.273
2	Mann	Fred	1884	CL5	AA	**18.400**	2	**9.200**	8	0	199	0.262
3	Smith	Pop	1884	CL5	AA	**17.836**	2	**8.918**	10	0	212	0.249
4	Richmond	John	1884	CL5	AA	**17.102**	2	**8.551**	3	0	209	0.267
5	Kuehne	Bill	1884	CL5	AA	**9.440**	2	**4.720**	6	0	183	0.232

Columbus Salons — **Top Career Batters**

Rank	Name	First	LYear	LTeam	Lg	PEVA-B	YRS	PER YR	HR	RBI	H	AVE
1	McTamany	Jim	1891	CL6	AA	**37.534**	3	**12.511**	8	135	342	0.263
2	Johnson	Spud	1890	CL6	AA	**30.007**	2	**15.003**	3	192	316	0.317
3	O'Connor	Jack	1891	CL6	AA	**23.383**	3	**7.794**	6	163	316	0.292
4	Crooks	Jack	1891	CL6	AA	**21.499**	3	**7.166**	1	115	248	0.237
5	Sneed	John	1891	CL6	AA	**17.891**	2	**8.946**	3	126	235	0.276
6	Marr	Lefty	1889	CL6	AA	**16.611**	1	**16.611**	1	75	167	0.306
7	Duffee	Charlie	1891	CL6	AA	**16.545**	1	**16.545**	10	90	166	0.301
8	Orr	Dave	1889	CL6	AA	**10.793**	1	**10.793**	4	87	183	0.327
9	Reilly	Charlie	1890	CL6	AA	**10.481**	2	**5.240**	7	83	152	0.275

Cleveland Indians — **Top Career Batters**

Rank	Name	First	LYear	LTeam	Lg	PEVA-B	YRS	PER YR	HR	RBI	H	AVE
1	Speaker	Tris	1926	CLE	AL	**268.546**	11	**24.413**	73	884	1965	0.354
2	Lajoie	Nap	1914	CLE	AL	**219.218**	13	**16.863**	34	919	2046	0.339
3	Averill	Earl	1939	CLE	AL	**183.365**	11	**16.670**	226	1084	1903	0.322
4	Boudreau	Lou	1950	CLE	AL	**166.252**	13	**12.789**	63	740	1706	0.296
5	Doby	Larry	1958	CLE	AL	**160.844**	10	**16.084**	215	776	1234	0.286
6	Thome	Jim	2002	CLE	AL	**158.767**	12	**13.231**	334	927	1332	0.287
7	Rosen	Al	1956	CLE	AL	**154.897**	10	**15.490**	192	717	1063	0.285
8	Belle	Albert	1996	CLE	AL	**134.867**	8	**16.858**	242	751	1014	0.295
9	Sewell	Joe	1930	CLE	AL	**131.698**	11	**11.973**	30	869	1800	0.320
10	Keltner	Ken	1949	CLE	AL	**129.826**	12	**10.819**	163	850	1561	0.276
11	Ramirez	Manny	2000	CLE	AL	**127.409**	8	**15.926**	236	804	1086	0.313

Rank	Name	First	LYear	LTeam	Lg	PEVA-B	YRS	PER YR	HR	RBI	H	AVE
12	Jackson	Joe	1915	CLE	AL	**124.016**	6	**20.669**	24	353	937	0.375
13	Trosky	Hal	1941	CLE	AL	**118.782**	9	**13.198**	216	911	1365	0.313
14	Flick	Elmer	1910	CLE	AL	**113.194**	9	**12.577**	19	376	1058	0.299
15	Bradley	Bill	1910	CLE	AL	**112.334**	10	**11.233**	26	473	1265	0.272
16	Lofton	Kenny	2007	CLE	AL	**107.726**	10	**10.773**	87	518	1512	0.300
17	Thornton	Andre	1987	CLE	AL	**101.651**	10	**10.165**	214	749	1095	0.254
18	Colavito	Rocky	1967	CLE	AL	**101.466**	8	**12.683**	190	574	851	0.267
19	Vizquel	Omar	2004	CLE	AL	**96.572**	11	**8.779**	60	584	1616	0.283
20	Heath	Jeff	1945	CLE	AL	**90.393**	10	**9.039**	122	619	1040	0.298

Cleveland Infants — **Top Career Batters**

Rank	Name	First	LYear	LTeam	Lg	PEVA-B	YRS	PER YR	HR	RBI	H	AVE
1	Browning	Pete	1890	CLP	PL	**29.964**	1	**29.964**	5	93	184	0.373
2	Larkin	Henry	1890	CLP	PL	**23.049**	1	**23.049**	5	112	167	0.330
3	Radford	Paul	1890	CLP	PL	**13.819**	1	**13.819**	2	62	136	0.292
4	Tebeau	Patsy	1890	CLP	PL	**10.414**	1	**10.414**	5	74	134	0.298
5	Delahanty	Ed	1890	CLP	PL	**9.191**	1	**9.191**	3	64	153	0.296

Cincinnati Kelly's Killers — **Top Career Batters**

Rank	Name	First	LYear	LTeam	Lg	PEVA-B	YRS	PER YR	HR	RBI	H	AVE
1	Seery	Emmett	1891	CN3	AA	**7.099**	1	**7.099**	4	36	106	0.285
2	Canavan	Jim	1891	CN3	AA	**6.859**	1	**6.859**	7	66	97	0.228
3	Kelly	King	1891	CN3	AA	**4.732**	1	**4.732**	1	53	84	0.297
4	Carney	John	1891	CN3	AA	**4.259**	1	**4.259**	3	43	102	0.278
5	Johnston	Dick	1891	CN3	AA	**3.338**	1	**3.338**	6	51	83	0.221

Cincinnati Outlaw Reds — **Top Career Batters**

Rank	Name	First	LYear	LTeam	Lg	PEVA-B	YRS	PER YR	HR	RBI	H	AVE
1	Glasscock	Jack	1884	CNU	UA	**6.534**	1	**6.534**	2	0	72	0.419
2	Burns	Dick	1884	CNU	UA	**5.365**	1	**5.365**	4	0	107	0.306
3	Harbidge	Bill	1884	CNU	UA	**5.304**	1	**5.304**	2	0	95	0.279
4	Sylvester	Lou	1884	CNU	UA	**4.341**	1	**4.341**	2	0	89	0.267
5	Hawes	Bill	1884	CNU	UA	**3.537**	1	**3.537**	4	0	97	0.278

Colorado Rockies — **Top Career Batters**

Rank	Name	First	LYear	LTeam	Lg	PEVA-B	YRS	PER YR	HR	RBI	H	AVE
1	Helton	Todd	2010	COL	NL	**184.243**	14	**13.160**	333	1239	2236	0.324
2	Walker	Larry	2004	COL	NL	**115.097**	10	**11.510**	258	848	1361	0.334
3	Bichette	Dante	1999	COL	NL	**80.981**	7	**11.569**	201	826	1278	0.316
4	Castilla	Vinny	2006	COL	NL	**78.028**	9	**8.670**	239	745	1206	0.294
5	Holliday	Matt	2008	COL	NL	**68.432**	5	**13.686**	128	483	848	0.319
6	Galarraga	Andres	1997	COL	NL	**55.236**	5	**11.047**	172	579	843	0.316
7	Atkins	Garrett	2009	COL	NL	**52.366**	7	**7.481**	98	479	805	0.289
8	Tulowitzki	Troy	2010	COL	NL	**49.664**	5	**9.933**	92	338	608	0.290
9	Hawpe	Brad	2010	COL	NL	**45.731**	7	**6.533**	118	464	749	0.280
10	Burks	Ellis	1998	COL	NL	**33.366**	5	**6.673**	115	337	558	0.306
11	Perez	Neifi	2001	COL	NL	**32.868**	6	**5.478**	43	281	769	0.282
12	Young	Eric	1997	COL	NL	**27.044**	5	**5.409**	30	227	626	0.295
13	Weiss	Walt	1997	COL	NL	**23.700**	4	**5.925**	14	143	469	0.266
14	Gonzalez	Carlos	2010	COL	NL	**22.458**	2	**11.229**	47	146	276	0.319
15	Barmes	Clint	2010	COL	NL	**20.389**	8	**2.549**	61	285	582	0.254
16	Cirillo	Jeff	2001	COL	NL	**20.168**	2	**10.084**	28	198	360	0.320
17	Hayes	Charlie	1994	COL	NL	**19.267**	2	**9.634**	35	148	297	0.298

Rank	Name	First	LYear	LTeam	Lg	PEVA-B	YRS	PER YR	HR	RBI	H	AVE
18	Wilson	Preston	2005	COL	NL	**18.149**	3	**6.050**	57	217	288	0.269
19	Pierre	Juan	2002	COL	NL	**14.945**	3	**4.982**	3	110	434	0.308
20	Girardi	Joe	1995	COL	NL	**13.862**	3	**4.621**	15	120	302	0.274

Detroit Tigers — **Top Career Batters**

Rank	Name	First	LYear	LTeam	Lg	PEVA-B	YRS	PER YR	HR	RBI	H	AVE
1	Cobb	Ty	1926	DET	AL	**553.520**	22	**25.160**	111	1804	3900	0.368
2	Kaline	Al	1974	DET	AL	**297.259**	22	**13.512**	399	1583	3007	0.297
3	Crawford	Sam	1917	DET	AL	**252.290**	15	**16.819**	70	1264	2466	0.309
4	Heilmann	Harry	1929	DET	AL	**235.849**	15	**15.723**	164	1442	2499	0.342
5	Gehringer	Charlie	1942	DET	AL	**233.202**	19	**12.274**	184	1427	2839	0.320
6	Greenberg	Hank	1946	DET	AL	**206.974**	12	**17.248**	306	1202	1528	0.319
7	Trammell	Alan	1996	DET	AL	**189.326**	20	**9.466**	185	1003	2365	0.285
8	Veach	Bobby	1923	DET	AL	**188.829**	12	**15.736**	59	1042	1859	0.311
9	Cash	Norm	1974	DET	AL	**173.654**	15	**11.577**	373	1087	1793	0.272
10	Whitaker	Lou	1995	DET	AL	**170.425**	19	**8.970**	244	1084	2369	0.276
11	Bush	Donie	1921	DET	AL	**149.331**	14	**10.667**	9	427	1745	0.250
12	Freehan	Bill	1976	DET	AL	**145.340**	15	**9.689**	200	758	1591	0.262
13	York	Rudy	1945	DET	AL	**124.435**	10	**12.444**	239	936	1317	0.282
14	Parrish	Lance	1986	DET	AL	**116.688**	10	**11.669**	212	700	1123	0.263
15	Fielder	Cecil	1996	DET	AL	**104.158**	7	**14.880**	245	758	947	0.258
16	Horton	Willie	1977	DET	AL	**102.064**	15	**6.804**	262	886	1490	0.276
17	McAuliffe	Dick	1973	DET	AL	**97.254**	14	**6.947**	192	672	1471	0.249
18	Fryman	Travis	1997	DET	AL	**92.871**	8	**11.609**	149	679	1176	0.274
19	Gibson	Kirk	1995	DET	AL	**92.140**	12	**7.678**	195	668	1140	0.273
20	Lemon	Chet	1990	DET	AL	**89.891**	9	**9.988**	142	536	1071	0.263

Detroit Wolverines — **Top Career Batters**

Rank	Name	First	LYear	LTeam	Lg	PEVA-B	YRS	PER YR	HR	RBI	H	AVE
1	Bennett	Charlie	1888	DTN	NL	**98.664**	8	**12.333**	37	353	654	0.278
2	Hanlon	Ned	1888	DTN	NL	**82.480**	8	**10.310**	27	342	879	0.261
3	Brouthers	Dan	1888	DTN	NL	**75.511**	3	**25.170**	32	239	510	0.338
4	Wood	George	1885	DTN	NL	**68.560**	5	**13.712**	27	165	558	0.281
5	Thompson	Sam	1888	DTN	NL	**49.145**	4	**12.286**	32	339	503	0.327
6	Richardson	Hardy	1888	DTN	NL	**33.801**	3	**11.267**	25	187	444	0.330

Elizabeth Resolutes — **Top Career Batters**

Rank	Name	First	LYear	LTeam	Lg	PEVA-B	YRS	PER YR	HR	RBI	H	AVE
1	Allison	Art	1873	ELI	NA	**0.869**	1	**0.869**	0	11	32	0.323
2	Allison	Doug	1873	ELI	NA	**0.637**	1	**0.637**	0	8	24	0.289
3	Booth	Eddie	1873	ELI	NA	**0.548**	1	**0.548**	0	4	21	0.292
4	Austin	Henry	1873	ELI	NA	**0.441**	1	**0.441**	0	11	25	0.248
5	Fleet	Frank	1873	ELI	NA	**0.350**	1	**0.350**	0	10	23	0.256

Florida Marlins — **Top Career Batters**

Rank	Name	First	LYear	LTeam	Lg	PEVA-B	YRS	PER YR	HR	RBI	H	AVE
1	Cabrera	Miguel	2007	FLO	NL	**96.341**	5	**19.268**	138	523	842	0.313
2	Ramirez	Hanley	2010	FLO	NL	**90.283**	5	**18.057**	124	389	934	0.313
3	Sheffield	Gary	1998	FLO	NL	**69.230**	6	**11.538**	122	380	538	0.288
4	Lowell	Mike	2005	FLO	NL	**68.068**	7	**9.724**	143	578	965	0.272
5	Uggla	Dan	2010	FLO	NL	**60.503**	5	**12.101**	154	465	771	0.263
5	Conine	Jeff	2005	FLO	NL	**58.808**	8	**7.351**	120	553	1005	0.290
6	Castillo	Luis	2005	FLO	NL	**54.996**	10	**5.500**	20	271	1273	0.293

7	Lee	Derrek	2003	FLO	NL	**46.603**	6	**7.767**	129	417	746	0.264
8	Floyd	Cliff	2002	FLO	NL	**45.583**	6	**7.597**	110	409	661	0.294
10	Wilson	Preston	2002	FLO	NL	**37.729**	5	**7.546**	104	329	549	0.262
11	Gonzalez	Alex	2005	FLO	NL	**36.315**	8	**4.539**	81	375	788	0.245
12	Pierre	Juan	2005	FLO	NL	**35.369**	3	**11.790**	6	137	606	0.303
13	Ross	Cody	2010	FLO	NL	**31.688**	5	**6.338**	80	297	502	0.265
14	Johnson	Charles	2002	FLO	NL	**29.358**	7	**4.194**	75	277	467	0.241
15	Encarnacion	Juan	2005	FLO	NL	**25.600**	4	**6.400**	46	223	414	0.271
16	Millar	Kevin	2002	FLO	NL	**23.757**	5	**4.751**	59	251	443	0.296
17	Willingham	Josh	2008	FLO	NL	**23.696**	5	**4.739**	63	219	378	0.266
18	Cantu	Jorge	2010	FLO	NL	**22.683**	3	**7.561**	55	249	441	0.278
19	Kotsay	Mark	2000	FLO	NL	**22.547**	4	**5.637**	31	179	463	0.280
20	Delgado	Carlos	2005	FLO	NL	**21.013**	1	**21.013**	33	115	157	0.301

Fort Wayne Kekiongas — **Top Career Batters**

Rank	Name	First	LYear	LTeam	Lg	PEVA-B	YRS	PER YR	HR	RBI	H	AVE
1	Foran	Jim	1871	FW1	NA	**1.456**	1	**1.456**	1	18	31	0.348
2	Mathews	Bobby	1871	FW1	NA	**0.589**	1	**0.589**	0	10	24	0.270
3	Goldsmith	Wally	1871	FW1	NA	**0.578**	1	**0.578**	0	12	18	0.205
4	Carey	Tom	1871	FW1	NA	**0.573**	1	**0.573**	0	10	20	0.230
5	Kelly	Bill	1871	FW1	NA	**0.484**	1	**0.484**	0	7	15	0.224

Hartford Dark Blues (National League) — **Top Career Batters**

Rank	Name	First	LYear	LTeam	Lg	PEVA-B	YRS	PER YR	HR	RBI	H	AVE
1	Cassidy	John	1877	HAR	NL	**22.978**	2	**11.489**	0	35	108	0.362
2	York	Tom	1877	HAR	NL	**21.181**	2	**10.591**	2	76	135	0.270
3	Start	Joe	1877	HAR	NL	**19.559**	1	**19.559**	1	21	90	0.332
4	Ferguson	Bob	1877	HAR	NL	**18.223**	2	**9.111**	0	67	147	0.261
5	Carey	Tom	1877	HAR	NL	**13.346**	2	**6.673**	1	46	148	0.263
6	Burdock	Jack	1877	HAR	NL	**12.431**	2	**6.215**	0	32	152	0.259
7	Higham	Dick	1876	HAR	NL	**11.623**	1	**11.623**	0	35	102	0.327

Houston Astros — **Top Career Batters**

Rank	Name	First	LYear	LTeam	Lg	PEVA-B	YRS	PER YR	HR	RBI	H	AVE
1	Bagwell	Jeff	2005	HOU	NL	**293.606**	15	**19.574**	449	1529	2314	0.297
2	Biggio	Craig	2007	HOU	NL	**226.358**	20	**11.318**	291	1175	3060	0.281
3	Berkman	Lance	2010	HOU	NL	**179.286**	12	**14.941**	326	1090	1648	0.296
4	Wynn	Jimmy	1973	HOU	NL	**172.930**	11	**15.721**	223	719	1291	0.255
5	Cruz	Jose	1987	HOU	NL	**168.825**	13	**12.987**	138	942	1937	0.292
6	Cedeno	Cesar	1981	HOU	NL	**156.475**	12	**13.040**	163	778	1659	0.289
7	Watson	Bob	1979	HOU	NL	**121.311**	14	**8.665**	139	782	1448	0.297
8	Puhl	Terry	1990	HOU	NL	**84.066**	14	**6.005**	62	432	1357	0.281
9	Morgan	Joe	1980	HOU	NL	**80.052**	10	**8.005**	72	327	972	0.261
10	Rader	Doug	1975	HOU	NL	**79.441**	9	**8.827**	128	600	1060	0.250
11	Doran	Bill	1990	HOU	NL	**69.842**	9	**7.760**	69	404	1139	0.267
12	Davis	Glenn	1990	HOU	NL	**67.815**	7	**9.688**	166	518	795	0.262
13	Caminiti	Ken	2000	HOU	NL	**67.204**	10	**6.720**	103	546	1037	0.264
14	Bass	Kevin	1994	HOU	NL	**61.995**	10	**6.200**	87	468	990	0.278
15	Staub	Rusty	1968	HOU	NL	**61.675**	6	**10.279**	57	370	792	0.273
16	Hidalgo	Richard	2004	HOU	NL	**57.353**	8	**7.169**	134	465	787	0.278
17	Cabell	Enos	1985	HOU	NL	**55.861**	8	**6.983**	45	405	1124	0.281
18	Aspromonte	Bob	1968	HOU	NL	**52.820**	7	**7.546**	51	385	925	0.258
19	Ausmus	Brad	2008	HOU	NL	**51.180**	10	**5.118**	41	386	970	0.246
20	Gonzalez	Luis	1997	HOU	NL	**47.156**	7	**6.737**	62	366	683	0.266

Hartford Dark Blues (National Association) — **Top Career Batters**

Rank	Name	First	LYear	LTeam	Lg	PEVA-B	YRS	PER YR	HR	RBI	H	AVE
1	York	Tom	1875	HR1	NA	**14.808**	1	**14.808**	0	37	111	0.296
2	Pike	Lip	1874	HR1	NA	**13.501**	1	**13.501**	1	51	83	0.355
3	Remsen	Jack	1875	HR1	NA	**11.122**	1	**11.122**	0	34	96	0.268
4	Ferguson	Bob	1875	HR1	NA	**10.643**	1	**10.643**	0	43	88	0.240
5	Burdock	Jack	1875	HR1	NA	**9.859**	1	**9.859**	0	35	103	0.294

Indianapolis Blues — **Top Career Batters**

Rank	Name	First	LYear	LTeam	Lg	PEVA-B	YRS	PER YR	HR	RBI	H	AVE
1	Shaffer	Orator	1878	IN1	NL	50.240	1	**50.240**	0	30	90	0.338
2	Clapp	John	1878	IN1	NL	20.427	1	**20.427**	0	29	80	0.304
3	McKelvy	Russ	1878	IN1	NL	10.612	1	**10.612**	2	36	57	0.225
4	Quest	Joe	1878	IN1	NL	6.698	1	**6.698**	0	13	57	0.205
5	Williamson	Ned	1878	IN1	NL	5.895	1	**5.895**	1	19	58	0.232

Indianapolis Hoosiers (American Association) — **Top Career Batters**

Rank	Name	First	LYear	LTeam	Lg	PEVA-B	YRS	PER YR	HR	RBI	H	AVE
1	Keenan	Jim	1884	IN2	AA	6.327	1	**6.327**	3	0	73	0.293
2	Phillips	Marr	1884	IN2	AA	5.336	1	**5.336**	0	0	111	0.269
3	Peltz	John	1884	IN2	AA	4.633	1	**4.633**	3	0	86	0.219
4	Kerins	John	1884	IN2	AA	3.299	1	**3.299**	6	0	78	0.214
5	Weihe	Podge	1884	IN2	AA	2.474	1	**2.474**	4	0	65	0.254

Indianapolis Hoosiers (National League) — **Top Career Batters**

Rank	Name	First	LYear	LTeam	Lg	PEVA-B	YRS	PER YR	HR	RBI	H	AVE
1	Glasscock	Jack	1889	IN3	NL	39.030	3	**13.010**	8	170	466	0.309
2	Denny	Jerry	1889	IN3	NL	38.772	3	**12.924**	41	272	465	0.288
3	Seery	Emmett	1889	IN3	NL	32.706	3	**10.902**	17	137	379	0.254
4	Hines	Paul	1889	IN3	NL	19.750	2	**9.875**	10	130	292	0.292
5	Bassett	Charley	1889	IN3	NL	13.730	3	**4.577**	7	175	337	0.239
6	McGeachy	Jack	1889	IN3	NL	13.203	3	**4.401**	3	149	350	0.252

Indianapolis Hoosiers & Newark Peppers (Federal League) (IND, NEW) — **Top Career Batters**

Rank	Name	First	LYear	LTeam	Lg	PEVA-B	YRS	PER YR	HR	RBI	H	AVE
1	Kauff	Benny	1914	IND	FL	**35.356**	1	**35.356**	8	95	211	0.370
2	Esmond	Jimmy	1915	NEW	FL	**22.057**	2	**11.029**	7	111	307	0.276
3	Rariden	Bill	1915	NEW	FL	**20.840**	2	**10.420**	0	87	213	0.254
4	Scheer	Al	1915	NEW	FL	**20.832**	2	**10.416**	5	105	257	0.283
5	LaPorte	Frank	1915	NEW	FL	**19.779**	2	**9.889**	6	163	296	0.281
6	Campbell	Vin	1915	NEW	FL	**19.521**	2	**9.760**	8	88	336	0.314
7	McKechnie	Bill	1915	NEW	FL	**19.064**	2	**9.532**	3	81	286	0.280
8	Roush	Edd	1915	NEW	FL	**16.881**	2	**8.440**	4	90	218	0.304

Kansas City Cowboys (American Association) — **Top Career Batters**

Rank	Name	First	LYear	LTeam	Lg	PEVA-B	YRS	PER YR	HR	RBI	H	AVE
1	Hamilton	Billy	1889	KC2	AA	**16.213**	2	**8.107**	3	88	195	0.294
2	Long	Herman	1889	KC2	AA	**13.249**	1	**13.249**	3	60	158	0.275
3	McTamany	Jim	1888	KC2	AA	**11.901**	1	**11.901**	4	41	127	0.246
4	Burns	Jim	1889	KC2	AA	**11.650**	2	**5.825**	5	101	196	0.304

Rank	Name	First	LYear	LTeam	Lg	PEVA-B	YRS	PER YR	HR	RBI	H	AVE
5	Stearns	Ecky	1889	KC2	AA	**8.475**	1	**8.475**	3	87	160	0.286

Kansas City Royals **Top Career Batters**

Rank	Name	First	LYear	LTeam	Lg	PEVA-B	YRS	PER YR	HR	RBI	H	AVE
1	Brett	George	1993	KCA	AL	**301.897**	21	**14.376**	317	1595	3154	0.305
2	Otis	Amos	1983	KCA	AL	**161.550**	14	**11.539**	193	992	1977	0.280
3	McRae	Hal	1987	KCA	AL	**125.000**	15	**8.333**	169	1012	1924	0.293
4	Wilson	Willie	1990	KCA	AL	**106.434**	15	**7.096**	40	509	1968	0.289
5	Mayberry	John	1977	KCA	AL	**94.598**	6	**15.766**	143	552	816	0.261
6	White	Frank	1990	KCA	AL	**93.076**	18	**5.171**	160	886	2006	0.255
7	Sweeney	Mike	2007	KCA	AL	**68.382**	13	**5.260**	197	837	1398	0.299
8	Tartabull	Danny	1991	KCA	AL	**67.175**	5	**13.435**	124	425	674	0.290
9	Beltran	Carlos	2004	KCA	AL	**62.766**	7	**8.967**	123	516	899	0.287
10	Porter	Darrell	1980	KCA	AL	**60.035**	4	**15.009**	61	301	514	0.271
11	Seitzer	Kevin	1991	KCA	AL	**59.378**	6	**9.896**	33	265	809	0.294
12	Randa	Joe	2004	KCA	AL	**55.767**	8	**6.971**	86	533	1084	0.288
13	Patek	Freddie	1979	KCA	AL	**55.332**	9	**6.148**	28	382	1036	0.241
14	DeJesus	David	2010	KCA	AL	**53.670**	8	**6.709**	61	390	971	0.289
15	Damon	Johnny	2000	KCA	AL	**49.516**	6	**8.253**	65	352	894	0.292
16	Cowens	Al	1979	KCA	AL	**45.245**	6	**7.541**	45	374	784	0.282
17	Macfarlane	Mike	1998	KCA	AL	**43.652**	11	**3.968**	103	398	717	0.256
18	Piniella	Lou	1973	KCA	AL	**41.461**	5	**8.292**	45	348	734	0.286
19	Dye	Jermaine	2001	KCA	AL	**38.031**	5	**7.606**	85	329	584	0.284
20	Aikens	Willie	1983	KCA	AL	**37.346**	4	**9.337**	77	297	499	0.282

Kansas City Packers **Top Career Batters**

Rank	Name	First	LYear	LTeam	Lg	PEVA-B	YRS	PER YR	HR	RBI	H	AVE
1	Kenworthy	Bill	1915	KCF	FL	**31.636**	2	**15.818**	18	143	291	0.309
2	Easterly	Ted	1915	KCF	FL	**22.776**	2	**11.388**	4	99	230	0.309
3	Chadbourne	Chet	1915	KCF	FL	**21.925**	2	**10.963**	2	72	294	0.252
4	Perring	George	1915	KCF	FL	**20.490**	2	**10.245**	9	136	281	0.268
5	Gilmore	Grover	1915	KCF	FL	**18.288**	2	**9.144**	2	79	269	0.286

Kansas City Cowboys (National League) **Top Career Batters**

Rank	Name	First	LYear	LTeam	Lg	PEVA-B	YRS	PER YR	HR	RBI	H	AVE
1	Myers	Al	1886	KCN	NL	**5.041**	1	**5.041**	4	51	131	0.277
2	Radford	Paul	1886	KCN	NL	**4.964**	1	**4.964**	0	20	113	0.229
3	McQuery	Mox	1886	KCN	NL	**4.197**	1	**4.197**	4	38	111	0.247
4	Bassett	Charley	1886	KCN	NL	**3.605**	1	**3.605**	2	32	89	0.260
5	Rowe	Dave	1886	KCN	NL	**3.549**	1	**3.549**	3	57	103	0.240

Kansas City Cowboys (Union League) **Top Career Batters**

Rank	Name	First	LYear	LTeam	Lg	PEVA-B	YRS	PER YR	HR	RBI	H	AVE
1	Black	Bob	1884	KCU	UA	**1.493**	1	**1.493**	1	0	36	0.247
2	Shafer	Taylor	1884	KCU	UA	**1.122**	1	**1.122**	0	0	28	0.171
3	McLaughlin	Barney	1884	KCU	UA	**0.992**	1	**0.992**	0	0	37	0.228
4	McLaughlin	Frank	1884	KCU	UA	**0.906**	1	**0.906**	1	0	28	0.228
5	Say	Lou	1884	KCU	UA	**0.774**	1	**0.774**	1	0	14	0.200

Keokuk Westerns **Top Career Batters**

Rank	Name	First	LYear	LTeam	Lg	PEVA-B	YRS	PER YR	HR	RBI	H	AVE
1	Hallinan	Jimmy	1875	KEO	NA	**0.886**	1	**0.886**	0	3	14	0.275

Rank	Name	First	LYear	LTeam	Lg	PEVA-B	YRS	PER YR	HR	RBI	H	AVE
2	Jones	Charley	1875	KEO	NA	**0.672**	1	**0.672**	0	10	13	0.277
3	Golden	Mike	1875	KEO	NA	**0.369**	1	**0.369**	0	1	6	0.130
4	Quinn	Paddy	1875	KEO	NA	**0.346**	1	**0.346**	0	5	14	0.326
5	Goldsmith	Wally	1875	KEO	NA	**0.200**	1	**0.200**	0	1	6	0.118

Los Angeles Dodgers (BR3, BRO, LAN) **Top Career Batters**

Rank	Name	First	LYear	LTeam	Lg	PEVA-B	YRS	PER YR	HR	RBI	H	AVE
1	Snider	Duke	1962	LAN	NL	**246.786**	16	**15.424**	389	1271	1995	0.300
2	Wheat	Zack	1926	BRO	NL	**211.353**	18	**11.742**	131	1210	2804	0.317
3	Hodges	Gil	1961	LAN	NL	**168.075**	16	**10.505**	361	1254	1884	0.274
4	Reese	Pee Wee	1958	LAN	NL	**166.690**	16	**10.418**	126	885	2170	0.269
5	Campanella	Roy	1957	BRO	NL	**162.658**	10	**16.266**	242	856	1161	0.276
6	Cey	Ron	1982	LAN	NL	**151.942**	12	**12.662**	228	842	1378	0.264
7	Garvey	Steve	1982	LAN	NL	**147.511**	14	**10.537**	211	992	1968	0.301
8	Robinson	Jackie	1956	BRO	NL	**146.209**	10	**14.621**	137	734	1518	0.311
9	Davis	Willie	1973	LAN	NL	**144.813**	14	**10.344**	154	849	2091	0.279
10	Piazza	Mike	1998	LAN	NL	**140.487**	7	**20.070**	177	563	896	0.331
11	Furillo	Carl	1960	LAN	NL	**125.828**	15	**8.389**	192	1058	1910	0.299
12	Walker	Dixie	1947	BRO	NL	**124.757**	9	**13.862**	67	725	1395	0.311
13	Guerrero	Pedro	1988	LAN	NL	**122.716**	11	**11.156**	171	585	1113	0.309
14	Griffin	Mike	1898	BRO	NL	**118.042**	8	**14.755**	29	477	1166	0.305
15	Gilliam	Jim	1966	LAN	NL	**109.763**	14	**7.840**	65	558	1889	0.265
16	Sheckard	Jimmy	1905	BRO	NL	**102.943**	8	**12.868**	36	420	966	0.295
17	Karros	Eric	2002	LAN	NL	**100.842**	12	**8.404**	270	976	1608	0.268
18	Wills	Maury	1972	LAN	NL	**95.928**	12	**7.994**	17	374	1732	0.281
19	Pinkney	George	1891	BRO	NL	**90.886**	7	**12.984**	20	436	1012	0.271
20	Baker	Dusty	1983	LAN	NL	**89.486**	8	**11.186**	144	586	1144	0.281

Louisville Grays **Top Career Batters**

Rank	Name	First	LYear	LTeam	Lg	PEVA-B	YRS	PER YR	HR	RBI	H	AVE
1	Hall	George	1877	LS1	NL	**14.527**	1	**14.527**	0	26	87	0.323
2	Gerhardt	Joe	1877	LS1	NL	**12.420**	2	**6.210**	3	53	152	0.280
3	Shaffer	Orator	1877	LS1	NL	**10.281**	1	**10.281**	3	34	74	0.285
4	Devlin	Jim	1877	LS1	NL	**9.716**	2	**4.858**	1	55	166	0.293
5	Hague	Bill	1877	LS1	NL	**9.284**	2	**4.642**	2	46	148	0.266

Louisville Eclipse & Colonels **Top Career Batters**

Rank	Name	First	LYear	LTeam	Lg	PEVA-B	YRS	PER YR	HR	RBI	H	AVE
1	Browning	Pete	1893	LS3	NL	**139.219**	10	**13.922**	34	451	1233	0.343
2	Wolf	Jimmy	1891	LS2	AA	**112.272**	10	**11.227**	18	592	1438	0.290
3	Clarke	Fred	1899	LS3	NL	**95.443**	6	**15.907**	34	393	1034	0.334
4	Weaver	Farmer	1894	LS3	NL	**39.297**	7	**5.614**	9	320	816	0.275
5	Wagner	Honus	1899	LS3	NL	**32.707**	3	**10.902**	19	257	448	0.321
6	Collins	Hub	1888	LS2	AA	**29.769**	3	**9.923**	3	126	340	0.297
7	Hoy	Dummy	1899	LS3	NL	**29.444**	2	**14.722**	11	115	371	0.305
8	Brown	Tom	1894	LS3	NL	**29.240**	3	**9.747**	16	156	413	0.239
9	Hecker	Guy	1889	LS2	AA	**28.731**	8	**3.591**	19	240	735	0.290
10	Kerins	John	1889	LS2	AA	**27.046**	5	**5.409**	14	202	460	0.263
11	Mack	Reddy	1888	LS2	AA	**23.014**	4	**5.754**	5	164	372	0.257
12	Clingman	Billy	1899	LS3	NL	**20.972**	4	**5.243**	6	178	423	0.246
13	McCreery	Tom	1897	LS3	NL	**20.362**	3	**6.787**	11	115	286	0.322
14	Taylor	Harry	1892	LS3	NL	**20.356**	3	**6.785**	2	124	402	0.287
15	White	Bill	1888	LS2	AA	**20.293**	3	**6.764**	4	175	327	0.258
16	Werrick	Joe	1888	LS2	AA	**20.183**	3	**6.728**	10	212	381	0.253

Rank	Name	First	LYear	LTeam	Lg	PEVA-B	YRS	PER YR	HR	RBI	H	AVE
17	Maskrey	Leech	1886	LS2	AA	**19.765**	5	**3.953**	2	84	341	0.227
18	Pfeffer	Fred	1895	LS3	NL	**18.135**	4	**4.534**	10	215	389	0.272
19	Dexter	Charlie	1899	LS3	NL	**14.563**	4	**3.641**	7	182	392	0.285
20	Cline	Monk	1891	LS2	AA	**14.336**	3	**4.779**	2	53	140	0.291

Middletown Mansfields — **Top Career Batters**

Rank	Name	First	LYear	LTeam	Lg	PEVA-B	YRS	PER YR	HR	RBI	H	AVE
1	Booth	Eddie	1872	MID	NA	**2.609**	1	**2.609**	0	12	38	0.325
2	Murnane	Tim	1872	MID	NA	**2.056**	1	**2.056**	0	13	42	0.359
3	Clapp	John	1872	MID	NA	**1.747**	1	**1.747**	1	10	28	0.289
4	O'Rourke	Jim	1872	MID	NA	**1.704**	1	**1.704**	0	12	31	0.307
5	McCarton	Frank	1872	MID	NA	**1.395**	1	**1.395**	0	10	28	0.329

Milwaukee Brewers (MIL, SE1) — **Top Career Batters**

Rank	Name	First	LYear	LTeam	Lg	PEVA-B	YRS	PER YR	HR	RBI	H	AVE
1	Yount	Robin	1993	ML4	AL	**291.976**	20	**14.599**	251	1406	3142	0.285
2	Molitor	Paul	1992	ML4	AL	**177.049**	15	**11.803**	160	790	2281	0.303
3	Cooper	Cecil	1987	ML4	AL	**143.483**	11	**13.044**	201	944	1815	0.302
4	Thomas	Gorman	1986	ML4	AL	**99.391**	11	**9.036**	208	605	815	0.230
5	Fielder	Prince	2010	MIL	NL	**98.457**	6	**16.410**	192	536	826	0.279
6	Oglivie	Ben	1986	ML4	AL	**98.016**	9	**10.891**	176	685	1144	0.277
7	Jenkins	Geoff	2007	MIL	NL	**79.013**	10	**7.901**	212	704	1221	0.277
8	Scott	George	1976	ML4	AL	**72.764**	5	**14.553**	115	463	851	0.283
9	Braun	Ryan	2010	MIL	NL	**72.509**	4	**18.127**	128	420	711	0.307
10	Money	Don	1983	ML4	AL	**69.513**	11	**6.319**	134	529	1168	0.270
11	Gantner	Jim	1992	ML4	AL	**66.603**	17	**3.918**	47	568	1696	0.274
12	Cirillo	Jeff	2006	MIL	NL	**65.226**	8	**8.153**	73	418	1000	0.307
13	Burnitz	Jeromy	2001	MIL	NL	**64.153**	6	**10.692**	165	525	714	0.258
14	Vaughn	Greg	1996	ML4	AL	**61.395**	8	**7.674**	169	566	799	0.246
15	Lezcano	Sixto	1980	ML4	AL	**58.995**	7	**8.428**	102	374	749	0.275
16	Surhoff	B.J.	1995	ML4	AL	**57.810**	9	**6.423**	57	524	1064	0.274
17	Simmons	Ted	1985	ML4	AL	**52.469**	5	**10.494**	66	394	666	0.262
18	May	Dave	1978	ML4	AL	**45.087**	6	**7.514**	69	287	652	0.259
19	Sexson	Richie	2003	MIL	NL	**44.362**	4	**11.091**	133	398	549	0.276
20	Deer	Rob	1990	ML4	AL	**43.296**	5	**8.659**	137	385	535	0.229

Milwaukee Grays — **Top Career Batters**

Rank	Name	First	LYear	LTeam	Lg	PEVA-B	YRS	PER YR	HR	RBI	H	AVE
1	Dalrymple	Abner	1878	ML2	NL	**21.930**	1	**21.930**	0	15	96	0.354
2	Peters	John	1878	ML2	NL	**4.881**	1	**4.881**	0	22	76	0.309
3	Foley	Will	1878	ML2	NL	**3.871**	1	**3.871**	0	22	62	0.271
4	Goodman	Jake	1878	ML2	NL	**3.847**	1	**3.847**	1	27	62	0.246
5	Golden	Mike	1878	ML2	NL	**1.479**	1	**1.479**	0	20	44	0.206

Milwaukee Brewers (American Association) — **Top Career Batters**

Rank	Name	First	LYear	LTeam	Lg	PEVA-B	YRS	PER YR	HR	RBI	H	AVE
1	Canavan	Jim	1891	ML3	AA	**2.286**	1	**2.286**	3	21	38	0.268
2	Shoch	George	1891	ML3	AA	**1.361**	1	**1.361**	1	16	40	0.315
3	Carney	John	1891	ML3	AA	**1.277**	1	**1.277**	3	23	33	0.300
4	Dalrymple	Abner	1891	ML3	AA	**0.887**	1	**0.887**	1	22	42	0.311
5	Vaughn	Farmer	1891	ML3	AA	**0.634**	1	**0.634**	0	9	33	0.333

Milwaukee Brewers (Union League) **Top Career Batters**

Rank	Name	First	LYear	LTeam	Lg	PEVA-B	YRS	PER YR	HR	RBI	H	AVE
1	Sexton	Tom	1884	MLU	UA	**0.804**	1	**0.804**	0	0	11	0.234
2	Baldwin	Lady	1884	MLU	UA	**0.200**	1	**0.200**	0	0	6	0.222
3	Behel	Steve	1884	MLU	UA	**0.200**	1	**0.200**	0	0	8	0.242
4	Broughton	Cal	1884	MLU	UA	**0.200**	1	**0.200**	0	0	12	0.308
5	Cushman	Ed	1884	MLU	UA	**0.200**	1	**0.200**	0	0	1	0.091

Minnesota Twins (WS1, MIN) **Top Career Batters**

Rank	Name	First	LYear	LTeam	Lg	PEVA-B	YRS	PER YR	HR	RBI	H	AVE
1	Killebrew	Harmon	1974	MIN	AL	**268.631**	21	**12.792**	559	1540	2024	0.258
2	Puckett	Kirby	1995	MIN	AL	**204.074**	12	**17.006**	207	1085	2304	0.318
3	Rice	Sam	1933	WS1	AL	**180.318**	19	**9.490**	33	1045	2889	0.323
4	Carew	Rod	1978	MIN	AL	**156.323**	12	**13.027**	74	733	2085	0.334
5	Oliva	Tony	1976	MIN	AL	**155.848**	15	**10.390**	220	947	1917	0.304
6	Yost	Eddie	1958	WS1	AL	**148.898**	14	**10.636**	101	550	1521	0.253
7	Vernon	Mickey	1955	WS1	AL	**141.926**	14	**10.138**	121	1026	1993	0.288
8	Goslin	Goose	1938	WS1	AL	**139.781**	12	**11.648**	127	931	1659	0.323
9	Milan	Clyde	1922	WS1	AL	**139.763**	16	**8.735**	17	617	2100	0.285
10	Allison	Bob	1970	MIN	AL	**135.670**	13	**10.436**	256	796	1281	0.255
11	Hrbek	Kent	1994	MIN	AL	**134.078**	14	**9.577**	293	1086	1749	0.282
12	Judge	Joe	1932	WS1	AL	**121.719**	18	**6.762**	71	1001	2291	0.299
13	Mauer	Joe	2010	MIN	AL	**119.391**	7	**17.056**	81	472	1011	0.327
14	Myer	Buddy	1941	WS1	AL	**107.428**	16	**6.714**	35	759	1828	0.303
15	Lewis	Buddy	1949	WS1	AL	**107.224**	11	**9.748**	71	607	1563	0.297
16	Gaetti	Gary	1990	MIN	AL	**102.866**	10	**10.287**	201	758	1276	0.256
17	Spence	Stan	1947	WS1	AL	**99.298**	5	**19.860**	66	427	852	0.296
18	Travis	Cecil	1947	WS1	AL	**97.425**	12	**8.119**	27	657	1544	0.314
19	Cronin	Joe	1934	WS1	AL	**94.086**	7	**13.441**	51	673	1090	0.304
20	Bluege	Ossie	1939	WS1	AL	**87.540**	18	**4.863**	43	848	1751	0.272

New Haven Elm Citys **Top Career Batters**

Rank	Name	First	LYear	LTeam	Lg	PEVA-B	YRS	PER YR	HR	RBI	H	AVE
1	Luff	Henry	1875	NH1	NA	**2.444**	1	**2.444**	2	18	45	0.271
2	McGinley	Tim	1875	NH1	NA	**1.847**	1	**1.847**	0	10	36	0.275
3	McKelvey	John	1875	NH1	NA	**1.415**	1	**1.415**	0	10	43	0.229
4	Geer	Billy	1875	NH1	NA	**1.325**	1	**1.325**	0	9	40	0.244
5	Gould	Charlie	1875	NH1	NA	**1.133**	1	**1.133**	0	8	29	0.266

New York Mutuals (National Association) **Top Career Batters**

Rank	Name	First	LYear	LTeam	Lg	PEVA-B	YRS	PER YR	HR	RBI	H	AVE
1	Start	Joe	1875	NY2	NA	**46.839**	5	**9.368**	8	187	387	0.295
2	Eggler	Dave	1873	NY2	NA	**46.499**	3	**15.500**	0	72	235	0.333
3	Hatfield	John	1875	NY2	NA	**39.408**	5	**7.882**	3	143	281	0.279
4	Hicks	Nat	1875	NY2	NA	**22.618**	3	**7.539**	1	69	178	0.271
5	Wolters	Rynie	1871	NY2	NA	**21.178**	1	**21.178**	0	44	51	0.370
6	Higham	Dick	1875	NY2	NA	**16.542**	4	**4.135**	1	90	223	0.303
7	Holdsworth	Jim	1875	NY2	NA	**11.953**	2	**5.976**	0	51	167	0.300
8	Bechtel	George	1872	NY2	NA	**11.150**	1	**11.150**	0	41	74	0.298
9	Pearce	Dickey	1872	NY2	NA	**10.719**	2	**5.360**	1	43	84	0.228

New York Mutuals (National League) **Top Career Batters**

Rank	Name	First	LYear	LTeam	Lg	PEVA-B	YRS	PER YR	HR	RBI	H	AVE

1	Hallinan	Jimmy	1876	NY3	NL	**4.913**	1	**4.913**	2	36	67	0.279
2	Treacey	Fred	1876	NY3	NL	**4.126**	1	**4.126**	0	18	54	0.211
3	Start	Joe	1876	NY3	NL	**3.497**	1	**3.497**	0	21	73	0.277
4	Holdsworth	Jim	1876	NY3	NL	**3.157**	1	**3.157**	0	19	64	0.266
5	Craver	Bill	1876	NY3	NL	**2.237**	1	**2.237**	0	22	55	0.224

New York Metropolitans **Top Career Batters**

Rank	Name	First	LYear	LTeam	Lg	PEVA-B	YRS	PER YR	HR	RBI	H	AVE
1	Orr	Dave	1887	NY4	AA	**81.804**	6	**13.634**	26	357	650	0.348
2	Nelson	Candy	1887	NY4	AA	**48.971**	5	**9.794**	2	77	500	0.258
3	Roseman	Chief	1887	NY4	AA	**39.418**	5	**7.884**	14	126	526	0.257
4	Brady	Steve	1886	NY4	AA	**30.906**	4	**7.727**	4	97	479	0.264
5	Esterbrook	Dude	1887	NY4	AA	**22.793**	3	**7.598**	1	7	270	0.274
6	Hankinson	Frank	1887	NY4	AA	**21.168**	3	**7.056**	5	178	344	0.246
7	Kennedy	Ed	1885	NY4	AA	**10.343**	3	**3.448**	5	21	221	0.204

New York Yankees (NYA, BLA) **Top Career Batters**

Rank	Name	First	LYear	LTeam	Lg	PEVA-B	YRS	PER YR	HR	RBI	H	AVE
1	Ruth	Babe	1934	NYA	AL	**566.741**	15	**37.783**	659	1975	2518	0.349
2	Gehrig	Lou	1939	NYA	AL	**479.522**	17	**28.207**	493	1995	2721	0.340
3	Mantle	Mickey	1968	NYA	AL	**455.611**	18	**25.312**	536	1509	2415	0.298
4	DiMaggio	Joe	1951	NYA	AL	**297.630**	13	**22.895**	361	1537	2214	0.325
5	Berra	Yogi	1963	NYA	AL	**251.424**	18	**13.968**	358	1430	2148	0.285
6	Jeter	Derek	2010	NYA	AL	**233.157**	16	**14.572**	234	1135	2926	0.314
7	Dickey	Bill	1946	NYA	AL	**202.680**	17	**11.922**	202	1209	1969	0.313
8	Williams	Bernie	2006	NYA	AL	**201.884**	16	**12.618**	287	1257	2336	0.297
9	Mattingly	Don	1995	NYA	AL	**193.054**	14	**13.790**	222	1099	2153	0.307
10	White	Roy	1979	NYA	AL	**164.619**	15	**10.975**	160	758	1803	0.271
11	Rodriguez	Alex	2010	NYA	AL	**159.965**	7	**22.852**	268	841	1137	0.296
12	Posada	Jorge	2010	NYA	AL	**156.632**	16	**9.790**	261	1021	1583	0.275
13	Winfield	Dave	1990	NYA	AL	**157.533**	9	**17.504**	205	818	1300	0.290
14	Keller	Charlie	1952	NYA	AL	**143.887**	11	**13.081**	184	723	1053	0.286
15	Munson	Thurman	1979	NYA	AL	**134.322**	11	**12.211**	113	701	1558	0.292
16	Combs	Earle	1935	NYA	AL	**131.405**	12	**10.950**	58	632	1866	0.325
17	Murcer	Bobby	1983	NYA	AL	**130.882**	13	**10.068**	175	687	1231	0.278
18	Nettles	Graig	1983	NYA	AL	**127.229**	11	**11.566**	250	834	1396	0.253
19	Lazzeri	Tony	1937	NYA	AL	**124.158**	12	**10.346**	169	1154	1784	0.293
20	O'Neill	Paul	2001	NYA	AL	**121.196**	9	**13.466**	185	858	1426	0.303

New York Mets **Top Career Batters**

Rank	Name	First	LYear	LTeam	Lg	PEVA-B	YRS	PER YR	HR	RBI	H	AVE
1	Strawberry	Darryl	1990	NYN	NL	**131.391**	8	**16.424**	252	733	1025	0.263
2	Wright	David	2010	NYN	NL	**127.862**	7	**18.266**	169	664	1149	0.309
3	Piazza	Mike	2005	NYN	NL	**105.622**	8	**13.203**	220	655	1028	0.296
4	Johnson	Howard	1993	NYN	NL	**94.784**	9	**10.532**	192	629	997	0.251
5	Beltran	Carlos	2010	NYN	NL	**83.466**	6	**13.911**	134	493	776	0.279
6	Hernandez	Keith	1989	NYN	NL	**81.195**	7	**11.599**	80	468	939	0.297
7	Alfonzo	Edgardo	2002	NYN	NL	**80.065**	8	**10.008**	120	538	1136	0.292
8	Jones	Cleon	1975	NYN	NL	**75.788**	12	**6.316**	93	521	1188	0.281
9	McReynolds	Kevin	1994	NYN	NL	**66.957**	6	**11.159**	122	456	791	0.272
10	Reyes	Jose	2010	NYN	NL	**62.024**	8	**7.753**	74	379	1119	0.286
11	Mazzilli	Lee	1989	NYN	NL	**61.912**	10	**6.191**	68	353	796	0.264
12	Staub	Rusty	1985	NYN	NL	**58.491**	9	**6.499**	75	399	709	0.276
13	Wilson	Mookie	1989	NYN	NL	**58.240**	10	**5.824**	60	342	1112	0.276

14	Hundley	Todd	1998	NYN	NL	**54.779**	9	**6.087**	124	397	612	0.240
15	Carter	Gary	1989	NYN	NL	**54.024**	5	**10.805**	89	349	542	0.249
16	Olerud	John	1999	NYN	NL	**53.907**	3	**17.969**	63	291	524	0.315
17	Kranepool	Ed	1979	NYN	NL	**52.975**	18	**2.943**	118	614	1418	0.261
18	Grote	Jerry	1977	NYN	NL	**52.327**	12	**4.361**	35	357	994	0.256
19	Harrelson	Bud	1977	NYN	NL	**50.685**	13	**3.899**	6	242	1029	0.234
20	Stearns	John	1984	NYN	NL	**47.579**	10	**4.758**	46	312	695	0.259

New York Giants (Pacific League) **Top Career Batters**

Rank	Name	First	LYear	LTeam	Lg	PEVA-B	YRS	PER YR	HR	RBI	H	AVE
1	Connor	Roger	1890	NYP	PL	**25.866**	1	**25.866**	14	103	169	0.349
2	O'Rourke	Jim	1890	NYP	PL	**16.853**	1	**16.853**	9	115	172	0.360
3	Ewing	Buck	1890	NYP	PL	**12.724**	1	**12.724**	8	72	119	0.338
4	Gore	George	1890	NYP	PL	**10.430**	1	**10.430**	10	55	127	0.318
5	Richardson	Danny	1890	NYP	PL	**7.003**	1	**7.003**	4	80	135	0.256

Oakland Athletics (PHA, KC1, OAK) **Top Career Batters**

Rank	Name	First	LYear	LTeam	Lg	PEVA-B	YRS	PER YR	HR	RBI	H	AVE
1	Foxx	Jimmie	1935	PHA	AL	**215.477**	11	**19.589**	302	1075	1492	0.339
2	Henderson	Rickey	1998	OAK	AL	**214.434**	14	**15.317**	167	648	1768	0.288
3	Johnson	Bob	1942	PHA	AL	**181.433**	10	**18.143**	252	1040	1617	0.298
4	McGwire	Mark	1997	OAK	AL	**169.899**	12	**14.158**	363	941	1157	0.260
5	Jackson	Reggie	1987	OAK	AL	**164.741**	10	**16.474**	269	776	1228	0.262
6	Simmons	Al	1944	PHA	AL	**163.521**	12	**13.627**	209	1178	1827	0.356
7	Collins	Eddie	1930	PHA	AL	**161.283**	13	**12.406**	16	496	1308	0.337
8	Bando	Sal	1976	OAK	AL	**158.997**	11	**14.454**	192	796	1311	0.255
9	Baker	Frank	1914	PHA	AL	**144.082**	7	**20.583**	48	612	1103	0.321
10	Cochrane	Mickey	1933	PHA	AL	**138.530**	9	**15.392**	108	680	1317	0.321
11	Canseco	Jose	1997	OAK	AL	**130.460**	9	**14.496**	254	793	1048	0.264
12	Giambi	Jason	2009	OAK	AL	**117.960**	8	**14.745**	198	715	1100	0.300
13	Davis	Harry	1917	PHA	AL	**113.195**	16	**7.075**	69	761	1500	0.279
14	Campaneris	Bert	1976	OAK	AL	**111.655**	13	**8.589**	70	529	1882	0.262
15	Murphy	Dwayne	1987	OAK	AL	**108.184**	10	**10.818**	153	563	999	0.247
16	Chapman	Sam	1951	PHA	AL	**99.023**	11	**9.002**	174	737	1273	0.268
17	Murphy	Danny	1913	PHA	AL	**98.297**	12	**8.191**	40	664	1489	0.290
18	Chavez	Eric	2010	OAK	AL	**97.259**	13	**7.481**	230	787	1276	0.267
19	Joost	Eddie	1954	PHA	AL	**92.011**	8	**11.501**	116	435	840	0.249
20	Hartsel	Topsy	1911	PHA	AL	**89.716**	10	**8.972**	21	266	1087	0.266

Philadelphia Athletics (National Association) **Top Career Batters**

Rank	Name	First	LYear	LTeam	Lg	PEVA-B	YRS	PER YR	HR	RBI	H	AVE
1	Anson	Cap	1875	PH1	NA	**53.497**	4	**13.374**	0	181	384	0.364
2	Meyerle	Levi	1872	PH1	NA	**28.933**	2	**14.466**	5	71	112	0.406
3	McGeary	Mike	1874	PH1	NA	**27.884**	3	**9.295**	0	88	251	0.326
4	Fisler	Wes	1875	PH1	NA	**27.606**	5	**5.521**	1	159	334	0.316
5	Sutton	Ezra	1875	PH1	NA	**27.018**	3	**9.006**	1	120	268	0.318
6	Cuthbert	Ned	1872	PH1	NA	**26.003**	2	**13.001**	4	77	125	0.305
7	Malone	Fergy	1872	PH1	NA	**20.018**	2	**10.009**	1	72	106	0.305
8	McBride	Dick	1875	PH1	NA	**18.444**	5	**3.689**	0	174	306	0.260
9	Force	Davy	1875	PH1	NA	**17.860**	1	**17.860**	0	49	120	0.311
10	Hall	George	1875	PH1	NA	**16.453**	1	**16.453**	4	62	107	0.299
11	Clapp	John	1875	PH1	NA	**15.941**	3	**5.314**	4	87	187	0.283
12	McMullin	John	1874	PH1	NA	**14.411**	2	**7.206**	2	61	152	0.312
13	Reach	Al	1875	PH1	NA	**11.160**	5	**2.232**	0	57	97	0.247

14	Craver	Bill	1875	PH1	NA	**10.479**	1	**10.479**	2	40	83	0.319
15	Mack	Denny	1872	PH1	NA	**10.386**	1	**10.386**	0	34	59	0.288

Philadelphia Athletics (American Association) **Top Career Batters**

Rank	Name	First	LYear	LTeam	Lg	PEVA-B	YRS	PER YR	HR	RBI	H	AVE
1	Stovey	Harry	1889	PH4	AA	**157.628**	7	**22.518**	76	544	1036	0.302
2	Larkin	Henry	1891	PH4	AA	**116.041**	7	**16.577**	36	555	1031	0.301
3	Lyons	Denny	1890	PH4	AA	**80.785**	5	**16.157**	28	351	658	0.329
4	Welch	Curt	1890	PH4	AA	**40.991**	3	**13.664**	3	140	401	0.274
5	Milligan	Jocko	1891	PH4	AA	**36.912**	5	**7.382**	23	240	476	0.286
6	O'Brien	Jack	1890	PH4	AA	**35.196**	6	**5.866**	10	273	505	0.273
7	Bierbauer	Lou	1889	PH4	AA	**31.800**	4	**7.950**	10	314	572	0.268
8	Knight	Lon	1885	PH4	AA	**21.689**	3	**7.230**	2	67	264	0.256
9	Purcell	Blondie	1890	PH4	AA	**20.622**	4	**5.155**	2	172	389	0.290
10	Houck	Sadie	1885	PH4	AA	**19.045**	2	**9.523**	0	54	239	0.278
11	Coleman	John	1889	PH4	AA	**18.954**	4	**4.739**	5	136	263	0.259
12	Birchall	Jud	1884	PH4	AA	**17.291**	3	**5.764**	1	51	254	0.252
13	Corey	Fred	1885	PH4	AA	**16.690**	3	**5.563**	7	78	292	0.260
14	Stricker	Cub	1885	PH4	AA	**16.474**	4	**4.118**	3	99	334	0.239
15	Moynahan	Mike	1884	PH4	AA	**16.146**	2	**8.073**	1	67	124	0.307
16	Wood	George	1891	PH4	AA	**15.229**	1	**15.229**	3	61	163	0.309
17	Poorman	Tom	1888	PH4	AA	**14.528**	2	**7.264**	6	105	242	0.250
18	Hallman	Bill	1891	PH4	AA	**10.783**	1	**10.783**	6	69	166	0.283
19	Robinson	Wilbert	1890	PH4	AA	**10.004**	5	**2.001**	7	155	330	0.227

Philadelphia Whites **Top Career Batters**

Rank	Name	First	LYear	LTeam	Lg	PEVA-B	YRS	PER YR	HR	RBI	H	AVE
1	Meyerle	Levi	1875	PH2	NA	**19.272**	2	**9.636**	4	112	178	0.330
2	Craver	Bill	1874	PH2	NA	**15.032**	1	**15.032**	0	56	91	0.343
3	Fulmer	Chick	1875	PH2	NA	**14.106**	3	**4.702**	1	99	203	0.257
4	Eggler	Dave	1874	PH2	NA	**12.715**	1	**12.715**	0	31	95	0.318
5	Malone	Fergy	1875	PH2	NA	**10.427**	2	**5.213**	0	53	103	0.270

Philadelphia Centennials **Top Career Batters**

Rank	Name	First	LYear	LTeam	Lg	PEVA-B	YRS	PER YR	HR	RBI	H	AVE
1	Craver	Bill	1875	PH3	NA	**2.620**	1	**2.620**	0	5	18	0.277
2	Bechtel	George	1875	PH3	NA	**0.891**	1	**0.891**	0	7	17	0.279
3	McGinley	Tim	1875	PH3	NA	**0.733**	1	**0.733**	0	5	12	0.231
4	Warner	Fred	1875	PH3	NA	**0.706**	1	**0.706**	0	2	14	0.246
5	Treacey	Fred	1875	PH3	NA	**0.574**	1	**0.574**	0	2	12	0.261

Philadelphia Phillies **Top Career Batters**

Rank	Name	First	LYear	LTeam	Lg	PEVA-B	YRS	PER YR	HR	RBI	H	AVE
1	Schmidt	Mike	1989	PHI	NL	**383.574**	18	**21.310**	548	1595	2234	0.267
2	Delahanty	Ed	1901	PHI	NL	**305.971**	13	**23.536**	87	1286	2213	0.348
3	Magee	Sherry	1914	PHI	NL	**196.074**	11	**17.825**	75	886	1647	0.299
4	Ashburn	Richie	1959	PHI	NL	**177.326**	12	**14.777**	22	499	2217	0.311
5	Abreu	Bobby	2006	PHI	NL	**156.932**	9	**17.437**	195	814	1474	0.303
6	Thompson	Sam	1898	PHI	NL	**156.862**	10	**15.686**	95	957	1469	0.333
7	Thomas	Roy	1911	PHI	NL	**155.658**	12	**12.972**	6	264	1364	0.295
8	Ennis	Del	1956	PHI	NL	**152.246**	11	**13.841**	259	1124	1812	0.286
9	Klein	Chuck	1944	PHI	NL	**149.923**	15	**9.995**	243	983	1705	0.326
10	Allen	Dick	1976	PHI	NL	**148.577**	9	**16.509**	204	655	1143	0.290

11	Callison	Johnny	1969	PHI	NL	**137.833**	10	**13.783**	185	666	1438	0.271
12	Hamilton	Billy	1895	PHI	NL	**133.809**	6	**22.302**	23	367	1079	0.361
13	Cravath	Gavvy	1920	PHI	NL	**129.501**	9	**14.389**	117	676	1054	0.291
14	Luzinski	Greg	1980	PHI	NL	**122.236**	11	**11.112**	223	811	1299	0.281
15	Titus	John	1912	PHI	NL	**115.622**	10	**11.562**	31	475	1209	0.278
16	Jones	Willie	1959	PHI	NL	**110.196**	13	**8.477**	180	753	1400	0.258
17	Rollins	Jimmy	2010	PHI	NL	**110.925**	11	**10.084**	154	662	1714	0.272
18	Howard	Ryan	2010	PHI	NL	**109.947**	7	**15.650**	253	748	902	0.279
19	Williams	Cy	1930	PHI	NL	**109.600**	13	**8.431**	217	795	1553	0.306
20	Flick	Elmer	1901	PHI	NL	**93.454**	4	**23.363**	29	377	683	0.338

Philadelphia Athletics (National League) — **Top Career Batters**

Rank	Name	First	LYear	LTeam	Lg	PEVA-B	YRS	PER YR	HR	RBI	H	AVE
1	Hall	George	1876	PHN	NL	**19.290**	1	**19.290**	5	45	98	0.366
2	Meyerle	Levi	1876	PHN	NL	**8.147**	1	**8.147**	0	34	87	0.340
3	Force	Davy	1876	PHN	NL	**5.165**	1	**5.165**	0	17	66	0.232
4	Fisler	Wes	1876	PHN	NL	**4.809**	1	**4.809**	1	30	80	0.288
5	Sutton	Ezra	1876	PHN	NL	**4.636**	1	**4.636**	1	31	70	0.297

Philadelphia Athletics (Pacific League) — **Top Career Batters**

Rank	Name	First	LYear	LTeam	Lg	PEVA-B	YRS	PER YR	HR	RBI	H	AVE
1	Shindle	Billy	1890	PHP	PL	**19.068**	1	**19.068**	10	90	189	0.324
2	Wood	George	1890	PHP	PL	**15.312**	1	**15.312**	9	102	156	0.289
3	Griffin	Mike	1890	PHP	PL	**12.624**	1	**12.624**	6	54	140	0.286
4	Mulvey	Joe	1890	PHP	PL	**9.179**	1	**9.179**	6	87	149	0.287
5	Farrar	Sid	1890	PHP	PL	**5.644**	1	**5.644**	1	69	123	0.256

Philadelphia Keystones — **Top Career Batters**

Rank	Name	First	LYear	LTeam	Lg	PEVA-B	YRS	PER YR	HR	RBI	H	AVE
1	Hoover	Buster	1884	PHU	UA	**8.132**	1	**8.132**	0	0	100	0.364
2	Kienzle	Bill	1884	PHU	UA	**4.253**	1	**4.253**	0	0	76	0.254
3	McCormick	Jerry	1884	PHU	UA	**3.702**	1	**3.702**	0	0	84	0.285
4	Clements	Jack	1884	PHU	UA	**2.903**	1	**2.903**	3	0	50	0.282
5	Flynn	Joe	1884	PHU	UA	**2.286**	1	**2.286**	4	0	52	0.249

Pittsburgh Pirates (PIT, PT1) — **Top Career Batters**

Rank	Name	First	LYear	LTeam	Lg	PEVA-B	YRS	PER YR	HR	RBI	H	AVE
1	Wagner	Honus	1917	PIT	NL	**418.765**	18	**23.265**	82	1475	2967	0.328
2	Clemente	Roberto	1972	PIT	NL	**264.635**	18	**14.702**	240	1305	3000	0.317
3	Stargell	Willie	1982	PIT	NL	**241.847**	21	**11.517**	475	1540	2232	0.282
4	Waner	Paul	1940	PIT	NL	**228.401**	15	**15.227**	109	1177	2868	0.340
5	Carey	Max	1926	PIT	NL	**183.198**	17	**10.776**	67	719	2416	0.287
6	Kiner	Ralph	1953	PIT	NL	**175.644**	8	**21.955**	301	801	1097	0.280
7	Vaughan	Arky	1941	PIT	NL	**162.788**	10	**16.279**	84	764	1709	0.324
8	Clarke	Fred	1915	PIT	NL	**159.569**	15	**10.638**	33	622	1638	0.299
9	Bonds	Barry	1992	PIT	NL	**148.339**	7	**21.191**	176	556	984	0.275
10	Traynor	Pie	1937	PIT	NL	**141.423**	17	**8.319**	58	1273	2416	0.320
11	Leach	Tommy	1918	PIT	NL	**125.144**	14	**8.939**	43	626	1603	0.271
12	Parker	Dave	1983	PIT	NL	**118.404**	11	**10.764**	166	758	1479	0.305
13	Oliver	Al	1977	PIT	NL	**106.312**	10	**10.631**	135	717	1490	0.296
14	Waner	Lloyd	1945	PIT	NL	**104.120**	17	**6.125**	27	577	2317	0.319
15	Kendall	Jason	2004	PIT	NL	**103.018**	9	**11.446**	67	471	1409	0.306
16	Van Slyke	Andy	1994	PIT	NL	**102.157**	8	**12.770**	117	564	1108	0.283

17	Beaumont	Ginger	1906	PIT	NL	**100.065**	8	**12.508**	31	421	1292	0.321
18	Smith	Elmer	1901	PIT	NL	**99.667**	7	**14.238**	30	467	958	0.324
19	Mazeroski	Bill	1972	PIT	NL	**98.836**	17	**5.814**	138	853	2016	0.260
20	Giles	Brian	2003	PIT	NL	**98.218**	5	**19.644**	165	506	782	0.308

Providence Grays — **Top Career Batters**

Rank	Name	First	LYear	LTeam	Lg	PEVA-B	YRS	PER YR	HR	RBI	H	AVE
1	Hines	Paul	1885	PRO	NL	**185.762**	8	**23.220**	23	323	964	0.309
2	York	Tom	1882	PRO	NL	**80.331**	5	**16.066**	5	181	414	0.285
3	Start	Joe	1885	PRO	NL	**70.375**	7	**10.054**	5	271	741	0.297
4	Farrell	Jack	1885	PRO	NL	**51.872**	7	**7.410**	15	225	563	0.251
5	Denny	Jerry	1885	PRO	NL	**40.825**	5	**8.165**	20	204	446	0.248
6	Ward	John	1882	PRO	NL	**33.736**	5	**6.747**	4	175	386	0.246
7	Higham	Dick	1878	PRO	NL	**27.094**	1	**27.094**	1	29	90	0.320
8	O'Rourke	Jim	1879	PRO	NL	**25.376**	1	**25.376**	1	46	126	0.348
9	Brown	Lew	1881	PRO	NL	**24.589**	3	**8.196**	3	91	151	0.276
10	Carroll	Cliff	1885	PRO	NL	**21.643**	4	**5.411**	5	116	285	0.246
11	Wright	George	1882	PRO	NL	**20.013**	2	**10.007**	1	51	137	0.239
12	Gross	Emil	1881	PRO	NL	**19.360**	3	**6.453**	2	82	186	0.281
13	Irwin	Arthur	1885	PRO	NL	**18.954**	3	**6.318**	2	102	252	0.245
14	Radbourn	Charley	1885	PRO	NL	**17.537**	5	**3.507**	5	167	386	0.243
15	Gilligan	Barney	1885	PRO	NL	**16.917**	5	**3.383**	1	120	263	0.220
16	Radford	Paul	1885	PRO	NL	**11.392**	2	**5.696**	1	61	160	0.220

Pittsburgh Rebels — **Top Career Batters**

Rank	Name	First	LYear	LTeam	Lg	PEVA-B	YRS	PER YR	HR	RBI	H	AVE
1	Oakes	Rebel	1915	PTF	FL	**26.174**	2	**13.087**	7	157	339	0.295
2	Lennox	Ed	1915	PTF	FL	**20.237**	2	**10.119**	12	93	150	0.311
3	Konetchy	Ed	1915	PTF	FL	**19.572**	1	**19.572**	10	93	181	0.314
4	Kelly	Jim	1915	PTF	FL	**11.361**	1	**11.361**	4	50	154	0.294
5	Mowrey	Mike	1915	PTF	FL	**11.159**	1	**11.159**	1	49	146	0.280

Pittsburgh Burghers — **Top Career Batters**

Rank	Name	First	LYear	LTeam	Lg	PEVA-B	YRS	PER YR	HR	RBI	H	AVE
1	Beckley	Jake	1890	PTP	PL	**23.357**	1	**23.357**	9	120	167	0.324
2	Fields	Jocko	1890	PTP	PL	**13.610**	1	**13.610**	9	86	148	0.281
3	Visner	Joe	1890	PTP	PL	**12.981**	1	**12.981**	3	71	139	0.267
4	Carroll	Fred	1890	PTP	PL	**12.170**	1	**12.170**	2	71	124	0.298
5	Hanlon	Ned	1890	PTP	PL	**10.532**	1	**10.532**	1	44	131	0.278

Rockford Forest Citys — **Top Career Batters**

Rank	Name	First	LYear	LTeam	Lg	PEVA-B	YRS	PER YR	HR	RBI	H	AVE
1	Anson	Cap	1871	RC1	NA	**5.003**	1	**5.003**	0	16	39	0.325
2	Stires	Gat	1871	RC1	NA	**3.837**	1	**3.837**	2	24	30	0.273
3	Hastings	Scott	1871	RC1	NA	**3.285**	1	**3.285**	0	20	30	0.254
4	Mack	Denny	1871	RC1	NA	**2.652**	1	**2.652**	0	17	30	0.246
5	Addy	Bob	1871	RC1	NA	**2.182**	1	**2.182**	0	13	32	0.271

Rochester Broncos — **Top Career Batters**

Rank	Name	First	LYear	LTeam	Lg	PEVA-B	YRS	PER YR	HR	RBI	H	AVE

Rank	Name	First	LYear	LTeam	Lg	PEVA-B	YRS	PER YR	HR	RBI	H	AVE
1	Knowles	Jimmy	1890	RC2	AA	**10.628**	1	**10.628**	5	84	138	0.281
2	Scheffler	Ted	1890	RC2	AA	**9.074**	1	**9.074**	3	34	109	0.245
3	Lyons	Harry	1890	RC2	AA	**8.265**	1	**8.265**	3	58	152	0.260
4	Griffin	Sandy	1890	RC2	AA	**8.181**	1	**8.181**	5	53	125	0.307
5	McGuire	Deacon	1890	RC2	AA	**5.883**	1	**5.883**	4	53	99	0.299

Richmond Virginians **Top Career Batters**

Rank	Name	First	LYear	LTeam	Lg	PEVA-B	YRS	PER YR	HR	RBI	H	AVE
1	Johnston	Dick	1884	RIC	AA	**1.770**	1	**1.770**	2	0	41	0.281
2	Nash	Billy	1884	RIC	AA	**1.547**	1	**1.547**	1	0	33	0.199
3	Glenn	Ed	1884	RIC	AA	**1.121**	1	**1.121**	1	0	43	0.246
4	Powell	Jim	1884	RIC	AA	**1.058**	1	**1.058**	0	0	37	0.245
5	Mansell	Mike	1884	RIC	AA	**0.943**	1	**0.943**	0	0	34	0.301

San Diego Padres **Top Career Batters**

Rank	Name	First	LYear	LTeam	Lg	PEVA-B	YRS	PER YR	HR	RBI	H	AVE
1	Gwynn	Tony	2001	SDN	NL	**238.944**	20	**11.947**	135	1138	3141	0.338
2	Winfield	Dave	1980	SDN	NL	**123.090**	8	**15.386**	154	626	1134	0.284
3	Gonzalez	Adrian	2010	SDN	NL	**99.263**	5	**19.853**	161	501	856	0.288
4	Giles	Brian	2009	SDN	NL	**74.100**	7	**10.586**	83	415	872	0.279
5	Colbert	Nate	1974	SDN	NL	**71.102**	6	**11.850**	163	481	780	0.253
6	Caminiti	Ken	1998	SDN	NL	**68.417**	4	**17.104**	121	396	592	0.295
7	Kennedy	Terry	1986	SDN	NL	**65.188**	6	**10.865**	76	424	817	0.274
8	Richards	Gene	1983	SDN	NL	**61.836**	7	**8.834**	26	251	994	0.291
9	Klesko	Ryan	2006	SDN	NL	**61.417**	7	**8.774**	133	493	786	0.279
10	Nevin	Phil	2005	SDN	NL	**60.293**	7	**8.613**	156	573	842	0.288
11	Finley	Steve	1998	SDN	NL	**50.714**	4	**12.679**	82	298	662	0.276
12	Tenace	Gene	1980	SDN	NL	**49.386**	4	**12.347**	68	239	384	0.237
13	Templeton	Garry	1991	SDN	NL	**47.904**	10	**4.790**	43	427	1135	0.252
14	Santiago	Benito	1992	SDN	NL	**47.661**	7	**6.809**	85	375	758	0.264
15	McGriff	Fred	1993	SDN	NL	**46.102**	3	**15.367**	84	256	382	0.281
16	McReynolds	Kevin	1986	SDN	NL	**40.028**	4	**10.007**	65	260	470	0.263
17	Gaston	Cito	1974	SDN	NL	**39.616**	6	**6.603**	77	316	672	0.257
18	Martinez	Carmelo	1989	SDN	NL	**34.515**	6	**5.753**	82	337	577	0.248
19	Smith	Ozzie	1981	SDN	NL	**33.585**	4	**8.396**	1	129	516	0.231
20	Garvey	Steve	1987	SDN	NL	**33.059**	5	**6.612**	61	316	631	0.275

Seattle Mariners **Top Career Batters**

Rank	Name	First	LYear	LTeam	Lg	PEVA-B	YRS	PER YR	HR	RBI	H	AVE
1	Martinez	Edgar	2004	SEA	AL	**225.436**	18	**12.524**	309	1261	2247	0.312
2	Griffey	Ken	2010	SEA	AL	**224.365**	13	**17.259**	417	1216	1843	0.292
3	Suzuki	Ichiro	2010	SEA	AL	**148.997**	10	**14.900**	90	558	2244	0.331
4	Buhner	Jay	2001	SEA	AL	**116.173**	14	**8.298**	307	951	1255	0.255
5	Rodriguez	Alex	2000	SEA	AL	**109.585**	7	**15.655**	189	595	966	0.309
6	Davis	Alvin	1991	SEA	AL	**104.099**	8	**13.012**	160	667	1163	0.281
7	Ibanez	Raul	2008	SEA	AL	**64.784**	10	**6.478**	127	547	967	0.284
8	Boone	Bret	2005	SEA	AL	**60.059**	7	**8.580**	143	535	863	0.277
9	Wilson	Dan	2005	SEA	AL	**54.699**	12	**4.558**	88	508	1071	0.262
10	Bradley	Phil	1987	SEA	AL	**53.524**	5	**10.705**	52	234	649	0.301
11	Olerud	John	2004	SEA	AL	**53.523**	5	**10.705**	72	405	709	0.285
12	Reynolds	Harold	1992	SEA	AL	**51.115**	10	**5.111**	17	295	1063	0.260
13	Presley	Jim	1989	SEA	AL	**49.017**	6	**8.170**	115	418	736	0.250
14	Beltre	Adrian	2009	SEA	AL	**48.640**	5	**9.728**	103	396	751	0.266
15	Cameron	Mike	2003	SEA	AL	**46.313**	4	**11.578**	87	344	554	0.256

16	Bochte	Bruce	1982	SEA	AL	**45.799**	5	**9.160**	58	329	697	0.290
17	Sexson	Richie	2008	SEA	AL	**34.758**	4	**8.690**	105	321	447	0.244
18	Winn	Randy	2005	SEA	AL	**34.029**	3	**11.343**	31	193	462	0.287
19	Valle	Dave	1993	SEA	AL	**33.708**	10	**3.371**	72	318	588	0.235
20	Jones	Ruppert	1979	SEA	AL	**32.630**	3	**10.877**	51	200	434	0.257

San Francisco Giants (NY1, SFN) **Top Career Batters**

Rank	Name	First	LYear	LTeam	Lg	PEVA-B	YRS	PER YR	HR	RBI	H	AVE
1	Mays	Willie	1972	SFN	NL	**516.157**	21	**24.579**	646	1859	3187	0.304
2	Bonds	Barry	2007	SFN	NL	**458.361**	15	**30.557**	586	1440	1951	0.312
3	Ott	Mel	1947	NY1	NL	**369.518**	22	**16.796**	511	1860	2876	0.304
4	McCovey	Willie	1980	SFN	NL	**241.730**	19	**12.723**	469	1388	1974	0.274
5	Connor	Roger	1894	NY1	NL	**204.857**	10	**20.486**	76	786	1388	0.319
6	Tiernan	Mike	1899	NY1	NL	**180.026**	13	**13.848**	106	851	1834	0.311
7	Terry	Bill	1936	NY1	NL	**157.226**	14	**11.230**	154	1078	2193	0.341
8	Burns	George	1921	NY1	NL	**143.471**	11	**13.043**	34	458	1541	0.290
9	Davis	George	1903	NY1	NL	**136.080**	10	**13.608**	53	816	1427	0.332
10	Clark	Will	1993	SFN	NL	**135.831**	8	**16.979**	176	709	1278	0.299
11	Van Haltren	George	1903	NY1	NL	**132.330**	10	**13.233**	29	604	1575	0.321
12	Cepeda	Orlando	1966	SFN	NL	**128.383**	9	**14.265**	226	767	1286	0.308
13	Bonds	Bobby	1974	SFN	NL	**123.893**	7	**17.699**	186	552	1106	0.273
14	Doyle	Larry	1920	NY1	NL	**123.810**	13	**9.524**	67	725	1751	0.292
15	Youngs	Ross	1926	NY1	NL	**110.998**	10	**11.100**	42	592	1491	0.322
16	Clark	Jack	1984	SFN	NL	**105.185**	10	**10.518**	163	595	1034	0.277
17	Williams	Matt	1996	SFN	NL	**104.459**	10	**10.446**	247	732	1092	0.264
18	Thomson	Bobby	1957	NY1	NL	**102.946**	9	**11.438**	189	704	1171	0.277
19	Jackson	Travis	1936	NY1	NL	**98.655**	15	**6.577**	135	929	1768	0.291
20	Ewing	Buck	1892	NY1	NL	**98.270**	9	**10.919**	47	459	905	0.306

St. Louis Red Stockings **Top Career Batters**

Rank	Name	First	LYear	LTeam	Lg	PEVA-B	YRS	PER YR	HR	RBI	H	AVE
1	Morgan	Pidgey	1875	SL1	NA	**1.023**	1	**1.023**	0	1	18	0.261
2	Hautz	Charlie	1875	SL1	NA	**1.022**	1	**1.022**	0	4	25	0.301
3	Redmon	Billy	1875	SL1	NA	**0.365**	1	**0.365**	0	1	16	0.195
4	Croft	Art	1875	SL1	NA	**0.228**	1	**0.228**	0	2	15	0.200
5	McSorley	Trick	1875	SL1	NA	**0.221**	1	**0.221**	0	2	11	0.212

St. Louis Brown Stockings (National Association) **Top Career Batters**

Rank	Name	First	LYear	LTeam	Lg	PEVA-B	YRS	PER YR	HR	RBI	H	AVE
1	Pike	Lip	1875	SL2	NA	**23.989**	1	**23.989**	0	44	108	0.346
2	Pearce	Dickey	1875	SL2	NA	**7.012**	1	**7.012**	0	29	77	0.248
3	Cuthbert	Ned	1875	SL2	NA	**6.283**	1	**6.283**	0	17	78	0.245
4	Battin	Joe	1875	SL2	NA	**4.445**	1	**4.445**	0	33	71	0.250
5	Dehlman	Herman	1875	SL2	NA	**4.138**	1	**4.138**	0	14	57	0.224

St. Louis Brown Stockings (National League) **Top Career Batters**

Rank	Name	First	LYear	LTeam	Lg	PEVA-B	YRS	PER YR	HR	RBI	H	AVE
1	Clapp	John	1877	SL3	NL	**33.035**	2	**16.518**	0	63	172	0.311
2	Pike	Lip	1876	SL3	NL	**17.351**	1	**17.351**	1	50	91	0.323
3	Battin	Joe	1877	SL3	NL	**14.171**	2	**7.085**	1	68	130	0.255
4	Dorgan	Mike	1877	SL3	NL	**13.349**	1	**13.349**	0	23	82	0.308
5	McGeary	Mike	1877	SL3	NL	**10.093**	2	**5.046**	0	50	137	0.257

St. Louis Maroons (SL5, SLU) — **Top Career Batters**

Rank	Name	First	LYear	LTeam	Lg	PEVA-B	YRS	PER YR	HR	RBI	H	AVE
1	Dunlap	Fred	1886	SL5	NL	**52.234**	3	**17.411**	18	57	375	0.324
2	Glasscock	Jack	1886	SL5	NL	**29.166**	2	**14.583**	4	80	283	0.304
3	Shaffer	Orator	1885	SL5	NL	**25.836**	2	**12.918**	2	18	218	0.301
4	McKinnon	Alex	1886	SL5	NL	**16.457**	2	**8.229**	9	116	269	0.298
5	Rowe	Dave	1885	SL5	NL	**11.570**	2	**5.785**	4	3	152	0.278
6	Gleason	Jack	1885	SL5	NL	**10.414**	2	**5.207**	4	0	129	0.321

St. Louis Terriers — **Top Career Batters**

Rank	Name	First	LYear	LTeam	Lg	PEVA-B	YRS	PER YR	HR	RBI	H	AVE
1	Tobin	Jack	1915	SLF	FL	**24.185**	2	**12.093**	13	86	327	0.283
2	Miller	Ward	1915	SLF	FL	**23.626**	2	**11.813**	5	113	282	0.301
3	Borton	Babe	1915	SLF	FL	**15.217**	1	**15.217**	3	83	157	0.286
4	Hartley	Grover	1915	SLF	FL	**10.126**	2	**5.063**	2	75	169	0.279
5	Drake	Delos	1915	SLF	FL	**8.980**	2	**4.490**	4	83	220	0.257

St. Louis Cardinals (SL4, SLN) — **Top Career Batters**

Rank	Name	First	LYear	LTeam	Lg	PEVA-B	YRS	PER YR	HR	RBI	H	AVE
1	Musial	Stan	1963	SLN	NL	**481.184**	22	**21.872**	475	1951	3630	0.331
2	Pujols	Albert	2010	SLN	NL	**308.345**	10	**30.835**	408	1230	1900	0.331
3	Hornsby	Rogers	1933	SLN	NL	**258.875**	13	**19.913**	193	1072	2110	0.359
4	Simmons	Ted	1980	SLN	NL	**184.755**	13	**14.212**	172	929	1704	0.298
5	Boyer	Ken	1965	SLN	NL	**164.783**	11	**14.980**	255	1001	1855	0.293
6	Slaughter	Enos	1953	SLN	NL	**164.390**	13	**12.645**	146	1148	2064	0.305
7	Brock	Lou	1979	SLN	NL	**162.510**	16	**10.157**	129	814	2713	0.297
8	Medwick	Joe	1948	SLN	NL	**130.327**	11	**11.848**	152	923	1590	0.335
9	Smith	Ozzie	1996	SLN	NL	**128.467**	15	**8.564**	27	664	1944	0.272
10	Edmonds	Jim	2007	SLN	NL	**124.000**	8	**15.500**	241	713	1033	0.285
11	Lankford	Ray	2004	SLN	NL	**123.023**	13	**9.463**	228	829	1479	0.273
12	Bottomley	Jim	1932	SLN	NL	**122.406**	11	**11.128**	181	1105	1727	0.325
13	Flood	Curt	1969	SLN	NL	**120.736**	12	**10.061**	84	633	1853	0.293
14	O'Neill	Tip	1891	SL4	AA	**118.748**	7	**16.964**	47	625	1092	0.343
15	Hernandez	Keith	1983	SLN	NL	**118.120**	10	**11.812**	81	595	1217	0.299
16	Schoendienst	Red	1963	SLN	NL	**110.289**	15	**7.353**	65	651	1980	0.289
17	Mize	Johnny	1941	SLN	NL	**101.247**	6	**16.874**	158	653	1048	0.336
18	Torre	Joe	1974	SLN	NL	**101.068**	6	**16.845**	98	558	1062	0.308
19	Konetchy	Ed	1913	SLN	NL	**95.329**	7	**13.618**	36	476	1013	0.283
20	McGwire	Mark	2001	SLN	NL	**91.288**	5	**18.258**	220	473	469	0.270

St. Paul Apostles — **Top Career Batters**

Rank	Name	First	LYear	LTeam	Lg	PEVA-B	YRS	PER YR	HR	RBI	H	AVE
1	Barnes	Bill	1884	SPU	UA	**0.200**	1	**0.200**	0	0	6	0.200
2	Galvin	Lou	1884	SPU	UA	**0.200**	1	**0.200**	0	0	2	0.222
3	Carroll	Scrappy	1884	SPU	UA	**0.200**	1	**0.200**	0	0	3	0.097
4	Dealy	Pat	1884	SPU	UA	**0.200**	1	**0.200**	0	0	2	0.133
5	Dunn	Steve	1884	SPU	UA	**0.200**	1	**0.200**	0	0	8	0.250

Syracuse Stars — **Top Career Batters**

Rank	Name	First	LYear	LTeam	Lg	PEVA-B	YRS	PER YR	HR	RBI	H	AVE
1	Farrell	Jack	1879	SR1	NL	**3.821**	1	**3.821**	1	21	73	0.303
2	Purcell	Blondie	1879	SR1	NL	**3.463**	1	**3.463**	0	25	72	0.260

Rank	Name	First	LYear	LTeam	Lg	PEVA-B	YRS	PER YR	HR	RBI	H	AVE
3	Dorgan	Mike	1879	SR1	NL	**2.974**	1	**2.974**	1	17	72	0.267
4	Richmond	John	1879	SR1	NL	**2.060**	1	**2.060**	1	23	54	0.213
5	Mansell	Mike	1879	SR1	NL	**1.971**	1	**1.971**	1	13	52	0.215

Syracuse Stars (American Association) **Top Career Batters**

Rank	Name	First	LYear	LTeam	Lg	PEVA-B	YRS	PER YR	HR	RBI	H	AVE
1	Childs	Cupid	1890	SR2	AA	**24.650**	1	**24.650**	2	89	170	0.345
2	McQuery	Mox	1890	SR2	AA	**7.634**	1	**7.634**	2	55	142	0.308
3	Wright	Rasty	1890	SR2	AA	**6.613**	1	**6.613**	0	27	106	0.305
4	Ely	Bones	1890	SR2	AA	**6.237**	1	**6.237**	0	64	130	0.262
5	O'Rourke	Tim	1890	SR2	AA	**3.882**	1	**3.882**	1	46	94	0.283

Tampa Bay Devil Rays **Top Career Batters**

Rank	Name	First	LYear	LTeam	Lg	PEVA-B	YRS	PER YR	HR	RBI	H	AVE
1	Crawford	Carl	2010	TBA	AL	**88.222**	9	**9.802**	104	592	1480	0.296
2	Pena	Carlos	2010	TBA	AL	**51.883**	4	**12.971**	144	407	461	0.238
3	Longoria	Evan	2010	TBA	AL	**51.341**	3	**17.114**	82	302	455	0.283
4	Huff	Aubrey	2006	TBA	AL	**46.522**	7	**6.646**	128	449	870	0.287
5	Upton	B.J.	2010	TBA	AL	**41.314**	6	**6.886**	67	288	633	0.260
6	McGriff	Fred	2004	TBA	AL	**36.431**	5	**7.286**	99	359	603	0.291
7	Lugo	Julio	2006	TBA	AL	**29.375**	4	**7.344**	40	212	550	0.287
8	Zobrist	Ben	2010	TBA	AL	**28.795**	5	**5.759**	52	223	384	0.253
9	Baldelli	Rocco	2008	TBA	AL	**25.841**	5	**5.168**	52	234	488	0.281
10	Winn	Randy	2002	TBA	AL	**21.673**	5	**4.335**	24	182	513	0.279
11	Bartlett	Jason	2010	TBA	AL	**20.954**	3	**6.985**	19	150	409	0.288
12	Hall	Toby	2006	TBA	AL	**20.404**	7	**2.915**	44	251	538	0.262
13	Grieve	Ben	2003	TBA	AL	**15.880**	3	**5.293**	34	153	302	0.254
14	Navarro	Dioner	2010	TBA	AL	**15.526**	5	**3.105**	29	157	367	0.243
15	Iwamura	Akinori	2009	TBA	AL	**15.525**	3	**5.175**	14	104	379	0.281
16	Flaherty	John	2002	TBA	AL	**14.726**	5	**2.945**	35	196	422	0.252
17	Vaughn	Greg	2002	TBA	AL	**14.717**	3	**4.906**	60	185	271	0.226
18	Cantu	Jorge	2007	TBA	AL	**14.386**	4	**3.597**	44	200	338	0.272
19	Gomes	Jonny	2008	TBA	AL	**14.364**	6	**2.394**	66	184	297	0.235
20	Lee	Travis	2006	TBA	AL	**13.198**	3	**4.399**	42	150	336	0.261

Texas Rangers (WS2, TEX) **Top Career Batters**

Rank	Name	First	LYear	LTeam	Lg	PEVA-B	YRS	PER YR	HR	RBI	H	AVE
1	Palmeiro	Rafael	2003	TEX	AL	**157.511**	10	**15.751**	321	1039	1692	0.290
2	Howard	Frank	1972	TEX	AL	**146.881**	8	**18.360**	246	701	1141	0.277
3	Gonzalez	Juan	2003	TEX	AL	**134.135**	13	**10.318**	372	1180	1595	0.293
4	Rodriguez	Ivan	2009	TEX	AL	**127.258**	13	**9.789**	217	842	1747	0.304
5	Young	Michael	2010	TEX	AL	**116.915**	11	**10.629**	158	811	1848	0.300
6	Sierra	Ruben	2003	TEX	AL	**100.532**	10	**10.053**	180	742	1281	0.280
7	Bell	Buddy	1989	TEX	AL	**99.017**	8	**12.377**	87	499	1060	0.293
8	Sundberg	Jim	1989	TEX	AL	**90.856**	12	**7.571**	60	480	1180	0.252
9	Harrah	Toby	1986	TEX	AL	**89.156**	11	**8.105**	124	568	1174	0.257
10	Greer	Rusty	2002	TEX	AL	**77.166**	9	**8.574**	119	614	1166	0.305
11	Rodriguez	Alex	2003	TEX	AL	**76.868**	3	**25.623**	156	395	569	0.305
12	Burroughs	Jeff	1976	TEX	AL	**65.070**	7	**9.296**	108	412	645	0.255
13	Teixeira	Mark	2007	TEX	AL	**63.627**	5	**12.725**	153	499	746	0.283
14	O'Brien	Pete	1988	TEX	AL	**60.585**	7	**8.655**	114	487	914	0.273
15	McMullen	Ken	1970	WS2	AL	**57.483**	6	**9.580**	86	327	709	0.251
16	Franco	Julio	1993	TEX	AL	**55.682**	5	**11.136**	55	331	725	0.307
17	Parrish	Larry	1988	TEX	AL	**54.765**	7	**7.824**	149	522	852	0.264

18	Hargrove	Mike	1978	TEX	AL	**51.842**	5	**10.368**	47	295	730	0.293
19	Blalock	Hank	2009	TEX	AL	**51.220**	8	**6.402**	152	535	943	0.269
20	Palmer	Dean	1997	TEX	AL	**51.196**	8	**6.400**	154	451	677	0.247

Toledo Blue Stockings **Top Career Batters**

Rank	Name	First	LYear	LTeam	Lg	PEVA-B	YRS	PER YR	HR	RBI	H	AVE
1	Barkley	Sam	1884	TL1	AA	**11.577**	1	**11.577**	1	0	133	0.306
2	Welch	Curt	1884	TL1	AA	**5.719**	1	**5.719**	0	0	95	0.224
3	Miller	Joe	1884	TL1	AA	**5.237**	1	**5.237**	1	0	101	0.239
4	Mullane	Tony	1884	TL1	AA	**4.851**	1	**4.851**	3	0	97	0.276
5	Poorman	Tom	1884	TL1	AA	**4.162**	1	**4.162**	0	0	89	0.233

Toledo Maumees **Top Career Batters**

Rank	Name	First	LYear	LTeam	Lg	PEVA-B	YRS	PER YR	HR	RBI	H	AVE
1	Swartwood	Ed	1890	TL2	AA	**17.677**	1	**17.677**	3	64	151	0.327
2	Werden	Perry	1890	TL2	AA	**13.212**	1	**13.212**	6	72	147	0.295
3	Nicholson	Parson	1890	TL2	AA	**6.869**	1	**6.869**	4	72	140	0.268
4	Scheibeck	Frank	1890	TL2	AA	**6.814**	1	**6.814**	1	49	117	0.241
5	Alvord	Billy	1890	TL2	AA	**5.814**	1	**5.814**	2	52	135	0.273

Toronto Blue Jays **Top Career Batters**

Rank	Name	First	LYear	LTeam	Lg	PEVA-B	YRS	PER YR	HR	RBI	H	AVE
1	Delgado	Carlos	2004	TOR	AL	**143.926**	12	**11.994**	336	1058	1413	0.282
2	Bell	George	1990	TOR	AL	**108.429**	9	**12.048**	202	740	1294	0.286
3	Wells	Vernon	2010	TOR	AL	**107.684**	12	**8.974**	223	813	1529	0.280
4	Moseby	Lloyd	1989	TOR	AL	**105.798**	10	**10.580**	149	651	1319	0.257
5	Fernandez	Tony	2001	TOR	AL	**95.931**	12	**7.994**	60	613	1583	0.297
6	Barfield	Jesse	1989	TOR	AL	**88.455**	9	**9.828**	179	527	919	0.265
7	Carter	Joe	1997	TOR	AL	**84.649**	7	**12.093**	203	736	1051	0.257
8	McGriff	Fred	1990	TOR	AL	**73.370**	5	**14.674**	125	305	540	0.278
9	Olerud	John	1996	TOR	AL	**69.764**	8	**8.721**	109	471	910	0.293
10	Whitt	Ernie	1989	TOR	AL	**65.164**	12	**5.430**	131	518	888	0.253
11	Alomar	Roberto	1995	TOR	AL	**59.926**	5	**11.985**	55	342	832	0.307
12	Upshaw	Willie	1987	TOR	AL	**59.524**	9	**6.614**	112	478	982	0.265
13	Gruber	Kelly	1992	TOR	AL	**58.353**	9	**6.484**	114	434	800	0.259
14	Stewart	Shannon	2008	TOR	AL	**52.918**	10	**5.292**	74	370	1082	0.298
15	Green	Shawn	1999	TOR	AL	**51.706**	7	**7.387**	119	376	718	0.286
16	White	Devon	1995	TOR	AL	**50.795**	5	**10.159**	72	274	733	0.270
17	Rios	Alexis	2009	TOR	AL	**47.865**	6	**7.977**	81	395	875	0.285
18	Sprague	Ed	1998	TOR	AL	**44.458**	8	**5.557**	113	418	773	0.245
19	Mulliniks	Rance	1992	TOR	AL	**44.028**	11	**4.003**	68	389	843	0.280
20	Molitor	Paul	1995	TOR	AL	**41.129**	3	**13.710**	51	246	508	0.315

Troy Trojans **Top Career Batters**

Rank	Name	First	LYear	LTeam	Lg	PEVA-B	YRS	PER YR	HR	RBI	H	AVE
1	Connor	Roger	1882	TRN	NL	**54.729**	3	**18.243**	9	120	335	0.317
2	Ferguson	Bob	1882	TRN	NL	**24.842**	4	**6.211**	1	93	296	0.266
3	Gillespie	Pete	1882	TRN	NL	**23.856**	3	**7.952**	4	98	262	0.264
4	Ewing	Buck	1882	TRN	NL	**15.012**	3	**5.004**	2	59	165	0.256
5	Cassidy	John	1882	TRN	NL	**14.100**	4	**3.525**	1	50	199	0.226
6	Caskin	Ed	1881	TRN	NL	**11.572**	3	**3.857**	0	70	206	0.237

Troy Haymakers — **Top Career Batters**

Rank	Name	First	LYear	LTeam	Lg	PEVA-B	YRS	PER YR	HR	RBI	H	AVE
1	King	Steve	1872	TRO	NA	**19.998**	2	**9.999**	0	55	96	0.353
2	Pike	Lip	1871	TRO	NA	**16.526**	1	**16.526**	4	39	49	0.377
3	Force	Davy	1872	TRO	NA	**10.863**	1	**10.863**	0	16	53	0.408
4	York	Tom	1871	TRO	NA	**7.002**	1	**7.002**	2	23	37	0.255
5	Flynn	Clipper	1871	TRO	NA	**6.865**	1	**6.865**	0	27	48	0.338

Washington Nationals (WAS, MON) — **Top Career Batters**

Rank	Name	First	LYear	LTeam	Lg	PEVA-B	YRS	PER YR	HR	RBI	H	AVE
1	Carter	Gary	1992	MON	NL	**174.379**	12	**14.532**	220	823	1427	0.269
2	Raines	Tim	2001	MON	NL	**160.252**	13	**12.327**	96	556	1622	0.301
3	Dawson	Andre	1986	MON	NL	**143.855**	11	**13.078**	225	838	1575	0.280
4	Wallach	Tim	1992	MON	NL	**128.113**	13	**9.855**	204	905	1694	0.259
5	Guerrero	Vladimir	2003	MON	NL	**103.212**	8	**12.902**	234	702	1215	0.323
6	Zimmerman	Ryan	2010	WAS	NL	**71.877**	6	**11.980**	116	449	833	0.288
7	Staub	Rusty	1979	MON	NL	**71.327**	4	**17.832**	81	284	531	0.295
8	Bailey	Bob	1975	MON	NL	**66.559**	7	**9.508**	118	466	791	0.264
9	Walker	Larry	1994	MON	NL	**56.370**	6	**9.395**	99	384	666	0.281
10	Parrish	Larry	1981	MON	NL	**56.177**	8	**7.022**	100	444	896	0.263
11	Cromartie	Warren	1983	MON	NL	**56.137**	9	**6.237**	60	371	1063	0.280
11	Vidro	Jose	2006	WAS	NL	**55.678**	10	**5.568**	115	550	1280	0.301
13	Galarraga	Andres	2002	MON	NL	**52.354**	8	**6.544**	115	473	906	0.269
14	Grissom	Marquis	1994	MON	NL	**49.187**	6	**8.198**	54	276	747	0.279
15	Singleton	Ken	1974	MON	NL	**49.019**	3	**16.340**	46	227	449	0.285
16	Alou	Moises	1996	MON	NL	**43.808**	6	**7.301**	84	373	626	0.292
17	Cabrera	Orlando	2004	MON	NL	**43.448**	8	**5.431**	66	381	877	0.267
18	Fairly	Ron	1974	MON	NL	**41.908**	6	**6.985**	86	331	615	0.276
19	White	Rondell	2000	MON	NL	**40.860**	8	**5.107**	101	384	808	0.293
20	Brooks	Hubie	1989	MON	NL	**39.992**	5	**7.998**	75	390	689	0.279

Wilmington Quicksteps — **Top Career Batters**

Rank	Name	First	LYear	LTeam	Lg	PEVA-B	YRS	PER YR	HR	RBI	H	AVE
1	Lynch	Tom	1884	WIL	UA	**1.002**	1	**1.002**	0	0	16	0.276
2	Bastian	Charlie	1884	WIL	UA	**0.366**	1	**0.366**	2	0	12	0.200
3	Say	Jimmy	1884	WIL	UA	**0.257**	1	**0.257**	0	0	13	0.220
4	McCloskey	Bill	1884	WIL	UA	**0.200**	1	**0.200**	0	0	3	0.100
5	Casey	Dan	1884	WIL	UA	**0.200**	1	**0.200**	0	0	1	0.167

Worcester Ruby Legs — **Top Career Batters**

Rank	Name	First	LYear	LTeam	Lg	PEVA-B	YRS	PER YR	HR	RBI	H	AVE
1	Stovey	Harry	1882	WOR	NL	**36.781**	3	**12.260**	13	84	290	0.275
2	Irwin	Arthur	1882	WOR	NL	**17.577**	3	**5.859**	1	89	219	0.246
3	Dickerson	Buttercup	1881	WOR	NL	**12.548**	2	**6.274**	1	51	155	0.310
4	Creamer	George	1882	WOR	NL	**10.746**	3	**3.582**	1	81	190	0.211
5	Richmond	Lee	1882	WOR	NL	**9.356**	3	**3.119**	2	90	197	0.250

Washington Olympics — **Top Career Batters**

Rank	Name	First	LYear	LTeam	Lg	PEVA-B	YRS	PER YR	HR	RBI	H	AVE
1	Waterman	Fred	1872	WS3	NA	**13.259**	2	**6.629**	0	23	67	0.330
2	Force	Davy	1871	WS3	NA	**13.071**	1	**13.071**	0	29	45	0.278
3	Mills	Everett	1871	WS3	NA	**7.910**	1	**7.910**	1	24	43	0.274
4	Allison	Doug	1871	WS3	NA	**7.859**	1	**7.859**	2	27	44	0.331

Rank	Name	First	LYear	LTeam	Lg	PEVA-B	YRS	PER YR	HR	RBI	H	AVE
5	Hall	George	1871	WS3	NA	**7.458**	1	**7.458**	2	17	40	0.294

Washington Nationals (National Association) **Top Career Batters**

Rank	Name	First	LYear	LTeam	Lg	PEVA-B	YRS	PER YR	HR	RBI	H	AVE
1	Bielaski	Oscar	1872	WS4	NA	**0.200**	1	**0.200**	0	3	9	0.196
2	Coughlin	Dennis	1872	WS4	NA	**0.200**	1	**0.200**	0	7	12	0.324
3	Doyle	Joe	1872	WS4	NA	**0.200**	1	**0.200**	0	9	12	0.293
4	Lennon	Bill	1872	WS4	NA	**0.200**	1	**0.200**	0	6	12	0.222
5	Hines	Paul	1872	WS4	NA	**0.200**	1	**0.200**	0	5	12	0.245

Washington Blue Legs **Top Career Batters**

Rank	Name	First	LYear	LTeam	Lg	PEVA-B	YRS	PER YR	HR	RBI	H	AVE
1	Hines	Paul	1873	WS5	NA	**3.798**	1	**3.798**	1	29	60	0.331
2	Beals	Tommy	1873	WS5	NA	**2.452**	1	**2.452**	0	24	46	0.272
3	Bielaski	Oscar	1873	WS5	NA	**2.134**	1	**2.134**	0	23	49	0.283
4	White	Warren	1873	WS5	NA	**1.811**	1	**1.811**	0	21	43	0.269
5	Glenn	John	1873	WS5	NA	**1.739**	1	**1.739**	1	21	49	0.265

Washington Nationals (National Association) **Top Career Batters**

Rank	Name	First	LYear	LTeam	Lg	PEVA-B	YRS	PER YR	HR	RBI	H	AVE
1	Allison	Art	1875	WS6	NA	**1.461**	1	**1.461**	0	3	24	0.214
2	Hollingshead	Holly	1875	WS6	NA	**0.719**	1	**0.719**	0	5	20	0.247
3	Stearns	Bill	1875	WS6	NA	**0.461**	1	**0.461**	0	7	20	0.256
4	Dailey	John	1875	WS6	NA	**0.415**	1	**0.415**	0	13	20	0.182
5	Ressler	Larry	1875	WS6	NA	**0.254**	1	**0.254**	0	5	21	0.194

Washington Nationals (American Association) **Top Career Batters**

Rank	Name	First	LYear	LTeam	Lg	PEVA-B	YRS	PER YR	HR	RBI	H	AVE
1	Fennelly	Frank	1884	WS7	AA	**10.940**	1	**10.940**	2	0	75	0.292
2	Humphries	John	1884	WS7	AA	**1.470**	1	**1.470**	0	0	34	0.176
3	Hawkes	Thorny	1884	WS7	AA	**1.209**	1	**1.209**	0	0	42	0.278
4	Olin	Frank	1884	WS7	AA	**1.179**	1	**1.179**	0	0	32	0.386
5	Gladman	Buck	1884	WS7	AA	**0.999**	1	**0.999**	1	0	35	0.156

Washington Nationals (National League) **Top Career Batters**

Rank	Name	First	LYear	LTeam	Lg	PEVA-B	YRS	PER YR	HR	RBI	H	AVE
1	Hoy	Dummy	1889	WS8	NL	**24.547**	2	**12.273**	2	68	277	0.274
2	Hines	Paul	1887	WS8	NL	**22.682**	2	**11.341**	19	128	299	0.310
3	Wilmot	Walt	1889	WS8	NL	**17.800**	2	**8.900**	13	100	231	0.255
4	O'Brien	Billy	1889	WS8	NL	**11.475**	3	**3.825**	28	139	245	0.248
5	Myers	Al	1889	WS8	NL	**8.266**	3	**2.755**	4	102	234	0.225

Washington Statesmen/Senators (WSN, WS9) **Top Career Batters**

Rank	Name	First	LYear	LTeam	Lg	PEVA-B	YRS	PER YR	HR	RBI	H	AVE
1	McGuire	Deacon	1899	WSN	NL	**66.121**	9	**7.347**	32	502	990	0.298
2	Selbach	Kip	1898	WSN	NL	**65.794**	5	**13.159**	26	345	736	0.310
3	Joyce	Bill	1896	WSN	NL	**51.923**	3	**17.308**	42	235	371	0.326
4	Abbey	Charlie	1897	WSN	NL	**25.580**	5	**5.116**	19	280	492	0.281
5	DeMontreville	Gene	1897	WSN	NL	**24.679**	3	**8.226**	11	179	386	0.337
6	Cartwright	Ed	1897	WSN	NL	**24.186**	4	**6.046**	16	273	472	0.295
7	Freeman	Buck	1899	WSN	NL	**23.261**	3	**7.754**	28	144	230	0.323

Rank	Name	First	LYear	LTeam	Lg	PEVA-B	YRS	PER YR	HR	RBI	H	AVE
8	Hoy	Dummy	1893	WSN	NL	**23.095**	2	**11.548**	3	120	304	0.263
9	Farrell	Duke	1899	WSN	NL	**19.711**	5	**3.942**	6	212	377	0.301
10	Radford	Paul	1894	WSN	NL	**16.461**	3	**5.487**	3	120	314	0.242
11	Brown	Tom	1898	WSN	NL	**16.340**	4	**4.085**	9	122	306	0.280
12	Anderson	John	1898	WSN	NL	**13.310**	1	**13.310**	9	71	131	0.305
13	Larkin	Henry	1893	WSN	NL	**12.679**	2	**6.340**	12	169	231	0.295
14	Hassamaer	Bill	1895	WSN	NL	**10.007**	2	**5.004**	5	150	259	0.304

Washington Nationals (Union League) — **Top Career Batters**

Rank	Name	First	LYear	LTeam	Lg	PEVA-B	YRS	PER YR	HR	RBI	H	AVE
1	Moore	Harry	1884	WSU	UA	**14.018**	1	**14.018**	1	0	155	0.336
2	Baker	Phil	1884	WSU	UA	**5.563**	1	**5.563**	1	0	107	0.288
3	Evers	Tom	1884	WSU	UA	**3.744**	1	**3.744**	0	0	99	0.232
4	Wise	Bill	1884	WSU	UA	**2.817**	1	**2.817**	2	0	79	0.233
5	Fulmer	Chris	1884	WSU	UA	**2.425**	1	**2.425**	0	0	50	0.276

Stat Geek Baseball Best Players (All Franchises Current and Past)

Top 20 Pitchers By Team/Franchise (1871-2009) - Regular Season

Altoona Mountain City

Rank	Name	First	LYear	LTeam	Lg	PEVA-P	YRS	PER YR	W	L	SV	IP
1	Murphy	John	1884	ALT	UA	**0.752**	1	**0.752**	5	6	0	111.7
2	Brown	Jim	1884	ALT	UA	**0.587**	1	**0.587**	1	9	0	74.0
3	Leary	Jack	1884	ALT	UA	**0.480**	1	**0.480**	0	3	0	24.0
4	Smith	Germany	1884	ALT	UA	**0.200**	1	**0.200**	0	0	0	1.0
5	Connors	Joe	1884	ALT	UA	**0.086**	1	**0.086**	0	1	0	9.0

Los Angeles Angels of Anaheim (ANA, CAL, LAA)

Rank	Name	First	LYear	LTeam	Lg	PEVA-P	YRS	PER YR	W	L	SV	IP
1	Finley	Chuck	1999	ANA	AL	**138.133**	14	**9.867**	165	140	0	2675.0
2	Ryan	Nolan	1979	CAL	AL	**104.056**	8	**13.007**	138	121	1	2181.3
3	Lackey	John	2009	LAA	AL	**93.923**	8	**11.740**	102	71	0	1501.0
4	Tanana	Frank	1980	CAL	AL	**89.677**	8	**11.210**	102	78	0	1615.3
5	Witt	Mike	1990	CAL	AL	**84.436**	10	**8.444**	109	107	6	1965.3
6	Langston	Mark	1997	ANA	AL	**78.413**	8	**9.802**	88	74	0	1445.3
7	Percival	Troy	2004	ANA	AL	**66.072**	10	**6.607**	29	38	316	586.7
8	Weaver	Jered	2010	LAA	AL	**64.182**	5	**12.836**	64	39	0	896.0
9	Chance	Dean	1966	CAL	AL	**62.961**	6	**10.493**	74	66	16	1236.7
10	Rodriguez	Francisco	2008	LAA	AL	**60.392**	7	**8.627**	23	17	208	451.7
11	Santana	Ervin	2010	LAA	AL	**57.275**	6	**9.546**	76	55	0	1069.0
12	Wright	Clyde	1973	CAL	AL	**48.718**	8	**6.090**	87	85	3	1403.3
13	Washburn	Jarrod	2005	LAA	AL	**47.648**	8	**5.956**	75	57	0	1153.3
14	Abbott	Jim	1996	CAL	AL	**46.499**	6	**7.750**	54	74	0	1073.7
15	McCaskill	Kirk	1991	CAL	AL	**44.068**	7	**6.295**	78	74	0	1221.0
16	Escobar	Kelvim	2009	LAA	AL	**44.054**	5	**8.811**	43	36	1	658.0
17	Messersmith	Andy	1972	CAL	AL	**42.182**	5	**8.436**	59	47	13	972.3
18	Shields	Scot	2010	LAA	AL	**38.686**	10	**3.869**	46	44	21	697.0
19	Harvey	Bryan	1992	CAL	AL	**34.750**	6	**5.792**	16	20	126	307.7
20	Saunders	Joe	2010	LAA	AL	**33.804**	6	**5.634**	54	32	0	692.0
21	Zahn	Geoff	1985	CAL	AL	**33.449**	5	**6.690**	52	42	0	830.0

Arizona Diamondbacks

Rank	Name	First	LYear	LTeam	Lg	PEVA-P	YRS	PER YR	W	L	SV	IP
1	Johnson	Randy	2008	ARI	NL	**223.198**	8	**27.900**	118	62	0	1630.3
2	Webb	Brandon	2009	ARI	NL	**142.694**	7	**20.385**	87	62	0	1319.7
3	Schilling	Curt	2003	ARI	NL	**94.007**	4	**23.502**	58	28	0	781.7
4	Haren	Danny	2010	ARI	NL	**64.598**	3	**21.533**	37	26	0	586.3
5	Batista	Miguel	2006	ARI	NL	**31.261**	4	**7.815**	40	34	0	723.7
6	Valverde	Jose	2007	ARI	NL	**29.829**	5	**5.966**	9	14	98	260.0
7	Kim	Byung-Hy	2007	ARI	NL	**28.739**	6	**4.790**	21	23	70	325.7
8	Anderson	Brian	2002	ARI	NL	**27.527**	5	**5.505**	41	42	1	840.7
9	Davis	Doug	2009	ARI	NL	**26.069**	3	**8.690**	28	34	0	542.0
10	Daal	Omar	2000	ARI	NL	**22.872**	3	**7.624**	26	31	0	473.3
11	Benes	Andy	1999	ARI	NL	**20.332**	2	**10.166**	27	25	0	429.7
12	Qualls	Chad	2010	ARI	NL	**17.351**	3	**5.784**	7	14	45	163.7
13	Mantei	Matt	2004	ARI	NL	**14.014**	6	**2.336**	8	11	74	173.7
14	Lyon	Brandon	2008	ARI	NL	**12.502**	4	**3.126**	11	15	42	232.0
15	Cruz	Juan	2008	ARI	NL	**12.096**	3	**4.032**	15	7	0	207.3
16	Koplove	Mike	2006	ARI	NL	**12.042**	6	**2.007**	15	7	2	248.7
17	Swindell	Greg	2002	ARI	NL	**11.394**	4	**2.848**	8	14	4	227.3
18	Pena	Tony	2009	ARI	NL	**11.368**	4	**2.842**	16	13	7	222.7
19	Reynoso	Armando	2002	ARI	NL	**11.311**	4	**2.828**	22	24	0	386.0
20	Hernandez	Livan	2007	ARI	NL	**10.904**	2	**5.452**	15	16	0	273.7

Atlanta Braves (ATL, BSN, ML1)

Rank	Name	First	LYear	LTeam	Lg	PEVA-P	YRS	PER YR	W	L	SV	IP
1	Maddux	Greg	2003	ATL	NL	**408.394**	11	**37.127**	194	88	0	2526.7
2	Spahn	Warren	1964	ML1	NL	**367.584**	20	**18.379**	356	229	29	5046.0
3	Nichols	Kid	1901	BSN	NL	**302.820**	12	**25.235**	329	183	16	4538.0
4	Smoltz	John	2008	ATL	NL	**300.465**	20	**15.023**	210	147	154	3395.0
5	Glavine	Tom	2008	ATL	NL	**269.702**	17	**15.865**	244	147	0	3408.0
6	Niekro	Phil	1987	ATL	NL	**242.904**	21	**11.567**	268	230	29	4622.7
7	Burdette	Lew	1963	ML1	NL	**126.672**	13	**9.744**	179	120	23	2638.0
8	Willis	Vic	1905	BSN	NL	**107.648**	8	**13.456**	151	147	5	2575.0
9	Bond	Tommy	1881	BSN	NL	**102.049**	5	**20.410**	149	87	0	2127.3
10	Sain	Johnny	1951	BSN	NL	**92.311**	7	**13.187**	104	91	11	1624.3
11	Clarkson	John	1892	BSN	NL	**78.099**	5	**15.620**	149	82	4	2092.7
12	Hudson	Tim	2010	ATL	NL	**76.163**	6	**12.694**	73	48	0	1047.7
13	Avery	Steve	1996	ATL	NL	**75.914**	7	**10.845**	72	62	0	1222.3
14	Rudolph	Dick	1927	BSN	NL	**71.596**	11	**6.509**	121	107	6	2035.0
15	Millwood	Kevin	2002	ATL	NL	**66.793**	6	**11.132**	75	46	0	1004.3
16	Whitney	Jim	1885	BSN	NL	**66.051**	5	**13.210**	133	121	2	2263.7
17	Buhl	Bob	1962	ML1	NL	**65.199**	10	**6.520**	109	72	5	1599.7
18	Mahler	Rick	1991	ATL	NL	**59.898**	11	**5.445**	79	89	2	1558.7
19	Brandt	Ed	1935	BSN	NL	**56.593**	8	**7.074**	94	119	13	1761.7
20	MacFayden	Danny	1943	BSN	NL	**53.591**	6	**8.932**	60	64	2	1097.0

Baltimore Orioles (BAL, MLA, SLA)

Rank	Name	First	LYear	LTeam	Lg	PEVA-P	YRS	PER YR	W	L	SV	IP
1	Palmer	Jim	1984	BAL	AL	**240.163**	19	**12.640**	268	152	4	3948.0
2	Mussina	Mike	2000	BAL	AL	**152.295**	10	**15.230**	147	81	0	2009.7
3	McNally	Dave	1974	BAL	AL	**110.675**	13	**8.513**	181	113	2	2652.7
4	Cuellar	Mike	1976	BAL	AL	**109.565**	8	**13.696**	143	88	1	2028.3
5	Shocker	Urban	1924	SLA	AL	**102.739**	7	**14.677**	126	80	20	1749.7

6	McGregor	Scott	1988	BAL	AL	**85.930**	13	**6.610**	138	108	5	2140.7
7	Flanagan	Mike	1992	BAL	AL	**85.686**	15	**5.712**	141	116	4	2317.7
8	Boddicker	Mike	1988	BAL	AL	**68.578**	9	**7.620**	79	73	0	1273.7
9	Pappas	Milt	1965	BAL	AL	**66.639**	9	**7.404**	110	74	4	1632.0
10	Martinez	Dennis	1986	BAL	AL	**60.275**	11	**5.480**	108	93	5	1775.0
11	Erickson	Scott	2002	BAL	AL	**56.698**	7	**8.100**	79	68	0	1287.7
12	Barber	Steve	1967	BAL	AL	**56.276**	8	**7.034**	95	75	4	1414.7
13	Wilhelm	Hoyt	1962	BAL	AL	**50.330**	5	**10.066**	43	39	40	616.3
14	Powell	Jack	1912	SLA	AL	**48.970**	10	**4.897**	117	143	11	2229.7
15	Hall	Dick	1971	BAL	AL	**47.719**	9	**5.302**	65	40	58	770.0
16	Gray	Dolly	1933	SLA	AL	**46.693**	6	**7.782**	67	82	14	1312.0
17	Howell	Harry	1910	SLA	AL	**45.638**	7	**6.520**	78	91	5	1580.7
18	McDonald	Ben	1995	BAL	AL	**44.919**	7	**6.417**	58	53	0	937.0
19	Blaeholder	George	1935	SLA	AL	**44.049**	10	**4.405**	90	111	12	1631.0
20	Stewart	Lefty	1932	SLA	AL	**43.384**	6	**7.231**	73	74	5	1236.7

Buffalo Bisons

Rank	Name	First	LYear	LTeam	Lg	PEVA-P	YRS	PER YR	W	L	SV	IP
1	Galvin	Pud	1885	BFN	NL	**102.507**	7	**14.644**	218	179	1	3547.7
2	Serad	Billy	1885	BFN	NL	**3.688**	2	**1.844**	23	41	0	549.3
3	McGunnigle	Bill	1880	BFN	NL	**2.218**	2	**1.109**	11	8	0	157.0
4	Daily	Hugh	1882	BFN	NL	**1.979**	1	**1.979**	15	14	0	255.7
5	Purcell	Blondie	1882	BFN	NL	**1.359**	2	**0.680**	6	2	0	92.7

Buffalo Bisons (Pacific Coast League)

Rank	Name	First	LYear	LTeam	Lg	PEVA-P	YRS	PER YR	W	L	SV	IP
1	Haddock	George	1890	BFP	PL	**1.851**	1	**1.851**	9	26	0	290.7
2	Cunningham	Bert	1890	BFP	PL	**1.524**	1	**1.524**	9	15	0	211.0
3	Keefe	George	1890	BFP	PL	**0.966**	1	**0.966**	6	16	0	196.0
4	Twitchell	Larry	1890	BFP	PL	**0.600**	1	**0.600**	5	7	0	104.3
5	Stafford	General	1890	BFP	PL	**0.578**	1	**0.578**	3	9	0	98.0

Baltimore Canaries

Rank	Name	First	LYear	LTeam	Lg	PEVA-P	YRS	PER YR	W	L	SV	IP
1	Cummings	Candy	1873	BL1	NA	**11.758**	1	**11.758**	28	14	0	382.0
2	Mathews	Bobby	1872	BL1	NA	**11.290**	1	**11.290**	25	18	0	405.3
3	Fisher	Cherokee	1872	BL1	NA	**5.672**	1	**5.672**	10	1	1	110.3
4	Brainard	Asa	1874	BL1	NA	**1.853**	2	**0.927**	10	29	0	347.7
5	Manning	Jack	1874	BL1	NA	**0.784**	1	**0.784**	4	16	0	179.7

Baltimore Orioles (American Association/National League) (BL2, BL3, BLN)

Rank	Name	First	LYear	LTeam	Lg	PEVA-P	YRS	PER YR	W	L	SV	IP
1	Kilroy	Matt	1889	BL2	AA	**65.201**	4	**16.300**	121	99	0	1974.0
2	McMahon	Sadie	1896	BLN	NL	**58.921**	7	**8.417**	130	91	3	1919.0
3	Hoffer	Bill	1898	BLN	NL	**42.819**	4	**10.705**	78	28	0	960.7
4	McGinnity	Joe	1899	BLN	NL	**34.184**	1	**34.184**	28	16	2	366.3
5	Kitson	Frank	1899	BLN	NL	**25.482**	2	**12.741**	30	21	0	447.0
6	Corbett	Joe	1897	BLN	NL	**25.055**	2	**12.528**	27	8	1	354.0
7	McJames	Doc	1898	BLN	NL	**23.828**	1	**23.828**	27	15	0	374.0
8	Nops	Jerry	1899	BLN	NL	**21.214**	4	**5.304**	55	27	0	736.7
9	Henderson	Hardie	1886	BL2	AA	**18.246**	4	**4.562**	65	105	0	1508.3
10	Pond	Arlie	1898	BLN	NL	**15.089**	4	**3.772**	35	19	2	496.0
11	Maul	Al	1898	BLN	NL	**13.778**	2	**6.889**	20	7	0	247.3

Rank	Name	First	LYear	LTeam	Lg	PEVA-P	YRS	PER YR	W	L	SV	IP
12	Foreman	Frank	1892	BLN	NL	**12.249**	3	**4.083**	25	25	0	466.0
13	Smith	Phenome	1888	BL2	AA	**12.058**	2	**6.029**	39	49	0	783.3
14	Emslie	Bob	1885	BL2	AA	**10.236**	3	**3.412**	44	40	0	763.7
15	Hemming	George	1896	BLN	NL	**10.126**	3	**3.375**	39	19	0	509.7

Baltimore Marylands

Rank	Name	First	LYear	LTeam	Lg	PEVA-P	YRS	PER YR	W	L	SV	IP
1	Stratton	Ed	1873	BL4	NA	**0.377**	1	**0.377**	0	3	0	27.0
2	French	Bill	1873	BL4	NA	**0.200**	1	**0.200**	0	1	0	9.0
3	Selman	Frank	1873	BL4	NA	**0.200**	1	**0.200**	0	1	0	9.0
4	McDoolan	NA	1873	BL4	NA	**0.200**	1	**0.200**	0	1	0	9.0

Baltimore Terrapins

Rank	Name	First	LYear	LTeam	Lg	PEVA-P	YRS	PER YR	W	L	SV	IP
1	Quinn	Jack	1915	BLF	FL	**21.400**	2	**10.700**	35	36	2	616.3
2	Suggs	George	1915	BLF	FL	**13.424**	2	**6.712**	35	31	7	552.0
3	Wilhelm	Kaiser	1915	BLF	FL	**3.470**	2	**1.735**	12	17	5	244.7
4	Bailey	Bill	1915	BLF	FL	**3.366**	2	**1.683**	13	28	0	319.0
5	Smith	Frank	1915	BLF	FL	**2.807**	2	**1.404**	14	12	2	263.3

Baltimore Monumentals

Rank	Name	First	LYear	LTeam	Lg	PEVA-P	YRS	PER YR	W	L	SV	IP
1	Sweeney	Bill	1884	BLU	UA	**12.261**	1	**12.261**	40	21	0	538.0
2	Robinson	Yank	1884	BLU	UA	**0.874**	1	**0.874**	3	3	0	75.0
3	Lee	Tom	1884	BLU	UA	**0.754**	1	**0.754**	5	8	0	122.0
4	Atkinson	Al	1884	BLU	UA	**0.720**	1	**0.720**	3	5	0	69.7
5	Ryan	John	1884	BLU	UA	**0.713**	1	**0.713**	3	2	0	51.0

Boston Red Sox

Rank	Name	First	LYear	LTeam	Lg	PEVA-P	YRS	PER YR	W	L	SV	IP
1	Clemens	Roger	1996	BOS	AL	**287.594**	13	**22.123**	192	111	0	2776.0
2	Martinez	Pedro	2004	BOS	AL	**205.515**	7	**29.359**	117	37	0	1383.7
3	Young	Cy	1908	BOS	AL	**191.940**	8	**23.992**	192	112	9	2728.3
4	Wakefield	Tim	2010	BOS	AL	**113.880**	16	**7.118**	179	160	22	2851.3
5	Grove	Lefty	1941	BOS	AL	**91.139**	8	**11.392**	105	62	4	1539.7
6	Tiant	Luis	1978	BOS	AL	**89.502**	8	**11.188**	122	81	3	1774.7
7	Parnell	Mel	1956	BOS	AL	**89.418**	10	**8.942**	123	75	10	1752.7
8	Wood	Joe	1915	BOS	AL	**89.336**	8	**11.167**	117	56	9	1418.0
9	Hughson	Tex	1949	BOS	AL	**88.733**	8	**11.092**	96	54	17	1375.7
10	Sullivan	Frank	1960	BOS	AL	**84.612**	8	**10.576**	90	80	6	1505.3
11	Lowe	Derek	2004	BOS	AL	**79.777**	8	**9.972**	70	55	85	1037.0
12	Kinder	Ellis	1955	BOS	AL	**76.624**	8	**9.578**	86	52	91	1142.3
13	Beckett	Josh	2010	BOS	AL	**72.256**	5	**14.451**	71	40	0	919.7
14	Papelbon	Jonathan	2010	BOS	AL	**65.202**	6	**10.867**	19	18	188	365.0
15	Eckersley	Dennis	1998	BOS	AL	**61.819**	8	**7.727**	88	71	1	1371.7
16	Dobson	Joe	1954	BOS	AL	**58.959**	9	**6.551**	106	72	9	1544.0
17	Lester	Jon	2010	BOS	AL	**58.299**	5	**11.660**	61	25	0	766.0
18	Schilling	Curt	2007	BOS	AL	**55.818**	4	**13.955**	53	29	9	675.0
19	Stanley	Bob	1989	BOS	AL	**54.531**	13	**4.195**	115	97	132	1707.0
20	Monbouquett	Bill	1965	BOS	AL	**54.268**	8	**6.784**	96	91	1	1622.0

Brooklyn Eckfords

Rank	Name	First	LYear	LTeam	Lg	PEVA-P	YRS	PER YR	W	L	SV	IP
1	Zettlein	George	1872	BR1	NA	**1.957**	1	**1.957**	1	8	0	75.3
2	Martin	Phonney	1872	BR1	NA	**0.477**	1	**0.477**	2	7	0	85.0
3	McDermott	Joe	1872	BR1	NA	**0.444**	1	**0.444**	0	7	0	63.0
4	Malone	Martin	1872	BR1	NA	**0.377**	1	**0.377**	0	3	0	27.0
5	O'Rourke	0.000	1872	BR1	NA	**0.200**	1	**0.200**	0	1	0	9.0

Brooklyn Atlantics

Rank	Name	First	LYear	LTeam	Lg	PEVA-P	YRS	PER YR	W	L	SV	IP
1	Britt	Jim	1873	BR2	NA	**10.101**	2	**5.050**	26	64	0	816.7
2	Bond	Tommy	1874	BR2	NA	**4.309**	1	**4.309**	22	32	0	497.0
3	Cassidy	John	1875	BR2	NA	**1.076**	1	**1.076**	1	20	0	214.7
4	Clinton	Jim	1875	BR2	NA	**1.039**	1	**1.039**	1	14	0	123.0
5	Ferguson	Bob	1874	BR2	NA	**0.400**	2	**0.200**	0	2	0	28.3

Brooklyn Gladiators

Rank	Name	First	LYear	LTeam	Lg	PEVA-P	YRS	PER YR	W	L	SV	IP
1	Daily	Ed	1890	BR4	AA	**3.555**	1	**3.555**	10	15	0	235.7
2	McCullough	Charlie	1890	BR4	AA	**1.142**	1	**1.142**	4	21	0	215.7
3	Mattimore	Mike	1890	BR4	AA	**0.955**	1	**0.955**	6	13	0	178.3
4	Toole	Steve	1890	BR4	AA	**0.831**	1	**0.831**	2	4	0	53.3
5	Murphy	Bob	1890	BR4	AA	**0.538**	1	**0.538**	3	9	0	96.0

Brooklyn Tip-Tops

Rank	Name	First	LYear	LTeam	Lg	PEVA-P	YRS	PER YR	W	L	SV	IP
1	Seaton	Tom	1915	BRF	FL	**12.763**	2	**6.381**	37	25	5	492.0
2	Lafitte	Ed	1915	BRF	FL	**8.538**	2	**4.269**	24	24	2	408.3
3	Finneran	Happy	1915	BRF	FL	**4.789**	2	**2.395**	22	23	3	390.7
4	Wiltse	Hooks	1915	BRF	FL	**3.540**	1	**3.540**	3	5	5	59.3
5	Marion	Dan	1915	BRF	FL	**3.292**	2	**1.646**	15	11	0	297.7

Brooklyn Ward's Wonders

Rank	Name	First	LYear	LTeam	Lg	PEVA-P	YRS	PER YR	W	L	SV	IP
1	Weyhing	Gus	1890	BRP	PL	**7.634**	1	**7.634**	30	16	0	390.0
2	Van Haltren	George	1890	BRP	PL	**1.564**	1	**1.564**	15	10	2	223.0
3	Sowders	John	1890	BRP	PL	**4.114**	1	**4.114**	19	16	0	309.0
4	Murphy	Con	1890	BRP	PL	**0.907**	1	**0.907**	4	10	2	139.0
5	Hemming	George	1890	BRP	PL	**0.761**	1	**0.761**	8	4	3	123.0

Boston Red Stockings

Rank	Name	First	LYear	LTeam	Lg	PEVA-P	YRS	PER YR	W	L	SV	IP
1	Spalding	Al	1875	BS1	NA	**144.902**	5	**28.980**	205	53	10	2351.0
2	Manning	Jack	1875	BS1	NA	**4.323**	1	**4.323**	15	2	7	139.7
3	Wright	Harry	1874	BS1	NA	**4.142**	4	**1.036**	4	4	8	99.3
4	McVey	Cal	1875	BS1	NA	**0.200**	1	**0.200**	1	0	1	11.0
5	Wright	George	1875	BS1	NA	**0.200**	1	**0.200**	0	1	0	4.0

Boston Reds (Pacific Coast League/American Association) (BS2, BSP)

Rank	Name	First	LYear	LTeam	Lg	PEVA-P	YRS	PER YR	W	L	SV	IP
1	Haddock	George	1891	BS2	AA	**17.269**	1	**17.269**	34	11	1	379.7
2	Buffinton	Charlie	1891	BS2	AA	**16.050**	1	**16.050**	29	9	1	363.7

3	Radbourn	Charley	1890	BSP	PL	**7.308**	1	**7.308**	27	12	0	343.0
4	Daley	Bill	1891	BS2	AA	**4.200**	2	**2.100**	26	13	4	361.7
5	Gumbert	Ad	1890	BSP	PL	**3.317**	1	**3.317**	23	12	0	277.3

Boston Reds (UA)

Rank	Name	First	LYear	LTeam	Lg	PEVA-P	YRS	PER YR	W	L	SV	IP
1	Shaw	Dupee	1884	BSU	UA	**7.504**	1	**7.504**	21	15	0	315.7
2	Burke	James	1884	BSU	UA	**3.920**	1	**3.920**	19	15	0	322.0
3	Tenney	Fred	1884	BSU	UA	**1.718**	1	**1.718**	3	1	0	35.0
4	Bond	Tommy	1884	BSU	UA	**1.492**	1	**1.492**	13	9	0	189.0
5	McCarthy	Tommy	1884	BSU	UA	**0.557**	1	**0.557**	0	7	0	56.0

Buffalo Buffeds & Blues

Rank	Name	First	LYear	LTeam	Lg	PEVA-P	YRS	PER YR	W	L	SV	IP
1	Ford	Russ	1915	BUF	FL	**14.482**	2	**7.241**	26	15	6	374.7
2	Anderson	Fred	1915	BUF	FL	**10.664**	2	**5.332**	32	28	0	500.3
3	Schulz	Al	1915	BUF	FL	**9.554**	2	**4.777**	30	26	2	480.7
4	Krapp	Gene	1915	BUF	FL	**8.421**	2	**4.210**	25	33	0	483.7
5	Bedient	Hugh	1915	BUF	FL	**8.280**	1	**8.280**	16	18	10	269.3

Chicago White Stockings (National Association) (CH1, CH2)

Rank	Name	First	LYear	LTeam	Lg	PEVA-P	YRS	PER YR	W	L	SV	IP
1	Zettlein	George	1875	CH2	NA	**45.460**	3	**15.153**	62	53	0	1038.3
2	Devlin	Jim	1875	CH2	NA	**1.544**	1	**1.544**	7	16	0	224.0
3	Golden	Mike	1875	CH2	NA	**0.884**	1	**0.884**	6	7	0	119.0
4	Force	Davy	1874	CH2	NA	**0.200**	1	**0.200**	0	0	0	7.0
5	Pinkham	Ed	1871	CH1	NA	**0.200**	1	**0.200**	1	0	1	10.3

Chicago White Sox

Rank	Name	First	LYear	LTeam	Lg	PEVA-P	YRS	PER YR	W	L	SV	IP
1	Walsh	Ed	1916	CHA	AL	**213.566**	13	**16.428**	195	125	34	2946.3
2	Lyons	Ted	1946	CHA	AL	**178.627**	21	**8.506**	260	230	23	4161.0
3	Faber	Red	1933	CHA	AL	**177.153**	20	**8.858**	254	213	28	4086.7
4	Pierce	Billy	1961	CHA	AL	**163.992**	13	**12.615**	186	152	19	2931.0
5	Buehrle	Mark	2010	CHA	AL	**148.328**	11	**13.484**	148	110	0	2271.3
6	Wood	Wilbur	1978	CHA	AL	**140.663**	12	**11.722**	163	148	57	2524.3
7	Cicotte	Eddie	1920	CHA	AL	**122.749**	9	**13.639**	156	102	21	2322.3
8	McDowell	Jack	1994	CHA	AL	**85.815**	7	**12.259**	91	58	0	1343.7
9	Horlen	Joe	1971	CHA	AL	**79.961**	11	**7.269**	113	113	3	1918.0
10	Lee	Thornton	1947	CHA	AL	**78.583**	11	**7.144**	104	104	6	1888.0
11	White	Doc	1913	CHA	AL	**76.793**	11	**6.981**	159	123	4	2498.3
12	Peters	Gary	1969	CHA	AL	**73.793**	11	**6.708**	91	78	3	1560.0
13	Thomas	Tommy	1932	CHA	AL	**72.167**	7	**10.310**	83	92	8	1557.3
14	Fernandez	Alex	1996	CHA	AL	**72.104**	7	**10.301**	79	63	0	1346.3
15	Garland	Jon	2007	CHA	AL	**65.908**	8	**8.238**	92	81	1	1428.7
16	Wilhelm	Hoyt	1968	CHA	AL	**62.366**	6	**10.394**	41	33	98	675.7
17	Smith	Frank	1910	CHA	AL	**60.157**	7	**8.594**	108	80	3	1717.3
18	Dotson	Richard	1989	CHA	AL	**59.722**	10	**5.972**	97	95	0	1606.0
19	Scott	Jim	1917	CHA	AL	**58.515**	9	**6.502**	107	113	9	1892.0
20	Donovan	Dick	1960	CHA	AL	**57.898**	6	**9.650**	73	50	3	1148.7

Chicago Chi-Feds

Rank	Name	First	LYear	LTeam	Lg	PEVA-P	YRS	PER YR	W	L	SV	IP
1	Hendrix	Claude	1915	CHF	FL	**32.400**	2	**16.200**	45	25	9	647.0
2	McConnell	George	1915	CHF	FL	**11.452**	1	**11.452**	25	10	1	303.0
3	Brown	Mordecai	1915	CHF	FL	**7.083**	1	**7.083**	17	8	4	236.3
4	Prendergast	Mike	1915	CHF	FL	**6.604**	2	**3.302**	19	21	0	389.7
5	Watson	Doc	1914	CHF	FL	**3.592**	1	**3.592**	9	8	1	172.0

Chicago Cubs

Rank	Name	First	LYear	LTeam	Lg	PEVA-P	YRS	PER YR	W	L	SV	IP
1	Jenkins	Fergie	1983	CHN	NL	**163.228**	10	**16.323**	167	132	6	2673.7
2	Maddux	Greg	2006	CHN	NL	**159.849**	10	**15.985**	133	112	0	2016.0
3	Brown	Mordecai	1916	CHN	NL	**151.146**	10	**15.115**	188	86	39	2329.0
4	Root	Charley	1941	CHN	NL	**143.727**	16	**8.983**	201	156	40	3137.3
5	Alexander	Pete	1926	CHN	NL	**130.911**	9	**14.546**	128	83	10	1884.3
6	Vaughn	Hippo	1921	CHN	NL	**118.611**	9	**13.179**	151	105	4	2216.3
7	Lee	Bill	1947	CHN	NL	**118.275**	11	**10.752**	139	123	9	2271.3
8	Passeau	Claude	1947	CHN	NL	**118.148**	9	**13.128**	124	94	15	1914.7
9	Zambrano	Carlos	2010	CHN	NL	**114.288**	10	**11.429**	116	74	0	1681.0
10	Hutchison	Bill	1895	CHN	NL	**110.650**	7	**15.807**	181	158	4	3021.0
11	Griffith	Clark	1900	CHN	NL	**107.861**	8	**13.483**	152	96	1	2188.7
12	Warneke	Lon	1945	CHN	NL	**106.352**	10	**10.635**	109	72	11	1624.7
13	Reuschel	Rick	1984	CHN	NL	**104.862**	12	**8.738**	135	127	3	2290.0
14	Clarkson	John	1887	CHN	NL	**100.405**	4	**25.101**	137	57	0	1730.7
15	Rush	Bob	1957	CHN	NL	**80.867**	10	**8.087**	110	140	7	2132.7
16	Bush	Guy	1934	CHN	NL	**75.071**	12	**6.256**	152	101	27	2201.7
17	Hands	Bill	1972	CHN	NL	**71.330**	7	**10.190**	92	86	9	1564.0
18	Malone	Pat	1934	CHN	NL	**66.936**	7	**9.562**	115	79	8	1632.0
19	Taylor	Jack	1907	CHN	NL	**66.889**	8	**8.361**	109	90	3	1801.0
20	Wood	Kerry	2008	CHN	NL	**65.632**	10	**6.563**	77	61	34	1219.3

Chicago Pirates

Rank	Name	First	LYear	LTeam	Lg	PEVA-P	YRS	PER YR	W	L	SV	IP
1	King	Silver	1890	CHP	PL	**18.015**	1	**18.015**	30	22	0	461.0
2	Baldwin	Mark	1890	CHP	PL	**13.556**	1	**13.556**	34	24	0	501.0
3	Bartson	Charlie	1890	CHP	PL	**1.003**	1	**1.003**	8	10	1	188.0
4	Dwyer	Frank	1890	CHP	PL	**0.543**	1	**0.543**	3	6	1	69.3

Chicago/Pittsburgh (Union League)

Rank	Name	First	LYear	LTeam	Lg	PEVA-P	YRS	PER YR	W	L	SV	IP
1	Daily	Hugh	1884	CHU	UA	**10.504**	1	**10.504**	27	27	0	484.7
2	Atkinson	Al	1884	CHU	UA	**1.447**	1	**1.447**	6	10	0	140.0
3	Horan	John	1884	CHU	UA	**0.984**	1	**0.984**	3	6	0	98.0
4	Cady	Charlie	1884	CHU	UA	**0.933**	1	**0.933**	3	1	0	35.0
5	Foreman	Frank	1884	CHU	UA	**0.514**	1	**0.514**	1	2	0	18.0

Cincinnati Reds (CN1, CN2, CIN)

Rank	Name	First	LYear	LTeam	Lg	PEVA-P	YRS	PER YR	W	L	SV	IP
1	Derringer	Paul	1933	CIN	NL	**175.613**	10	**17.561**	161	150	17	2615.3
2	Walters	Bucky	1938	CIN	NL	**160.759**	11	**14.614**	160	107	4	2355.7
3	Rixey	Eppa	1924	CIN	NL	**157.828**	13	**12.141**	179	148	8	2890.7
4	Luque	Dolf	1929	CIN	NL	**145.848**	12	**12.154**	154	152	10	2668.7
5	Rijo	Jose	2001	CIN	NL	**129.993**	10	**12.999**	97	61	0	1478.0
6	Hahn	Noodles	1904	CIN	NL	**110.989**	7	**15.856**	127	92	0	1987.3

Rank	Name	First	LYear	LTeam	Lg	PEVA-P	YRS	PER YR	W	L	SV	IP
7	Donohue	Pete	1923	CIN	NL	**110.489**	10	**11.049**	127	110	11	1996.3
8	Soto	Mario	1984	CIN	NL	**109.458**	12	**9.121**	100	92	4	1730.3
9	White	Will	1886	CN2	AA	**108.243**	8	**13.530**	227	163	0	3497.7
10	Blackwell	Ewell	1949	CIN	NL	**98.446**	8	**12.306**	79	77	8	1281.3
11	Browning	Tom	1984	CIN	NL	**89.064**	11	**8.097**	123	88	0	1911.0
12	Vander Meer	Johnny	1948	CIN	NL	**88.777**	11	**8.071**	116	116	1	2028.0
13	Purkey	Bob	1964	CIN	NL	**79.132**	7	**11.305**	103	76	3	1588.0
14	Lucas	Red	1928	CIN	NL	**77.958**	8	**9.745**	109	99	6	1768.7
15	Arroyo	Bronson	2010	CIN	NL	**75.870**	5	**15.174**	70	60	0	1087.3
16	Nolan	Gary	1972	CIN	NL	**74.062**	10	**7.406**	110	67	0	1656.3
17	Seaver	Tom	1977	CIN	NL	**73.586**	6	**12.264**	75	46	0	1085.7
18	Raffensberge	Ken	1947	CIN	NL	**73.018**	8	**9.127**	89	99	7	1490.0
19	Maloney	Jim	1963	CIN	NL	**72.308**	11	**6.573**	134	81	4	1818.7
20	Harang	Aaron	2010	CIN	NL	**71.915**	8	**8.989**	75	80	0	1343.0

Cleveland Forest Citys

Rank	Name	First	LYear	LTeam	Lg	PEVA-P	YRS	PER YR	W	L	SV	IP
1	Pratt	Al	1872	CL1	NA	**7.439**	2	**3.720**	12	26	0	329.3
2	Pabor	Charlie	1872	CL1	NA	**0.577**	2	**0.289**	1	3	0	47.3
3	Wolters	Rynie	1872	CL1	NA	**0.486**	1	**0.486**	3	6	0	76.3

Cleveland Blues

Rank	Name	First	LYear	LTeam	Lg	PEVA-P	YRS	PER YR	W	L	SV	IP
1	McCormick	Jim	1884	CL2	NL	**85.338**	6	**14.223**	174	162	1	3026.7
2	Daily	Hugh	1883	CL2	NL	**6.392**	1	**6.392**	23	19	1	378.7
3	Harkins	John	1884	CL2	NL	**2.934**	1	**2.934**	12	32	0	391.0
4	Hankinson	Frank	1880	CL2	NL	**2.879**	1	**2.879**	1	1	1	25.0
5	Sawyer	Will	1883	CL2	NL	**2.182**	1	**2.182**	4	10	0	141.0

Cleveland Blues & Spiders (CL3, CL4)

Rank	Name	First	LYear	LTeam	Lg	PEVA-P	YRS	PER YR	W	L	SV	IP
1	Young	Cy	1898	CL4	NL	**234.004**	9	**26.000**	241	135	7	3351.0
2	Cuppy	Nig	1898	CL4	NL	**68.450**	7	**9.779**	139	80	4	1914.0
3	Powell	Jack	1898	CL4	NL	**22.120**	2	**11.060**	38	25	0	567.0
4	Clarkson	John	1894	CL4	NL	**15.430**	3	**5.143**	41	37	1	689.0
5	Wilson	Zeke	1898	CL4	NL	**14.760**	4	**3.690**	49	39	1	803.0
6	Bakely	Jersey	1889	CL4	NL	**14.743**	2	**7.371**	37	55	0	837.0
7	Beatin	Ed	1891	CL4	NL	**11.523**	3	**3.841**	42	48	0	821.0

Columbus Buckeyes

Rank	Name	First	LYear	LTeam	Lg	PEVA-P	YRS	PER YR	W	L	SV	IP
1	Morris	Ed	1884	CL5	AA	**15.285**	1	**15.285**	34	13	0	429.7
2	Mountain	Frank	1884	CL5	AA	**11.148**	2	**5.574**	49	50	1	863.7
3	Dundon	Ed	1884	CL5	AA	**1.446**	2	**0.723**	9	20	0	247.7
4	Valentine	John	1883	CL5	AA	**0.859**	1	**0.859**	2	10	0	102.0
5	Sullivan	Tom	1884	CL5	AA	**0.481**	1	**0.481**	2	2	0	31.0

Columbus Salons

Rank	Name	First	LYear	LTeam	Lg	PEVA-P	YRS	PER YR	W	L	SV	IP
1	Gastright	Hank	1891	CL6	AA	**16.582**	3	**5.527**	52	49	0	907.7
2	Knell	Phil	1891	CL6	AA	**14.684**	1	**14.684**	28	27	0	462.0
3	Baldwin	Mark	1889	CL6	AA	**10.311**	1	**10.311**	27	34	1	513.7

4	Knauss	Frank	1890	CL6	AA	**8.110**	1	**8.110**	17	12	2	275.7
5	Easton	Jack	1891	CL6	AA	**4.289**	4	**1.072**	21	26	2	424.0

Cleveland Indians

Rank	Name	First	LYear	LTeam	Lg	PEVA-P	YRS	PER YR	W	L	SV	IP
1	Feller	Bob	1956	CLE	AL	**246.330**	18	**13.685**	266	162	21	3827.0
2	Lemon	Bob	1958	CLE	AL	**165.840**	13	**12.757**	207	128	22	2850.0
3	Coveleski	Stan	1924	CLE	AL	**140.733**	9	**15.637**	172	123	20	2502.3
4	Wynn	Early	1963	CLE	AL	**136.070**	10	**13.607**	164	102	10	2286.7
5	Garcia	Mike	1959	CLE	AL	**123.307**	12	**10.276**	142	96	21	2138.0
6	Joss	Addie	1910	CLE	AL	**121.395**	9	**13.488**	160	97	5	2327.0
7	Sabathia	C.C.	2008	CLE	AL	**120.902**	8	**15.113**	106	71	0	1528.7
8	Harder	Mel	1947	CLE	AL	**116.071**	20	**5.804**	223	186	23	3426.3
9	McDowell	Sam	1971	CLE	AL	**104.898**	11	**9.536**	122	109	11	2109.7
10	Lee	Cliff	2009	CLE	AL	**94.650**	8	**11.831**	83	48	0	1117.0
11	Nagy	Charles	2002	CLE	AL	**92.023**	13	**7.079**	129	103	0	1942.3
12	Perry	Gaylord	1975	CLE	AL	**84.762**	4	**21.190**	70	57	1	1130.7
13	Uhle	George	1936	CLE	AL	**83.512**	11	**7.592**	147	119	15	2200.3
14	Hudlin	Willis	1940	CLE	AL	**81.421**	15	**5.428**	157	151	31	2557.7
15	Bagby	Jim	1922	CLE	AL	**72.142**	7	**10.306**	122	85	26	1735.7
16	Ferrell	Wes	1933	CLE	AL	**71.005**	7	**10.144**	102	62	12	1321.3
17	Colon	Bartolo	2002	CLE	AL	**58.031**	6	**9.672**	75	45	0	1029.7
18	Candiotti	Tom	1999	CLE	AL	**55.224**	7	**7.889**	73	66	0	1201.7
19	Blyleven	Bert	1985	CLE	AL	**54.199**	5	**10.840**	48	37	0	760.7
20	Westbrook	Jake	2010	CLE	AL	**49.376**	9	**5.486**	69	69	0	1191.3

Cleveland Infants

Rank	Name	First	LYear	LTeam	Lg	PEVA-P	YRS	PER YR	W	L	SV	IP
1	Gruber	Henry	1890	CLP	PL	**4.143**	1	**4.143**	22	23	0	383.3
2	Bakely	Jersey	1890	CLP	PL	**2.445**	1	**2.445**	12	25	0	326.3
3	O'Brien	Darby	1890	CLP	PL	**1.719**	1	**1.719**	8	16	0	206.3
4	McGill	Willie	1890	CLP	PL	**1.069**	1	**1.069**	11	9	0	183.7
5	Gleason	Bill	1890	CLP	PL	**0.200**	1	**0.200**	0	1	0	4.0

Cincinnati Kelly's Killers

Rank	Name	First	LYear	LTeam	Lg	PEVA-P	YRS	PER YR	W	L	SV	IP
1	Crane	Ed	1891	CN3	AA	**5.311**	1	**5.311**	14	14	0	250.0
2	Dwyer	Frank	1891	CN3	AA	**3.102**	1	**3.102**	13	19	0	289.0
3	Mains	Willard	1891	CN3	AA	**2.998**	1	**2.998**	12	12	0	204.0
4	McGill	Willie	1891	CN3	AA	**1.142**	1	**1.142**	2	5	0	65.0
5	Kilroy	Matt	1891	CN3	AA	**0.973**	1	**0.973**	1	4	0	45.3

Cincinnati Outlaw Reds

Rank	Name	First	LYear	LTeam	Lg	PEVA-P	YRS	PER YR	W	L	SV	IP
1	Bradley	George	1884	CNU	UA	**6.254**	1	**6.254**	25	15	0	342.0
2	Burns	Dick	1884	CNU	UA	**5.938**	1	**5.938**	23	15	0	329.7
3	McCormick	Jim	1884	CNU	UA	**5.734**	1	**5.734**	21	3	0	210.0
4	Sylvester	Lou	1884	CNU	UA	**0.904**	1	**0.904**	0	1	1	32.7

Colorado Rockies

Rank	Name	First	LYear	LTeam	Lg	PEVA-P	YRS	PER YR	W	L	SV	IP
1	Jimenez	Ubaldo	2010	COL	NL	**65.562**	5	**13.112**	50	36	0	728.0

Rank	Name	First	LYear	LTeam	Lg	PEVA-P	YRS	PER YR	W	L	SV	IP
2	Cook	Aaron	2010	COL	NL	**45.294**	9	**5.033**	69	58	0	1215.3
3	Fuentes	Brian	2008	COL	NL	**41.071**	7	**5.867**	16	26	115	410.3
4	Francis	Jeff	2008	COL	NL	**39.044**	5	**7.809**	51	44	0	778.3
5	Jennings	Jason	2006	COL	NL	**37.351**	6	**6.225**	58	56	0	941.0
7	Astacio	Pedro	2001	COL	NL	**34.466**	5	**6.893**	53	48	0	827.3
8	Reed	Steve	2004	COL	NL	**27.747**	7	**3.964**	33	29	15	499.0
9	Ritz	Kevin	1998	COL	NL	**23.826**	5	**4.765**	39	38	2	576.3
10	Jimenez	Jose	2003	COL	NL	**20.928**	4	**5.232**	15	23	102	300.7
11	Ruffin	Bruce	1997	COL	NL	**18.840**	5	**3.768**	17	18	60	321.0
12	Holmes	Darren	1997	COL	NL	**17.893**	5	**3.579**	23	13	46	328.0
13	Wright	Jamey	2005	COL	NL	**17.827**	6	**2.971**	35	52	0	791.7
14	Reynoso	Armando	1996	COL	NL	**17.434**	4	**4.358**	30	31	0	503.0
15	de la Rosa	Jorge	2010	COL	NL	**17.133**	3	**5.711**	34	24	0	436.7
16	Corpas	Manuel	2010	COL	NL	**17.030**	5	**3.406**	12	16	34	286.0
17	Thomson	John	2002	COL	NL	**15.909**	5	**3.182**	27	43	0	611.0
18	Leskanic	Curt	1999	COL	NL	**15.534**	7	**2.219**	31	20	20	470.0
19	Street	Huston	2010	COL	NL	**15.457**	2	**7.729**	8	5	55	109.0
20	Bohanon	Brian	2001	COL	NL	**14.973**	3	**4.991**	29	30	0	471.3

Detroit Tigers

Rank	Name	First	LYear	LTeam	Lg	PEVA-P	YRS	PER YR	W	L	SV	IP
1	Newhouser	Hal	1953	DET	AL	**214.336**	15	**14.289**	200	148	19	2944.0
2	Morris	Jack	1990	DET	AL	**155.657**	14	**11.118**	198	150	0	3042.7
3	Lolich	Mickey	1975	DET	AL	**129.756**	13	**9.981**	207	175	10	3361.7
4	Trout	Dizzy	1952	DET	AL	**126.383**	14	**9.027**	161	153	34	2591.7
5	Bunning	Jim	1963	DET	AL	**124.530**	9	**13.837**	118	87	12	1867.3
6	Bridges	Tommy	1946	DET	AL	**117.328**	16	**7.333**	194	138	10	2826.3
7	Lary	Frank	1964	DET	AL	**116.661**	11	**10.606**	123	110	7	2008.7
8	Mullin	George	1913	DET	AL	**94.732**	12	**7.894**	209	179	6	3394.0
9	McLain	Denny	1970	DET	AL	**91.872**	8	**11.484**	117	62	1	1593.0
10	Dauss	Hooks	1926	DET	AL	**89.926**	15	**5.995**	222	182	40	3390.7
11	Verlander	Justin	2010	DET	AL	**89.645**	6	**14.941**	83	52	0	1064.3
12	Petry	Dan	1991	DET	AL	**73.495**	11	**6.681**	119	93	0	1843.0
13	Trucks	Virgil	1956	DET	AL	**69.230**	12	**5.769**	114	96	13	1800.7
14	Whitehill	Earl	1932	DET	AL	**67.701**	10	**6.770**	133	119	7	2171.3
15	Hiller	John	1980	DET	AL	**63.837**	15	**4.256**	87	76	125	1242.0
16	Hutchinson	Fred	1953	DET	AL	**59.270**	10	**5.927**	95	71	7	1464.0
17	Rowe	Schoolbo	1942	DET	AL	**54.275**	10	**5.427**	105	62	8	1445.0
18	Tanana	Frank	1992	DET	AL	**52.880**	8	**6.610**	96	82	1	1551.3
19	Foytack	Paul	1963	DET	AL	**49.068**	10	**4.907**	81	81	7	1425.3
20	Coleman	Joe	1976	DET	AL	**48.914**	6	**8.152**	88	73	0	1407.7

Detroit Wolverines

Rank	Name	First	LYear	LTeam	Lg	PEVA-P	YRS	PER YR	W	L	SV	IP
1	Baldwin	Lady	1888	DTN	NL	**39.645**	4	**9.911**	69	35	1	930.3
2	Getzein	Charlie	1888	DTN	NL	**25.902**	5	**5.180**	95	86	0	1634.7
3	Derby	George	1882	DTN	NL	**25.471**	2	**12.736**	46	46	0	856.7
4	Wiedman	Stump	1887	DTN	NL	**19.243**	6	**3.207**	84	101	2	1654.0
5	Conway	Pete	1888	DTN	NL	**17.549**	3	**5.850**	44	28	0	628.0

Elizabeth Resolutes

Rank	Name	First	LYear	LTeam	Lg	PEVA-P	YRS	PER YR	W	L	SV	IP
1	Campbell	Hugh	1873	ELI	NA	**1.710**	1	**1.710**	2	16	0	165.0
2	Fleet	Frank	1873	ELI	NA	**0.200**	1	**0.200**	0	3	0	24.0

Rank	Name	First	LYear	LTeam	Lg	PEVA-P	YRS	PER YR	W	L	SV	IP
3	Lovett	Len	1873	ELI	NA	**0.200**	1	**0.200**	0	1	0	9.0
4	Wolters	Rynie	1873	ELI	NA	**0.200**	1	**0.200**	0	1	0	9.0

Florida Marlins

Rank	Name	First	LYear	LTeam	Lg	PEVA-P	YRS	PER YR	W	L	SV	IP
1	Willis	Dontrelle	2007	FLO	NL	**70.888**	5	**14.178**	68	54	0	1022.7
2	Brown	Kevin	1997	FLO	NL	**62.698**	2	**31.349**	33	19	0	470.3
3	Johnson	Josh	2010	FLO	NL	**56.672**	6	**9.445**	45	22	0	665.0
4	Burnett	A.J.	2005	FLO	NL	**36.538**	7	**5.220**	49	50	0	853.7
5	Nolasco	Ricky	2010	FLO	NL	**34.894**	5	**6.979**	54	39	0	716.3
6	Penny	Brad	2004	FLO	NL	**32.724**	5	**6.545**	48	42	0	781.7
7	Nen	Robb	1997	FLO	NL	**32.683**	5	**6.537**	20	16	108	314.0
8	Dempster	Ryan	2002	FLO	NL	**30.382**	5	**6.076**	42	43	0	759.7
9	Pavano	Carl	2004	FLO	NL	**28.334**	3	**9.445**	33	23	0	485.0
10	Beckett	Josh	2005	FLO	NL	**26.235**	5	**5.247**	41	34	0	609.0
11	Rapp	Pat	1997	FLO	NL	**25.886**	5	**5.177**	37	43	0	665.7
12	Olsen	Scott	2008	FLO	NL	**24.917**	4	**6.229**	31	37	0	579.3
13	Sanchez	Anibal	2010	FLO	NL	**22.337**	5	**4.467**	31	29	0	477.0
14	Leiter	Al	2005	FLO	NL	**21.615**	3	**7.205**	30	28	0	446.7
15	Fernandez	Alex	2000	FLO	NL	**21.268**	3	**7.089**	28	24	0	414.0
16	Alfonseca	Antonio	2005	FLO	NL	**19.039**	6	**3.173**	19	25	102	333.0
17	Burkett	John	1996	FLO	NL	**17.392**	2	**8.696**	20	24	0	342.3
18	Harvey	Bryan	1995	FLO	NL	**16.533**	3	**5.511**	1	5	51	79.3
19	Hammond	Chris	1998	FLO	NL	**16.353**	5	**3.271**	29	32	0	520.0
20	Looper	Braden	2003	FLO	NL	**16.352**	5	**3.270**	19	16	46	388.0

Fort Wayne Kekiongas

Rank	Name	First	LYear	LTeam	Lg	PEVA-P	YRS	PER YR	W	L	SV	IP
1	Mathews	Bobby	1871	FW1	NA	**1.663**	1	**1.663**	6	11	0	169.0

Hartford Dark Blues (National League)

Rank	Name	First	LYear	LTeam	Lg	PEVA-P	YRS	PER YR	W	L	SV	IP
1	Bond	Tommy	1876	HAR	NL	**19.434**	1	**19.434**	31	13	0	408.0
2	Larkin	Terry	1877	HAR	NL	**18.656**	1	**18.656**	29	25	0	501.0
3	Cummings	Candy	1876	HAR	NL	**4.764**	1	**4.764**	16	8	0	216.0
4	Ferguson	Bob	1877	HAR	NL	**0.629**	1	**0.629**	1	1	0	25.0
5	Cassidy	John	1877	HAR	NL	**0.200**	1	**0.200**	1	1	0	18.0

Houston Astros

Rank	Name	First	LYear	LTeam	Lg	PEVA-P	YRS	PER YR	W	L	SV	IP
1	Oswalt	Roy	2010	HOU	NL	**175.479**	10	**17.548**	143	82	0	1932.3
2	Scott	Mike	1991	HOU	NL	**131.199**	9	**14.578**	110	81	0	1704.0
3	Niekro	Joe	1985	HOU	NL	**122.547**	11	**11.141**	144	116	9	2270.0
4	Ryan	Nolan	1988	HOU	NL	**110.250**	9	**12.250**	106	94	0	1854.7
5	Richard	J.R.	1980	HOU	NL	**99.479**	10	**9.948**	107	71	0	1606.0
6	Reynolds	Shane	2002	HOU	NL	**91.426**	11	**8.311**	103	86	0	1622.3
7	Dierker	Larry	1976	HOU	NL	**81.453**	13	**6.266**	137	117	1	2294.3
8	Knepper	Bob	1989	HOU	NL	**80.903**	9	**8.989**	93	100	1	1738.0
9	Wagner	Billy	2003	HOU	NL	**74.263**	9	**8.251**	26	29	225	504.3
10	Smith	Dave	1990	HOU	NL	**68.474**	11	**6.225**	53	47	199	762.0
11	Hampton	Mike	2009	HOU	NL	**63.325**	7	**9.046**	76	50	0	1138.0
12	Clemens	Roger	2006	HOU	NL	**61.294**	3	**20.431**	38	18	0	539.0
13	Kile	Darryl	1997	HOU	NL	**55.820**	7	**7.974**	71	65	0	1200.0

Rank	Name	First	LYear	LTeam	Lg	PEVA-P	YRS	PER YR	W	L	SV	IP
14	Wilson	Don	1974	HOU	NL	**53.550**	9	**5.950**	104	92	2	1748.3
15	Forsch	Ken	1980	HOU	NL	**49.929**	11	**4.539**	78	81	50	1493.7
16	Pettitte	Andy	2006	HOU	NL	**46.901**	3	**15.634**	37	26	0	519.7
17	Rodriguez	Wandy	2010	HOU	NL	**45.220**	6	**7.537**	62	64	0	985.0
18	Lima	Jose	2001	HOU	NL	**45.043**	5	**9.009**	46	42	2	804.0
19	Harnisch	Pete	1994	HOU	NL	**44.498**	4	**11.125**	45	33	0	736.0
20	Sambito	Joe	1984	HOU	NL	**42.534**	8	**5.317**	33	32	72	536.0

Hartford Dark Blues (National Association)

Rank	Name	First	LYear	LTeam	Lg	PEVA-P	YRS	PER YR	W	L	SV	IP
1	Cummings	Candy	1875	HR1	NA	**19.393**	1	**19.393**	35	12	0	417.0
2	Bond	Tommy	1875	HR1	NA	**13.404**	1	**13.404**	19	16	0	352.0
3	Fisher	Cherokee	1874	HR1	NA	**2.732**	1	**2.732**	14	23	0	317.0
4	Stearns	Bill	1874	HR1	NA	**0.877**	1	**0.877**	2	14	1	164.0
5	Ferguson	Bob	1875	HR1	NA	**0.200**	1	**0.200**	0	0	0	2.0

Indianapolis Blues

Rank	Name	First	LYear	LTeam	Lg	PEVA-P	YRS	PER YR	W	L	SV	IP
1	Nolan	The Only	1878	IN1	NL	**3.014**	1	**3.014**	13	22	0	347.0
2	McCormick	Jim	1878	IN1	NL	**2.216**	1	**2.216**	5	8	0	117.0
3	Healey	Tom	1878	IN1	NL	**0.470**	1	**0.470**	6	4	1	89.0
4	McKelvy	Russ	1878	IN1	NL	**0.456**	1	**0.456**	0	2	0	25.0

Indianapolis Hoosiers (American Association)

Rank	Name	First	LYear	LTeam	Lg	PEVA-P	YRS	PER YR	W	L	SV	IP
1	McKeon	Larry	1884	IN2	AA	**5.217**	1	**5.217**	18	41	0	512.0
2	Barr	Bob	1884	IN2	AA	**1.032**	1	**1.032**	3	11	0	132.0
3	Aydelott	Jake	1884	IN2	AA	**1.005**	1	**1.005**	5	7	0	106.0
4	McCauley	Al	1884	IN2	AA	**0.613**	1	**0.613**	2	7	0	76.0
5	MacArthur	Mac	1884	IN2	AA	**0.599**	1	**0.599**	1	5	0	52.0

Indianapolis Hoosiers (National League)

Rank	Name	First	LYear	LTeam	Lg	PEVA-P	YRS	PER YR	W	L	SV	IP
1	Boyle	Henry	1889	IN3	NL	**12.800**	3	**4.267**	49	69	0	1029.7
2	Healy	John	1888	IN3	NL	**4.937**	2	**2.468**	24	53	0	662.3
3	Getzein	Charlie	1889	IN3	NL	**3.074**	1	**3.074**	18	22	1	349.0
4	Shreve	Lev	1889	IN3	NL	**2.869**	3	**0.956**	16	36	0	435.3
5	Burdick	Bill	1889	IN3	NL	**1.880**	2	**0.940**	12	14	1	221.7

Indianapolis Hoosiers & Newark Peppers (Federal League) (IND, NEW)

Rank	Name	First	LYear	LTeam	Lg	PEVA-P	YRS	PER YR	W	L	SV	IP
1	Falkenberg	Cy	1915	NEW	FL	**25.787**	2	**12.894**	34	27	4	549.3
2	Moseley	Earl	1915	NEW	FL	**15.566**	2	**7.783**	34	33	2	584.7
3	Kaiserling	George	1915	NEW	FL	**12.402**	2	**6.201**	32	25	2	536.7
4	Reulbach	Ed	1915	NEW	FL	**8.067**	1	**8.067**	21	10	1	270.0
5	Mullin	George	1915	NEW	FL	**3.735**	2	**1.867**	16	12	2	235.3

Kansas City Cowboys (American Association)

Rank	Name	First	LYear	LTeam	Lg	PEVA-P	YRS	PER YR	W	L	SV	IP
1	Conway	Jim	1889	KC2	AA	**7.431**	1	**7.431**	19	19	0	335.0
2	Porter	Henry	1889	KC2	AA	**5.673**	2	**2.837**	18	40	0	497.0

Rank	Name	First	LYear	LTeam	Lg	PEVA-P	YRS	PER YR	W	L	SV	IP
3	Swartzel	Park	1889	KC2	AA	**4.020**	1	**4.020**	19	27	1	410.3
4	Sullivan	Tom	1889	KC2	AA	**1.998**	2	**0.999**	10	24	0	302.0
5	Hoffman	Frank	1888	KC2	AA	**1.120**	1	**1.120**	3	9	0	104.0

Kansas City Royals

Rank	Name	First	LYear	LTeam	Lg	PEVA-P	YRS	PER YR	W	L	SV	IP
1	Saberhagen	Bret	1991	KCA	AL	**138.736**	8	**17.342**	110	78	1	1660.3
2	Appier	Kevin	2004	KCA	AL	**127.556**	13	**9.812**	115	92	0	1843.7
3	Leonard	Dennis	1986	KCA	AL	**104.156**	12	**8.680**	144	106	1	2187.0
4	Quisenberry	Dan	1988	KCA	AL	**100.751**	10	**10.075**	51	44	238	920.3
5	Gubicza	Mark	1996	KCA	AL	**96.379**	13	**7.414**	132	135	2	2218.7
6	Greinke	Zack	2010	KCA	AL	**78.498**	7	**11.214**	60	67	1	1108.0
7	Gura	Larry	1985	KCA	AL	**76.835**	10	**7.683**	111	78	12	1701.3
8	Montgomery	Jeff	1999	KCA	AL	**76.504**	12	**6.375**	44	50	304	849.3
9	Splittorff	Paul	1984	KCA	AL	**72.582**	15	**4.839**	166	143	1	2554.7
10	Leibrandt	Charlie	1989	KCA	AL	**55.642**	6	**9.274**	76	61	0	1257.0
11	Soria	Joakim	2010	KCA	AL	**46.569**	4	**11.642**	8	10	132	255.0
12	Gordon	Tom	1995	KCA	AL	**42.358**	8	**5.295**	79	71	3	1149.7
13	Black	Bud	1988	KCA	AL	**41.994**	7	**5.999**	56	57	10	977.7
14	Busby	Steve	1980	KCA	AL	**41.250**	8	**5.156**	70	54	0	1060.7
15	Cone	David	1994	KCA	AL	**38.388**	3	**12.796**	27	19	0	448.3
16	Belcher	Tim	1998	KCA	AL	**32.914**	3	**10.971**	42	37	0	686.0
17	Fitzmorris	Al	1976	KCA	AL	**32.562**	8	**4.070**	70	48	7	1098.0
18	Suppan	Jeff	2002	KCA	AL	**32.457**	5	**6.491**	39	51	0	864.7
19	Meche	Gil	2010	KCA	AL	**28.720**	4	**7.180**	29	39	0	617.0
20	Drago	Dick	1973	KCA	AL	**27.004**	5	**5.401**	61	70	1	1134.0

Kansas City Packers

Rank	Name	First	LYear	LTeam	Lg	PEVA-P	YRS	PER YR	W	L	SV	IP
1	Cullop	Nick	1915	KCF	FL	**17.620**	2	**8.810**	36	30	3	598.0
2	Packard	Gene	1915	KCF	FL	**14.384**	2	**7.192**	40	26	7	583.7
3	Johnson	Chief	1915	KCF	FL	**7.673**	2	**3.837**	26	27	2	415.3
4	Main	Alex	1915	KCF	FL	**4.588**	1	**4.588**	13	14	3	230.0
5	Henning	Pete	1915	KCF	FL	**2.873**	2	**1.436**	14	25	4	345.0

Kansas City Cowboys (National League)

Rank	Name	First	LYear	LTeam	Lg	PEVA-P	YRS	PER YR	W	L	SV	IP
1	Wiedman	Stump	1886	KCN	NL	**3.572**	1	**3.572**	12	36	0	427.7
2	Whitney	Jim	1886	KCN	NL	**3.297**	1	**3.297**	12	32	0	393.0
3	Conway	Pete	1886	KCN	NL	**1.078**	1	**1.078**	5	15	0	180.0
4	King	Silver	1886	KCN	NL	**0.763**	1	**0.763**	1	3	0	39.0
5	Lillie	Jim	1886	KCN	NL	**0.200**	1	**0.200**	0	0	0	6.0
6	McKeon	Larry	1886	KCN	NL	**0.111**	1	**0.111**	0	2	0	21.0

Kansas City Cowboys (Union League)

Rank	Name	First	LYear	LTeam	Lg	PEVA-P	YRS	PER YR	W	L	SV	IP
1	Veach	Peek-A-B	1884	KCU	UA	**2.327**	1	**2.327**	3	9	0	104.0
2	Crothers	Doug	1884	KCU	UA	**1.348**	1	**1.348**	1	2	0	25.0
3	Black	Bob	1884	KCU	UA	**1.201**	1	**1.201**	4	9	0	123.0
4	Hickman	Ernie	1884	KCU	UA	**0.793**	1	**0.793**	4	13	0	137.3
5	Blaisdell	Dick	1884	KCU	UA	**0.668**	1	**0.668**	0	3	0	26.0

Keokuk Westerns

Rank	Name	First	LYear	LTeam	Lg	PEVA-P	YRS	PER YR	W	L	SV	IP
1	Golden	Mike	1875	KEO	NA	**0.832**	1	**0.832**	1	12	0	112.0

Los Angeles Dodgers (BR3, BRO, LAN)

Rank	Name	First	LYear	LTeam	Lg	PEVA-P	YRS	PER YR	W	L	SV	IP
1	Drysdale	Don	1969	LAN	NL	**228.992**	14	**16.357**	209	166	6	3432.0
2	Vance	Dazzy	1935	BRO	NL	**208.994**	12	**17.416**	190	131	7	2757.7
3	Koufax	Sandy	1966	LAN	NL	**206.787**	12	**17.232**	165	87	9	2324.3
4	Sutton	Don	1988	LAN	NL	**191.643**	16	**11.978**	233	181	5	3816.3
5	Hershiser	Orel	2000	LAN	NL	**158.331**	13	**12.179**	135	107	5	2180.7
6	Valenzuela	Fernando	1990	LAN	NL	**143.987**	11	**13.090**	141	116	2	2348.7
7	Newcombe	Don	1958	LAN	NL	**122.137**	8	**15.267**	123	66	4	1662.7
8	Grimes	Burleigh	1926	BRO	NL	**109.893**	9	**12.210**	158	121	5	2426.0
9	Osteen	Claude	1973	LAN	NL	**98.204**	9	**10.912**	147	126	0	2396.7
10	Welch	Bob	1987	LAN	NL	**96.199**	10	**9.620**	115	86	8	1820.7
11	Hooton	Burt	1984	LAN	NL	**93.717**	10	**9.372**	112	84	6	1861.3
12	Wyatt	Whit	1944	BRO	NL	**93.449**	6	**15.575**	80	45	1	1072.3
13	Martinez	Ramon	1998	LAN	NL	**92.094**	11	**8.372**	123	77	0	1731.7
14	Reuss	Jerry	1987	LAN	NL	**91.391**	9	**10.155**	86	69	8	1407.7
15	Brown	Kevin	2003	LAN	NL	**82.913**	5	**16.583**	58	32	0	872.7
16	Clark	Watty	1937	BRO	NL	**82.849**	11	**7.532**	106	88	16	1659.0
17	Mungo	Van	1941	BRO	NL	**81.720**	11	**7.429**	102	99	14	1739.3
18	Podres	Johnny	1966	LAN	NL	**80.285**	13	**6.176**	136	104	6	2029.3
19	Roe	Preacher	1954	BRO	NL	**77.500**	7	**11.071**	93	37	4	1277.3
20	Kennedy	Brickyard	1901	BRO	NL	**77.246**	10	**7.725**	177	149	9	2857.0

Louisville Grays

Rank	Name	First	LYear	LTeam	Lg	PEVA-P	YRS	PER YR	W	L	SV	IP
1	Devlin	Jim	1877	LS1	NL	**72.489**	2	**36.245**	65	60	0	1181.0
2	Clinton	Jim	1876	LS1	NL	**0.200**	1	**0.200**	0	1	0	9.0
3	Pearce	Frank	1876	LS1	NL	**0.200**	1	**0.200**	0	0	0	4.0
4	Ryan	Johnny	1876	LS1	NL	**0.200**	1	**0.200**	0	0	0	8.0

Louisville Eclipse & Colonels

Rank	Name	First	LYear	LTeam	Lg	PEVA-P	YRS	PER YR	W	L	SV	IP
1	Hecker	Guy	1889	LS2	AA	**78.200**	8	**9.775**	171	137	1	2786.3
2	Ramsey	Toad	1889	LS2	AA	**65.347**	5	**13.069**	87	106	0	1711.0
3	Stratton	Scott	1894	LS3	NL	**49.144**	7	**7.021**	87	104	1	1715.7
4	Cunningham	Bert	1899	LS3	NL	**31.479**	5	**6.296**	77	75	1	1340.7
5	Ehret	Red	1898	LS3	NL	**20.056**	4	**5.014**	51	63	2	1032.7
6	Phillippe	Deacon	1899	LS3	NL	**17.432**	1	**17.432**	21	17	1	321.0
7	Fraser	Chick	1898	LS3	NL	**13.381**	3	**4.460**	34	63	1	838.7
8	Mullane	Tony	1882	LS2	AA	**13.305**	1	**13.305**	30	24	0	460.3
9	Dowling	Pete	1899	LS3	NL	**12.291**	3	**4.097**	27	39	0	601.3
10	Hemming	George	1897	LS3	NL	**11.691**	4	**2.923**	36	42	2	728.3
11	Chamberlain	Elton	1888	LS2	AA	**10.544**	3	**3.515**	32	28	0	536.3

Middletown Mansfields

Rank	Name	First	LYear	LTeam	Lg	PEVA-P	YRS	PER YR	W	L	SV	IP
1	Bentley	Cy	1872	MID	NA	**0.693**	1	**0.693**	2	15	0	154.0

2	Buttery	Frank	1872	MID	NA	**0.630**	1	**0.630**	3	2	0	49.0
3	Brainard	Asa	1872	MID	NA	**0.045**	1	**0.045**	0	2	0	8.0

Milwaukee Brewers (MIL, ML4, SE1)

Rank	Name	First	LYear	LTeam	Lg	PEVA-P	YRS	PER YR	W	L	SV	IP
1	Sheets	Ben	2008	MIL	NL	**90.234**	8	**11.279**	86	83	0	1428.0
2	Higuera	Teddy	1994	ML4	AL	**79.331**	9	**8.815**	94	64	0	1380.0
3	Caldwell	Mike	1984	ML4	AL	**64.597**	8	**8.075**	102	80	2	1604.7
4	Wegman	Bill	1995	ML4	AL	**55.455**	11	**5.041**	81	90	2	1482.7
5	Slaton	Jim	1983	ML4	AL	**50.213**	12	**4.184**	117	121	11	2025.3
6	Bosio	Chris	1992	ML4	AL	**49.803**	7	**7.115**	67	62	8	1190.0
7	Eldred	Cal	1999	MIL	NL	**44.631**	9	**4.959**	64	65	0	1078.7
8	Haas	Moose	1985	ML4	AL	**44.160**	10	**4.416**	91	79	2	1542.0
9	Plesac	Dan	1992	ML4	AL	**40.054**	7	**5.722**	29	37	133	524.3
10	Navarro	Jaime	2000	MIL	NL	**38.957**	7	**5.565**	62	64	1	1061.7
11	Fingers	Rollie	1985	ML4	AL	**38.411**	4	**9.603**	13	17	97	259.0
12	Davis	Doug	2006	MIL	NL	**35.578**	4	**8.895**	37	36	0	685.7
13	Bush	David	2010	MIL	NL	**34.641**	5	**6.928**	46	53	0	870.0
14	Bones	Ricky	1996	ML4	AL	**34.402**	5	**6.880**	47	56	0	883.0
15	Karl	Scott	1999	MIL	NL	**32.889**	5	**6.578**	50	51	0	914.7
16	Capuano	Chris	2007	MIL	NL	**32.358**	4	**8.090**	40	44	0	678.7
17	Sanders	Ken	1972	ML4	AL	**27.662**	3	**9.221**	14	23	61	321.0
18	Sorensen	Lary	1980	ML4	AL	**26.678**	4	**6.670**	52	46	2	854.0
19	Travers	Bill	1980	ML4	AL	**26.143**	7	**3.735**	65	67	1	1068.3
20	Colborn	Jim	1976	ML4	AL	**26.011**	5	**5.202**	57	60	3	1118.0

Milwaukee Grays

Rank	Name	First	LYear	LTeam	Lg	PEVA-P	YRS	PER YR	W	L	SV	IP
1	Weaver	Sam	1878	ML2	NL	**12.351**	1	**12.351**	12	31	0	383.0
2	Golden	Mike	1878	ML2	NL	**0.822**	1	**0.822**	3	13	0	161.0
3	Ellick	Joe	1878	ML2	NL	**0.200**	1	**0.200**	0	1	0	3.0

Milwaukee Brewers (American Association)

Rank	Name	First	LYear	LTeam	Lg	PEVA-P	YRS	PER YR	W	L	SV	IP
1	Davies	George	1891	ML3	AA	**4.217**	1	**4.217**	7	5	0	102.0
2	Killen	Frank	1891	ML3	AA	**3.860**	1	**3.860**	7	4	0	96.7
3	Dwyer	Frank	1891	ML3	AA	**0.923**	1	**0.923**	6	4	0	86.0
4	Hughey	Jim	1891	ML3	AA	**0.200**	1	**0.200**	1	0	0	15.0
5	Mains	Willard	1891	ML3	AA	**0.147**	1	**0.147**	0	2	0	10.0

Milwaukee Brewers (Union League)

Rank	Name	First	LYear	LTeam	Lg	PEVA-P	YRS	PER YR	W	L	SV	IP
1	Baldwin	Lady	1884	MLU	UA	**0.200**	1	**0.200**	1	1	0	17.0
2	Porter	Henry	1884	MLU	UA	**1.144**	1	**1.144**	3	3	0	51.0
3	Cushman	Ed	1884	MLU	UA	**5.306**	1	**5.306**	4	0	0	36.0

Minnesota Twins (WS1, MIN)

Rank	Name	First	LYear	LTeam	Lg	PEVA-P	YRS	PER YR	W	L	SV	IP
1	Johnson	Walter	1927	WS1	AL	**479.881**	21	**22.851**	417	279	34	5914.7
2	Santana	Johan	2007	MIN	AL	**153.720**	8	**19.215**	93	44	1	1308.7
3	Blyleven	Bert	1988	MIN	AL	**144.556**	11	**13.141**	149	138	0	2566.7
4	Radke	Brad	2006	MIN	AL	**130.309**	12	**10.859**	148	139	0	2451.0

Rank	Name	First	LYear	LTeam	Lg	PEVA-P	YRS	PER YR	W	L	SV	IP
5	Kaat	Jim	1973	MIN	AL	**128.252**	15	**8.550**	190	159	6	3014.3
6	Viola	Frank	1989	MIN	AL	**115.982**	8	**14.498**	112	93	0	1772.7
7	Pascual	Camilo	1966	MIN	AL	**110.796**	13	**8.523**	145	141	10	2465.0
8	Marberry	Firpo	1936	WS1	AL	**99.087**	11	**9.008**	117	71	96	1654.0
9	Nathan	Joe	2009	MIN	AL	**93.770**	6	**15.628**	22	12	246	418.7
10	Leonard	Dutch	1946	WS1	AL	**93.004**	9	**10.334**	118	101	2	1899.3
11	Perry	Jim	1972	MIN	AL	**80.186**	10	**8.019**	128	90	5	1883.3
12	Tapani	Kevin	1995	MIN	AL	**64.625**	7	**9.232**	75	63	0	1171.3
13	Crowder	Alvin	1934	WS1	AL	**59.546**	7	**8.507**	98	69	12	1331.0
14	Aguilera	Rick	1999	MIN	AL	**58.838**	11	**5.349**	40	47	254	694.0
15	Goltz	Dave	1979	MIN	AL	**54.265**	8	**6.783**	96	79	3	1638.0
16	Ramos	Pedro	1961	MIN	AL	**49.523**	7	**7.075**	78	112	12	1544.3
17	Hadley	Bump	1935	WS1	AL	**48.246**	7	**6.892**	68	71	10	1299.0
18	Erickson	Scott	1995	MIN	AL	**47.705**	6	**7.951**	61	60	0	979.3
19	Zachary	Tom	1928	WS1	AL	**44.702**	9	**4.967**	96	103	8	1589.0
20	Shaw	Jim	1921	WS1	AL	**41.703**	9	**4.634**	84	98	17	1600.3

New Haven Elm Citys

Rank	Name	First	LYear	LTeam	Lg	PEVA-P	YRS	PER YR	W	L	SV	IP
1	Nichols	Tricky	1875	NH1	NA	**1.569**	1	**1.569**	4	29	0	288.0
2	Luff	Henry	1875	NH1	NA	**0.789**	1	**0.789**	1	6	0	69.7
3	Ryan	Johnny	1875	NH1	NA	**0.535**	1	**0.535**	1	5	0	59.3
4	Knight	George	1875	NH1	NA	**0.200**	1	**0.200**	1	0	0	9.0

New York Mutuals (National Association)

Rank	Name	First	LYear	LTeam	Lg	PEVA-P	YRS	PER YR	W	L	SV	IP
1	Mathews	Bobby	1875	NY2	NA	**63.252**	3	**21.084**	100	83	0	1647.7
2	Cummings	Candy	1872	NY2	NA	**26.548**	1	**26.548**	33	20	0	497.0
3	Wolters	Rynie	1871	NY2	NA	**21.840**	1	**21.840**	16	16	0	283.0
4	Martin	Phonney	1873	NY2	NA	**0.652**	1	**0.652**	0	1	0	34.0
5	Hatfield	John	1874	NY2	NA	**0.200**	1	**0.200**	0	1	0	8.0

New York Mutuals (National League)

Rank	Name	First	LYear	LTeam	Lg	PEVA-P	YRS	PER YR	W	L	SV	IP
1	Booth	Eddie	1876	NY3	NL	**0.200**	1	**0.200**	0	0	0	5.0
2	Larkin	Terry	1876	NY3	NL	**0.200**	1	**0.200**	0	1	0	9.0
3	Mathews	Bobby	1876	NY3	NL	**5.545**	1	**5.545**	21	34	0	516.0

New York Metropolitans

Rank	Name	First	LYear	LTeam	Lg	PEVA-P	YRS	PER YR	W	L	SV	IP
1	Keefe	Tim	1884	NY4	AA	**42.664**	2	**21.332**	78	44	0	1102.0
2	Lynch	Jack	1887	NY4	AA	**23.697**	5	**4.739**	100	95	0	1749.7
3	Cushman	Ed	1887	NY4	AA	**7.498**	3	**2.499**	35	49	0	736.7
4	Mays	Al	1887	NY4	AA	**7.452**	2	**3.726**	28	62	0	791.3
5	Shaffer	John	1887	NY4	AA	**4.122**	2	**2.061**	7	14	0	181.0

New York Yankees (NYA, BLA)

Rank	Name	First	LYear	LTeam	Lg	PEVA-P	YRS	PER YR	W	L	SV	IP
1	Rivera	Mariano	2010	NYA	AL	**215.110**	16	**13.444**	74	55	559	1150.0
2	Ford	Whitey	1967	NYA	AL	**200.553**	16	**12.535**	236	106	10	3170.3
3	Pettitte	Andy	2010	NYA	AL	**151.286**	13	**11.637**	203	112	0	2535.7
4	Guidry	Ron	1988	NYA	AL	**144.658**	14	**10.333**	170	91	4	2392.0

Rank	Name	First	LYear	LTeam	Lg	PEVA-P	YRS	PER YR	W	L	SV	IP
5	Ruffing	Red	1946	NYA	AL	**137.320**	15	**9.155**	231	124	8	3168.7
6	Gomez	Lefty	1942	NYA	AL	**135.369**	13	**10.413**	189	101	9	2498.3
7	Mussina	Mike	2008	NYA	AL	**124.700**	8	**15.587**	123	72	0	1553.0
8	Stottlemyre	Mel	1974	NYA	AL	**105.938**	11	**9.631**	164	139	1	2661.3
9	Hoyt	Waite	1930	NYA	AL	**102.845**	10	**10.284**	157	98	28	2272.3
10	Chandler	Spud	1947	NYA	AL	**99.944**	11	**9.086**	109	43	6	1485.0
11	Pennock	Herb	1933	NYA	AL	**96.556**	11	**8.778**	162	90	20	2203.3
12	Shawkey	Bob	1927	NYA	AL	**88.534**	13	**6.810**	168	131	26	2488.7
13	Chesbro	Jack	1909	NYA	AL	**84.045**	7	**12.006**	128	93	2	1952.0
14	Peterson	Fritz	1974	NYA	AL	**83.654**	9	**9.295**	109	106	1	1857.3
15	Reynolds	Allie	1954	NYA	AL	**82.812**	8	**10.352**	131	60	41	1700.0
16	Raschi	Vic	1953	NYA	AL	**76.467**	8	**9.558**	120	50	3	1537.0
17	Righetti	Dave	1990	NYA	AL	**72.977**	11	**6.634**	74	61	224	1136.7
18	Lopat	Ed	1955	NYA	AL	**72.869**	8	**9.109**	113	59	2	1497.3
19	John	Tommy	1989	NYA	AL	**67.807**	8	**8.476**	91	60	0	1367.0
20	Gossage	Rich	1989	NYA	AL	**65.037**	7	**9.291**	42	28	151	533.0

New York Mets

Rank	Name	First	LYear	LTeam	Lg	PEVA-P	YRS	PER YR	W	L	SV	IP
1	Seaver	Tom	1983	NYN	NL	**235.428**	12	**19.619**	198	124	1	3045.3
2	Gooden	Dwight	1994	NYN	NL	**176.442**	11	**16.040**	157	85	1	2169.7
3	Koosman	Jerry	1978	NYN	NL	**102.649**	12	**8.554**	140	137	5	2544.7
4	Leiter	Al	2004	NYN	NL	**85.459**	7	**12.208**	95	67	0	1360.0
5	Fernandez	Sid	1993	NYN	NL	**84.206**	10	**8.421**	98	78	1	1584.7
6	Cone	David	2003	NYN	NL	**74.974**	7	**10.711**	81	51	1	1209.3
7	Darling	Ron	1991	NYN	NL	**67.663**	9	**7.518**	99	70	0	1620.0
8	Matlack	Jon	1977	NYN	NL	**62.324**	7	**8.903**	82	81	0	1448.0
9	Franco	John	2004	NYN	NL	**59.854**	14	**4.275**	48	56	276	702.7
10	Glavine	Tom	2007	NYN	NL	**55.345**	5	**11.069**	61	56	0	1005.3
11	Jones	Bobby	2000	NYN	NL	**54.087**	8	**6.761**	74	56	0	1215.7
12	Reed	Rick	2001	NYN	NL	**53.269**	5	**10.654**	59	36	0	888.7
13	Santana	Johan	2010	NYN	NL	**51.512**	3	**17.171**	40	25	0	600.0
14	Trachsel	Steve	2006	NYN	NL	**44.517**	6	**7.420**	66	59	0	956.3
15	Ojeda	Bob	1990	NYN	NL	**43.710**	5	**8.742**	51	40	0	764.0
16	Saberhagen	Bret	1995	NYN	NL	**42.807**	4	**10.702**	29	21	0	524.3
17	Orosco	Jesse	1987	NYN	NL	**41.499**	8	**5.187**	47	47	107	595.7
18	Benitez	Armando	2003	NYN	NL	**39.664**	5	**7.933**	18	14	160	347.0
19	Viola	Frank	1991	NYN	NL	**38.753**	3	**12.918**	38	32	0	566.3
20	Swan	Craig	1984	NYN	NL	**35.859**	12	**2.988**	59	71	2	1230.7

New York Giants (Pacific League)

Rank	Name	First	LYear	LTeam	Lg	PEVA-P	YRS	PER YR	W	L	SV	IP
1	Keefe	Tim	1890	NYP	PL	**3.709**	1	**3.709**	17	11	0	229.0
2	O'Day	Hank	1890	NYP	PL	**3.441**	1	**3.441**	22	13	3	329.0
3	Crane	Ed	1890	NYP	PL	**2.641**	1	**2.641**	16	19	0	330.3
4	Ewing	John	1890	NYP	PL	**2.579**	1	**2.579**	18	12	2	267.3
5	Hatfield	Gil	1890	NYP	PL	**0.200**	1	**0.200**	1	1	1	7.7

Oakland Athletics (PHA, KC1, OAK)

Rank	Name	First	LYear	LTeam	Lg	PEVA-P	YRS	PER YR	W	L	SV	IP
1	Grove	Lefty	1933	PHA	AL	**251.486**	9	**27.943**	195	79	51	2401.0
2	Plank	Eddie	1914	PHA	AL	**148.985**	14	**10.642**	284	162	16	3860.7
3	Blue	Vida	1977	OAK	AL	**120.100**	9	**13.344**	124	86	2	1945.7
4	Hunter	Catfish	1974	OAK	AL	**119.807**	10	**11.981**	161	113	1	2456.3

5	Rommel	Eddie	1932	PHA	AL	**107.495**	13	**8.269**	171	119	29	2556.3
6	Zito	Barry	2006	OAK	AL	**106.442**	7	**15.206**	102	63	0	1430.3
7	Bender	Chief	1914	PHA	AL	**99.665**	12	**8.305**	193	102	28	2602.0
8	Stewart	Dave	1995	OAK	AL	**97.757**	8	**12.220**	119	78	0	1717.3
9	Waddell	Rube	1907	PHA	AL	**95.353**	6	**15.892**	131	82	0	1869.3
10	Hudson	Tim	2004	OAK	AL	**93.684**	6	**15.614**	92	39	0	1240.7
11	Eckersley	Dennis	1995	OAK	AL	**92.686**	9	**10.298**	41	31	320	637.0
12	Walberg	Rube	1933	PHA	AL	**78.745**	11	**7.159**	134	114	27	2186.7
13	Mulder	Mark	2004	OAK	AL	**67.653**	5	**13.531**	81	42	0	1003.0
14	Earnshaw	George	1933	PHA	AL	**66.290**	6	**11.048**	98	58	10	1353.7
15	Coombs	Jack	1914	PHA	AL	**61.912**	9	**6.879**	115	67	8	1629.7
16	Fingers	Rollie	1976	OAK	AL	**61.804**	9	**6.867**	67	61	136	1016.0
17	Welch	Bob	1994	OAK	AL	**61.494**	7	**8.785**	96	60	0	1271.3
18	Moore	Mike	1992	OAK	AL	**53.509**	4	**13.377**	66	46	0	874.0
19	Holtzman	Ken	1975	OAK	AL	**52.037**	4	**13.009**	77	55	0	1084.3
20	Haren	Danny	2007	OAK	AL	**49.646**	3	**16.549**	43	34	0	662.7

Philadelphia Athletics (National Association)

Rank	Name	First	LYear	LTeam	Lg	PEVA-P	YRS	PER YR	W	L	SV	IP
1	McBride	Dick	1875	PH1	NA	**67.696**	5	**13.539**	149	74	0	2048.7
2	Fisher	Cherokee	1873	PH1	NA	**4.479**	1	**4.479**	3	4	1	84.3
3	Knight	Lon	1875	PH1	NA	**1.367**	1	**1.367**	6	5	0	107.0
4	Bechtel	George	1875	PH1	NA	**0.975**	2	**0.488**	4	3	0	62.0
5	Sutton	Ezra	1875	PH1	NA	**0.200**	1	**0.200**	0	0	0	6.0

Philadelphia Athletics (American Association)

Rank	Name	First	LYear	LTeam	Lg	PEVA-P	YRS	PER YR	W	L	SV	IP
1	Weyhing	Gus	1891	PH4	AA	**43.174**	4	**10.793**	115	87	0	1769.3
2	Seward	Ed	1890	PH4	AA	**36.855**	4	**9.214**	87	71	0	1463.3
3	Mathews	Bobby	1887	PH4	AA	**32.970**	5	**6.594**	106	61	0	1489.7
4	McMahon	Sadie	1890	PH4	AA	**17.088**	2	**8.544**	43	30	1	652.0
5	Taylor	Billy	1887	PH4	AA	**9.066**	3	**3.022**	20	17	0	321.3

Philadelphia Whites

Rank	Name	First	LYear	LTeam	Lg	PEVA-P	YRS	PER YR	W	L	SV	IP
1	Zettlein	George	1875	PH2	NA	**21.736**	2	**10.868**	48	23	0	640.3
2	Fisher	Cherokee	1875	PH2	NA	**8.534**	1	**8.534**	22	19	0	356.7
3	Cummings	Candy	1874	PH2	NA	**6.770**	1	**6.770**	28	26	0	482.0
4	Borden	Joe	1875	PH2	NA	**3.528**	1	**3.528**	2	4	0	66.0
5	Bechtel	George	1874	PH2	NA	**0.832**	2	**0.416**	1	5	0	58.0

Philadelphia Centennials

Rank	Name	First	LYear	LTeam	Lg	PEVA-P	YRS	PER YR	W	L	SV	IP
1	Bechtel	George	1875	PH3	NA	**1.200**	1	**1.200**	2	12	0	126.0

Philadelphia Phillies

Rank	Name	First	LYear	LTeam	Lg	PEVA-P	YRS	PER YR	W	L	SV	IP
1	Carlton	Steve	1986	PHI	NL	**278.981**	15	**18.599**	241	161	0	3697.3
2	Roberts	Robin	1961	PHI	NL	**275.929**	14	**19.709**	234	199	24	3739.3
3	Alexander	Pete	1930	PHI	NL	**207.112**	8	**25.889**	190	91	15	2513.7
4	Schilling	Curt	2000	PHI	NL	**126.818**	9	**14.091**	101	78	2	1659.3
5	Bunning	Jim	1971	PHI	NL	**100.697**	6	**16.783**	89	73	4	1520.7

6	Short	Chris	1972	PHI	NL	**80.283**	14	**5.734**	132	127	16	2253.0
7	Simmons	Curt	1960	PHI	NL	**73.304**	13	**5.639**	115	110	4	1939.7
8	Hamels	Cole	2010	PHI	NL	**72.490**	5	**14.498**	60	45	0	945.3
9	Myers	Brett	2009	PHI	NL	**50.608**	8	**6.326**	73	63	21	1183.7
10	Reed	Ron	1983	PHI	NL	**48.807**	8	**6.101**	57	38	90	809.3
11	Orth	Al	1901	PHI	NL	**48.441**	7	**6.920**	100	72	4	1504.7
12	Mulholland	Terry	1996	PHI	NL	**48.326**	6	**8.054**	62	57	0	1070.3
13	Denny	John	1985	PHI	NL	**46.283**	4	**11.571**	37	29	0	650.0
14	Ferguson	Charlie	1887	PHI	NL	**45.301**	4	**11.325**	99	64	4	1514.7
15	Wolf	Randy	2006	PHI	NL	**44.760**	8	**5.595**	69	60	0	1175.0
16	McGraw	Tug	1984	PHI	NL	**44.635**	10	**4.464**	49	37	94	722.0
17	Rixey	Eppa	1920	PHI	NL	**44.172**	8	**5.521**	87	103	6	1604.0
18	Halladay	Roy	2010	PHI	NL	**43.340**	1	**43.340**	21	10	0	250.7
19	Christenson	Larry	1983	PHI	NL	**42.223**	11	**3.838**	83	71	4	1402.7
20	Sparks	Tully	1910	PHI	NL	**41.974**	9	**4.664**	95	95	7	1698.0

Philadelphia Athletics (National League)

Rank	Name	First	LYear	LTeam	Lg	PEVA-P	YRS	PER YR	W	L	SV	IP
1	Knight	Lon	1876	PHN	NL	**1.878**	1	**1.878**	10	22	0	282.0
2	Zettlein	George	1876	PHN	NL	**1.477**	1	**1.477**	4	20	2	234.0
3	Coon	William	1876	PHN	NL	**0.200**	1	**0.200**	0	0	0	7.0
4	Meyerle	Levi	1876	PHN	NL	**0.200**	1	**0.200**	0	2	0	18.0
5	Lafferty	Flip	1876	PHN	NL	**0.200**	1	**0.200**	0	1	0	9.0

Philadelphia Athletics (Pacific League)

Rank	Name	First	LYear	LTeam	Lg	PEVA-P	YRS	PER YR	W	L	SV	IP
1	Sanders	Ben	1890	PHP	PL	**4.233**	1	**4.233**	19	18	1	346.7
2	Knell	Phil	1890	PHP	PL	**3.221**	1	**3.221**	22	11	0	286.7
3	Buffinton	Charlie	1890	PHP	PL	**2.990**	1	**2.990**	19	15	1	283.3
4	Husted	Bill	1890	PHP	PL	**0.841**	1	**0.841**	5	10	0	129.0
5	Cunningham	Bert	1890	PHP	PL	**0.785**	1	**0.785**	3	9	0	108.7

Philadelphia Keystones

Rank	Name	First	LYear	LTeam	Lg	PEVA-P	YRS	PER YR	W	L	SV	IP
1	Bakely	Jersey	1884	PHU	UA	**3.380**	1	**3.380**	14	25	0	344.7
2	Fisher	J.	1884	PHU	UA	**0.919**	1	**0.919**	1	7	0	70.7
3	Weaver	Sam	1884	PHU	UA	**0.903**	1	**0.903**	5	10	0	136.0
4	Gallagher	Bill	1884	PHU	UA	**0.462**	1	**0.462**	1	2	0	25.0
5	Maul	Al	1884	PHU	UA	**0.200**	1	**0.200**	0	1	0	8.0

Pittsburgh Pirates (PIT, PT1)

Rank	Name	First	LYear	LTeam	Lg	PEVA-P	YRS	PER YR	W	L	SV	IP
1	Friend	Bob	1965	PIT	NL	**175.551**	15	**11.703**	191	218	10	3480.3
2	Cooper	Wilbur	1924	PIT	NL	**158.666**	13	**12.205**	202	159	14	3199.0
3	Adams	Babe	1926	PIT	NL	**158.236**	18	**8.791**	194	139	15	2991.3
4	Candelaria	John	1993	PIT	NL	**107.249**	12	**8.937**	124	87	16	1873.0
5	Law	Vern	1967	PIT	NL	**103.746**	16	**6.484**	162	147	13	2672.0
6	Kremer	Ray	1933	PIT	NL	**101.477**	10	**10.148**	143	85	10	1954.7
7	Drabek	Doug	1992	PIT	NL	**99.151**	6	**16.525**	92	62	0	1362.7
8	Leever	Sam	1910	PIT	NL	**96.852**	13	**7.450**	194	100	13	2660.7
9	Phillippe	Deacon	1911	PIT	NL	**96.644**	12	**8.054**	168	92	11	2286.0
10	Face	Roy	1968	PIT	NL	**88.354**	15	**5.890**	100	93	188	1314.7
11	Tannehill	Jesse	1902	PIT	NL	**80.132**	6	**13.355**	116	58	5	1499.0

Rank	Name	First	LYear	LTeam	Lg	PEVA-P	YRS	PER YR	W	L	SV	IP
12	Sewell	Rip	1949	PIT	NL	**77.735**	12	**6.478**	143	97	15	2108.7
13	Veale	Bob	1972	PIT	NL	**76.671**	11	**6.970**	116	91	6	1868.7
14	Tekulve	Kent	1985	PIT	NL	**76.544**	12	**6.379**	70	61	158	1017.0
15	Killen	Frank	1898	PIT	NL	**72.982**	6	**12.164**	112	82	0	1661.3
16	Rhoden	Rick	1986	PIT	NL	**72.112**	8	**9.014**	79	73	1	1448.0
17	Morris	Ed	1889	PIT	NL	**72.078**	5	**14.416**	129	102	1	2104.0
18	French	Larry	1934	PIT	NL	**68.367**	6	**11.395**	87	83	9	1502.7
19	Blanton	Cy	1939	PIT	NL	**66.710**	6	**11.118**	58	51	4	955.3
20	Swift	Bill	1939	PIT	NL	**61.922**	8	**7.740**	91	79	18	1555.0

Providence Grays

Rank	Name	First	LYear	LTeam	Lg	PEVA-P	YRS	PER YR	W	L	SV	IP
1	Radbourn	Charley	1885	PRO	NL	**105.640**	5	**21.128**	193	89	2	2556.0
2	Ward	John	1882	PRO	NL	**68.865**	5	**13.773**	145	86	3	2124.0
3	Sweeney	Charlie	1884	PRO	NL	**13.469**	2	**6.734**	24	15	1	367.7
4	Shaw	Dupee	1885	PRO	NL	**6.378**	1	**6.378**	23	26	0	399.7
5	Bradley	George	1880	PRO	NL	**4.649**	1	**4.649**	13	8	1	196.0

Pittsburgh Rebels

Rank	Name	First	LYear	LTeam	Lg	PEVA-P	YRS	PER YR	W	L	SV	IP
1	Knetzer	Elmer	1915	PTF	FL	**13.033**	2	**6.517**	38	26	4	551.0
2	Allen	Frank	1915	PTF	FL	**9.704**	2	**4.852**	24	13	0	290.3
3	Barger	Cy	1915	PTF	FL	**6.809**	2	**3.405**	19	24	7	381.3
4	Rogge	Clint	1915	PTF	FL	**5.847**	1	**5.847**	17	11	0	254.3
5	Camnitz	Howie	1915	PTF	FL	**4.741**	2	**2.371**	14	19	1	282.0

Pittsburgh Burghers

Rank	Name	First	LYear	LTeam	Lg	PEVA-P	YRS	PER YR	W	L	SV	IP
1	Staley	Harry	1890	PTP	PL	**10.092**	1	**10.092**	21	25	0	387.7
2	Maul	Al	1890	PTP	PL	**2.071**	1	**2.071**	16	12	0	246.7
3	Galvin	Pud	1890	PTP	PL	**1.564**	1	**1.564**	12	13	0	217.0
4	Morris	Ed	1890	PTP	PL	**0.778**	1	**0.778**	8	7	0	144.3
5	Tener	John	1890	PTP	PL	**0.631**	1	**0.631**	3	11	0	117.0

Rockford Forest Citys

Rank	Name	First	LYear	LTeam	Lg	PEVA-P	YRS	PER YR	W	L	SV	IP
1	Fisher	Cherokee	1871	RC1	NA	**2.834**	1	**2.834**	4	16	0	213.0
2	Mack	Denny	1871	RC1	NA	**0.200**	1	**0.200**	0	1	0	13.0

Rochester Broncos

Rank	Name	First	LYear	LTeam	Lg	PEVA-P	YRS	PER YR	W	L	SV	IP
1	Barr	Bob	1890	RC2	AA	**10.690**	1	**10.690**	28	24	0	493.3
2	Calihan	Will	1890	RC2	AA	**4.396**	1	**4.396**	18	15	0	296.3
3	Titcomb	Cannonbı	1890	RC2	AA	**1.121**	1	**1.121**	10	9	0	168.7
4	Miller	Bob	1890	RC2	AA	**0.942**	1	**0.942**	3	7	1	92.3
5	Fitzgerald	John	1890	RC2	AA	**0.825**	1	**0.825**	3	8	0	78.0

Richmond Virginians

Rank	Name	First	LYear	LTeam	Lg	PEVA-P	YRS	PER YR	W	L	SV	IP
1	Meegan	Pete	1884	RIC	AA	**1.107**	1	**1.107**	7	12	0	179.0
2	Dugan	Ed	1884	RIC	AA	**0.933**	1	**0.933**	5	14	0	166.3

Rank	Name	First	LYear	LTeam	Lg	PEVA-P	YRS	PER YR	W	L	SV	IP
3	Firth	Ted	1884	RIC	AA	**0.200**	1	**0.200**	0	1	0	9.0
4	Curry	Wes	1884	RIC	AA	**0.200**	1	**0.200**	0	2	0	16.0

San Diego Padres

Rank	Name	First	LYear	LTeam	Lg	PEVA-P	YRS	PER YR	W	L	SV	IP
1	Hoffman	Trevor	2008	SDN	NL	**151.267**	16	**9.454**	54	64	552	952.3
2	Peavy	Jake	2009	SDN	NL	**107.744**	8	**13.468**	92	68	0	1342.7
3	Jones	Randy	1980	SDN	NL	**88.355**	8	**11.044**	92	105	2	1766.0
4	Benes	Andy	1995	SDN	NL	**68.733**	7	**9.819**	69	75	0	1235.0
5	Whitson	Ed	1991	SDN	NL	**66.290**	8	**8.286**	77	72	1	1354.3
6	Ashby	Andy	2004	SDN	NL	**65.177**	8	**8.147**	70	62	0	1212.0
7	Show	Eric	1990	SDN	NL	**62.499**	10	**6.250**	100	87	7	1603.3
8	Hurst	Bruce	1993	SDN	NL	**55.228**	5	**11.046**	55	38	0	911.7
9	Hamilton	Joey	1998	SDN	NL	**43.336**	5	**8.667**	55	44	0	934.7
10	Brown	Kevin	1998	SDN	NL	**42.712**	1	**42.712**	18	7	0	257.0
11	Bell	Heath	2010	SDN	NL	**38.731**	4	**9.683**	24	15	91	311.3
12	Lawrence	Brian	2005	SDN	NL	**37.932**	5	**7.586**	49	61	0	934.0
13	Dravecky	Dave	1987	SDN	NL	**34.872**	6	**5.812**	53	50	10	900.3
14	Lefferts	Craig	1992	SDN	NL	**34.087**	7	**4.870**	42	40	64	659.0
15	Williams	Woody	2006	SDN	NL	**33.849**	5	**6.770**	51	45	0	826.3
16	Hawkins	Andy	1988	SDN	NL	**33.432**	7	**4.776**	60	58	0	1102.7
17	Young	Chris	2010	SDN	NL	**31.955**	5	**6.391**	33	25	0	550.7
18	Harris	Greg	1993	SDN	NL	**30.337**	6	**5.056**	41	39	15	673.3
19	Fingers	Rollie	1980	SDN	NL	**30.243**	4	**7.561**	34	40	108	426.3
20	Perry	Gaylord	1979	SDN	NL	**27.401**	2	**13.700**	33	17	0	493.3

Seattle Mariners

Rank	Name	First	LYear	LTeam	Lg	PEVA-P	YRS	PER YR	W	L	SV	IP
1	Johnson	Randy	1998	SEA	AL	**145.762**	10	**14.576**	130	74	2	1838.3
2	Moyer	Jamie	2006	SEA	AL	**125.477**	11	**11.407**	145	87	0	2093.0
3	Hernandez	Felix	2010	SEA	AL	**95.601**	6	**15.934**	71	53	0	1154.7
4	Garcia	Freddy	2004	SEA	AL	**63.854**	6	**10.642**	76	50	0	1096.3
5	Langston	Mark	1989	SEA	AL	**59.270**	6	**9.878**	74	67	0	1197.7
6	Hanson	Erik	1993	SEA	AL	**42.617**	6	**7.103**	56	54	0	967.3
7	Putz	J.J.	2008	SEA	AL	**42.423**	6	**7.071**	22	15	101	323.0
8	Moore	Mike	1988	SEA	AL	**40.810**	7	**5.830**	66	96	2	1457.0
9	Pineiro	Joel	2006	SEA	AL	**38.578**	7	**5.511**	58	55	1	996.0
10	Rhodes	Arthur	2008	SEA	AL	**30.835**	5	**6.167**	28	16	9	283.0
11	Fassero	Jeff	1999	SEA	AL	**29.545**	3	**9.848**	33	35	0	598.0
12	Sasaki	Kazuhiro	2003	SEA	AL	**28.187**	4	**7.047**	7	16	129	223.3
13	Sele	Aaron	2005	SEA	AL	**28.057**	3	**9.352**	38	27	0	542.7
14	Young	Matt	1990	SEA	AL	**27.166**	5	**5.433**	45	66	14	864.3
15	Bannister	Floyd	1982	SEA	AL	**26.981**	4	**6.745**	40	50	0	768.3
16	Swift	Bill	1998	SEA	AL	**26.796**	7	**3.828**	41	49	24	903.7
17	Beattie	Jim	1986	SEA	AL	**26.725**	7	**3.818**	43	72	1	944.7
18	Franklin	Ryan	2005	SEA	AL	**25.105**	6	**4.184**	35	50	0	811.3
19	Meche	Gil	2006	SEA	AL	**24.657**	6	**4.110**	55	44	0	815.3
20	Washburn	Jarrod	2009	SEA	AL	**23.589**	4	**5.897**	31	49	1	667.3

San Francisco Giants (NY1, SFN)

Rank	Name	First	LYear	LTeam	Lg	PEVA-P	YRS	PER YR	W	L	SV	IP
1	Hubbell	Carl	1943	NY1	NL	**301.955**	16	**18.872**	253	154	33	3590.3
2	Mathewson	Christy	1916	NY1	NL	**290.825**	17	**17.107**	372	188	28	4771.7
3	Marichal	Juan	1973	SFN	NL	**230.597**	14	**16.471**	238	140	2	3444.0

Rank	Name	First	LYear	LTeam	Lg	PEVA-P	YRS	PER YR	W	L	SV	IP
4	Rusie	Amos	1898	NY1	NL	**207.458**	8	**25.932**	233	163	5	3522.7
5	Perry	Gaylord	1971	SFN	NL	**120.599**	10	**12.060**	134	109	10	2294.7
6	Jansen	Larry	1954	NY1	NL	**118.311**	8	**14.789**	120	86	9	1731.0
7	Schumacher	Hal	1946	NY1	NL	**113.603**	13	**8.739**	158	121	7	2482.3
8	Fitzsimmons	Freddie	1937	NY1	NL	**105.185**	13	**8.091**	170	114	9	2514.3
9	Keefe	Tim	1891	NY1	NL	**101.563**	6	**16.927**	174	82	1	2265.0
10	McGinnity	Joe	1908	NY1	NL	**97.683**	7	**13.955**	151	88	21	2151.3
11	Lincecum	Tim	2010	SFN	NL	**92.775**	4	**23.194**	56	27	0	811.0
12	Welch	Mickey	1892	NY1	NL	**89.061**	10	**8.906**	238	146	4	3579.0
13	Schmidt	Jason	2006	SFN	NL	**86.771**	6	**14.462**	78	37	0	1069.7
14	Maglie	Sal	1955	NY1	NL	**78.428**	7	**11.204**	95	42	8	1297.7
15	Cain	Matt	2010	SFN	NL	**76.859**	6	**12.810**	57	62	0	1095.7
16	McCormick	Mike	1970	SFN	NL	**71.853**	9	**7.984**	104	94	11	1741.3
17	Rueter	Kirk	2005	SFN	NL	**71.357**	10	**7.136**	105	80	0	1614.0
18	Melton	Cliff	1944	NY1	NL	**69.463**	8	**8.683**	86	80	16	1453.7
19	Antonelli	Johnny	1957	NY1	NL	**67.046**	4	**16.762**	67	54	4	964.7
20	Nen	Robb	2002	SFN	NL	**64.914**	5	**12.983**	24	25	206	378.3

St. Louis Red Stockings

Rank	Name	First	LYear	LTeam	Lg	PEVA-P	YRS	PER YR	W	L	SV	IP
1	Morgan	Pidgey	1875	SL1	NA	**1.981**	1	**1.981**	1	3	0	42.0
2	Blong	Joe	1875	SL1	NA	**1.374**	1	**1.374**	3	12	0	129.0

St. Louis Brown Stockings (National Association)

Rank	Name	First	LYear	LTeam	Lg	PEVA-P	YRS	PER YR	W	L	SV	IP
1	Bradley	George	1875	SL2	NA	**11.903**	1	**11.903**	33	26	0	535.7
2	Galvin	Pud	1875	SL2	NA	**2.934**	1	**2.934**	4	2	1	62.0
3	Fleet	Frank	1875	SL2	NA	**0.419**	1	**0.419**	2	1	0	27.0
4	Pearce	Dickey	1875	SL2	NA	**0.200**	1	**0.200**	0	0	0	5.3

St. Louis Brown Stockings (National League)

Rank	Name	First	LYear	LTeam	Lg	PEVA-P	YRS	PER YR	W	L	SV	IP
1	Bradley	George	1876	SL3	NL	**36.677**	1	**36.677**	45	19	0	573.0
2	Nichols	Tricky	1877	SL3	NL	**4.735**	1	**4.735**	18	23	0	350.0
3	Blong	Joe	1877	SL3	NL	**1.778**	2	**0.889**	10	9	0	191.3
4	Battin	Joe	1877	SL3	NL	**0.200**	1	**0.200**	0	0	0	3.7

St. Louis Maroons (SL5, SLU)

Rank	Name	First	LYear	LTeam	Lg	PEVA-P	YRS	PER YR	W	L	SV	IP
1	Sweeney	Charlie	1886	SL5	NL	**17.594**	3	**5.865**	40	34	0	639.0
2	Boyle	Henry	1886	SL5	NL	**15.422**	3	**5.141**	40	42	1	726.7
3	Taylor	Billy	1884	SLU	UA	**8.136**	1	**8.136**	25	4	4	263.0
4	Healy	John	1886	SL5	NL	**6.032**	2	**3.016**	18	30	0	419.7
5	Kirby	John	1886	SL5	NL	**3.752**	2	**1.876**	16	34	0	454.3

St. Louis Terriers

Rank	Name	First	LYear	LTeam	Lg	PEVA-P	YRS	PER YR	W	L	SV	IP
1	Davenport	Dave	1915	SLF	FL	**25.304**	2	**12.652**	30	31	5	608.3
2	Plank	Eddie	1915	SLF	FL	**16.422**	1	**16.422**	21	11	3	268.3
3	Crandall	Doc	1915	SLF	FL	**10.974**	2	**5.487**	34	24	1	508.7
4	Groom	Bob	1915	SLF	FL	**7.151**	2	**3.575**	24	31	3	489.7
5	Watson	Doc	1915	SLF	FL	**2.649**	2	**1.324**	12	13	0	191.7

St. Louis Cardinals (SL4, SLN)

Rank	Name	First	LYear	LTeam	Lg	PEVA-P	YRS	PER YR	W	L	SV	IP
1	Gibson	Bob	1975	SLN	NL	**225.768**	17	**13.280**	251	174	6	3884.3
2	Dean	Dizzy	1937	SLN	NL	**152.623**	7	**21.803**	134	75	30	1737.3
3	Carpenter	Chris	2010	SLN	NL	**128.778**	7	**18.397**	84	33	0	1094.3
4	Cooper	Mort	1945	SLN	NL	**118.364**	8	**14.795**	105	50	12	1480.3
5	Haines	Jesse	1937	SLN	NL	**116.240**	18	**6.458**	210	158	10	3203.7
6	Brecheen	Harry	1952	SLN	NL	**107.820**	11	**9.802**	128	79	17	1790.3
7	Forsch	Bob	1988	SLN	NL	**93.912**	15	**6.261**	163	127	3	2658.7
8	Wainwright	Adam	2010	SLN	NL	**90.090**	6	**15.015**	66	35	3	874.3
9	Morris	Matt	2005	SLN	NL	**84.692**	8	**10.586**	101	62	4	1377.3
10	King	Silver	1889	SL4	AA	**84.369**	3	**28.123**	111	50	2	1433.7
11	Andujar	Joaquin	1985	SLN	NL	**81.975**	5	**16.395**	68	53	1	1077.0
12	Jackson	Larry	1962	SLN	NL	**80.627**	8	**10.078**	101	86	20	1672.3
13	Lanier	Max	1951	SLN	NL	**79.002**	12	**6.583**	101	69	12	1454.7
14	Pollet	Howie	1951	SLN	NL	**78.541**	9	**8.727**	97	65	11	1401.7
15	Sherdel	Bill	1932	SLN	NL	**75.363**	14	**5.383**	153	131	25	2450.7
16	Tudor	John	1990	SLN	NL	**72.894**	5	**14.579**	62	26	0	881.7
17	Doak	Bill	1929	SLN	NL	**72.526**	13	**5.579**	144	136	13	2387.0
18	Tewksbury	Bob	1994	SLN	NL	**72.388**	6	**12.065**	67	46	1	968.7
19	Breitenstein	Ted	1901	SLN	NL	**62.958**	7	**8.994**	94	125	3	1925.3
20	Young	Cy	1900	SLN	NL	**62.390**	2	**31.195**	45	35	1	690.7

St. Paul Apostles

Rank	Name	First	LYear	LTeam	Lg	PEVA-P	YRS	PER YR	W	L	SV	IP
1	Galvin	Lou	1884	SPU	UA	**0.666**	1	**0.666**	0	2	0	25.0
2	Brown	Jim	1884	SPU	UA	**0.286**	1	**0.286**	1	4	0	36.0
3	O'Brien	Billy	1884	SPU	UA	**0.200**	1	**0.200**	1	0	0	10.0

Syracuse Stars

Rank	Name	First	LYear	LTeam	Lg	PEVA-P	YRS	PER YR	W	L	SV	IP
1	McCormick	Harry	1879	SR1	NL	**3.295**	1	**3.295**	18	33	0	457.3
2	Purcell	Blondie	1879	SR1	NL	**0.704**	1	**0.704**	4	15	0	179.7
3	Dorgan	Mike	1879	SR1	NL	**0.200**	1	**0.200**	0	0	0	12.0

Syracuse Stars (American Association)

Rank	Name	First	LYear	LTeam	Lg	PEVA-P	YRS	PER YR	W	L	SV	IP
1	Casey	Dan	1890	SR2	AA	**3.522**	1	**3.522**	19	22	0	360.7
2	Keefe	John	1890	SR2	AA	**3.209**	1	**3.209**	17	24	0	352.3
3	Mars	Ed	1890	SR2	AA	**0.971**	1	**0.971**	9	5	0	121.3
4	Morrison	Mike	1890	SR2	AA	**0.667**	1	**0.667**	6	9	0	127.0
5	Sullivan	Bill	1890	SR2	AA	**0.390**	1	**0.390**	1	4	0	42.0

Tampa Bay Devil Rays

Rank	Name	First	LYear	LTeam	Lg	PEVA-P	YRS	PER YR	W	L	SV	IP
1	Shields	James	2010	TBA	AL	**56.011**	5	**11.202**	56	51	0	977.7
2	Kazmir	Scott	2009	TBA	AL	**46.339**	6	**7.723**	55	44	0	834.0
3	Garza	Matt	2010	TBA	AL	**29.796**	3	**9.932**	34	31	1	592.3
4	Price	David	2010	TBA	AL	**22.932**	3	**7.644**	29	13	0	351.0
4	Hernandez	Roberto	2000	TBA	AL	**18.374**	3	**6.125**	8	16	101	218.0
5	Lopez	Albie	2001	TBA	AL	**16.280**	4	**4.070**	26	31	4	453.7

Rank	Name	First	LYear	LTeam	Lg	PEVA-P	YRS	PER YR	W	L	SV	IP
6	Zambrano	Victor	2004	TBA	AL	**15.939**	4	**3.985**	35	27	3	481.7
7	Niemann	Jeff	2010	TBA	AL	**15.177**	3	**5.059**	27	16	0	371.0
8	Balfour	Grant	2010	TBA	AL	**15.147**	4	**3.787**	14	7	8	203.0
9	Wheeler	Dan	2010	TBA	AL	**15.087**	7	**2.155**	13	25	18	268.7
10	Howell	J.P.	2009	TBA	AL	**14.924**	4	**3.731**	15	15	20	249.3
11	Arrojo	Rolando	1999	TBA	AL	**14.832**	2	**7.416**	21	24	0	342.7
12	Soriano	Rafael	2010	TBA	AL	**14.495**	1	**14.495**	3	2	45	62.3
13	Hendrickson	Mark	2006	TBA	AL	**13.932**	3	**4.644**	25	31	0	451.3
14	Jackson	Edwin	2008	TBA	AL	**13.059**	3	**4.353**	19	26	0	380.7
15	Sonnanstine	Andy	2010	TBA	AL	**13.472**	4	**3.368**	28	29	1	504.7
16	Yan	Esteban	2002	TBA	AL	**12.145**	5	**2.429**	26	30	42	418.7
17	Sturtze	Tanyon	2002	TBA	AL	**11.929**	3	**3.976**	19	30	1	472.0
18	Baez	Danys	2005	TBA	AL	**11.405**	2	**5.703**	9	8	71	140.3

Texas Rangers (WS2, TEX)

Rank	Name	First	LYear	LTeam	Lg	PEVA-P	YRS	PER YR	W	L	SV	IP
1	Hough	Charlie	1990	TEX	AL	**104.223**	11	**9.475**	139	123	1	2308.0
2	Rogers	Kenny	2005	TEX	AL	**96.190**	12	**8.016**	133	96	28	1909.0
3	Jenkins	Fergie	1981	TEX	AL	**71.298**	6	**11.883**	93	72	0	1410.3
4	Brown	Kevin	1994	TEX	AL	**59.377**	8	**7.422**	78	64	0	1278.7
5	Witt	Bobby	1998	TEX	AL	**50.054**	11	**4.550**	104	104	0	1680.7
6	Ryan	Nolan	1993	TEX	AL	**48.016**	5	**9.603**	51	39	0	840.0
7	Helling	Rick	2001	TEX	AL	**47.706**	8	**5.963**	68	51	0	1008.0
8	Bosman	Dick	1973	TEX	AL	**39.545**	8	**4.943**	59	64	2	1103.3
9	Russell	Jeff	1996	TEX	AL	**39.226**	10	**3.923**	42	40	134	752.7
10	Perry	Gaylord	1980	TEX	AL	**38.047**	4	**9.512**	48	43	0	827.3
11	Millwood	Kevin	2009	TEX	AL	**36.162**	4	**9.040**	48	46	0	755.0
12	Wetteland	John	2000	TEX	AL	**35.347**	4	**8.837**	20	12	150	253.0
13	Guzman	Jose	1992	TEX	AL	**34.705**	6	**5.784**	66	62	0	1013.7
14	Cordero	Francisco	2006	TEX	AL	**34.114**	7	**4.873**	21	20	117	397.0
15	Matlack	Jon	1983	TEX	AL	**30.682**	6	**5.114**	43	45	3	915.0
16	Wilson	C.J.	2010	TEX	AL	**29.395**	6	**4.899**	27	28	52	484.7
17	Darwin	Danny	1995	TEX	AL	**28.432**	8	**3.554**	55	52	15	872.0
18	Oliver	Darren	2001	TEX	AL	**28.158**	8	**3.520**	54	47	2	842.7
19	Kline	Ron	1966	WS2	AL	**26.053**	4	**6.513**	26	25	83	364.7
20	Tanana	Frank	1985	TEX	AL	**25.483**	4	**6.371**	31	49	0	677.7

Toledo Blue Stockings

Rank	Name	First	LYear	LTeam	Lg	PEVA-P	YRS	PER YR	W	L	SV	IP
1	Mullane	Tony	1884	TL1	AA	**15.319**	1	**15.319**	36	26	0	567.0
2	O'Day	Hank	1884	TL1	AA	**2.252**	1	**2.252**	9	28	1	326.7
3	Brown	Ed	1884	TL1	AA	**0.200**	1	**0.200**	0	1	0	9.0
4	Kent	Ed	1884	TL1	AA	**0.200**	1	**0.200**	0	1	0	9.0
5	McSorley	Trick	1884	TL1	AA	**0.200**	1	**0.200**	0	0	0	2.0

Toledo Maumees

Rank	Name	First	LYear	LTeam	Lg	PEVA-P	YRS	PER YR	W	L	SV	IP
1	Healy	John	1890	TL2	AA	**14.100**	1	**14.100**	22	21	0	389.0
2	Smith	Fred	1890	TL2	AA	**5.225**	1	**5.225**	19	13	0	286.0
3	Cushman	Ed	1890	TL2	AA	**3.263**	1	**3.263**	17	21	1	315.7
4	Sprague	Charlie	1890	TL2	AA	**1.399**	1	**1.399**	9	5	0	122.7
5	Doty	Babe	1890	TL2	AA	**0.200**	1	**0.200**	1	0	0	9.0

Toronto Blue Jays

Rank	Name	First	LYear	LTeam	Lg	PEVA-P	YRS	PER YR	W	L	SV	IP
1	Halladay	Roy	2009	TOR	AL	**214.098**	12	**17.841**	148	76	1	2046.7
2	Stieb	Dave	1998	TOR	AL	**173.801**	15	**11.587**	175	134	3	2873.0
3	Key	Jimmy	1992	TOR	AL	**105.710**	9	**11.746**	116	81	10	1695.7
4	Hentgen	Pat	2004	TOR	AL	**88.578**	10	**8.858**	107	85	0	1636.0
5	Clancy	Jim	1988	TOR	AL	**84.708**	12	**7.059**	128	140	1	2204.7
6	Clemens	Roger	1998	TOR	AL	**80.252**	2	**40.126**	41	13	0	498.7
7	Henke	Tom	1992	TOR	AL	**74.479**	8	**9.310**	29	29	217	563.0
8	Guzman	Juan	1998	TOR	AL	**65.588**	8	**8.199**	76	62	0	1215.7
9	Wells	David	2000	TOR	AL	**61.483**	8	**7.685**	84	55	13	1148.7
10	Ward	Duane	1995	TOR	AL	**52.059**	9	**5.784**	32	36	121	650.7
11	Alexander	Doyle	1986	TOR	AL	**43.563**	4	**10.891**	46	26	0	750.0
12	Stottlemyre	Todd	1994	TOR	AL	**34.472**	7	**4.925**	69	70	1	1139.0
13	Eichhorn	Mark	1993	TOR	AL	**31.812**	6	**5.302**	29	19	15	493.0
14	Escobar	Kelvim	2003	TOR	AL	**30.725**	7	**4.389**	58	55	58	849.0
15	Leal	Luis	1985	TOR	AL	**30.387**	6	**5.064**	51	58	1	946.0
16	Burnett	A.J.	2008	TOR	AL	**28.502**	3	**9.501**	38	26	0	522.7
17	Quantrill	Paul	2001	TOR	AL	**26.825**	6	**4.471**	30	34	15	517.7
18	Downs	Scott	2010	TOR	AL	**26.265**	6	**4.378**	20	18	16	407.7
19	Carpenter	Chris	2002	TOR	AL	**25.937**	6	**4.323**	49	50	0	870.7
20	Lilly	Ted	2006	TOR	AL	**24.752**	3	**8.251**	37	34	0	505.3

Troy Trojans

Rank	Name	First	LYear	LTeam	Lg	PEVA-P	YRS	PER YR	W	L	SV	IP
1	Welch	Mickey	1882	TRN	NL	**16.961**	3	**5.654**	69	64	0	1223.0
2	Keefe	Tim	1882	TRN	NL	**12.792**	3	**4.264**	41	59	0	882.0
3	Goldsmith	Fred	1879	TRN	NL	**4.900**	1	**4.900**	2	4	0	63.0
4	Bradley	George	1879	TRN	NL	**2.892**	1	**2.892**	13	40	0	487.0
5	Salisbury	Harry	1879	TRN	NL	**1.642**	1	**1.642**	4	6	0	89.0

Troy Haymakers

Rank	Name	First	LYear	LTeam	Lg	PEVA-P	YRS	PER YR	W	L	SV	IP
1	McMullin	John	1871	TRO	NA	**6.373**	1	**6.373**	12	15	0	249.0
2	Zettlein	George	1872	TRO	NA	**4.876**	1	**4.876**	14	8	0	187.7
3	Martin	Phonney	1872	TRO	NA	**0.209**	1	**0.209**	1	2	0	37.3
4	Flowers	Dickie	1871	TRO	NA	**0.200**	1	**0.200**	0	0	0	1.0

Washington Nationals (WAS, MON)

Rank	Name	First	LYear	LTeam	Lg	PEVA-P	YRS	PER YR	W	L	SV	IP
1	Rogers	Steve	1985	MON	NL	**159.201**	13	**12.246**	158	152	2	2837.7
2	Martinez	Dennis	1993	MON	NL	**118.829**	8	**14.854**	100	72	1	1609.0
3	Martinez	Pedro	1997	MON	NL	**80.240**	4	**20.060**	55	33	1	797.3
4	Vazquez	Javier	2003	MON	NL	**75.095**	6	**12.516**	64	68	0	1229.3
5	Smith	Bryn	1989	MON	NL	**68.521**	9	**7.613**	81	71	6	1400.3
6	Hernandez	Livan	2010	WAS	NL	**67.430**	6	**11.238**	62	59	0	1141.7
7	Fassero	Jeff	1996	MON	NL	**63.444**	6	**10.574**	58	48	10	850.0
8	Gullickson	Bill	1985	MON	NL	**56.505**	7	**8.072**	72	61	0	1186.3
9	Burke	Tim	1991	MON	NL	**42.498**	7	**6.071**	43	26	101	600.3
10	Hill	Ken	1994	MON	NL	**41.059**	3	**13.686**	41	21	0	556.3
11	Rojas	Mel	1999	MON	NL	**39.056**	8	**4.882**	29	23	109	512.3
12	Sanderson	Scott	1983	MON	NL	**39.053**	6	**6.509**	56	47	2	883.0
13	Reardon	Jeff	1986	MON	NL	**38.090**	6	**6.348**	32	37	152	506.3
14	Lea	Charlie	1987	MON	NL	**38.025**	6	**6.337**	55	41	0	793.3

Rank	Name	First	LYear	LTeam	Lg	PEVA-P	YRS	PER YR	W	L	SV	IP
15	Hermanson	Dustin	2000	MON	NL	**36.488**	4	**9.122**	43	47	4	759.7
16	Wetteland	John	1994	MON	NL	**35.557**	3	**11.852**	17	13	105	232.3
17	Cordero	Chad	2008	WAS	NL	**35.248**	6	**5.875**	20	14	128	320.7
18	Urbina	Ugueth	2001	MON	NL	**30.955**	7	**4.422**	31	26	125	406.7
19	Fryman	Woodie	1983	MON	NL	**30.931**	8	**3.866**	51	52	52	721.7
20	Perez	Pascual	1989	MON	NL	**30.691**	3	**10.230**	28	21	0	456.7

Wilmington Quicksteps

Rank	Name	First	LYear	LTeam	Lg	PEVA-P	YRS	PER YR	W	L	SV	IP
1	Nolan	The Only	1884	WIL	UA	**0.999**	1	**0.999**	1	4	0	40.0
2	Tenney	Fred	1884	WIL	UA	**0.393**	1	**0.393**	0	1	0	8.0
3	Murphy	John	1884	WIL	UA	**0.323**	1	**0.323**	0	6	0	48.0
4	Bastian	Charlie	1884	WIL	UA	**0.200**	1	**0.200**	0	0	0	6.0
5	Casey	Dan	1884	WIL	UA	**0.200**	1	**0.200**	1	1	0	18.0

Worcester Ruby Legs

Rank	Name	First	LYear	LTeam	Lg	PEVA-P	YRS	PER YR	W	L	SV	IP
1	Richmond	Lee	1882	WOR	NL	**24.438**	3	**8.146**	71	91	3	1464.0
2	Corey	Fred	1882	WOR	NL	**2.878**	3	**0.959**	15	37	2	476.0
3	Mountain	Frank	1882	WOR	NL	**0.765**	2	**0.383**	2	16	0	144.0
4	McCormick	Harry	1881	WOR	NL	**0.459**	1	**0.459**	1	8	0	78.3
5	Clarkson	John	1882	WOR	NL	**0.200**	1	**0.200**	1	2	0	24.0

Washington Olympics

Rank	Name	First	LYear	LTeam	Lg	PEVA-P	YRS	PER YR	W	L	SV	IP
1	Brainard	Asa	1872	WS3	NA	**7.743**	2	**3.872**	14	22	0	343.0
2	Stearns	Bill	1871	WS3	NA	**0.200**	1	**0.200**	2	0	0	18.0

Washington Nationals (National Association)

Rank	Name	First	LYear	LTeam	Lg	PEVA-P	YRS	PER YR	W	L	SV	IP
1	Stearns	Bill	1872	WS4	NA	**0.668**	1	**0.668**	0	11	0	99.0

Washington Blue Legs

Rank	Name	First	LYear	LTeam	Lg	PEVA-P	YRS	PER YR	W	L	SV	IP
1	Stearns	Bill	1873	WS5	NA	**2.029**	1	**2.029**	7	25	0	283.0
2	Greason	John	1873	WS5	NA	**0.482**	1	**0.482**	1	6	0	63.0

Washington Nationals (National Association)

Rank	Name	First	LYear	LTeam	Lg	PEVA-P	YRS	PER YR	W	L	SV	IP
1	Stearns	Bill	1875	WS6	NA	**1.014**	1	**1.014**	1	14	0	141.0
2	Parks	Bill	1875	WS6	NA	**0.639**	1	**0.639**	4	8	0	106.7
3	Mason	Charlie	1875	WS6	NA	**0.200**	1	**0.200**	0	0	0	2.0
4	Witherow	Charles	1875	WS6	NA	**0.200**	1	**0.200**	0	1	0	1.0

Washington Nationals (American Association)

Rank	Name	First	LYear	LTeam	Lg	PEVA-P	YRS	PER YR	W	L	SV	IP
1	Barr	Bob	1884	WS7	AA	**2.196**	1	**2.196**	9	23	0	281.0
2	Hamill	John	1884	WS7	AA	**0.855**	1	**0.855**	2	17	0	156.7
3	Trumbull	Ed	1884	WS7	AA	**0.622**	1	**0.622**	1	9	0	84.0
4	Smith	Edgar (EE	1884	WS7	AA	**0.200**	1	**0.200**	0	2	0	22.0

Washington Nationals (National League)

Rank	Name	First	LYear	LTeam	Lg	PEVA-P	YRS	PER YR	W	L	SV	IP
1	Whitney	Jim	1888	WS8	NL	**18.697**	2	**9.349**	42	42	0	729.7
2	O'Day	Hank	1889	WS8	NL	**8.680**	4	**2.170**	28	61	0	814.7
3	Gilmore	Frank	1888	WS8	NL	**5.022**	3	**1.674**	12	33	0	405.3
4	Shaw	Dupee	1888	WS8	NL	**4.986**	3	**1.662**	20	47	0	592.0
5	Keefe	George	1889	WS8	NL	**3.108**	4	**0.777**	14	29	0	383.3

Washington Statesmen/Senators (WSN, WS9)

Rank	Name	First	LYear	LTeam	Lg	PEVA-P	YRS	PER YR	W	L	SV	IP
1	Mercer	Win	1899	WSN	NL	**49.924**	6	**8.321**	94	116	8	1766.3
2	McJames	Doc	1897	WSN	NL	**16.405**	3	**5.468**	28	44	3	621.0
3	Weyhing	Gus	1899	WSN	NL	**13.857**	2	**6.928**	32	47	0	695.7
4	Maul	Al	1897	WSN	NL	**11.712**	5	**2.342**	38	44	0	701.3
5	Killen	Frank	1899	WSN	NL	**10.502**	3	**3.501**	35	37	0	600.0

Washington Nationals (Union League)

Rank	Name	First	LYear	LTeam	Lg	PEVA-P	YRS	PER YR	W	L	SV	IP
1	Wise	Bill	1884	WSU	UA	**4.409**	1	**4.409**	23	18	0	364.3
2	Geggus	Charlie	1884	WSU	UA	**3.031**	1	**3.031**	10	9	0	177.3
3	Voss	Alex	1884	WSU	UA	**1.105**	1	**1.105**	5	14	0	186.3
4	Powell	Abner	1884	WSU	UA	**0.990**	1	**0.990**	6	12	0	134.0
5	Lockwood	Milo	1884	WSU	UA	**0.614**	1	**0.614**	1	9	0	67.7

Note1: LYEAR, LTEAM, LG reflects last year with team.

Note2: Franchises with less than 5 players above 10.000 PEVA are listed with the five highest players.

Chapter 7 - BEST POSTSEASON SEASONS EVER

PEVA Player Rating Boxscore - Postseason			
Regular Season	Explanation		Postseason PEVA
64.000	Maximum	Highest PEVA Player Rating Available	6.400
32.000	Fantastic	MVP/CY Young Candidate	3.200
20.000	Great	All-League	2.000
15.000	Very Good	All-Star Caliber	1.500
10.000	Good	Plus Starter	1.000
3.500	Average	Bench Player	3.500
0.200	Minimum	Lowest PEVA Value	0.200

Ranked by Postseason PEVA Player Ratings, Postseaon Grades reflect 10% of Regular Season Values.

Players listed below reflect the Best Postseasons Ever for batters and pitchers. They include players from the three distinct eras of postseason professional Major League Baseball; the pre1900 era, noted by the Championship Series (CS) from 1884-1890 and 1892; the World Series (WS) only round postseasons between 1903 and 1968 (with no postseason held in 1904) between the National League regular season champ and the American League regular season champ; and the Multiple Round Playoffs (PL) beginning in 1969 through today, except for the strike season of 1994.

Because it's a whole lot easier to dominate a short amount of games than a full season of them, four position players and one pitcher have received a perfect (in relative terms) score for their postseason exploits in the Best Postseasons Ever Ratings. And it is a diverse group, not only in when but who. There are two catchers, one in the Hall of Fame, the other not, a Hall of Fame first baseman and recently added Hall of Fame outfielder, and a pitcher who is still in the prime of his career. There's a player from 1914, 1932, 1956, 1989, and 2007, which means they've covered two of the three playoff types, leaving out the pre-1900 mixture of the eight Championship Series years, although their Best Ever Batter came in at #18, not too far down the line, and their pitcher at #4. Anybody remember that 1884 series and the pitching of Charley Radbourn. Boy, was he great for the Providence Grays or what!

But there will be a whole lot of players that you'll remember, those enshrined already in Cooperstown or waiting to get there, plus those with highlights in World Series and playoffs not too far past. And while it seems to be much harder to push out a top PEVA score in the years after three playoff rounds became commonplace in 1995, that's not necessarily true. Since durability remains a factor in the PEVA equation; it's imperative that a player plays in all three rounds, therefore, it's the same dynamic no matter if you were a Championship Series player, a World Series only player, or one in the Multiple Playoff Round timeframe.

On to the players who created our World Series memories with postseaons of All-Time stats. And let's start out with the Best Position Players in a Postseason and those four men who performed at the highest PEVA level, a top score of 64.000; Yogi Berra in 1956, Lou Gehrig in 1932, Hank Gowdy in 1912, and Rickey Henderson in 1989.

Who Entered the Best Postseasons Ever List in 2010

Nobody jumped into the top ten, but two people, by the names of Nelson Cruz and Cody Ross, did move into pretty high spots on the list. Cruz cracked the Top 100 at #98. Ross just below that at #104. Sure, they weren't exactly household names at the beginning of the series between the Rangers and Giants, but by the end, everyone knew who they were. Just sneaking into the Top 400 was another Ranger who had a very good postseason in 2010, Ian Kinsler. No, Edgar Renteria didn't quite make it, although that last game blast sure put him into the never to forget moments in San Francisco history, that's for sure.

Best Batting Postseasons Ever

1. Yogi Berra, New York Yankees, 1956

He called the perfect game of Don Larsen, and in many ways that feat overshadowed Yogi's accomplishments. Berra caught an almost perfect series behind the plate, rating a top level 2.10 Field Value for the postseason. He knocked in ten runs and batted 0.360. Berra caught five consecutive games in which the starting pitcher of the Yankees pitched a complete game. Think that had nothing to do with Berra. In a seven game series between the Brooklyn Dodgers and the Bombers, won by the Yankees, Yogi Berra was the glue that held the entire team together, both in the field and at the plate. But Berra did not even win the World Series MVP, that went to Larson, and it's understandable to honor the only perfect game and no hitter in World Series history, we know, even though since Larson only pitched 10.7 innings in 1956, it didn't raise him to perfect status for us. So here's to you, Yogi, a bit late we know. As far as we're concerned, you stand at the top of the list, in the company of three other men as having one heck of a World Series, good enough to stand at #1 on the Best Postseason by a Position Player list.

Stat Dominance Chart - Postseason			
Yogi Berra 64.000		**Pct. of**	
Stats for Year		**PYr Ave**	**PAve**
RPR	15	273.7%	5.48
OBP	0.448	146.9%	0.305
SLG	0.800	220.4%	0.363
Field Value	2.10		

1. Lou Gehrig, New York Yankees, 1932

It's funny what a World Series becomes known for, whether it be the perfect game in 1956 or the shot called round the world in 1932. Yes, this was the series of vibrant taunting and the famous called Home Run shot by Babe Ruth. But this year it wasn't Ruth who did most of the talking with his bat, although he certainly did his share, coming in at #194, behind another teammate Earle Combs, too. This year it was Gehrig. Lou Gehrig hit three home runs and knocked in eight during the four game sweep of the Cubs. His OPS was off the charts at 1.718. But as time has gone by, it's a bit like Berra in 1956 falling behind Larson in baseball lore, Gehrig gets second mention to the Babe once again. One forgotten fact about that 5th Inning Game 3 shot by the Babe is the stat that another home run followed up that home run, hit by, you guessed it, Lou Gehrig.

Stat Dominance Chart - Postseason			
Lou Gehrig 64.000		**Pct. of**	
Stats for Year		**PYr Ave**	**PAve**
RPR	17	299.3%	5.68
OBP	0.600	163.9%	0.366
SLG	1.118	243.6%	0.459
Field Value	1.08		

1. Hank Gowdy, Boston Braves, 1914

They were the Miracle Braves, in last place on Independence Day, then the World Series champs in October in a four game sweep of the favored Athletics. And the main reason for their playoff run was a catcher in his first year of major action, Hank Gowdy. If you thought the OPS of Gehrig in the 1932 World Series was gawdy, just look at Hank's. 1.961, 434.5% of the average for the series. Gowdy's exploits are credited for winning two games, which was held in Fenway Park as the Bravos waited for the construction of their new field to be completed. And Gowdy was the man in Fenway in 1914. He would appear in two more

World Series later in his career and in many circles is more noted as the first baseball player to enlist in World War I, as well as enlisting again in World War II. Patriotic and the man who had one of the best Postseasons Ever, right up there with his three better known counterparts.

Stat Dominance Chart - Postseason			
Hank Gowdy 64.000		Pct. of	
Stats for Year		PYr Ave	PAve
RPR	6	229.9%	2.61
OBP	0.688	240.6%	0.286
SLG	1.273	434.5%	0.293
Field Value	2.10		

1. Rickey Henderson, Oakland A's, 1989

Eleven stolen bases and a run production of twenty in just nine games. That's the legacy of Rickey Henderson in the World Series that was once again noted for more than baseball itself. On October 17, in the 4th inning of Game 3, the 1989 earthquake hit California and the Bay Area. It was supposed to be a celebration of a bridge series, but this time that idea took on more serious consequences as the ground shook during play in Candlestick Park, causing a suspension of the series for ten days. When play resumed, the A's continued their domination, led by the man with power and disruptive speed. Henderson would not become the MVP of the series, just like Berra had not in 1956, losing again to a pitcher who had great performances, Dave Stewart. Stewart had performed well, too, coming in at #29 on the Best Postseason List for a pitcher. But it really was Rickey who led the charge from the top of the order and did not let go. Oakland led for every moment of the World Series, never once allowing San Francisco to get ahead of them in any game, the first time in history that had happened. And Rickey Henderson led the way.

Stat Dominance Chart - Postseason			
R. Henderson 64.000		Pct. of	
Stats for Year		PYr Ave	PAve
RPR	20	470.6%	4.25
OBP	0.568	170.6%	0.333
SLG	0.941	213.9%	0.440
Field Value	1.31		

5. Harry Hooper, Boston Red Sox, 1915

Okay, so you don't know who Harry Hooper is. Not many baseball fans do. But it was Hooper in 1915 that took the World Series, one with limited offensive production and led his Red Sox team to victory, 4 to 1, over the Philadelphia Phillies led by ace Grover Cleveland Alexander. He batted 0.350 with a slugging percentage of 0.650 when the average for the series was 0.303. It was a series of firsts and a couple odd twists. Woodrow Wilson became the first president to attend a World Series game, the first. Fenway Park, even though used during the regular season, was not used during the World Series as the owners of the Red Sox chose the new Boston Braves field as their home in the series because it was bigger. They did the same thing again when the Red Sox made it back to the series the next year, but did use Fenway Park as their World Series home in 1918. Can you imagine Red Sox nation putting up with that one today? Even odder was the affect that extra outfield seats had in a game at the A's Baker Bowl, making the small field even smaller and allowing Hooper to bounce two balls into the stands for home runs. Ground rule doubles were ground rule home runs then.

Stat Dominance Chart - Postseason			
Harry Hooper 62.837		Pct. of	
Stats for Year		PYr Ave	PAve
RPR	7	266.2%	2.63
OBP	0.435	155.4%	0.280

SLG	0.650	214.5%	0.303
Field Value	1.12		

6. Mel Ott, New York Giants, 1933

Mel Ott was the toast of the Giants at the time, a man who would hit 511 home runs, knock in 1,860, and bat 0.304 during his career, and in the 1933 World Series, he would not disappoint the fans of the Polo Grounds in postseason play, leading the Giants to a four to one victory over the Washington Senators. Ott hit two home runs and knocked in four with a slugging percentage of 0.722. The Giants would get back to the World Series two more times in 1936 and 1937 during Ott's reign in New York, but he would not lead his team to victory over the crosstown Yanks either season. It was a series of lasts for Washington, too. Washington, D.C. would never again host a World Series game after 1933. Some day, perhaps, the Nationals will end that streak.

Stat Dominance Chart - Postseason			
Mel Ott 62.642		**Pct. of**	
Stats for Year		**PYr Ave**	**PAve**
RPR	7	233.3%	3.00
OBP	0.500	172.4%	0.290
SLG	0.722	233.7%	0.309
Field Value	1.45		

7. Carl Yastrzemski, Boston Red Sox, 1967

It was a marvelous sight to see, Yastremski patrolling in front of the Green Monster, playing the caroms with perfection. 1967 had been a magical season for Carl, winning the Triple Crown, the American League MVP Award, and leading his Red Sox to the World Series. But Boston ran into a buzzsaw in Bob Gibson that postseason and could not win the ultimate crown, succumbing the fans of Fenway to wait nearly forty more years to get their elusive championship. It was hardly Yaz's fault as he slugged to the tune of 0.840 with three homers and five RBI's while playing stellar outfield. One footnote to the series was its odd winning ways. Each visitor team won every game of the series, with the Red Sox taking all games in Busch Stadium, and the Cardinals winning all four games in Fenway Park.

Stat Dominance Chart - Postseason			
Yastrzemski 62.624		**Pct. of**	
Stats for Year		**PYr Ave**	**PAve**
RPR	9	236.2%	3.81
OBP	0.500	181.8%	0.275
SLG	0.840	235.3%	0.357
Field Value	1.70		

8. Alex Rodriguez, New York Yankees, 2009

Up until the 2009 playoffs when you looked up the record of Alex Rodriguez in postseason play you might have seen the picture of a goat. But oh, how that would change. Rodriguez would plow through the postseason of 2009, particularly in the Division and League Championship rounds, hitting bomb after bomb into the seats of new Yankee Stadium and opposing parks, six of them in all. With an OPS of 1.308 and good defensive play at Third Base, Alex Rodriguez would lead the Yankees past the Minnesota Twins, the Los Angeles Angels of Anaheim, and then the Philadelphia Phillies during the three playoff rounds. His Run Production (Runs plus Runs Batted In) at 33 would be 505.4% higher than the average run production for a postseason player. No more livestock next to the Rodriguez name and postseason play after this postseason, that's for sure.

Stat Dominance Chart - Postseason		
A. Rodriguez 62.492		**Pct. of**

Stats for Year		PYr Ave	PAve
RPR	33	505.4%	6.53
OBP	0.500	152.9%	0.327
SLG	0.808	201.5%	0.401
Field Value	1.57		

9. Frank Robinson, Baltimore Orioles, 1966

Frank Robinson was the consummate professional baseball player in his tenure of play first in Cincinnati, then Baltimore, and later as a manager, but there was never a better moment in his career, one that included two MVP Awards and 12 All-Star selections, than his performance in the 1966 World Series when he won the Most Valuable Player Award and led the Orioles to a victory sweep. It was a series again dominated by pitching by both staffs, led by aces Dave McNally, Jim Palmer, Sandy Koufax and Don Drysdale stifling the other team's batters to an OBP% of 0.246 and SLG% of 0.267. Boy, that's really low. The only hitter who was effective, who solved the pitching riddle of the series, was Frank, slugging at a pace 321% above the average. The Orioles won the last two games of the series 1-0, winning the final game on a solo home run by Frank Robinson. This was the first championship in franchise history.

Stat Dominance Chart - Postseason			
F. Robinson 62.344		**Pct. of**	
Stats for Year		**PYr Ave**	**PAve**
RPR	7	285.7%	2.45
OBP	0.375	152.4%	0.246
SLG	0.857	321.0%	0.267
Field Value	1.49		

10. Babe Ruth, New York Yankees, 1926

Seems odd that Babe Ruth's best postseason was in a losing effort. That wasn't usually the case. But with his 1.448 OPS and 4 Home Runs in 1926, it slightly edged out a winning turn in 1923, which comes in at #11, with 1928 not too far behind as well at #28. All told, Ruth goes on to appear a total of five times on the batting list. St. Louis would win the 1926 World Series in seven games, the first of their ten championships to date.

Stat Dominance Chart - Postseason			
Babe Ruth 62.088		**Pct. of**	
Stats for Year		**PYr Ave**	**PAve**
RPR	11	237.1%	4.64
OBP	0.548	172.3%	0.318
SLG	0.900	245.9%	0.366
Field Value	1.47		

Note: Field Value Maximums: Catcher 2.10, SS 1.75, OF/3B 1.70, 2B 1.50, 1B 1.40.

Note: PAve: Average for Postseason Year, RPR - Run Production (Runs + RBI) Ave. for Non-Zero players

Best Batting Postseasons Ever (Through 2010 Postseason)

													PEVA-B		
Rank	**Year**	**Name**	**First**	**Type**	**Team**	**Lg**	**G**	**HR**	**RBI**	**AVG**	**PA**	**RPR**	**FV**	**REG**	**POST**
1	1956	Berra	Yogi	WS	NYA	AL	7	3	10	0.360	29	15	2.10	**64.000**	**6.400**
1	1932	Gehrig	Lou	WS	NYA	AL	4	3	8	0.529	20	17	1.08	**64.000**	**6.400**
1	1914	Gowdy	Hank	WS	BSN	NL	4	1	3	0.545	16	6	2.10	**64.000**	**6.400**
1	1989	Henderson	Rickey	PL	OAK	AL	9	3	8	0.441	44	20	1.31	**64.000**	**6.400**
5	1915	Hooper	Harry	WS	BOS	AL	5	2	3	0.350	23	7	1.12	**62.837**	**6.284**
6	1933	Ott	Mel	WS	NY1	NL	5	2	4	0.389	22	7	1.45	**62.642**	**6.264**
7	1967	Yastrzemski	Carl	WS	BOS	AL	7	3	5	0.400	30	9	1.70	**62.624**	**6.262**
8	2009	Rodriguez	Alex	PL	NYA	AL	15	6	18	0.365	68	33	1.57	**62.492**	**6.249**

9	1966	Robinson	Frank	WS	BAL	AL	4	2	3	0.286	16	7	1.49	**62.344**	**6.234**
10	1926	Ruth	Babe	WS	NYA	AL	7	4	5	0.300	31	11	1.47	**62.088**	**6.209**
11	1923	Ruth	Babe	WS	NYA	AL	6	3	3	0.368	27	11	1.23	**61.557**	**6.156**
12	1905	Bresnahan	Roger	WS	NY1	NL	5	0	1	0.313	22	4	1.91	**61.259**	**6.126**
13	1915	Lewis	Duffy	WS	BOS	AL	5	1	5	0.444	20	6	1.70	**61.085**	**6.109**
14	1911	Baker	Frank	WS	PHA	AL	6	2	5	0.375	25	12	1.49	**60.480**	**6.048**
15	1939	Keller	Charlie	WS	NYA	AL	4	3	6	0.438	17	14	1.37	**60.460**	**6.046**
16	2002	Bonds	Barry	PL	SFN	NL	17	8	16	0.356	74	34	1.22	**60.380**	**6.038**
17	1983	Matthews	Gary	PL	PHI	NL	9	4	9	0.333	34	14	1.70	**60.342**	**6.034**
18	1892	Duffy	Hugh	CS	BSN	NL	6	1	9	0.462	27	12	1.46	**60.186**	**6.019**
19	1927	Ruth	Babe	WS	NYA	AL	4	2	7	0.400	18	11	1.53	**60.021**	**6.002**
20	1960	Mantle	Mickey	WS	NYA	AL	7	3	11	0.400	33	19	1.52	**59.978**	**5.998**
21	1928	Ruth	Babe	WS	NYA	AL	4	3	4	0.625	17	13	1.53	**59.503**	**5.950**
22	1928	Gehrig	Lou	WS	NYA	AL	4	4	9	0.545	17	14	1.30	**57.470**	**5.747**
23	1913	Collins	Eddie	WS	PHA	AL	5	0	3	0.421	22	8	1.50	**56.762**	**5.676**
24	1930	Simmons	Al	WS	PHA	AL	6	2	4	0.364	24	8	1.70	**56.665**	**5.667**
25	1931	Martin	Pepper	WS	SLN	NL	7	1	5	0.500	26	10	1.66	**56.065**	**5.606**
26	1967	Brock	Lou	WS	SLN	NL	7	1	3	0.414	31	11	1.54	**56.032**	**5.603**
27	1944	McQuinn	George	WS	SLA	AL	6	1	5	0.438	24	7	1.03	**55.103**	**5.510**
28	1963	Skowron	Bill	WS	LAN	NL	4	1	3	0.385	14	5	1.40	**54.172**	**5.417**
29	2003	Rodriguez	Ivan	PL	FLO	NL	17	3	17	0.313	77	27	1.88	**53.829**	**5.383**
30	1931	Simmons	Al	WS	PHA	AL	7	2	8	0.333	30	12	1.70	**53.555**	**5.356**
31	1978	Jackson	Reggie	PL	NYA	AL	10	4	14	0.417	45	21	1.44	**53.527**	**5.353**
32	1977	Jackson	Reggie	PL	NYA	AL	11	5	9	0.306	42	20	1.69	**53.085**	**5.309**
33	1884	Gilligan	Barney	CS	PRO	NL	3	0	2	0.444	9	5	2.07	**53.062**	**5.306**
34	1952	Snider	Duke	WS	BRO	NL	7	4	8	0.345	31	13	1.65	**52.801**	**5.280**
35	1985	Brett	George	PL	KCA	AL	14	3	6	0.360	61	17	1.59	**52.225**	**5.223**
36	1950	DiMaggio	Joe	WS	NYA	AL	4	1	2	0.308	17	4	1.45	**52.107**	**5.211**
37	1913	Baker	Frank	WS	PHA	AL	5	1	7	0.450	21	9	1.27	**51.709**	**5.171**
38	1993	Molitor	Paul	PL	TOR	AL	12	3	13	0.447	55	30	1.17	**51.398**	**5.140**
39	1972	Bench	Johnny	PL	CIN	NL	12	2	3	0.293	48	10	2.00	**50.909**	**5.091**
40	1915	Luderus	Fred	WS	PHI	NL	5	1	6	0.438	18	7	1.12	**50.662**	**5.066**
41	1948	Elliott	Bob	WS	BSN	NL	6	2	5	0.333	23	9	1.19	**50.216**	**5.022**
42	1957	Aaron	Hank	WS	ML1	NL	7	3	7	0.393	29	12	1.44	**50.155**	**5.016**
43	2000	Piazza	Mike	PL	NYN	NL	14	4	8	0.302	62	19	2.09	**49.615**	**4.961**
44	1941	Keller	Charlie	WS	NYA	AL	5	0	5	0.389	21	10	1.55	**49.450**	**4.945**
45	1970	Robinson	Brooks	PL	BAL	AL	8	2	8	0.485	34	16	1.70	**49.434**	**4.943**
46	1962	Tresh	Tom	WS	NYA	AL	7	1	4	0.321	30	9	1.56	**49.420**	**4.942**
47	1930	Cochrane	Mickey	WS	PHA	AL	6	2	4	0.222	24	9	1.84	**48.284**	**4.828**
48	1990	Hatcher	Billy	PL	CIN	NL	8	1	4	0.519	31	12	1.70	**48.270**	**4.827**
49	1945	Greenberg	Hank	WS	DET	AL	7	2	7	0.304	31	14	1.13	**48.038**	**4.804**
50	1962	Hiller	Chuck	WS	SFN	NL	7	1	5	0.269	29	9	1.33	**47.984**	**4.798**

Best Batting Postseasons Ever (Through 2010 Postseason)

														PEVA-B	
Rank	**Year**	**Name**	**First**	**Type**	**Team**	**Lg**	**G**	**HR**	**RBI**	**AVG**	**PA**	**RPR**	**FV**	**REG**	**POST**
51	1953	Martin	Billy	WS	NYA	AL	6	2	8	0.500	25	13	1.41	**47.498**	**4.750**
52	1981	Garvey	Steve	PL	LAN	NL	16	3	6	0.359	66	15	1.40	**47.466**	**4.747**
53	2004	Beltran	Carlos	PL	HOU	NL	12	8	14	0.435	56	35	1.54	**47.219**	**4.722**
54	2000	Jeter	Derek	PL	NYA	AL	16	4	9	0.317	75	22	1.65	**46.858**	**4.686**
55	1936	Powell	Jake	WS	NYA	AL	6	1	5	0.455	26	13	1.48	**46.763**	**4.676**
56	1925	Harris	Joe	WS	WS1	AL	7	3	6	0.440	28	11	1.44	**46.667**	**4.667**
57	1984	Trammell	Alan	PL	DET	AL	8	3	9	0.419	37	16	1.40	**46.577**	**4.658**
58	1964	McCarver	Tim	WS	SLN	NL	7	1	5	0.478	29	9	2.07	**46.461**	**4.646**
59	1951	Dark	Alvin	WS	NY1	NL	6	1	4	0.417	26	9	1.41	**45.968**	**4.597**
60	1996	Williams	Bernie	PL	NYA	AL	15	6	15	0.345	69	29	1.64	**44.716**	**4.472**
61	1989	Clark	Will	PL	SFN	NL	9	2	8	0.472	39	18	1.40	**44.640**	**4.464**

62	1927	Koenig	Mark	WS	NYA	AL	4	0	2	0.500	18	7	1.60	**44.321**	**4.432**
63	1974	Garvey	Steve	PL	LAN	NL	9	2	6	0.385	40	12	1.19	**44.304**	**4.430**
64	1968	Brock	Lou	WS	SLN	NL	7	2	5	0.464	31	11	1.27	**44.197**	**4.420**
65	1998	Brosius	Scott	PL	NYA	AL	13	4	15	0.383	51	21	1.54	**44.197**	**4.420**
66	1920	Speaker	Tris	WS	CLE	AL	7	0	1	0.320	28	7	1.62	**43.495**	**4.349**
67	1990	Sabo	Chris	PL	CIN	NL	10	3	8	0.368	42	11	1.70	**43.443**	**4.344**
68	1981	Carter	Gary	PL	MON	NL	10	2	6	0.429	41	12	2.10	**43.353**	**4.335**
69	1973	Staub	Rusty	PL	NYN	NL	11	4	11	0.341	46	16	1.43	**43.309**	**4.331**
70	1954	Thompson	Hank	WS	NY1	NL	4	0	2	0.364	18	8	1.70	**43.225**	**4.322**
71	1973	Campaneris	Bert	PL	OAK	AL	12	3	6	0.308	57	15	1.69	**43.006**	**4.301**
72	1976	Bench	Johnny	PL	CIN	NL	7	3	7	0.444	28	14	2.10	**42.608**	**4.261**
73	1886	O'Neill	Tip	WS	SL4	AA	6	2	5	0.400	24	9	1.59	**42.090**	**4.209**
74	1890	Wolf	Jimmy	WS	LS2	AA	7	0	8	0.360	28	12	1.37	**41.638**	**4.164**
75	1965	Fairly	Ron	WS	LAN	NL	7	2	6	0.379	29	13	1.44	**41.621**	**4.162**
76	1970	Powell	Boog	PL	BAL	AL	8	3	11	0.355	36	19	1.30	**40.833**	**4.083**
77	1959	Kluszewski	Ted	WS	CHA	AL	6	3	10	0.391	25	15	1.40	**40.658**	**4.066**
78	1964	Mantle	Mickey	WS	NYA	AL	7	3	8	0.333	30	16	1.00	**40.433**	**4.043**
79	1971	Robertson	Bob	PL	PIT	NL	11	6	11	0.317	45	20	1.16	**40.364**	**4.036**
80	1979	Stargell	Willie	PL	PIT	NL	10	5	13	0.415	46	22	1.13	**40.097**	**4.010**
81	1945	Cavarretta	Phil	WS	CHN	NL	7	1	5	0.423	31	12	1.23	**39.955**	**3.995**
82	1906	Rohe	George	WS	CHA	AL	6	0	4	0.333	25	6	1.47	**39.810**	**3.981**
83	2004	Pujols	Albert	PL	SLN	NL	15	6	14	0.414	67	29	1.40	**39.515**	**3.951**
84	1941	Gordon	Joe	WS	NYA	AL	5	1	5	0.500	21	7	1.14	**39.391**	**3.939**
85	1947	Lindell	Johnny	WS	NYA	AL	6	0	7	0.500	24	10	1.32	**38.878**	**3.888**
86	2005	Berkman	Lance	PL	HOU	NL	14	2	14	0.333	60	20	1.52	**38.818**	**3.882**
87	2007	Ramirez	Manny	PL	BOS	AL	14	4	16	0.348	63	27	1.61	**38.517**	**3.852**
88	1885	Anson	Cap	WS	CHN	NL	7	0	0	0.423	28	8	1.31	**38.108**	**3.811**
89	1908	Schulte	Frank	WS	CHN	NL	5	0	2	0.389	22	6	1.70	**38.101**	**3.810**
90	1996	Lopez	Javy	PL	ATL	NL	15	3	8	0.365	61	20	2.10	**38.068**	**3.807**
91	1907	Steinfeldt	Harry	WS	CHN	NL	5	0	2	0.471	21	4	1.47	**38.023**	**3.802**
92	1884	Denny	Jerry	WS	PRO	NL	3	1	2	0.444	9	5	1.49	**38.006**	**3.801**
93	1938	Gordon	Joe	WS	NYA	AL	4	1	6	0.400	16	9	1.25	**37.632**	**3.763**
94	1984	Gibson	Kirk	PL	DET	AL	8	3	9	0.367	37	15	1.31	**37.620**	**3.762**
95	1939	Dickey	Bill	WS	NYA	AL	4	2	5	0.267	16	7	2.10	**37.019**	**3.702**
96	1972	Tenace	Gene	PL	OAK	AL	12	4	10	0.225	45	16	1.73	**36.996**	**3.700**
97	1952	Mantle	Mickey	WS	NYA	AL	7	2	3	0.345	32	8	1.52	**36.978**	**3.698**
98	2010	Cruz	Nelson	PL	TEX	AL	16	6	11	0.317	63	24	1.63	**36.923**	**3.692**
99	1885	Kelly	King	WS	CHN	NL	7	0	0	0.346	28	9	1.45	**36.670**	**3.667**
100	2004	Ortiz	David	PL	BOS	AL	14	5	19	0.400	68	32	1.36	**36.660**	**3.666**

Best Batting Postseasons Ever (Through 2010 Postseason)

														PEVA-B	
Rank	**Year**	**Name**	**First**	**Type**	**Team**	**Lg**	**G**	**HR**	**RBI**	**AVG**	**PA**	**RPR**	**FV**	**REG**	**POST**
101	1993	Dykstra	Lenny	PL	PHI	NL	12	6	10	0.313	60	24	1.70	**36.399**	**3.640**
102	1922	Groh	Heinie	WS	NY1	NL	5	0	0	0.474	21	4	1.70	**36.327**	**3.633**
103	1979	Garner	Phil	PL	PIT	NL	10	1	6	0.472	41	14	1.42	**36.032**	**3.603**
104	2010	Ross	Cody	PL	SFN	NL	15	5	10	0.294	59	21	1.52	**35.975**	**3.598**
105	1884	Hines	Paul	WS	PRO	NL	3	0	1	0.250	11	6	1.57	**35.971**	**3.597**
106	1949	Brown	Bobby	WS	NYA	AL	4	0	5	0.500	14	9	1.57	**35.957**	**3.596**
107	1946	York	Rudy	WS	BOS	AL	7	2	5	0.261	30	11	1.00	**35.709**	**3.571**
108	1997	Sheffield	Gary	PL	FLO	NL	16	3	7	0.320	71	20	1.43	**35.476**	**3.548**
109	1914	Evers	Johnny	WS	BSN	NL	4	0	2	0.438	18	4	1.44	**34.849**	**3.485**
110	2007	Ortiz	David	PL	BOS	AL	14	3	10	0.370	63	26	0.85	**34.699**	**3.470**
111	1916	Hooper	Harry	WS	BOS	AL	5	0	1	0.333	24	7	1.64	**34.628**	**3.463**
112	1972	Rose	Pete	PL	CIN	NL	12	1	4	0.313	54	8	1.69	**34.616**	**3.462**
113	1991	Puckett	Kirby	PL	MIN	AL	12	4	9	0.333	54	17	1.70	**34.611**	**3.461**
114	1935	Gehringer	Charlie	WS	DET	AL	6	0	4	0.375	27	8	1.50	**34.526**	**3.453**

														PEVA-B	
Rank	Year	Name	First	Type	Team	Lg	G	HR	RBI	AVG	PA	RPR	FV	REG	POST
115	1957	Mathews	Eddie	WS	ML1	NL	7	1	4	0.227	31	8	1.70	**34.409**	**3.441**
116	1913	Schang	Wally	WS	PHA	AL	4	1	7	0.357	16	9	1.64	**34.310**	**3.431**
117	1980	Aikens	Willie	PL	KCA	AL	9	4	10	0.387	37	15	1.06	**34.232**	**3.423**
118	1971	Clemente	Roberto	PL	PIT	NL	11	2	8	0.383	50	13	1.64	**34.168**	**3.417**
119	1976	Chambliss	Chris	PL	NYA	AL	9	2	9	0.432	39	15	1.15	**34.099**	**3.410**
120	1919	Jackson	Joe	WS	CHA	AL	8	1	6	0.375	33	11	1.50	**34.072**	**3.407**
121	1961	Skowron	Bill	WS	NYA	AL	5	1	5	0.353	20	8	1.40	**33.875**	**3.387**
122	1988	Hatcher	Mickey	PL	LAN	NL	11	2	8	0.300	44	17	1.17	**33.856**	**3.386**
123	1958	Bauer	Hank	WS	NYA	AL	7	4	8	0.323	31	14	1.39	**33.818**	**3.382**
124	1974	Wynn	Jimmy	PL	LAN	NL	9	1	4	0.192	40	9	1.49	**33.753**	**3.375**
125	1968	Kaline	Al	WS	DET	AL	7	2	8	0.379	30	14	1.56	**33.588**	**3.359**
126	2006	Molina	Yadier	PL	SLN	NL	16	2	8	0.358	59	13	1.80	**33.577**	**3.358**
127	1905	Donlin	Mike	WS	NY1	NL	5	0	1	0.263	22	5	1.60	**33.566**	**3.357**
128	1912	Herzog	Buck	WS	NY1	NL	8	0	5	0.400	34	11	1.37	**33.397**	**3.340**
129	2009	Ruiz	Carlos	PL	PHI	NL	15	2	9	0.341	57	17	2.10	**33.283**	**3.328**
130	1967	Maris	Roger	WS	SLN	NL	7	1	7	0.385	30	10	1.31	**32.928**	**3.293**
131	1924	Goslin	Goose	WS	WS1	AL	7	3	7	0.344	32	11	1.67	**32.752**	**3.275**
132	2007	Youkilis	Kevin	PL	BOS	AL	14	4	10	0.388	59	26	1.08	**32.721**	**3.272**
133	1910	Murphy	Danny	WS	PHA	AL	5	1	9	0.400	22	15	1.45	**32.655**	**3.266**
134	1938	Crosetti	Frankie	WS	NYA	AL	4	1	6	0.250	19	7	1.75	**32.496**	**3.250**
135	1958	Mantle	Mickey	WS	NYA	AL	7	2	3	0.250	31	7	1.51	**31.869**	**3.187**
136	1955	Snider	Duke	WS	BRO	NL	7	4	7	0.320	28	12	1.47	**31.339**	**3.134**
137	1955	Berra	Yogi	WS	NYA	AL	7	1	2	0.417	28	7	2.08	**31.329**	**3.133**
138	1921	Meusel	Irish	WS	NY1	NL	8	1	7	0.345	31	11	1.64	**31.235**	**3.123**
139	1980	Otis	Amos	PL	KCA	AL	9	3	7	0.429	38	13	1.60	**31.203**	**3.120**
140	2009	Werth	Jayson	PL	PHI	NL	15	7	13	0.275	62	26	1.64	**31.167**	**3.117**
141	1986	Barrett	Marty	PL	BOS	AL	14	0	9	0.400	68	14	1.50	**31.151**	**3.115**
142	1937	Gehrig	Lou	WS	NYA	AL	5	1	3	0.294	22	7	1.40	**31.046**	**3.105**
143	1969	Agee	Tommie	PL	NYN	NL	8	3	5	0.250	36	10	1.62	**30.886**	**3.089**
144	1934	Martin	Pepper	WS	SLN	NL	7	0	4	0.355	34	12	1.24	**30.692**	**3.069**
145	1930	Foxx	Jimmie	WS	PHA	AL	6	1	3	0.333	23	6	1.40	**30.616**	**3.062**
146	2001	Finley	Steve	PL	ARI	NL	16	1	9	0.365	59	16	1.56	**30.486**	**3.049**
147	1956	Mantle	Mickey	WS	NYA	AL	7	3	4	0.250	30	10	1.66	**30.463**	**3.046**
148	2006	Pujols	Albert	PL	SLN	NL	16	3	6	0.288	65	17	1.19	**30.436**	**3.044**
149	1989	Henderson	Dave	PL	OAK	AL	9	3	5	0.281	39	15	1.61	**30.386**	**3.039**
150	1910	Baker	Frank	WS	PHA	AL	5	0	4	0.409	24	10	1.57	**30.320**	**3.032**

Best Batting Postseasons Ever (Through 2010 Postseason)

														PEVA-B	
Rank	**Year**	**Name**	**First**	**Type**	**Team**	**Lg**	**G**	**HR**	**RBI**	**AVG**	**PA**	**RPR**	**FV**	**REG**	**POST**
151	1929	Foxx	Jimmie	WS	PHA	AL	5	2	5	0.350	21	10	1.34	**30.250**	**3.025**
152	1937	Lazzeri	Tony	WS	NYA	AL	5	1	2	0.400	19	5	1.50	**30.239**	**3.024**
153	1962	Boyer	Clete	WS	NYA	AL	7	1	4	0.318	24	6	1.58	**30.142**	**3.014**
154	1958	McDougald	Gil	WS	NYA	AL	7	2	4	0.321	30	9	1.50	**29.989**	**2.999**
155	1909	Wagner	Honus	WS	PIT	NL	7	0	6	0.333	30	10	1.69	**29.815**	**2.982**
156	2006	Delgado	Carlos	PL	NYN	NL	10	4	11	0.351	43	19	1.00	**29.785**	**2.979**
157	1954	Wertz	Vic	WS	CLE	AL	4	1	3	0.500	18	5	1.03	**29.719**	**2.972**
158	2001	Posada	Jorge	PL	NYA	AL	17	2	3	0.273	66	12	1.91	**29.713**	**2.971**
159	1908	Chance	Frank	WS	CHN	NL	5	0	2	0.421	22	6	1.40	**29.657**	**2.966**
160	1976	Munson	Thurman	PL	NYA	AL	9	0	5	0.475	40	10	1.95	**29.558**	**2.956**
161	1927	Gehrig	Lou	WS	NYA	AL	4	0	4	0.308	18	6	1.40	**29.514**	**2.951**
162	1943	Dickey	Bill	WS	NYA	AL	5	1	4	0.278	20	5	2.10	**29.418**	**2.942**
163	1917	Collins	Eddie	WS	CHA	AL	6	0	2	0.409	24	6	1.50	**29.380**	**2.938**
164	1995	Jones	Chipper	PL	ATL	NL	14	3	8	0.364	65	18	1.69	**29.148**	**2.915**
165	1978	Lopes	Davey	PL	LAN	NL	10	5	12	0.341	46	22	1.35	**29.103**	**2.910**
166	1982	Porter	Darrell	PL	SLN	NL	10	1	6	0.351	43	10	2.10	**29.073**	**2.907**
167	1932	Combs	Earle	WS	NYA	AL	4	1	4	0.375	20	12	1.55	**29.051**	**2.905**

168	1981	Watson	Bob	PL	NYA	AL	14	2	9	0.340	55	13	1.00	**29.045**	**2.904**
169	1997	Alomar	Sandy	PL	CLE	AL	18	5	19	0.274	76	31	1.86	**28.987**	**2.899**
170	1962	Maris	Roger	WS	NYA	AL	7	1	5	0.174	28	9	1.70	**28.962**	**2.896**
171	1992	Borders	Pat	PL	TOR	AL	12	2	6	0.381	47	11	1.87	**28.960**	**2.896**
172	1982	Yount	Robin	PL	ML4	AL	12	1	6	0.356	52	13	1.52	**28.905**	**2.891**
173	2009	Jeter	Derek	PL	NYA	AL	15	3	6	0.344	74	20	1.75	**28.788**	**2.879**
174	2008	Upton	B.J.	PL	TBA	AL	16	7	16	0.288	72	32	1.63	**28.663**	**2.866**
175	1981	Randolph	Willie	PL	NYA	AL	14	3	6	0.240	61	13	1.43	**28.571**	**2.857**
176	1934	Greenberg	Hank	WS	DET	AL	7	1	7	0.321	32	11	1.00	**28.531**	**2.853**
177	1995	Griffey	Ken	PL	SEA	AL	11	6	9	0.364	52	20	1.28	**28.427**	**2.843**
178	2002	Glaus	Troy	PL	ANA	AL	16	7	13	0.344	69	28	1.30	**28.406**	**2.841**
179	1930	Dykes	Jimmie	WS	PHA	AL	6	1	5	0.222	25	7	1.32	**28.270**	**2.827**
180	1957	Berra	Yogi	WS	NYA	AL	7	1	2	0.320	29	7	1.83	**28.216**	**2.822**
181	1970	Johnson	Davey	PL	BAL	AL	8	2	6	0.333	34	12	1.50	**27.922**	**2.792**
182	1975	Fisk	Carlton	PL	BOS	AL	10	2	7	0.297	44	16	1.72	**27.838**	**2.784**
183	1944	Cooper	Walker	WS	SLN	NL	6	0	2	0.318	26	3	2.10	**27.790**	**2.779**
184	1917	Robertson	Dave	WS	NY1	NL	6	0	1	0.500	23	4	1.37	**27.721**	**2.772**
185	1888	Tiernan	Mike	WS	NY1	NL	10	1	6	0.342	46	14	1.48	**27.668**	**2.767**
186	1940	Greenberg	Hank	WS	DET	AL	7	1	6	0.357	30	11	1.26	**27.648**	**2.765**
187	1950	Coleman	Jerry	WS	NYA	AL	4	0	3	0.286	16	5	1.50	**27.639**	**2.764**
188	2001	Williams	Bernie	PL	NYA	AL	17	3	11	0.220	71	21	1.29	**27.365**	**2.736**
189	1929	Simmons	Al	WS	PHA	AL	5	2	5	0.300	22	11	1.35	**27.167**	**2.717**
190	1935	Herman	Billy	WS	CHN	NL	6	1	6	0.333	25	9	1.05	**27.044**	**2.704**
191	2003	Williams	Bernie	PL	NYA	AL	17	2	10	0.318	76	23	1.24	**26.760**	**2.676**
192	1905	McGann	Dan	WS	NY1	NL	5	0	4	0.235	20	5	1.30	**26.689**	**2.669**
193	1910	Collins	Eddie	WS	PHA	AL	5	0	3	0.429	24	8	1.50	**26.520**	**2.652**
194	1946	Slaughter	Enos	WS	SLN	NL	7	1	2	0.320	30	7	1.70	**26.399**	**2.640**
195	2003	Jeter	Derek	PL	NYA	AL	17	2	5	0.314	78	15	1.74	**26.396**	**2.640**
196	1932	Ruth	Babe	WS	NYA	AL	4	2	6	0.333	20	12	1.12	**26.376**	**2.638**
197	1916	Lewis	Duffy	WS	BOS	AL	5	0	1	0.353	23	4	1.42	**26.349**	**2.635**
198	1986	Henderson	Dave	PL	BOS	AL	12	3	9	0.324	41	18	1.56	**26.334**	**2.633**
199	1925	Carey	Max	WS	PIT	NL	7	0	2	0.458	31	8	1.19	**26.240**	**2.624**
200	1927	Waner	Lloyd	WS	PIT	NL	4	0	0	0.400	18	5	1.23	**26.230**	**2.623**

Best Batting Postseasons Ever (Through 2010 Postseason)

														PEVA-B	
Rank	**Year**	**Name**	**First**	**Type**	**Team**	**Lg**	**G**	**HR**	**RBI**	**AVG**	**PA**	**RPR**	**FV**	**REG**	**POST**
201	1909	Clarke	Fred	WS	PIT	NL	7	2	7	0.211	30	14	1.45	**26.084**	**2.608**
202	1993	White	Devon	PL	TOR	AL	12	2	9	0.373	56	20	1.66	**26.070**	**2.607**
203	1961	Robinson	Frank	WS	CIN	NL	5	1	4	0.200	20	7	1.44	**26.055**	**2.605**
204	1983	Murray	Eddie	PL	BAL	AL	9	3	6	0.257	39	13	1.26	**25.999**	**2.600**
205	1963	Davis	Tommy	WS	LAN	NL	4	0	2	0.400	15	2	1.46	**25.998**	**2.600**
206	1946	Walker	Harry	WS	SLN	NL	7	0	6	0.412	22	9	1.46	**25.991**	**2.599**
207	1995	McGriff	Fred	PL	ATL	NL	14	4	9	0.333	65	23	1.39	**25.797**	**2.580**
208	1903	Parent	Freddy	WS	BOS	AL	8	0	4	0.290	34	12	1.69	**25.794**	**2.579**
209	1951	Rizzuto	Phil	WS	NYA	AL	6	1	3	0.320	28	8	1.75	**25.778**	**2.578**
210	1929	Cochrane	Mickey	WS	PHA	AL	5	0	0	0.400	22	5	2.03	**25.737**	**2.574**
211	1998	Ramirez	Manny	PL	CLE	AL	10	4	7	0.343	43	11	1.38	**25.631**	**2.563**
212	1987	Pena	Tony	PL	SLN	NL	14	0	4	0.395	49	11	2.10	**25.620**	**2.562**
213	1981	Cey	Ron	PL	LAN	NL	11	1	9	0.316	45	13	1.70	**25.586**	**2.559**
214	1944	Sanders	Ray	WS	SLN	NL	6	1	1	0.286	26	6	1.31	**25.586**	**2.559**
215	1975	Rose	Pete	PL	CIN	NL	10	1	4	0.366	47	10	1.54	**25.513**	**2.551**
216	1950	Berra	Yogi	WS	NYA	AL	4	1	2	0.200	17	4	2.06	**25.493**	**2.549**
217	1909	Leach	Tommy	WS	PIT	NL	7	0	2	0.360	30	10	1.68	**25.490**	**2.549**

218	1925	Goslin	Goose	WS	WS1	AL	7	3	6	0.308	29	12	1.36	**25.440**	**2.544**
219	1922	Frisch	Frankie	WS	NY1	NL	5	0	2	0.471	20	5	1.32	**25.396**	**2.540**
220	1920	O'Neill	Steve	WS	CLE	AL	7	0	2	0.333	25	3	1.78	**25.367**	**2.537**
221	1926	Thevenow	Tommy	WS	SLN	NL	7	1	4	0.417	26	9	1.51	**25.365**	**2.537**
222	1933	Schulte	Fred	WS	WS1	AL	5	1	4	0.333	22	5	1.39	**25.361**	**2.536**
223	1981	Guerrero	Pedro	PL	LAN	NL	16	4	10	0.211	63	14	1.62	**25.245**	**2.524**
224	1982	Molitor	Paul	PL	ML4	AL	12	2	8	0.340	54	17	1.39	**25.201**	**2.520**
225	1992	Justice	David	PL	ATL	NL	13	3	9	0.227	56	18	1.60	**25.156**	**2.516**
226	1983	Dempsey	Rick	PL	BAL	AL	9	1	2	0.280	28	6	2.10	**25.080**	**2.508**
227	1981	Gamble	Oscar	PL	NYA	AL	10	2	6	0.409	30	11	1.47	**25.030**	**2.503**
228	2004	Matsui	Hideki	PL	NYA	AL	11	3	13	0.412	57	25	1.70	**25.005**	**2.501**
229	1938	Dickey	Bill	WS	NYA	AL	4	1	2	0.400	16	4	2.10	**24.877**	**2.488**
230	1961	Howard	Elston	WS	NYA	AL	5	1	1	0.250	22	6	2.07	**24.873**	**2.487**
231	1887	Thompson	Sam	WS	DTN	NL	15	2	7	0.362	61	15	1.46	**24.839**	**2.484**
232	1990	O'Neill	Paul	PL	CIN	NL	9	1	5	0.310	36	8	1.70	**24.754**	**2.475**
233	1923	Stengel	Casey	WS	NY1	NL	6	2	4	0.417	16	7	1.30	**24.621**	**2.462**
234	1924	Harris	Bucky	WS	WS1	AL	7	2	7	0.333	34	12	1.47	**24.614**	**2.461**
235	1960	Richardson	Bobby	WS	NYA	AL	7	1	12	0.367	31	20	1.09	**24.412**	**2.441**
236	1939	DiMaggio	Joe	WS	NYA	AL	4	1	3	0.313	17	6	1.57	**24.357**	**2.436**
237	1989	Lansford	Carney	PL	OAK	AL	7	1	8	0.444	32	15	1.60	**24.324**	**2.432**
238	1923	Dugan	Joe	WS	NYA	AL	6	1	5	0.280	28	10	1.70	**24.218**	**2.422**
239	1977	Baker	Dusty	PL	LAN	NL	10	3	13	0.316	41	21	1.19	**24.195**	**2.420**
240	1943	Sanders	Ray	WS	SLN	NL	5	1	2	0.294	20	5	1.05	**24.179**	**2.418**
241	1954	Mays	Willie	WS	NY1	NL	4	0	3	0.286	18	7	1.70	**24.161**	**2.416**
242	1940	Campbell	Bruce	WS	DET	AL	7	1	5	0.360	30	9	1.41	**24.151**	**2.415**
243	2003	Conine	Jeff	PL	FLO	NL	17	1	5	0.367	71	15	1.15	**24.115**	**2.411**
244	1999	Garciaparra	Nomar	PL	BOS	AL	10	4	9	0.406	38	17	1.36	**24.093**	**2.409**
245	1973	Jones	Cleon	PL	NYN	NL	12	1	4	0.292	55	12	1.34	**24.070**	**2.407**
246	2009	Utley	Chase	PL	PHI	NL	15	6	10	0.296	66	25	1.12	**24.069**	**2.407**
247	1973	Rudi	Joe	PL	OAK	AL	12	1	7	0.289	53	11	1.70	**23.981**	**2.398**
248	1990	Larkin	Barry	PL	CIN	NL	10	0	2	0.300	45	10	1.73	**23.922**	**2.392**
249	1953	Berra	Yogi	WS	NYA	AL	6	1	4	0.429	26	7	2.03	**23.876**	**2.388**
250	1987	Brunansky	Tom	PL	MIN	AL	12	2	11	0.286	50	21	1.56	**23.731**	**2.373**

Best Batting Postseasons Ever (Through 2010 Postseason)

														PEVA-B	
Rank	**Year**	**Name**	**First**	**Type**	**Team**	**Lg**	**G**	**HR**	**RBI**	**AVG**	**PA**	**RPR**	**FV**	**REG**	**POST**
251	1981	Milbourne	Larry	PL	NYA	AL	14	0	4	0.327	58	14	1.46	**23.692**	**2.369**
252	1961	Lopez	Hector	WS	NYA	AL	4	1	7	0.333	12	10	1.46	**23.613**	**2.361**
253	1977	Munson	Thurman	PL	NYA	AL	11	2	8	0.304	49	15	2.10	**23.575**	**2.358**
254	2003	Pierre	Juan	PL	FLO	NL	17	0	7	0.301	83	19	1.61	**23.485**	**2.349**
255	1949	Campanella	Roy	WS	BRO	NL	5	1	2	0.267	18	4	2.10	**23.460**	**2.346**
256	1981	Piniella	Lou	PL	NYA	AL	12	2	8	0.367	30	13	1.00	**23.414**	**2.341**
257	1975	Perez	Tony	PL	CIN	NL	10	4	11	0.250	44	18	1.30	**23.374**	**2.337**
258	1942	Kurowski	Whitey	WS	SLN	NL	5	1	5	0.267	18	8	1.22	**23.371**	**2.337**
259	1975	Yastrzemski	Carl	PL	BOS	AL	10	1	6	0.350	46	17	1.52	**23.364**	**2.336**
260	1951	Irvin	Monte	WS	NY1	NL	6	0	2	0.458	26	5	1.28	**23.164**	**2.316**
261	1940	Werber	Billy	WS	CIN	NL	7	0	2	0.370	31	7	1.60	**23.084**	**2.308**
262	1996	Jones	Andruw	PL	ATL	NL	14	3	9	0.345	37	16	1.52	**23.065**	**2.306**
263	2003	Alou	Moises	PL	CHN	NL	12	2	8	0.388	52	15	1.50	**23.010**	**2.301**
264	1996	McGriff	Fred	PL	ATL	NL	16	5	16	0.281	69	27	1.34	**22.901**	**2.290**
265	1936	Gehrig	Lou	WS	NYA	AL	6	2	7	0.292	28	12	1.37	**22.901**	**2.290**
266	2006	Monroe	Craig	PL	DET	AL	13	5	9	0.240	54	20	1.60	**22.876**	**2.288**
267	1906	Donahue	Jiggs	WS	CHA	AL	6	0	4	0.278	24	4	1.40	**22.861**	**2.286**
268	2007	Lowell	Mike	PL	BOS	AL	14	2	15	0.353	61	25	1.62	**22.848**	**2.285**
269	2002	Aurilia	Rich	PL	SFN	NL	17	6	17	0.265	77	30	1.68	**22.813**	**2.281**
270	1935	Fox	Pete	WS	DET	AL	6	0	4	0.385	26	5	1.43	**22.703**	**2.270**

271	1912	Murray	Red	WS	NY1	NL	8	0	4	0.323	34	9	1.70	**22.592**	**2.259**
272	1981	Nettles	Graig	PL	NYA	AL	11	1	10	0.282	46	14	1.60	**22.554**	**2.255**
273	1908	Cobb	Ty	WS	DET	AL	5	0	4	0.368	21	7	1.00	**22.542**	**2.254**
274	1987	Gaetti	Gary	PL	MIN	AL	12	3	9	0.277	53	18	1.70	**22.510**	**2.251**
275	1888	O'Neill	Tip	WS	SL4	AA	10	2	11	0.243	43	19	1.42	**22.489**	**2.249**
276	1987	Gladden	Dan	PL	MIN	AL	12	1	12	0.314	58	20	1.56	**22.363**	**2.236**
277	1885	Pfeffer	Fred	WS	CHN	NL	7	1	0	0.407	27	5	1.26	**22.306**	**2.231**
278	2004	Walker	Larry	PL	SLN	NL	15	6	11	0.293	67	25	1.56	**22.288**	**2.229**
279	1889	Ward	John	WS	NY1	NL	9	0	7	0.417	41	17	1.50	**22.248**	**2.225**
280	2006	Edmonds	Jim	PL	SLN	NL	16	2	10	0.250	62	18	1.70	**22.216**	**2.222**
281	2002	Spiezio	Scott	PL	ANA	AL	16	3	19	0.327	66	29	1.40	**22.134**	**2.213**
282	1956	Hodges	Gil	WS	BRO	NL	7	1	8	0.304	27	13	1.40	**22.076**	**2.208**
283	1926	Southworth	Billy	WS	SLN	NL	7	1	4	0.345	30	10	1.64	**22.034**	**2.203**
284	1983	Ripken	Cal	PL	BAL	AL	9	0	2	0.273	39	9	1.74	**21.907**	**2.191**
285	1974	Ferguson	Joe	PL	LAN	NL	9	1	4	0.172	40	9	1.66	**21.864**	**2.186**
286	1940	Higgins	Pinky	WS	DET	AL	7	1	6	0.333	27	8	1.36	**21.791**	**2.179**
287	1968	McCarver	Tim	WS	SLN	NL	7	1	4	0.333	30	7	2.10	**21.697**	**2.170**
288	1980	Brett	George	PL	KCA	AL	9	3	7	0.343	38	13	1.64	**21.596**	**2.160**
289	1968	Horton	Willie	WS	DET	AL	7	1	3	0.304	29	9	1.51	**21.545**	**2.154**
290	1956	Snider	Duke	WS	BRO	NL	7	1	4	0.304	30	9	1.63	**21.400**	**2.140**
291	1985	Sundberg	Jim	PL	KCA	AL	14	1	7	0.208	55	16	2.10	**21.325**	**2.132**
292	1947	Johnson	Billy	WS	NYA	AL	7	0	2	0.269	30	10	1.70	**21.286**	**2.129**
293	2000	Martinez	Tino	PL	NYA	AL	16	1	7	0.364	71	17	1.19	**21.282**	**2.128**
294	1999	Jeter	Derek	PL	NYA	AL	12	1	4	0.375	53	14	1.59	**21.224**	**2.122**
295	1995	Buhner	Jay	PL	SEA	AL	11	4	8	0.383	51	15	1.52	**21.216**	**2.122**
296	1943	Marion	Marty	WS	SLN	NL	5	1	2	0.357	18	3	1.16	**21.200**	**2.120**
297	1965	Wills	Maury	WS	LAN	NL	7	0	3	0.367	31	6	1.75	**21.048**	**2.105**
298	1950	Woodling	Gene	WS	NYA	AL	4	0	1	0.429	16	3	1.02	**21.029**	**2.103**
299	1986	Rice	Jim	PL	BOS	AL	14	2	6	0.241	65	20	1.70	**21.019**	**2.102**
300	1981	Jackson	Reggie	PL	NYA	AL	10	3	6	0.278	40	14	1.07	**21.002**	**2.100**

Best Batting Postseasons Ever (Through 2010 Postseason)

														PEVA-B	
Rank	**Year**	**Name**	**First**	**Type**	**Team**	**Lg**	**G**	**HR**	**RBI**	**AVG**	**PA**	**RPR**	**FV**	**REG**	**POST**
301	1986	Evans	Dwight	PL	BOS	AL	14	3	13	0.259	61	19	1.37	**20.938**	**2.094**
302	1991	Olson	Greg	PL	ATL	NL	14	1	5	0.275	60	11	2.10	**20.908**	**2.091**
303	1918	Whiteman	George	WS	BOS	AL	6	0	1	0.250	23	3	1.45	**20.840**	**2.084**
304	1980	Schmidt	Mike	PL	PHI	NL	11	2	8	0.289	51	15	1.70	**20.757**	**2.076**
305	1957	Bauer	Hank	WS	NYA	AL	7	2	6	0.258	32	9	1.43	**20.690**	**2.069**
306	1955	Campanella	Roy	WS	BRO	NL	7	2	4	0.259	31	8	2.08	**20.668**	**2.067**
307	1996	Jones	Chipper	PL	ATL	NL	16	1	9	0.345	67	20	1.46	**20.628**	**2.063**
308	1988	Sax	Steve	PL	LAN	NL	12	0	3	0.280	56	13	1.49	**20.572**	**2.057**
309	1887	Rowe	Jack	WS	DTN	NL	15	0	7	0.333	65	19	1.57	**20.561**	**2.056**
310	1906	Hofman	Solly	WS	CHN	NL	6	0	2	0.304	27	5	1.66	**20.533**	**2.053**
311	2003	Ramirez	Aramis	PL	CHN	NL	12	4	10	0.250	52	16	1.67	**20.511**	**2.051**
312	1917	Felsch	Happy	WS	CHA	AL	6	1	3	0.273	23	7	1.70	**20.511**	**2.051**
313	1972	Perez	Tony	PL	CIN	NL	12	0	4	0.326	48	7	1.40	**20.459**	**2.046**
314	1988	Hassey	Ron	PL	OAK	AL	9	1	4	0.375	20	6	1.98	**20.422**	**2.042**
315	1931	Foxx	Jimmie	WS	PHA	AL	7	1	3	0.348	29	6	1.00	**20.406**	**2.041**
316	1884	Farrell	Jack	WS	PRO	NL	3	0	0	0.444	9	3	1.30	**20.388**	**2.039**
317	1952	Reese	Pee Wee	WS	BRO	NL	7	1	4	0.345	32	8	1.44	**20.374**	**2.037**
318	1908	Evers	Johnny	WS	CHN	NL	5	0	2	0.350	22	7	1.30	**20.369**	**2.037**
319	1969	Robinson	Frank	PL	BAL	AL	8	2	3	0.250	35	6	1.23	**20.297**	**2.030**
320	1959	Neal	Charlie	WS	LAN	NL	6	2	6	0.370	27	10	1.41	**20.259**	**2.026**
321	2005	Konerko	Paul	PL	CHA	AL	12	5	15	0.265	52	21	1.40	**20.231**	**2.023**
322	1983	Roenicke	Gary	PL	BAL	AL	6	1	4	0.273	17	8	1.70	**20.227**	**2.023**
323	1916	Hoblitzel	Dick	WS	BOS	AL	5	0	2	0.235	23	5	1.40	**20.219**	**2.022**

324	1934	Medwick	Joe	WS	SLN	NL	7	1	5	0.379	30	9	1.31	**20.186**	**2.019**
325	1978	White	Roy	PL	NYA	AL	10	2	5	0.325	46	19	1.53	**20.166**	**2.017**
326	2003	Cabrera	Miguel	PL	FLO	NL	17	4	12	0.265	74	23	1.26	**20.116**	**2.012**
327	1910	Sheckard	Jimmy	WS	CHN	NL	5	0	1	0.357	22	6	1.47	**20.102**	**2.010**
328	1907	Rossman	Claude	WS	DET	AL	5	0	2	0.474	21	3	1.05	**20.072**	**2.007**
329	1884	Irwin	Arthur	WS	PRO	NL	3	0	2	0.222	9	5	1.57	**20.010**	**2.001**
330	1940	Ripple	Jimmy	WS	CIN	NL	7	1	6	0.333	25	9	1.20	**19.890**	**1.989**
331	1892	Childs	Cupid	CS	CL4	NL	6	0	0	0.409	27	3	1.26	**19.871**	**1.987**
332	1906	Isbell	Frank	WS	CHA	AL	6	0	4	0.308	26	8	1.10	**19.794**	**1.979**
333	2005	Pierzynski	A.J.	PL	CHA	AL	12	3	9	0.262	46	18	1.85	**19.793**	**1.979**
334	2005	Pujols	Albert	PL	SLN	NL	9	2	8	0.375	37	15	1.40	**19.781**	**1.978**
335	1936	Bartell	Dick	WS	NY1	NL	6	1	3	0.381	27	8	1.10	**19.742**	**1.974**
336	1997	Williams	Matt	PL	CLE	AL	18	2	8	0.288	80	21	1.63	**19.723**	**1.972**
337	2005	Crede	Joe	PL	CHA	AL	12	4	11	0.289	47	17	1.70	**19.708**	**1.971**
338	1948	Salkeld	Bill	WS	BSN	NL	5	1	1	0.222	15	3	1.89	**19.700**	**1.970**
339	1961	Blanchard	Johnny	WS	NYA	AL	4	2	3	0.400	12	7	1.49	**19.608**	**1.961**
340	1919	Roush	Edd	WS	CIN	NL	8	0	7	0.214	34	13	1.64	**19.553**	**1.955**
341	1916	Gardner	Larry	WS	BOS	AL	5	2	6	0.176	19	8	1.57	**19.495**	**1.949**
342	1986	Dykstra	Lenny	PL	NYN	NL	13	3	6	0.300	56	13	1.46	**19.334**	**1.933**
343	1892	McCarthy	Tommy	CS	BSN	NL	6	0	2	0.381	27	4	1.40	**19.287**	**1.929**
344	1903	Stahl	Chick	WS	BOS	AL	8	0	3	0.303	34	9	1.50	**19.264**	**1.926**
345	1973	Bando	Sal	PL	OAK	AL	12	2	4	0.205	53	11	1.69	**19.261**	**1.926**
346	1936	Selkirk	George	WS	NYA	AL	6	2	3	0.333	28	9	1.08	**19.255**	**1.925**
347	1969	Jones	Cleon	PL	NYN	NL	8	1	4	0.273	35	10	1.54	**19.249**	**1.925**
348	1927	Combs	Earle	WS	NYA	AL	4	0	2	0.313	18	8	1.64	**19.216**	**1.922**
349	1999	Perez	Eddie	PL	ATL	NL	13	2	8	0.341	49	11	2.10	**19.066**	**1.907**
350	1943	Johnson	Billy	WS	NYA	AL	5	0	3	0.300	20	6	1.33	**18.974**	**1.897**

Best Batting Postseasons Ever (Through 2009 Postseason)

														PEVA-B	
Rank	**Year**	**Name**	**First**	**Type**	**Team**	**Lg**	**G**	**HR**	**RBI**	**AVG**	**PA**	**RPR**	**FV**	**REG**	**POST**
351	1951	DiMaggio	Joe	WS	NYA	AL	6	1	5	0.261	25	8	1.58	**18.825**	**1.882**
352	1936	Rolfe	Red	WS	NYA	AL	6	0	4	0.400	29	9	1.70	**18.781**	**1.878**
353	1961	Post	Wally	WS	CIN	NL	5	1	2	0.333	19	5	1.55	**18.742**	**1.874**
354	1956	Slaughter	Enos	WS	NYA	AL	6	1	4	0.350	25	10	1.67	**18.693**	**1.869**
355	1978	Munson	Thurman	PL	NYA	AL	10	1	9	0.302	46	16	2.10	**18.681**	**1.868**
356	1945	Cramer	Doc	WS	DET	AL	7	0	4	0.379	31	11	1.58	**18.666**	**1.867**
357	1972	Morgan	Joe	PL	CIN	NL	12	2	4	0.186	51	13	1.47	**18.654**	**1.865**
358	2009	Howard	Ryan	PL	PHI	NL	15	3	17	0.278	64	28	1.34	**18.616**	**1.862**
359	1981	Cerone	Rick	PL	NYA	AL	14	2	8	0.224	55	12	2.10	**18.559**	**1.856**
360	1948	Torgeson	Earl	WS	BSN	NL	5	0	1	0.389	20	3	1.40	**18.533**	**1.853**
361	1889	Collins	Hub	WS	BR3	AA	9	1	2	0.371	42	15	1.21	**18.533**	**1.853**
362	1914	Maranville	Rabbit	WS	BSN	NL	4	0	3	0.308	16	4	1.57	**18.526**	**1.853**
363	2009	Matsui	Hideki	PL	NYA	AL	15	4	13	0.349	52	18	0.85	**18.480**	**1.848**
364	1974	Bando	Sal	PL	OAK	AL	9	2	4	0.138	37	11	1.66	**18.433**	**1.843**
365	1965	Johnson	Lou	WS	LAN	NL	7	2	4	0.296	29	7	1.56	**18.375**	**1.837**
366	1923	Ward	Aaron	WS	NYA	AL	6	1	2	0.417	25	6	1.50	**18.335**	**1.833**
367	1943	Gordon	Joe	WS	NYA	AL	5	1	2	0.235	20	4	1.50	**18.311**	**1.831**
368	1966	Powell	Boog	WS	BAL	AL	4	0	1	0.357	15	2	1.35	**18.286**	**1.829**
369	1982	Moore	Charlie	PL	ML4	AL	12	0	2	0.385	44	8	1.68	**18.202**	**1.820**
370	1925	Traynor	Pie	WS	PIT	NL	7	1	4	0.346	30	6	1.70	**18.191**	**1.819**
371	1952	Woodling	Gene	WS	NYA	AL	7	1	1	0.348	26	5	1.22	**18.187**	**1.819**
372	1948	Boudreau	Lou	WS	CLE	AL	6	0	3	0.273	24	4	1.56	**18.124**	**1.812**
373	1999	Valentin	John	PL	BOS	AL	10	4	17	0.333	48	26	1.38	**18.099**	**1.810**
374	1974	Jackson	Reggie	PL	OAK	AL	9	1	2	0.231	36	5	1.24	**17.981**	**1.798**
375	1947	DiMaggio	Joe	WS	NYA	AL	7	2	5	0.231	32	9	1.68	**17.924**	**1.792**
376	1889	O'Rourke	Jim	WS	NY1	NL	9	2	7	0.389	38	14	1.24	**17.862**	**1.786**

377	1999	Jordan	Brian	PL	ATL	NL	14	3	13	0.255	65	19	1.50	**17.797**	**1.780**
378	2000	Edmonds	Jim	PL	SLN	NL	8	3	13	0.361	38	19	1.35	**17.761**	**1.776**
379	2006	Guillen	Carlos	PL	DET	AL	13	1	4	0.362	53	10	1.31	**17.707**	**1.771**
380	1962	Pagan	Jose	WS	SFN	NL	7	1	2	0.368	21	4	1.10	**17.660**	**1.766**
381	2008	Ramirez	Manny	PL	LAN	NL	8	4	10	0.520	36	19	1.28	**17.631**	**1.763**
382	1964	Tresh	Tom	WS	NYA	AL	7	2	7	0.273	29	11	1.27	**17.614**	**1.761**
383	1999	Boone	Bret	PL	ATL	NL	14	0	5	0.370	58	11	1.50	**17.577**	**1.758**
384	1892	McKean	Ed	CS	CL4	NL	6	0	6	0.440	26	8	1.14	**17.569**	**1.757**
385	1884	Esterbrook	Dude	WS	NY4	AA	3	0	0	0.300	10	0	1.58	**17.493**	**1.749**
386	1965	Versalles	Zoilo	WS	MIN	AL	7	1	4	0.286	30	7	1.64	**17.460**	**1.746**
387	1915	Speaker	Tris	WS	BOS	AL	5	0	0	0.294	21	2	1.56	**17.441**	**1.744**
388	2001	Gonzalez	Luis	PL	ARI	NL	17	3	10	0.246	74	19	1.39	**17.403**	**1.740**
389	1970	May	Lee	PL	CIN	NL	8	2	10	0.300	32	16	1.40	**17.392**	**1.739**
390	1950	Hamner	Granny	WS	PHI	NL	4	0	0	0.429	15	1	1.10	**17.325**	**1.733**
391	1942	Keller	Charlie	WS	NYA	AL	5	2	5	0.200	21	7	1.58	**17.300**	**1.730**
392	1965	Parker	Wes	WS	LAN	NL	7	1	2	0.304	29	5	1.36	**17.214**	**1.721**
393	1966	Robinson	Brooks	WS	BAL	AL	4	1	1	0.214	15	3	1.69	**17.210**	**1.721**
394	2010	Kinsler	Ian	PL	TEX	AL	16	3	9	0.296	64	16	1.14	**17.207**	**1.721**
395	1908	Hofman	Solly	WS	CHN	NL	5	0	4	0.316	20	6	1.70	**17.204**	**1.720**
396	1903	Freeman	Buck	WS	BOS	AL	8	0	4	0.290	34	10	1.35	**17.159**	**1.716**
397	1942	Rizzuto	Phil	WS	NYA	AL	5	1	1	0.381	23	3	1.48	**17.158**	**1.716**
398	1986	Owen	Spike	PL	BOS	AL	14	0	5	0.366	51	12	1.44	**17.138**	**1.714**
399	1989	Mitchell	Kevin	PL	SFN	NL	9	3	9	0.324	38	16	1.53	**17.114**	**1.711**
400	2005	Podsednik	Scott	PL	CHA	AL	12	2	6	0.286	56	15	1.41	**17.102**	**1.710**
401	1999	Olerud	John	PL	NYN	NL	10	3	12	0.349	48	19	1.22	**17.086**	**1.709**
402	1974	Fosse	Ray	PL	OAK	AL	9	2	4	0.231	29	6	2.10	**17.069**	**1.707**
403	1958	Aaron	Hank	WS	ML1	NL	7	0	2	0.333	31	5	1.49	**17.057**	**1.706**

Best Pitching Postseasons Ever

For most, if you go by the amount of pitchers who get the World Series MVP award versus a position player, a pitcher who dominates the playoffs is more valuable. We're not sure we agree, although we'll admit that it's true that in any baseball game, a pitcher has the most direct impact on the outcome of that contest. They just don't play as many games, that's all. But the point is really, ah, pointless, because both pitchers and hitters have a direct impact on the outcome of a World Series or playoff round and the players below who constitute the Best of the Best are a special group we all know well.

And at the head of the pitching postseason class is a pitcher from only two years ago who put together a postseason run through all three rounds of the playoffs that is the only "perfect" Postseason by a Pitcher in the PEVA rating system galaxy. And how do you achieve that, you ask? Be the best pitcher in every statistical category that the PEVA pitching system uses and do it in a scenario where the pitching park factors of the playoff parks you played in were no higher than League Average.

Who Entered the Best Postseasons Ever Pitching List in 2010

Well, there was no Cliff Lee in the Top Ten moment in 2010 like there was in 2009, however, there was a Cliff Lee sighting. Even though Lee lost that last game, his overall postseason in 2010 ranked the highest among the three stellar performances this year. Coming in at #78, Lee added to his playoff repertoire, just a few spots ahead of Tim Lincecum, who bested him in the victory column, but not the final PEVA numbers for the entire postseason. Lincecum finished at #102 with teammate Brian Wilson, and his 0.00 ERA, coming in at #147.

1. Josh Beckett, Boston Red Sox, 2007

Through three rounds of the playoffs, Josh Beckett had taken the pitching mound at Fenway Park in dominant fashion, as if he had channeled 2004 hero Curt Schilling every time he toed the rubber. But he actually was better. Less dramatic perhaps. Less long suffering as Red Sox nation had already broken the curse under Schilling's gaze. However, Beckett was having the most overwhelming pitching postseason anyone had ever witnessed. An ERA over three runs under the playoff average, in a home hitter's park, at 1.20 just 27.3% of that average, too. He walked almost nobody, as a 17.50 SO/W ratio suggests, and he didn't give up hits either over the 30 innings he pitched. The Red Sox beat the Rockies in four straight to take the title again. Beckett won game one; he had been the ALCS MVP the round before. And he was becoming known as one great big game, big playoff series game pitcher.

Stat Dominance Chart - Postseason			
J. Beckett 64.000		**Pct. of**	
Stats for Year		**PYr Ave**	**PAve**
ERA	1.20	27.3%	4.39
WHIP9	6.30	50.2%	12.54
SO/W	17.50	833.3%	2.10

2. Christy Mathewson, New York Giants, 1905

One sentence is really all you have to know about the pitching prowess of Christy Mathewson in leading his Giants team to World Series victory. He pitched three shutout victories over a six day span. All at the age of 25. It was a series of pitching magnificance, at first order by Mathewson, but by his competitors and teammates, too. Four Hall of Fame pitchers pitched, including Eddie Plank, Chief Bender, and Joe McGinnity as well as Mathewson. The ERA of the entire World Series was 0.82. In the end, there was too much Christy for the Athletics to overcome and the Giants won four games to one. Should this postseason rank as #1 and supplant Mr. Beckett? Perhaps. But it was an era of limited runs while Beckett pitches in a time, even post steroid era, where runs are plentiful. To each his own. But there's no doubt that Christy Mathewson's postseason of 1905 ranks high in the Top Ten. No doubt at all.

Stat Dominance Chart - Postseason			
Mathewson 63.468		**Pct. of**	
Stats for Year		**PYr Ave**	**PAve**
ERA	0.00	0.0%	0.82
WHIP9	5.00	62.7%	7.98
SO/W	18.00	705.9%	2.55

3. Whitey Ford, New York Yankees, 1961

Whitey Ford was always pitching in the World Series, at least it seemed that way for the fans of other teams. He was the Game One pitcher in eight such classics, including this one, and won the MVP Award for his two victories over the Reds in the Yankees march to the World Series crown. Ford gave up no runs over 14 innings and his WHIP% of the World Series average was 40.8%, lowest among the Top Ten Postseasons. It was a season of marvelous achievements for Yankees players with the home run chase by Maris and Mantle taking most of the attention away from Ford, and some say, even their World Series victory. But in the end, in the championship moment, it was Whitey Ford who was taking center stage, and he was exiting with the #3 Postseason in Pitching history.

Stat Dominance Chart - Postseason			
Whitey Ford 63.199		**Pct. of**	
Stats for Year		**PYr Ave**	**PAve**
ERA	0.00	0.0%	3.24

WHIP9	4.50	40.8%	11.02
SO/W	7.00	429.4%	1.63

4. Charley Radbourn, Providence Grays, 1884

It was the first world championship in baseball history, a history that had only begun in a professional way a little over a decade before. So when the Providence Grays of the National League met the New York Metropolitans of the American Association at the Polo Grounds for a three game series to prove who was the best of the two leagues, it was really a novel idea. And Charley Radbourn was the reason why Providence would win, pitching all 22 innings of the three games and giving up no runs. Radbourne had won 59 games during the regular season, a record that still stands, and dominated each game, including the third, which was actually an unnecessary contest since Providence had already won the first two. Baseball was new and the World Series even newer, and the success of Providence on the field was not seen at the gate. They would fold one year later.

Stat Dominance Chart - Postseason			
Radbourn 61.571		**Pct. of**	
Stats for Year		**PYr Ave**	**PAve**
ERA	0.00	0.0%	2.72
WHIP9	4.50	61.4%	7.33
SO/W	17.00	283.3%	6.00

5. Sandy Koufax, Los Angeles Dodgers, 1965

It ended up a seven game series with the Dodgers prevailing, but not after both Koufax and Drysdale lost a game apiece in the first two games against Minnesota. The tandem was too tough to overcome, though, as Koufax pitched to the #5 spot in the All-Time Postseason Season list. The last two games Koufax pitched against the Twins were complete game shutouts, the last on two days rest. Overall his ERA for the series was a minuscule 0.38 and he combined that with command of the strike zone that had eluded Koufax earlier in his career. Koufax would retire from baseball after the 1966 season and be elected to the Hall of Fame at 36 years of age, the youngest of any player.

Stat Dominance Chart - Postseason			
S. Koufax 61.427		**Pct. of**	
Stats for Year		**PYr Ave**	**PAve**
ERA	0.38	14.4%	2.63
WHIP9	6.75	65.2%	10.35
SO/W	5.80	218.0%	2.66

6. Ed Lopat, New York Giants, 1951

The Yankees would win the series and the names of lore include DiMaggio, in his last World Series, plus Mays and Mantle in their first, but it was Ed Lopat who threw down the gauntlet as he pushed his Giants to six games before succumbing to the men from the Bronx. Lopat pitched Games 2 and 5, winning both, giving up only one-half run per game. But it wasn't enough to hold off their cross-town rival, cross-town at least for the several more seasons that the Giants remained in New York.

Stat Dominance Chart - Postseason			
Ed Lopat 58.207		**Pct. of**	
Stats for Year		**PYr Ave**	**PAve**
ERA	0.50	15.3%	3.26
WHIP9	6.50	52.0%	12.51
SO/W	1.33	151.1%	0.88

7. Bob Gibson, St. Louis Cardinals, 1967

It's really pretty ironic when you think about it, the #7 Best Pitching Postseason Ever in Bob Gibson faced the #7 Best Hitter Ever in the Postseason in Carl Yastzremski in the 1967 World Series. And even though Yaz won some of those contests, the pitcher won the MVP Award and the Cardinals won the series. Gibson was a monster on the mount in the 1967 postseason, pitching three complete games and giving up only three runs. He was a dominant, intimidating pitcher who owned the inside part of the plate and was willing to remind you of that fact. It's hard to imagine anyone better, unless you gave up nothing, which some of those above him on this list did. Either way, for many who have watched baseball over the last fifty years, there was nothing so dominating as watching Gibson in the 1967 World Series, and he takes the #7 Spot on our list.

Stat Dominance Chart - Postseason			
Bob Gibson 55.887		**Pct. of**	
Stats for Year		**PYr Ave**	**PAve**
ERA	1.00	33.1%	3.02
WHIP9	6.33	64.5%	9.81
SO/W	5.20	224.1%	2.32

8. Orel Hershiser, Los Angeles Dodgers, 1988

Orel Hershiser won the Series Most Valuable Player Award and Kurt Gibson had his moment as the Los Angeles Dodgers overcame the Oakland Athletics in six games to take the 1988 World Series. Hershiser seems to us as one of those pitchers who should be thought better of. He would win the Cy Young Award and Series MVP in 1988, break the record for the consecutive scoreless innings streak this same year, and overall won three games and had a save in the two playoff rounds. And by the way, he also won the MVP Award in the National League Championship Series. Oh, how the tenacious bulldog could compete, pitch, and win. Seven years later Hershiser would be back again in the playoffs and add the #11 best Postseason while hurling for Cleveland.

Stat Dominance Chart - Postseason			
O. Hershiser 55.625		**Pct. of**	
Stats for Year		**PYr Ave**	**PAve**
ERA	1.05	30.8%	3.41
WHIP9	8.02	71.8%	11.17
SO/W	2.46	118.3%	2.08

9. Bobo Newsom, Detroit Tigers, 1940

War raged in Europe, but it would be another year before the USA was involved and Major League Baseball still had it's full complement of players when Bobo Newsom took the mound for the Tigers in the 1940 World Series. But despite his effort, they would lose in seven games to the Reds. Newsom pitched 26 innings with an ERA of 1.38, 2 runs below the series average.

Stat Dominance Chart - Postseason			
B. Newsom 55.224		**Pct. of**	
Stats for Year		**PYr Ave**	**PAve**
ERA	1.38	41.2%	3.35
WHIP9	7.62	64.4%	11.83
SO/W	4.25	319.5%	1.33

10. Cliff Lee, Philadelphia Phillies, 2009

Cliff Lee was so good in the 2009 playoffs that at the end of the World Series they traded him for an upgrade. What's up with that? It was pretty hard to have been better than Lee was for the Phillies after coming over in a midseason trade. He got them to the playoffs, won four games while there, all with an ERA of 1.56, a WHIP9 at 61.3% of playoff pitcher average over 40.3 innings. Guess #10 on the All-Time Postseason Pitching Seasons list wasn't good enough for them. They were looking for higher. Maybe that will happen with Mr. Halladay, who essentially replaced him on the staff, but boy, oh, boy, was Mr. Lee good in 2009 and gonna be hard for anyone to beat. For Phillies fans, that's the plan.

Stat Dominance Chart - Postseason			
Cliff Lee 55.124		**Pct. of**	
Stats for Year		**PYr Ave**	**PAve**
ERA	1.56	37.9%	4.12
WHIP9	7.36	61.3%	12.01
SO/W	5.50	270.9%	2.03

Note: Field Value Maximums: Catcher 2.10, SS 1.75, OF/3B 1.70, 2B 1.50, 1B 1.40.

Note: PAve: Average for Postseason Year, RPR - Run Production (Runs + RBI) Ave. for Non-Zero players

Best Pitching Postseasons Ever (Through 2010 Postseason)

Rank	Year	Name	First	Type	Team	Lg	W	L	SV	IP	ERA	WHIP	SO/W	PEVA-P REG	POST
1	2007	Beckett	Josh	PL	BOS	AL	4	0	0	30.0	1.20	6.30	17.50	**64.000**	**6.400**
2	1905	Mathewson	Christy	WS	NY1	NL	3	0	0	27.0	0.00	5.00	18.00	**63.468**	**6.347**
3	1961	Ford	Whitey	WS	NYA	AL	2	0	0	14.0	0.00	4.50	7.00	**63.199**	**6.320**
4	1884	Radbourn	Charley	CS	PRO	NL	3	0	0	22.0	0.00	4.50	17.00	**61.571**	**6.157**
5	1965	Koufax	Sandy	WS	LAN	NL	2	1	0	24.0	0.38	6.75	5.80	**61.427**	**6.143**
6	1951	Lopat	Ed	WS	NYA	AL	2	0	0	18.0	0.50	6.50	1.33	**58.207**	**5.821**
7	1967	Gibson	Bob	WS	SLN	NL	3	0	0	27.0	1.00	6.33	5.20	**55.887**	**5.589**
8	1988	Hershiser	Orel	PL	LAN	NL	3	0	1	42.7	1.05	8.02	2.46	**55.625**	**5.562**
9	1940	Newsom	Bobo	WS	DET	AL	2	1	0	26.0	1.38	7.62	4.25	**55.224**	**5.522**
10	2009	Lee	Cliff	PL	PHI	NL	4	0	0	40.3	1.56	7.36	5.50	**55.124**	**5.512**
11	1995	Hershiser	Orel	PL	CLE	AL	4	1	0	35.3	1.53	7.39	3.89	**54.868**	**5.487**
12	1926	Alexander	Pete	WS	SLN	NL	2	0	1	20.3	1.33	7.08	4.25	**53.897**	**5.390**
13	2008	Hamels	Cole	PL	PHI	NL	4	0	0	35.0	1.80	8.23	3.33	**53.813**	**5.381**
14	1920	Coveleski	Stan	WS	CLE	AL	3	0	0	27.0	0.67	5.67	4.00	**52.358**	**5.236**
15	1968	Gibson	Bob	WS	SLN	NL	2	1	0	27.0	1.67	7.33	8.75	**51.441**	**5.144**
16	2001	Schilling	Curt	PL	ARI	NL	4	0	0	48.3	1.12	5.77	9.33	**50.112**	**5.011**
17	1996	Smoltz	John	PL	ATL	NL	4	1	0	38.0	0.95	8.29	2.54	**49.648**	**4.965**
18	1930	Earnshaw	George	WS	PHA	AL	2	0	0	25.0	0.72	7.20	2.71	**49.429**	**4.943**
19	1926	Pennock	Herb	WS	NYA	AL	2	0	0	22.0	1.23	6.95	2.00	**48.487**	**4.849**
20	2005	Contreras	Jose	PL	CHA	AL	3	1	0	32.0	3.09	7.88	7.00	**46.121**	**4.612**
21	1955	Podres	Johnny	WS	BRO	NL	2	0	0	18.0	1.00	9.50	2.50	**46.109**	**4.611**
22	1892	Stivetts	Jack	CS	BSN	NL	2	0	0	29.0	0.93	8.69	2.43	**45.255**	**4.525**
23	1886	Clarkson	John	WS	CHN	NL	2	2	0	31.0	2.03	10.74	2.33	**45.227**	**4.523**
24	1996	Maddux	Greg	PL	ATL	NL	3	2	0	37.0	1.70	8.51	7.33	**44.475**	**4.448**
25	1914	Rudolph	Dick	WS	BSN	NL	2	0	0	18.0	0.50	8.00	3.75	**42.295**	**4.229**
26	1888	Keefe	Tim	WS	NY1	NL	4	0	0	35.0	0.51	6.94	3.33	**40.508**	**4.051**
27	1979	McGregor	Scott	PL	BAL	AL	2	1	0	26.0	2.08	8.65	4.00	**39.644**	**3.964**
28	1968	Lolich	Mickey	WS	DET	AL	3	0	0	27.0	1.67	8.67	3.50	**39.581**	**3.958**
29	1989	Stewart	Dave	PL	OAK	AL	4	0	0	32.0	2.25	7.88	4.60	**38.982**	**3.898**
30	1950	Raschi	Vic	WS	NYA	AL	1	0	0	9.0	0.00	3.00	5.00	**38.180**	**3.818**

31	1933	Hubbell	Carl	WS	NY1	NL	2	0	0	20.0	0.00	8.55	2.50	**37.790**	**3.779**
32	1960	Ford	Whitey	WS	NYA	AL	2	0	0	18.0	0.00	6.50	4.00	**37.723**	**3.772**
33	1946	Brecheen	Harry	WS	SLN	NL	3	0	0	20.0	0.45	8.55	2.20	**37.674**	**3.767**
34	1935	Warneke	Lon	WS	CHN	NL	2	0	0	16.7	0.54	7.02	1.25	**36.835**	**3.683**
35	1908	Overall	Orval	WS	CHN	NL	2	0	0	18.3	0.98	6.87	2.14	**36.550**	**3.655**
36	1962	Terry	Ralph	WS	NYA	AL	2	1	0	25.0	1.80	6.84	8.00	**36.062**	**3.606**
37	1927	Pennock	Herb	WS	NYA	AL	1	0	0	9.0	1.00	3.00	1.00	**35.579**	**3.558**
38	1925	Johnson	Walter	WS	WS1	AL	2	1	0	26.0	2.08	10.38	3.75	**35.389**	**3.539**
39	1957	Burdette	Lew	WS	ML1	NL	3	0	0	27.0	0.67	8.33	3.25	**35.279**	**3.528**
40	1934	Dean	Dizzy	WS	SLN	NL	2	1	0	26.0	1.73	8.65	3.40	**34.494**	**3.449**
41	1943	Chandler	Spud	WS	NYA	AL	2	0	0	18.0	0.50	10.00	3.33	**34.346**	**3.435**
42	1937	Gomez	Lefty	WS	NYA	AL	2	0	0	18.0	1.50	9.00	4.00	**34.318**	**3.432**
43	1921	Hoyt	Waite	WS	NYA	AL	2	1	0	27.0	0.00	9.67	1.64	**34.000**	**3.400**
44	1972	Odom	Blue Moo	PL	OAK	AL	2	1	0	27.7	0.65	5.86	2.25	**33.944**	**3.394**
45	2006	Rogers	Kenny	PL	DET	AL	3	0	0	23.0	0.00	6.26	2.71	**33.864**	**3.386**
46	2001	Johnson	Randy	PL	ARI	NL	5	1	0	41.3	1.52	7.19	5.88	**33.682**	**3.368**
47	1974	Sutton	Don	PL	LAN	NL	3	0	0	30.0	1.50	6.30	5.00	**33.551**	**3.355**
48	1916	Shore	Ernie	WS	BOS	AL	2	0	0	17.7	1.53	8.15	2.25	**33.135**	**3.314**
49	1956	Larsen	Don	WS	NYA	AL	1	0	0	10.7	0.00	4.22	1.75	**32.383**	**3.238**
50	1927	Moore	Wilcy	WS	NYA	AL	1	0	1	10.7	0.84	10.97	1.00	**32.151**	**3.215**

Best Pitching Postseasons Ever (Through 2010 Postseason)

														PEVA-P	
Rank	**Year**	**Name**	**First**	**Type**	**Team**	**Lg**	**W**	**L**	**SV**	**IP**	**ERA**	**WHIP**	**SO/W**	**REG**	**POST**
51	2004	Lowe	Derek	PL	BOS	AL	3	0	0	19.3	1.86	6.52	3.33	**31.833**	**3.183**
52	1928	Hoyt	Waite	WS	NYA	AL	2	0	0	18.0	1.50	10.00	2.33	**31.657**	**3.166**
53	1993	Schilling	Curt	PL	PHI	NL	1	1	0	31.3	2.59	9.77	2.80	**31.541**	**3.154**
54	1984	Morris	Jack	PL	DET	AL	3	0	0	25.0	1.80	7.92	4.25	**31.473**	**3.147**
55	1947	Casey	Hugh	WS	BRO	NL	2	0	1	10.3	0.87	5.23	3.00	**31.376**	**3.138**
56	1980	Carlton	Steve	PL	PHI	NL	3	0	0	27.3	2.30	13.83	1.35	**31.348**	**3.135**
57	1918	Vaughn	Hippo	WS	CHN	NL	1	2	0	27.0	1.00	7.33	3.40	**31.062**	**3.106**
58	1938	Ruffing	Red	WS	NYA	AL	2	0	0	18.0	1.50	9.50	5.50	**30.831**	**3.083**
59	1911	Bender	Chief	WS	PHA	AL	2	1	0	26.0	1.04	8.31	2.50	**30.444**	**3.044**
60	1959	Sherry	Larry	WS	LAN	NL	2	0	2	12.7	0.71	7.11	2.50	**30.134**	**3.013**
61	2000	Pettitte	Andy	PL	NYA	AL	2	0	0	31.7	2.84	13.64	2.25	**29.942**	**2.994**
62	1913	Plank	Eddie	WS	PHA	AL	1	1	0	19.0	0.95	5.68	2.33	**29.594**	**2.959**
63	1969	Cuellar	Mike	PL	BAL	AL	1	0	0	24.00	1.50	7.88	4.00	**29.579**	**2.958**
64	2003	Beckett	Josh	PL	FLO	NL	2	2	0	42.7	2.11	6.96	3.92	**29.260**	**2.926**
65	1949	Reynolds	Allie	WS	NYA	AL	1	0	1	12.3	0.00	4.38	3.50	**29.198**	**2.920**
66	1973	Matlack	Jon	PL	NYN	NL	2	2	0	25.7	1.40	7.01	2.50	**29.076**	**2.908**
67	1947	Shea	Spec	WS	NYA	AL	2	0	0	15.3	2.35	10.57	1.25	**28.787**	**2.879**
68	1999	Hernandez	Orlando	PL	NYA	AL	3	0	0	30.0	1.20	8.70	1.93	**28.749**	**2.875**
69	1952	Reynolds	Allie	WS	NYA	AL	2	1	1	20.3	1.77	7.97	3.00	**28.711**	**2.871**
70	1983	Boddicker	Mike	PL	BAL	AL	2	0	0	18.0	0.00	5.50	6.67	**28.661**	**2.866**
71	1990	Stewart	Dave	PL	OAK	AL	2	2	0	29.0	2.17	8.07	1.13	**28.522**	**2.852**
72	1958	Spahn	Warren	WS	ML1	NL	2	1	0	28.7	2.20	8.48	2.25	**28.162**	**2.816**
73	1975	Tiant	Luis	PL	BOS	AL	3	0	0	34.0	2.65	10.32	1.82	**28.156**	**2.816**
74	1981	Hooton	Burt	PL	LAN	NL	4	1	0	33.0	0.82	10.91	0.67	**28.055**	**2.806**
75	1978	John	Tommy	PL	LAN	NL	2	0	0	23.7	1.90	9.13	1.67	**27.872**	**2.787**
76	1970	Carroll	Clay	PL	CIN	NL	1	0	1	10.3	0.00	7.84	6.50	**27.774**	**2.777**
77	1948	Sain	Johnny	WS	BSN	NL	1	1	0	17.0	1.06	4.76	9.00	**27.566**	**2.757**
78	2010	Lee	Cliff	PL	TEX	AL	3	2	0	35.7	2.78	7.31	25.50	**27.384**	**2.738**
79	1964	Gibson	Bob	WS	SLN	NL	2	1	0	27.0	3.00	10.33	3.88	**27.373**	**2.737**
80	1903	Phillippe	Deacon	WS	PIT	NL	3	2	0	44.0	3.07	8.39	7.33	**27.165**	**2.717**
81	1929	Ehmke	Howard	WS	PHA	AL	1	0	0	12.7	1.42	12.08	4.33	**26.750**	**2.675**
82	1997	Mussina	Mike	PL	BAL	AL	2	0	0	29.0	1.24	5.59	5.86	**26.646**	**2.665**
83	1921	Mays	Carl	WS	NYA	AL	1	2	0	26.0	1.73	6.92	9.00	**26.608**	**2.661**

84	1992	Smoltz	John	PL	ATL	NL	3	0	0	33.7	2.67	11.76	1.82	**26.577**	**2.658**
85	2000	Clemens	Roger	PL	NYA	AL	2	2	0	28.0	3.21	8.36	3.40	**26.437**	**2.644**
86	2007	Corpas	Manuel	PL	COL	NL	1	0	5	10.3	0.87	5.23	7.00	**26.173**	**2.617**
87	1981	Gossage	Rich	PL	NYA	AL	0	0	6	14.3	0.00	6.28	3.75	**26.157**	**2.616**
88	2002	Rodriguez	Francisc	PL	ANA	AL	5	1	0	18.7	1.93	7.23	5.60	**26.103**	**2.610**
89	2003	Rivera	Mariano	PL	NYA	AL	1	0	5	16.0	0.56	3.94	14.00	**25.887**	**2.589**
90	1926	Haines	Jesse	WS	SLN	NL	2	0	0	16.7	1.08	11.88	0.56	**25.770**	**2.577**
91	1989	Moore	Mike	PL	OAK	AL	3	0	0	20.0	1.35	7.65	2.60	**25.568**	**2.557**
92	1886	Caruthers	Bob	WS	SL4	AA	2	1	0	26.0	2.42	8.31	2.00	**25.517**	**2.552**
93	1969	McNally	Dave	PL	BAL	AL	1	1	0	27.00	1.67	8.00	2.40	**25.500**	**2.550**
94	1998	Rivera	Mariano	PL	NYA	AL	0	0	6	13.3	0.00	5.40	5.50	**25.271**	**2.527**
95	1955	Ford	Whitey	WS	NYA	AL	2	0	0	17.0	2.12	11.12	1.25	**24.784**	**2.478**
96	1976	Gullett	Don	PL	CIN	NL	2	0	0	15.3	1.17	7.63	1.33	**24.696**	**2.470**
97	1973	Hunter	Catfish	PL	OAK	AL	3	0	0	29.7	1.82	9.71	1.33	**24.567**	**2.457**
98	2000	Rivera	Mariano	PL	NYA	AL	0	0	6	15.7	1.72	6.32	10.00	**24.541**	**2.454**
99	2004	Lidge	Brad	PL	HOU	NL	1	0	3	12.3	0.73	5.84	6.67	**24.475**	**2.448**
100	1931	Grove	Lefty	WS	PHA	AL	2	1	0	26.0	2.42	10.38	8.00	**24.472**	**2.447**

Best Pitching Postseasons Ever (Through 2010 Postseason)

														PEVA-P	
Rank	Year	Name	First	Type	Team	Lg	W	L	SV	IP	ERA	WHIP	SO/W	REG	POST
101	1984	Lefferts	Craig	PL	SDN	NL	2	0	1	10.0	0.00	4.50	4.00	**24.300**	**2.430**
102	2010	Lincecum	Tim	PL	SFN	NL	4	1	0	37.0	2.43	8.27	4.78	**24.101**	**2.410**
103	2003	Pettitte	Andy	PL	NYA	AL	3	1	0	34.3	2.10	11.53	3.09	**23.615**	**2.361**
104	1924	Zachary	Tom	WS	WS1	AL	2	0	0	17.7	2.04	8.15	1.00	**23.571**	**2.357**
105	1913	Mathewson	Christy	WS	NY1	NL	1	1	0	19.0	0.95	7.58	3.50	**23.479**	**2.348**
106	1936	Hubbell	Carl	WS	NY1	NL	1	1	0	16.0	2.25	9.56	5.00	**23.424**	**2.342**
107	2009	Sabathia	C.C.	PL	NYA	AL	3	1	0	36.3	1.98	9.17	3.56	**23.401**	**2.340**
108	1977	Lyle	Sparky	PL	NYA	AL	3	0	0	14.0	1.29	5.79	5.00	**23.395**	**2.339**
109	1978	Gossage	Rich	PL	NYA	AL	2	0	1	10.0	1.80	4.50	7.00	**23.250**	**2.325**
110	1986	Hurst	Bruce	PL	BOS	AL	3	0	0	38.0	2.13	10.18	3.57	**22.913**	**2.291**
111	1990	Rijo	Jose	PL	CIN	NL	3	0	0	27.7	2.28	10.08	2.42	**22.884**	**2.288**
112	1985	Jackson	Danny	PL	KCA	AL	2	1	0	26.0	1.04	8.65	3.17	**22.805**	**2.281**
113	1909	Adams	Babe	WS	PIT	NL	3	0	0	27.0	1.33	8.00	1.83	**22.779**	**2.278**
114	1995	Glavine	Tom	PL	ATL	NL	2	0	0	28.0	1.61	8.04	2.11	**22.722**	**2.272**
115	1995	Maddux	Greg	PL	ATL	NL	3	1	0	38.0	2.84	9.95	2.71	**22.691**	**2.269**
116	1999	Rivera	Mariano	PL	NYA	AL	2	0	6	12.3	0.00	7.30	9.00	**22.666**	**2.267**
117	1990	Myers	Randy	PL	CIN	NL	0	0	4	8.7	0.00	7.27	3.33	**22.617**	**2.262**
118	1993	Guzman	Juan	PL	TOR	AL	2	1	0	25.0	2.88	12.60	1.24	**22.557**	**2.256**
119	1969	Taylor	Ron	PL	NYN	NL	1	0	2	5.67	0.00	6.35	7.00	**22.515**	**2.252**
120	1991	Morris	Jack	PL	MIN	AL	4	0	0	36.3	2.23	11.15	2.20	**22.323**	**2.232**
121	1926	Hoyt	Waite	WS	NYA	AL	1	1	0	15.0	1.20	12.00	10.00	**22.143**	**2.214**
122	1991	Avery	Steve	PL	ATL	NL	2	0	0	29.3	1.53	7.36	5.00	**22.069**	**2.207**
123	1930	Grove	Lefty	WS	PHA	AL	2	1	0	19.0	1.42	8.53	3.33	**22.025**	**2.203**
124	1890	Lovett	Tom	WS	BRO	NL	2	2	0	35.0	2.83	9.00	2.33	**21.922**	**2.192**
125	2000	Stanton	Mike	PL	NYA	AL	3	0	0	8.7	1.04	6.23	10.00	**21.856**	**2.186**
126	2006	Carpenter	Chris	PL	SLN	NL	3	1	0	32.3	2.78	10.02	2.88	**21.823**	**2.182**
127	1950	Ford	Whitey	WS	NYA	AL	1	0	0	8.7	0.00	8.31	7.00	**21.742**	**2.174**
128	1903	Dinneen	Bill	WS	BOS	AL	3	1	0	35.0	2.06	9.51	3.50	**21.732**	**2.173**
129	1967	Lonborg	Jim	WS	BOS	AL	2	1	0	24.0	2.63	6.00	5.50	**21.673**	**2.167**
130	1982	Andujar	Joaquin	PL	SLN	NL	3	0	0	20.0	1.80	8.55	2.67	**21.618**	**2.162**
131	1960	Law	Vern	WS	PIT	NL	2	0	0	18.3	3.44	12.27	2.67	**21.536**	**2.154**
132	1887	Caruthers	Bob	WS	SL4	AA	4	4	0	71.0	2.15	9.63	1.58	**21.267**	**2.127**
133	1948	Lemon	Bob	WS	CLE	AL	2	0	0	16.3	1.65	12.67	0.86	**21.256**	**2.126**
134	1889	Crane	Ed	WS	NY1	NL	4	1	0	38.0	3.79	14.45	0.59	**21.207**	**2.121**
135	1917	Faber	Red	WS	CHA	AL	3	1	0	27.0	2.33	8.00	3.00	**21.142**	**2.114**
136	1921	Douglas	Phil	WS	NY1	NL	2	1	0	26.0	2.08	8.65	3.40	**21.019**	**2.102**

137	1907	Overall	Orval	WS	CHN	NL	1	0	0	18.0	1.00	9.00	2.75	**21.012**	**2.101**
138	1927	Pipgras	George	WS	NYA	AL	1	0	0	9.0	2.00	8.00	2.00	**20.983**	**2.098**
139	1983	Carlton	Steve	PL	PHI	NL	2	1	0	20.3	1.33	11.51	2.50	**20.896**	**2.090**
140	1942	Beazley	Johnny	WS	SLN	NL	2	0	0	18.0	2.50	10.00	2.00	**20.756**	**2.076**
141	1922	Nehf	Art	WS	NY1	NL	1	0	0	16.0	2.25	7.88	2.00	**20.735**	**2.074**
142	1929	Earnshaw	George	WS	PHA	AL	1	1	0	13.7	2.63	13.17	2.83	**20.733**	**2.073**
143	1914	James	Bill	WS	BSN	NL	2	0	0	11.0	0.00	6.55	1.50	**20.714**	**2.071**
144	2009	Rivera	Mariano	PL	NYA	AL	0	0	5	16.0	0.56	8.44	2.80	**20.687**	**2.069**
145	2004	Schilling	Curt	PL	BOS	AL	3	1	0	22.7	3.57	11.12	2.60	**20.659**	**2.066**
146	1944	Cooper	Mort	WS	SLN	NL	1	1	0	16.0	1.13	7.88	3.20	**20.594**	**2.059**
147	2010	Wilson	Brian	PL	SFN	NL	1	0	6	11.7	0.00	6.92	4.00	**20.552**	**2.055**
148	1946	Dobson	Joe	WS	BOS	AL	1	0	0	12.7	0.00	4.97	3.33	**20.541**	**2.054**
149	1885	McCormick	Jim	WS	CHN	NL	3	2	0	36.0	2.00	8.25	3.17	**20.463**	**2.046**
150	1944	Donnelly	Blix	WS	SLN	NL	1	0	0	6.0	0.00	4.50	9.00	**20.397**	**2.040**

Best Pitching Postseasons Ever (Through 2010 Postseason)

														PEVA-P	
Rank	**Year**	**Name**	**First**	**Type**	**Team**	**Lg**	**W**	**L**	**SV**	**IP**	**ERA**	**WHIP**	**SO/W**	**REG**	**POST**
151	2008	Lidge	Brad	PL	PHI	NL	0	0	7	9.3	0.96	8.68	4.33	**20.328**	**2.033**
152	1956	Maglie	Sal	WS	BRO	NL	1	1	0	17.0	2.65	10.59	2.50	**20.325**	**2.033**
153	1983	Holland	Al	PL	PHI	NL	0	0	2	6.7	0.00	2.70	8.00	**20.314**	**2.031**
154	2005	Garcia	Freddy	PL	CHA	AL	3	0	0	21.0	2.14	9.86	1.63	**20.140**	**2.014**
155	1979	Blyleven	Bert	PL	PIT	NL	2	0	0	19.0	1.42	9.00	4.33	**20.052**	**2.005**
156	2006	Weaver	Jeff	PL	SLN	NL	3	2	0	29.7	2.43	10.31	2.11	**19.936**	**1.994**
157	2006	Wainwright	Adam	PL	SLN	NL	1	0	4	9.7	0.00	8.38	7.50	**19.927**	**1.993**
158	2004	Rivera	Mariano	PL	NYA	AL	1	0	2	12.7	0.71	7.11	4.00	**19.852**	**1.985**
159	2001	Rivera	Mariano	PL	NYA	AL	2	1	5	16.0	1.13	7.88	7.00	**19.845**	**1.985**
160	2005	Oswalt	Roy	PL	HOU	NL	3	0	0	27.3	3.29	10.87	2.00	**19.812**	**1.981**
161	1970	Palmer	Jim	PL	BAL	AL	2	0	0	24.7	3.28	10.95	1.75	**19.709**	**1.971**
162	1996	Glavine	Tom	PL	ATL	NL	2	2	0	26.7	1.69	8.44	4.00	**19.615**	**1.961**
163	1923	Bush	Joe	WS	NYA	AL	1	1	0	16.7	1.08	5.94	1.25	**19.589**	**1.959**
164	1940	Walters	Bucky	WS	CIN	NL	2	0	0	18.0	1.50	7.00	1.00	**19.499**	**1.950**
165	1948	Bearden	Gene	WS	CLE	AL	1	0	1	10.7	0.00	5.91	4.00	**19.433**	**1.943**
166	1945	Newhouser	Hal	WS	DET	AL	2	1	0	20.7	6.10	12.63	5.50	**19.211**	**1.921**
167	1939	Pearson	Monte	WS	NYA	AL	1	0	0	9.0	0.00	3.00	8.00	**19.143**	**1.914**
168	1908	Brown	Mordeca	WS	CHN	NL	2	0	0	11.0	0.00	5.73	5.00	**19.116**	**1.912**
169	1963	Koufax	Sandy	WS	LAN	NL	2	0	0	18.0	1.50	7.50	7.67	**19.095**	**1.909**
170	1985	Dayley	Ken	PL	SLN	NL	1	0	2	12.0	0.00	5.25	2.00	**18.928**	**1.893**
171	1918	Mays	Carl	WS	BOS	AL	2	0	0	18.0	1.00	6.50	1.67	**18.750**	**1.875**
172	2005	Buehrle	Mark	PL	CHA	AL	2	0	1	23.3	3.47	8.10	12.00	**18.747**	**1.875**
173	1931	Earnshaw	George	WS	PHA	AL	1	2	0	24.0	1.88	6.00	5.00	**18.608**	**1.861**
174	1944	Galehouse	Denny	WS	SLA	AL	1	1	0	18.0	1.50	9.00	3.00	**18.595**	**1.860**
175	1985	Tudor	John	PL	SLN	NL	3	2	0	30.7	2.93	10.27	2.20	**18.543**	**1.854**
176	1962	Sanford	Jack	WS	SFN	NL	1	2	0	23.3	1.93	9.26	2.38	**18.490**	**1.849**
177	1990	Dibble	Rob	PL	CIN	NL	1	0	1	9.7	0.00	4.66	7.00	**18.419**	**1.842**
178	1989	Eckersley	Dennis	PL	OAK	AL	0	0	4	7.3	1.23	4.91	2.00	**18.297**	**1.830**
179	2007	Schilling	Curt	PL	BOS	AL	3	0	0	24.0	3.00	10.50	5.33	**18.249**	**1.825**
180	1941	Murphy	Johnny	WS	NYA	AL	1	0	0	6.0	0.00	4.50	3.00	**18.235**	**1.823**
181	1909	Mullin	George	WS	DET	AL	2	1	0	32.0	2.25	8.72	2.50	**18.229**	**1.823**
182	1995	Charlton	Norm	PL	SEA	AL	2	0	2	13.3	1.35	6.08	3.50	**18.135**	**1.813**
183	1995	Charlton	Norm	PL	SEA	AL	2	0	2	13.3	1.35	6.08	3.50	**18.135**	**1.813**
184	1998	Wells	David	PL	NYA	AL	4	0	0	30.7	2.93	8.51	6.20	**17.887**	**1.789**
185	1942	Chandler	Spud	WS	NYA	AL	0	1	1	8.3	1.08	6.48	3.00	**17.817**	**1.782**
186	1910	Coombs	Jack	WS	PHA	AL	3	0	0	27.0	3.33	12.67	1.21	**17.583**	**1.758**
187	1971	Giusti	Dave	PL	PIT	NL	0	0	1	10.7	0.00	6.75	1.75	**17.571**	**1.757**
188	1983	McGregor	Scott	PL	BAL	AL	1	2	0	23.7	1.14	7.61	2.80	**17.528**	**1.753**
189	2004	Clemens	Roger	PL	HOU	NL	2	1	0	25.0	3.60	11.52	2.10	**17.452**	**1.745**

190	1944	Kramer	Jack	WS	SLA	AL	1	0	0	11.0	0.00	10.64	3.00	**17.392**	**1.739**
191	1985	Saberhagen	Bret	PL	KCA	AL	2	0	0	25.3	2.13	9.24	5.33	**17.384**	**1.738**
192	1931	Hallahan	Bill	WS	SLN	NL	2	0	1	18.3	0.49	9.82	1.50	**17.346**	**1.735**
193	2004	Foulke	Keith	PL	BOS	AL	1	0	3	14.0	0.64	9.64	2.38	**17.206**	**1.721**
194	1992	Guzman	Juan	PL	TOR	AL	2	0	0	21.0	1.71	11.14	3.00	**17.196**	**1.720**
195	1991	Aguilera	Rick	PL	MIN	AL	1	1	5	8.3	1.08	8.64	6.00	**17.166**	**1.717**
196	1919	Kerr	Dickey	WS	CHA	AL	2	0	0	19.0	1.42	8.05	2.00	**17.126**	**1.713**
197	1955	Byrne	Tommy	WS	NYA	AL	1	1	0	14.3	1.88	10.05	1.00	**17.079**	**1.708**
198	1981	Valenzuela	Fernand	PL	LAN	NL	3	1	0	40.7	2.21	9.74	1.73	**17.062**	**1.706**
199	1980	Gura	Larry	PL	KCA	AL	1	0	0	21.3	2.11	9.28	2.00	**17.056**	**1.706**
200	2004	Martinez	Pedro	PL	BOS	AL	2	1	0	27.0	4.00	12.00	2.00	**17.035**	**1.704**

Best Pitching Postseasons Ever (Through 2010 Postseason)

														PEVA-P	
Rank	**Year**	**Name**	**First**	**Type**	**Team**	**Lg**	**W**	**L**	**SV**	**IP**	**ERA**	**WHIP**	**SO/W**	**REG**	**POST**
201	1940	Derringer	Paul	WS	CIN	NL	2	1	0	19.3	2.79	12.57	0.60	**16.916**	**1.692**
202	2008	Papelbon	Jonathar	PL	BOS	AL	1	0	3	10.3	0.00	4.35	6.50	**16.848**	**1.685**
203	2008	Lester	Jon	PL	BOS	AL	1	2	0	26.7	2.36	9.79	5.20	**16.842**	**1.684**
204	1890	Ehret	Red	WS	LS2	AA	2	0	1	20.0	1.35	8.10	2.17	**16.790**	**1.679**
205	1977	Torrez	Mike	PL	NYA	AL	2	1	0	29.0	3.10	11.48	2.00	**16.779**	**1.678**
206	1950	Reynolds	Allie	WS	NYA	AL	1	0	1	10.3	0.87	9.58	1.75	**16.775**	**1.678**
207	1987	Viola	Frank	PL	MIN	AL	3	1	0	31.3	4.31	11.20	3.13	**16.761**	**1.676**
208	1981	Rogers	Steve	PL	MON	NL	3	1	0	27.7	0.98	9.11	2.75	**16.759**	**1.676**
209	1993	Stewart	Dave	PL	TOR	AL	2	1	0	25.3	4.26	12.08	1.00	**16.672**	**1.667**
210	1974	Holtzman	Ken	PL	OAK	AL	2	0	0	21.0	0.86	10.29	2.17	**16.659**	**1.666**
211	1915	Foster	Rube	WS	BOS	AL	2	0	0	18.0	2.00	7.00	6.50	**16.474**	**1.647**
212	1977	Sutton	Don	PL	LAN	NL	2	0	0	25.0	2.88	9.72	10.00	**16.392**	**1.639**
213	2004	Suppan	Jeff	PL	SLN	NL	2	2	0	23.7	3.80	9.89	1.88	**16.384**	**1.638**
214	1999	Martinez	Pedro	PL	BOS	AL	2	0	0	17.0	0.00	5.82	3.83	**16.383**	**1.638**
215	1979	Tekulve	Kent	PL	PIT	NL	0	1	3	12.3	2.92	8.03	2.40	**16.373**	**1.637**
216	1936	Gomez	Lefty	WS	NYA	AL	2	0	0	15.3	4.70	14.67	0.82	**16.274**	**1.627**
217	1988	Honeycutt	Rick	PL	OAK	AL	2	0	0	5.3	0.00	3.37	2.50	**16.167**	**1.617**
218	1975	Eastwick	Rawly	PL	CIN	NL	3	0	1	11.7	1.54	10.03	1.00	**16.147**	**1.615**
219	2009	Pettitte	Andy	PL	NYA	AL	4	0	0	30.7	3.52	10.86	2.27	**16.081**	**1.608**
220	2000	Hampton	Mike	PL	NYN	NL	2	2	0	27.3	2.96	11.52	1.50	**16.036**	**1.604**
221	2005	Jenks	Bobby	PL	CHA	AL	0	0	4	8.0	2.25	7.88	2.67	**15.948**	**1.595**
222	1981	Reuss	Jerry	PL	LAN	NL	2	2	0	36.7	2.21	8.84	1.89	**15.939**	**1.594**
223	1912	Mathewson	Christy	WS	NY1	NL	0	2	0	28.7	0.94	8.79	2.00	**15.927**	**1.593**
224	1972	Hunter	Catfish	PL	OAK	AL	2	0	0	31.3	2.01	9.48	1.82	**15.894**	**1.589**
225	1952	Raschi	Vic	WS	NYA	AL	2	0	0	17.0	1.59	10.59	2.25	**15.811**	**1.581**
226	1979	Jackson	Grant	PL	PIT	NL	2	0	0	6.7	0.00	6.75	1.33	**15.734**	**1.573**
227	1945	Borowy	Hank	WS	CHN	NL	2	2	0	18.0	4.00	13.50	1.33	**15.689**	**1.569**
228	1999	Millwood	Kevin	PL	ATL	NL	2	1	1	24.7	3.65	9.12	6.67	**15.414**	**1.541**
229	1987	Blyleven	Bert	PL	MIN	AL	3	1	0	26.3	3.42	10.25	4.20	**15.353**	**1.535**
230	1996	Eckersley	Dennis	PL	SLN	NL	1	0	4	7.0	0.00	6.43	6.00	**15.317**	**1.532**
231	1923	Pennock	Herb	WS	NYA	AL	2	0	1	17.3	3.63	10.38	8.00	**15.302**	**1.530**
232	1966	Drabowsky	Moe	WS	BAL	AL	1	0	0	6.7	0.00	4.05	5.50	**15.292**	**1.529**
233	1934	Dean	Paul	WS	SLN	NL	2	0	0	18.0	1.00	11.00	1.57	**15.269**	**1.527**
234	1972	Fingers	Rollie	PL	OAK	AL	2	1	2	15.7	1.72	7.47	2.60	**15.260**	**1.526**
235	1915	Alexander	Pete	WS	PHI	NL	1	1	0	17.7	1.53	9.17	2.50	**15.211**	**1.521**
236	1977	Guidry	Ron	PL	NYA	AL	2	0	0	20.3	3.10	8.41	2.50	**15.181**	**1.518**
237	1956	Kucks	Johnny	WS	NYA	AL	1	0	0	11.0	0.82	7.36	0.67	**15.178**	**1.518**
238	2005	Carpenter	Chris	PL	SLN	NL	2	0	0	21.0	2.14	10.29	1.71	**15.166**	**1.517**
239	1973	Seaver	Tom	PL	NYN	NL	1	2	0	31.7	1.99	9.66	4.38	**15.147**	**1.515**
240	1975	Drago	Dick	PL	BOS	AL	0	1	2	8.7	1.04	7.27	1.50	**15.129**	**1.513**
241	2007	Papelbon	Jonathar	PL	BOS	AL	1	0	4	10.7	0.00	7.59	1.75	**15.099**	**1.510**
242	1921	Nehf	Art	WS	NY1	NL	1	2	0	26.0	1.38	9.00	0.62	**15.098**	**1.510**

243	1998	Brown	Kevin	PL	SDN	NL	2	2	0	39.3	2.52	9.38	2.71	**15.095**	**1.510**
244	1991	Smoltz	John	PL	ATL	NL	2	0	0	29.7	1.52	9.40	6.50	**15.035**	**1.504**
245	1935	Bridges	Tommy	WS	DET	AL	2	0	0	18.0	2.50	11.00	2.25	**14.957**	**1.496**
246	1992	Key	Jimmy	PL	TOR	AL	2	0	0	12.0	0.75	7.50	3.50	**14.946**	**1.495**
247	1979	Flanagan	Mike	PL	BAL	AL	2	1	0	22.0	3.68	11.05	5.00	**14.897**	**1.490**
248	1932	Gomez	Lefty	WS	NYA	AL	1	0	0	9.0	1.00	10.00	8.00	**14.824**	**1.482**
249	1966	McNally	Dave	WS	BAL	AL	1	0	0	11.3	1.59	10.32	0.71	**14.809**	**1.481**
250	1995	Johnson	Randy	PL	SEA	AL	2	1	0	25.3	2.49	8.88	3.63	**14.802**	**1.480**
251	1995	Johnson	Randy	PL	SEA	AL	2	1	0	25.3	2.49	8.88	3.63	**14.802**	**1.480**
252	1971	McNally	Dave	PL	BAL	AL	3	1	0	20.7	2.61	10.02	2.83	**14.716**	**1.472**
253	1965	Grant	Mudcat	WS	MIN	AL	2	1	0	23.0	2.74	9.39	6.00	**14.691**	**1.469**

Abbreviation Code: Type: CS (Championship Series) - Prior to 1900; WS - World Series Only; PL - 1969 to Present.

RPR - Run Production (Runs Scored plus Runs Batted In)

FV - Field Value, Defensive value for player during postseason. Maximum, average, and minimum values differ for each position.

WHIP9 - Walks and Hits Per 9 Innings Pitched

SO/W - Strike Outs to Walks Ratio

Note: Postseason PEVA-B (Batting) and PEVA-P (Pitching) currently valued at 10% of what would represent regular season PEVA values.

Chapter 8 - BEST POSTSEASON CAREERS EVER

Total Postseason Rankings

To have a great postseason career, you have to have one thing. It's more than talent. It's more than being able to rise to the occasion. It's inherently a dependent function. You must play on winning teams, teams that make the playoffs, and teams that make it to the World Series. You can't be Ernie Banks. Not that a great player does not have an impact on making the postseason and that the even greater ones don't rise to that occasion. They do. But there are a whole lot of Yankees on these lists. They play in the World Series. They've won a whole lot of them.

It's even harder to compare total postseason numbers, because of the expansion in the number of rounds and teams, yet it seems unfair to discount the additional rounds as well. For PEVA calcuations, it adjusts for these things, but for the gross amount of Home Runs hit or Wins, the list includes all rounds; take all that into account as much as you'd like.

Listed below are PEVA postseason rankings and totals that reflect the 10% level. It includes a per season number that may even be better to use as a comparison. But the folks near the top are those that we relate to as great World Series players. They're in the lexicon of moments that play on and on in our minds when we recall October classics from the first in 1884 (okay, not too many recall that) through last year.

We don't think there are many surprises on these lists. The Babe, a Ford, some Mariano, and a dash of Berra show up high. And that's the way it should be I guess, our postseason heros of the diamond near the pinnacle of any Best Ever list.

Best Postseason Careers by Batters

1. Babe Ruth

Babe Ruth appeared in ten World Series. His teams won seven of them, including three with Boston before that trade, even though we'll admit he didn't have that much to do with those victories. He pitched a bit, but only batted 12 times during those series. Still explains a bit of the intense feelings in the rivalry between the Yanks and the Sox. Even though Babe Ruth ranks at the top of the total list, it's perfectly understandable to consider Lou Gehrig an equal to his World Series exploits. While Ruth had an OPS of 1.211 in those ten playoffs, Gehrig had a 1.208 in his seven, pretty hard to pick between the two. Ruth had 15 HR; Gehrig 10. Ruth had 33 RBI; Gehrig 35. Once again, one great tandem. About all you can say is that any club today would be glad to have either one as they make their playoff run.

Babe Ruth - Postseason (10 Years)			
Category			**Rank**
TOTAL POSTSEASON PEVA		**28.065**	**1**
POST PEVA PER YEAR*		**2.806**	**8**
POST OBP	0.467		
POST SLG	0.744		

*Note: Per Year Rank with at least 3 postseasons

2. Yogi Berra

When you look at Berra's postseason batting stats, they don't quite measure up to those of Ruth or Gehrig, but that would be missing a good part of the point. It was his defense and leadership that made up a good part of the difference, because over those fourteen seasons where Berra squated behind the plate, his team would win ten championships. Not that his offense was bad. He had an OPS of 0.811, hit 12 homers, and knocked in 39. Ten of those years he had a regular PEVA over 10.000 (Post 1.000), including that magic season of 1956 when he hit the perfect 64.000 (Post 6.400). To put those numbers into context, he had the best postseason ever and five of the top 400. Pretty fine line for Yogi.

Yogi Berra - Postseason (14 Years)			
Category			**Rank**
TOTAL POSTSEASON PEVA		**22.935**	**2**
POST PEVA PER YEAR		**1.638**	**42**
POST OBP	0.359		
POST SLG	0.452		

3. Mickey Mantle

Mickey Mantle was productive in the World Series, but not consistent. It was a bit of an all or nothing roll. His 18 HR and 40 RBI over the 12 postseasons rank high on the All-Time list, but he hit just 0.257 to get there. He had five great postseasons, topping out with the 1960 campaign, six lousy postseasons, and one that some would say was fair. The Bombers won seven of them either way, when he was hot, cold, and indifferent. So while Mantle ranks high on this list because of the amount of years he played, with some playing well, he probably wasn't the third best postseason batter ever, if you put everything into context. Fans of Charlie Keller and Al Simmons, just outside the Top Ten in Total, but way up there in Per Season Averages, would argue they were better. But they don't have as many rings, now do they.

Mickey Mantle - Postseason (12 Years)			
Category			**Rank**
TOTAL POSTSEASON PEVA		**22.596**	**3**
POST PEVA PER YEAR		**1.883**	**30**
POST OBP	0.374		
POST SLG	0.535		

4. Lou Gehrig

He came onto the postseason scene a decade after Ruth and teamed with the Babe for four World Series, winning three of them, but when Gehrig led the Yankees to three more series after Ruth retired, all victories, it became apparent just how special Gehrig was outside the shadow of the Big Bambino. Overall, Gehrig batted 0.361 during his seven years in the playoffs and his PEVA per year ranking at #4, which is tops among the players in the Top Ten Total race, pays testiment to his consistency over those years. As we said above, you'd have to flip a coin to pick between the two, Ruth or Gehrig, as the best postseason position player ever taking everything into consideration. You sure couldn't go wrong either way.

Lou Gehrig - Postseason (7 Years)			
Category			**Rank**
TOTAL POSTSEASON PEVA		**22.452**	**4**
POST PEVA PER YEAR		**3.207**	**4**
POST OBP	0.477		
POST SLG	0.731		

5. Derek Jeter

We all have our favorite moments, those indelible images of watching Derek Jeter perform in the playoffs from the Division Series to Championship Series to the World Series, and for us, despite his heroics with the bat, many of them come on the defensive side. And that just proves how special Jeter is, that you could go past the OPS of 0.850, the 185 hits, and 0.309 Batting Average, and choose a flip or a dive. Since Jeter came onto the Bronx scene, the Yankees have won five World Series. That's all you need to say.

Derek Jeter - Postseason (13 Years)			
Category			**Rank**

TOTAL POSTSEASON PEVA		18.800	5
POST PEVA PER YEAR		1.343	65
POST OBP	0.377		
POST SLG	0.472		

6. Reggie Jackson

Okay, if you were a Yankees fan, and you had to pick one man to be up at the plate for that important World Series moment, would you pick Ruth, Gehrig, Jeter, or Jackson. There'd be more than a few who'd go with Mr. October. Reggie played in four postseasons with the Yankees, winning two titles; five with Oakland, winning three; and two with California, the Angels didn't make the Series either time. In 1977 and 1978, during both Yankee championships, Jackson carried the club, producing 41 runs on 9 homers and 23 Runs Batted In. He was brash, bold, powerful, and timely. Okay, we'd still go with Ruth or Gehrig first, but what a good third choice Jackson would make.

Reggie Jackson - Postseason (11 Years)			
Category			**Rank**
TOTAL POSTSEASON PEVA		**17.015**	**6**
POST PEVA PER YEAR		**1.547**	**51**
POST OBP	0.358		
POST SLG	0.527		

7. Joe DiMaggio

Ninety percent of the time Joe DiMaggio laced up his cleats for the New York Yankees in the World Series his team won the championship. Nine out of ten. Now that's one fantastic feat, and he even missed the 1943 playoffs due to his war commitment or it would have been 10 for 11. DiMaggio was not always at his best in the Series with his most significant playoffs coming late in his career in 1950 when he batted 0.308 during the postseason sweep of the Phillies, his performance ranking #36 All-Time. So his average postseason lags a bit in the per year rankings at #43, but as with most of these Yankee exploits, it's so hard to argue with success. Particularly when it's at a rate of ninety percent.

Joe DiMaggio - Postseason (10 Years)			
Category			**Rank**
TOTAL POSTSEASON PEVA		**16.143**	**7**
POST PEVA PER YEAR		**1.614**	**43**
POST OBP	0.338		
POST SLG	0.422		

8. Bernie Williams

A bit of a surprise that Bernie accumulated this much PEVA during those twelve postseasons, and he may be the man who benefited most from the counting stats over quality in the Top Ten. So we may think that Williams gets too much credit for his postseasons, but too little for his regular seasons when you glean through his stats and accomplishments. Look through his regular season stats some day. More than pretty good. Bernie Williams was a member of a gaggle of great Yankee teams and contributing mightily to their success, both in the field and at the plate. He was at his best in the postseason of 1996 when he hit 6 HR and knocked in 15 to lead the Yankees to one of the four rings he would garner during his career.

Bernie Williams - Postseason (12 Years)			
Category			**Rank**
TOTAL POSTSEASON PEVA		**15.676**	**8**
POST PEVA PER YEAR		**1.306**	**71**

POST OBP	0.371
POST SLG	0.480

9. Frank Baker

Baker hit 0.363 during his six seasons in the spotlight for the Phladelphia Athletics and New York Yankees from 1910 to 1922. He was a member of three World Champions, all with the A's. During those three seasons, Frank hit 0.409, 0.375, and 0.450 and ranked as the #14, #37, and #148 on the All-Time Best Postseason Year list. Overall, his OPS was 0.952, higher than all other Top Ten players on this list outside of Ruth and Gehrig.

Frank Baker - Postseason (6 Years)			
Category			**Rank**
TOTAL POSTSEASON PEVA		**15.186**	**9**
POST PEVA PER YEAR		**2.531**	**16**
POST OBP	0.392		
POST SLG	0.560		

10. Bill Dickey

He plied his trade with the tools of catching ignorance and played during the times of Gehrig and DiMaggio in the Bronx. His main contributions, at least for most, was behind the plate. During his best postseasons, 1939, 1943, and 1938, his Field Value of 2.10 was at the top of the charts. He added timely hits at the plate as well, 145 in total over those eight seasons, but would hit only 0.245 overall during those World Series stints. Dickey would win seven World Series rings in his eight tries.

Bill Dickey - Postseason (8 Years)			
Category			**Rank**
TOTAL POSTSEASON PEVA		**14.432**	**10**
POST PEVA PER YEAR		**1.804**	**34**
POST OBP	0.329		
POST SLG	0.379		

Best Batting Postseason Careers
From 1884-2010

Rank	LPostYr	Name	First	Final PTeam	Postseason Career PEVA-B	Total Yrs	Career PerYr	HR	RBI	H	AB	AVG
1	1932	Ruth	Babe	NYA	**28.065**	10	**2.806**	15	33	42	129	0.326
2	1963	Berra	Yogi	NYA	**22.935**	14	**1.638**	12	39	71	259	0.274
3	1964	Mantle	Mickey	NYA	**22.596**	12	**1.883**	18	40	59	230	0.257
4	1938	Gehrig	Lou	NYA	**22.452**	7	**3.207**	10	35	43	119	0.361
5	2010	Jeter	Derek	NYA	**18.800**	14	**1.343**	20	57	185	599	0.309
6	1986	Jackson	Reggie	CAL	**17.015**	11	**1.547**	18	48	78	281	0.278
7	1951	DiMaggio	Joe	NYA	**16.143**	10	**1.614**	8	30	54	199	0.271
8	2006	Williams	Bernie	NYA	**15.676**	12	**1.306**	22	80	128	465	0.275
9	1922	Baker	Frank	NYA	**15.186**	6	**2.531**	3	18	33	91	0.363
10	1943	Dickey	Bill	NYA	**14.432**	8	**1.804**	5	24	37	145	0.255
11	1943	Keller	Charlie	NYA	**13.928**	4	**3.482**	5	18	22	72	0.306
12	1939	Simmons	Al	CIN	**13.759**	4	**3.440**	6	17	24	73	0.329
13	1984	Garvey	Steve	SDN	**13.368**	5	**2.674**	11	31	75	222	0.338
14	1971	Robinson	Frank	BAL	**13.183**	5	**2.637**	10	19	30	126	0.238

15	1919	Collins	Eddie	CHA	**13.154**	6	**2.192**	0	11	42	128	0.328
16	2009	Ramirez	Manny	LAN	**12.793**	11	**1.163**	29	78	117	410	0.285
17	1963	Skowron	Bill	LAN	**12.652**	8	**1.581**	8	29	39	133	0.293
18	1979	Bench	Johnny	CIN	**12.180**	6	**2.030**	10	20	45	169	0.266
19	1959	Snider	Duke	LAN	**12.125**	6	**2.021**	11	26	38	133	0.286
20	1918	Hooper	Harry	BOS	**11.353**	4	**2.838**	2	6	27	92	0.293
21	1983	Rose	Pete	PHI	**10.690**	8	**1.336**	5	22	86	268	0.321
22	1968	Brock	Lou	SLN	**10.684**	3	**3.561**	4	13	34	87	0.391
23	1948	Gordon	Joe	CLE	**10.642**	6	**1.774**	4	16	25	103	0.243
24	1945	Greenberg	Hank	DET	**10.574**	4	**2.644**	5	22	27	85	0.318
25	1935	Cochrane	Mickey	DET	**10.233**	5	**2.047**	2	7	27	110	0.245
26	2010	Posada	Jorge	NYA	**9.835**	14	**0.703**	11	42	97	402	0.241
27	2000	Henderson	Rickey	SEA	**9.648**	8	**1.206**	5	20	63	222	0.284
28	2010	Rodriguez	Alex	NYA	**9.495**	9	**1.055**	13	38	67	231	0.290
29	2005	Jones	Chipper	ATL	**9.474**	11	**0.861**	13	47	96	333	0.288
30	2009	Pujols	Albert	SLN	**9.369**	6	**1.562**	13	36	64	199	0.322
31	2002	Justice	David	OAK	**9.205**	10	**0.921**	14	63	89	398	0.224
32	1985	Brett	George	KCA	**9.179**	7	**1.311**	10	23	56	166	0.337
33	1916	Lewis	Duffy	BOS	**9.120**	3	**3.040**	1	7	20	67	0.299
34	1934	Martin	Pepper	SLN	**8.696**	3	**2.899**	1	9	23	55	0.418
35	1975	Yastrzemski	Carl	BOS	**8.599**	2	**4.299**	4	11	24	65	0.369
36	1974	Robinson	Brooks	BAL	**8.597**	6	**1.433**	5	22	44	145	0.303
37	1967	Howard	Elston	BOS	**8.580**	10	**0.858**	5	19	42	171	0.246
38	1935	Goslin	Goose	DET	**8.444**	5	**1.689**	7	19	37	129	0.287
39	1968	Maris	Roger	SLN	**8.404**	7	**1.201**	6	18	33	152	0.217
40	1931	Foxx	Jimmie	PHA	**8.127**	3	**2.709**	4	11	22	64	0.344
41	1960	McDougald	Gil	NYA	**8.052**	8	**1.007**	7	24	45	190	0.237
42	1955	Rizzuto	Phil	NYA	**8.046**	9	**0.894**	2	8	45	183	0.246
43	2001	O'Neill	Paul	NYA	**7.977**	8	**0.997**	11	39	85	299	0.284
44	1964	Tresh	Tom	NYA	**7.954**	3	**2.651**	4	13	18	65	0.277
45	1969	Aaron	Hank	ATL	**7.915**	3	**2.638**	6	16	25	69	0.362
46	1993	Molitor	Paul	TOR	**7.901**	3	**2.634**	6	22	43	117	0.368
47	2009	Ortiz	David	BOS	**7.791**	7	**1.113**	12	47	69	244	0.283
48	1888	O'Neill	Tip	SL4	**7.688**	4	**1.922**	5	25	35	146	0.240
49	1958	Bauer	Hank	NYA	**7.612**	9	**0.846**	7	24	46	188	0.245
50	1978	McCarver	Tim	PHI	**7.566**	6	**1.261**	2	12	24	88	0.273

Best Batting Postseason Careers

From 1884-2010

					Postseason							
				Final	Career	Total	Career					
Rank	**LPostYr**	**Name**	**First**	**PTeam**	**PEVA-B**	**Yrs**	**PerYr**	**HR**	**RBI**	**H**	**AB**	**AVG**
51	1953	Woodling	Gene	NYA	**7.503**	5	**1.501**	3	6	27	85	0.318
52	1920	Speaker	Tris	CLE	**7.464**	3	**2.488**	0	3	22	72	0.306
53	1984	Matthews	Gary	CHN	**7.439**	3	**2.480**	7	15	21	65	0.323
54	1937	Ott	Mel	NY1	**7.438**	3	**2.479**	4	10	18	61	0.295
55	2003	Bonds	Barry	SFN	**7.351**	7	**1.050**	9	24	37	151	0.245
56	1910	Schulte	Frank	CHN	**7.318**	4	**1.829**	0	9	26	81	0.321
57	1990	Henderson	Dave	OAK	**7.314**	4	**1.829**	7	20	36	121	0.298
58	1934	Frisch	Frankie	SLN	**7.306**	8	**0.913**	0	10	58	197	0.294
59	1978	Munson	Thurman	NYA	**7.181**	3	**2.394**	3	22	46	129	0.357
60	1923	Schang	Wally	NYA	**7.147**	6	**1.191**	1	9	27	94	0.287
61	1956	Martin	Billy	NYA	**7.095**	5	**1.419**	5	19	33	99	0.333
62	1974	Powell	Boog	BAL	**7.086**	6	**1.181**	6	18	33	126	0.262
63	1924	Gowdy	Hank	NY1	**7.039**	3	**2.346**	1	4	13	42	0.310
64	1956	Reese	Pee Wee	BRO	**7.008**	7	**1.001**	2	16	46	169	0.272
65	1993	Dykstra	Lenny	PHI	**6.996**	3	**2.332**	10	19	36	112	0.321
66	1984	Cey	Ron	CHN	**6.702**	5	**1.340**	7	27	42	161	0.261
67	1914	Evers	Johnny	BSN	**6.617**	4	**1.654**	0	6	24	76	0.316

68	1943	Crosetti	Frankie	NYA	**6.606**	7	**0.944**	1	11	20	115	0.174
69	2006	Rodriguez	Ivan	DET	**6.542**	5	**1.308**	4	25	39	153	0.255
70	2008	Edmonds	Jim	CHN	**6.413**	7	**0.916**	13	43	63	230	0.274
71	2006	Beltran	Carlos	NYN	**6.405**	2	**3.203**	11	19	30	82	0.366
72	1932	Combs	Earle	NYA	**6.372**	4	**1.593**	1	9	21	60	0.350
73	2001	Brosius	Scott	NYA	**6.362**	4	**1.590**	8	30	48	196	0.245
74	2005	Jones	Andruw	ATL	**6.320**	10	**0.632**	10	33	65	238	0.273
75	1986	Lopes	Davey	HOU	**6.276**	6	**1.046**	6	22	43	181	0.238
76	1938	Lazzeri	Tony	CHN	**6.270**	7	**0.896**	4	19	28	107	0.262
77	1956	Campanella	Roy	BRO	**6.262**	5	**1.252**	4	12	27	114	0.237
78	1979	Stargell	Willie	PIT	**6.202**	6	**1.034**	7	20	37	133	0.278
79	1954	Dark	Alvin	NY1	**6.189**	3	**2.063**	1	4	21	65	0.323
80	1905	Bresnahan	Roger	NY1	**6.126**	1	**6.126**	0	1	5	16	0.313
81	1988	Carter	Gary	NYN	**6.119**	3	**2.040**	4	21	33	118	0.280
82	1958	Slaughter	Enos	NYA	**6.070**	5	**1.214**	3	8	23	79	0.291
83	2000	Clark	Will	SLN	**6.039**	5	**1.208**	5	16	39	117	0.333
84	1892	Duffy	Hugh	BSN	**6.019**	1	**6.019**	1	9	12	26	0.462
85	2005	Martinez	Tino	NYA	**5.923**	9	**0.658**	9	38	83	356	0.233
86	2010	Berkman	Lance	HOU	**5.880**	4	**1.470**	7	30	39	122	0.320
87	1959	Hodges	Gil	LAN	**5.871**	7	**0.839**	5	21	35	131	0.267
88	1942	Rolfe	Red	NYA	**5.842**	6	**0.974**	0	6	33	116	0.284
89	1997	McGriff	Fred	ATL	**5.826**	5	**1.165**	10	37	57	188	0.303
90	1947	McQuinn	George	NYA	**5.746**	2	**2.873**	1	6	10	39	0.256
91	1979	Campaneris	Bert	CAL	**5.735**	6	**0.956**	3	11	35	144	0.243
92	2009	Matsui	Hideki	NYA	**5.722**	6	**0.954**	10	39	64	205	0.312
93	2003	Lopez	Javy	ATL	**5.705**	9	**0.634**	10	28	57	205	0.278
94	1981	Bando	Sal	ML4	**5.692**	6	**0.949**	5	13	39	159	0.245
95	1910	Steinfeldt	Harry	CHN	**5.638**	4	**1.410**	0	8	19	73	0.260
96	1964	Richardson	Bobby	NYA	**5.602**	7	**0.800**	1	15	40	131	0.305
97	1924	Meusel	Irish	NY1	**5.522**	4	**1.381**	3	17	24	87	0.276
98	1951	Brown	Bobby	NYA	**5.513**	4	**1.378**	0	9	18	41	0.439
99	1983	Baker	Dusty	LAN	**5.415**	4	**1.354**	5	21	42	149	0.282
100	1940	Gehringer	Charlie	DET	**5.357**	3	**1.786**	1	7	26	81	0.321

Best Batting Postseason Careers

From 1884-2010

Rank	**LPostYr**	**Name**	**First**	**Final PTeam**	**Postseason Career PEVA-B**	**Total Yrs**	**Career PerYr**	**HR**	**RBI**	**H**	**AB**	**AVG**
101	1983	Perez	Tony	PHI	**5.320**	6	**0.887**	6	25	41	172	0.238
102	1884	Gilligan	Barney	PRO	**5.306**	1	**5.306**	0	2	4	9	0.444
103	2006	Piazza	Mike	SDN	**5.275**	5	**1.055**	6	15	29	120	0.242
104	1945	Cavarretta	Phil	CHN	**5.269**	3	**1.756**	1	5	20	63	0.317
105	1936	Koenig	Mark	NY1	**5.253**	5	**1.051**	0	5	18	76	0.237
106	1982	Tenace	Gene	SLN	**5.219**	6	**0.870**	4	14	18	114	0.158
107	1996	Murray	Eddie	BAL	**5.157**	4	**1.289**	9	25	41	159	0.258
108	2010	Werth	Jayson	PHI	**5.107**	5	**1.021**	13	26	41	153	0.268
109	2007	Lofton	Kenny	CLE	**5.105**	11	**0.464**	7	34	97	392	0.247
110	1931	Dykes	Jimmie	PHA	**5.091**	3	**1.697**	1	11	17	59	0.288
111	1975	Rudi	Joe	OAK	**5.075**	5	**1.015**	3	15	37	140	0.264
112	1915	Luderus	Fred	PHI	**5.066**	1	**5.066**	1	6	7	16	0.438
113	1990	Hatcher	Billy	CIN	**5.031**	2	**2.515**	2	6	21	52	0.404
114	1948	Elliott	Bob	BSN	**5.022**	1	**5.022**	2	5	7	21	0.333
115	1948	Sanders	Ray	BSN	**5.016**	4	**1.254**	2	3	11	40	0.275
116	1927	Groh	Heinie	PIT	**5.007**	5	**1.001**	0	4	19	72	0.264
117	1984	Nettles	Graig	SDN	**5.000**	7	**0.714**	5	27	41	182	0.225
118	1983	Morgan	Joe	PHI	**4.968**	7	**0.710**	5	13	33	181	0.182
119	1988	Gibson	Kirk	LAN	**4.945**	3	**1.648**	7	21	22	78	0.282
120	2008	Varitek	Jason	BOS	**4.906**	7	**0.701**	11	33	54	228	0.237

121	1927	Harris	Joe	PIT	**4.864**	2	**2.432**	3	7	14	40	0.350
122	1957	Coleman	Jerry	NYA	**4.808**	6	**0.801**	0	9	19	69	0.275
123	1962	Hiller	Chuck	SFN	**4.798**	1	**4.798**	1	5	7	26	0.269
124	1969	Boyer	Clete	ATL	**4.797**	6	**0.799**	2	14	21	95	0.221
125	1944	Cooper	Walker	SLN	**4.742**	3	**1.581**	0	6	18	60	0.300
126	1938	Powell	Jake	NYA	**4.716**	3	**1.572**	1	5	10	23	0.435
127	1911	Murphy	Danny	PHA	**4.696**	3	**1.565**	1	12	18	59	0.305
128	1987	Trammell	Alan	DET	**4.681**	2	**2.341**	3	11	17	51	0.333
129	1919	Jackson	Joe	CHA	**4.658**	2	**2.329**	1	8	19	55	0.345
130	2010	Ruiz	Carlos	PHI	**4.652**	4	**1.163**	4	15	35	125	0.280
131	2005	Olerud	John	BOS	**4.642**	8	**0.580**	9	34	66	237	0.278
132	1990	Randolph	Willie	OAK	**4.640**	6	**0.773**	4	14	36	162	0.222
133	1983	Russell	Bill	LAN	**4.625**	5	**0.925**	0	18	57	194	0.294
134	2006	Sheffield	Gary	NYA	**4.602**	6	**0.767**	6	19	40	161	0.248
135	1910	Chance	Frank	CHN	**4.589**	4	**1.147**	0	6	21	70	0.300
136	1954	Thompson	Hank	NY1	**4.528**	2	**2.264**	0	2	6	25	0.240
137	1942	Selkirk	George	NYA	**4.454**	6	**0.742**	2	10	18	68	0.265
138	1977	Johnson	Davey	PHI	**4.449**	5	**0.890**	2	12	25	111	0.225
139	2001	Alomar	Roberto	CLE	**4.405**	7	**0.629**	4	33	72	230	0.313
140	1966	Fairly	Ron	LAN	**4.394**	4	**1.098**	2	6	12	40	0.300
141	1928	Meusel	Bob	NYA	**4.386**	6	**0.731**	1	17	29	129	0.225
142	2005	Finley	Steve	LAA	**4.347**	7	**0.621**	1	22	41	165	0.248
143	1888	Latham	Arlie	SL4	**4.347**	4	**1.087**	0	8	38	143	0.266
144	1990	Sabo	Chris	CIN	**4.344**	1	**4.344**	3	8	14	38	0.368
145	1973	Jones	Cleon	NYN	**4.332**	2	**2.166**	2	8	23	81	0.284
146	1973	Staub	Rusty	NYN	**4.331**	1	**4.331**	4	11	14	41	0.341
147	1910	Hofman	Solly	CHN	**4.320**	3	**1.440**	0	8	17	57	0.298
148	1946	York	Rudy	BOS	**4.319**	3	**1.440**	3	10	17	77	0.221
149	1997	White	Devon	FLO	**4.316**	5	**0.863**	3	20	56	189	0.296
150	1991	Puckett	Kirby	MIN	**4.314**	2	**2.157**	5	15	30	97	0.309

Best Batting Postseason Careers
From 1884-2010

Rank	LPostYr	Name	First	Final PTeam	Postseason Career PEVA-B	Total Yrs	Career PerYr	HR	RBI	H	AB	AVG
151	1924	Youngs	Ross	NY1	**4.292**	4	**1.073**	1	10	26	91	0.286
152	1972	Clemente	Roberto	PIT	**4.279**	4	**1.070**	3	14	34	107	0.318
153	1946	Musial	Stan	SLN	**4.269**	4	**1.067**	1	8	22	86	0.256
154	1968	Mathews	Eddie	DET	**4.267**	3	**1.422**	1	7	10	50	0.200
155	1946	Kurowski	Whitey	SLN	**4.256**	4	**1.064**	1	9	21	83	0.253
156	1963	Kubek	Tony	NYA	**4.203**	6	**0.700**	2	10	35	146	0.240
157	1946	Marion	Marty	SLN	**4.199**	4	**1.050**	1	11	18	78	0.231
158	1949	Henrich	Tommy	NYA	**4.179**	4	**1.045**	4	8	22	84	0.262
159	1917	Herzog	Buck	NY1	**4.178**	4	**1.045**	0	7	23	94	0.245
160	1886	Anson	Cap	CHN	**4.175**	2	**2.088**	0	1	16	47	0.340
161	1890	Wolf	Jimmy	LS2	**4.164**	1	**4.164**	0	8	9	25	0.360
162	1982	Chambliss	Chris	ATL	**4.132**	4	**1.033**	3	15	32	114	0.281
163	1992	Smith	Lonnie	ATL	**4.127**	6	**0.688**	4	17	57	205	0.278
164	1996	McGee	Willie	SLN	**4.120**	6	**0.687**	4	23	53	192	0.276
165	1975	Robertson	Bob	PIT	**4.116**	5	**0.823**	6	12	15	53	0.283
166	1956	Robinson	Jackie	BRO	**4.093**	6	**0.682**	2	12	32	137	0.234
167	1985	Porter	Darrell	SLN	**4.093**	5	**0.819**	1	9	32	120	0.267
168	1950	Johnson	Billy	NYA	**4.080**	4	**1.020**	0	5	14	59	0.237
169	1992	Lansford	Carney	OAK	**4.078**	5	**0.816**	2	18	39	128	0.305
170	2002	Williams	Matt	ARI	**4.069**	5	**0.814**	6	28	47	190	0.247
171	1959	Kluszewski	Ted	CHA	**4.066**	1	**4.066**	3	10	9	23	0.391
172	1966	Gilliam	Jim	LAN	**4.065**	7	**0.581**	2	12	31	147	0.211
173	1892	Kelly	King	BSN	**4.064**	3	**1.355**	1	1	14	58	0.241

174	1973	Mays	Willie	NYN	**4.032**	5	**0.806**	1	10	22	89	0.247
175	1909	Leach	Tommy	PIT	**4.031**	2	**2.015**	0	9	18	58	0.310
176	1978	Blair	Paul	NYA	**4.027**	8	**0.503**	3	15	38	146	0.260
177	2010	Utley	Chase	PHI	**4.016**	4	**1.004**	10	24	36	148	0.243
178	1949	Lindell	Johnny	NYA	**4.010**	3	**1.337**	0	7	11	34	0.324
179	1920	Gardner	Larry	CLE	**4.001**	4	**1.000**	3	13	17	86	0.198
180	1906	Rohe	George	CHA	**3.981**	1	**3.981**	0	4	7	21	0.333
181	1997	Ripken	Cal	BAL	**3.980**	3	**1.327**	1	8	37	110	0.336
182	2009	Youkilis	Kevin	BOS	**3.973**	4	**0.993**	6	17	34	111	0.306
183	1915	Barry	Jack	BOS	**3.938**	5	**0.788**	0	7	21	87	0.241
184	1889	Tiernan	Mike	NY1	**3.931**	2	**1.966**	2	11	24	76	0.316
185	1981	Piniella	Lou	NYA	**3.858**	5	**0.772**	3	18	42	140	0.300
186	2003	Grissom	Marquis	SFN	**3.840**	4	**0.960**	5	20	69	218	0.317
187	2010	Thome	Jim	MIN	**3.834**	9	**0.426**	17	37	47	217	0.217
188	1884	Denny	Jerry	PRO	**3.801**	1	**3.801**	1	2	4	9	0.444
189	1928	Dugan	Joe	NYA	**3.794**	5	**0.759**	1	8	24	90	0.267
190	1969	Roseboro	Johnny	MIN	**3.784**	5	**0.757**	1	7	12	75	0.160
191	1971	Buford	Don	BAL	**3.781**	3	**1.260**	5	11	22	86	0.256
192	2009	Molina	Yadier	SLN	**3.726**	4	**0.931**	2	11	34	108	0.315
193	1983	Schmidt	Mike	PHI	**3.698**	6	**0.616**	4	16	33	140	0.236
194	2010	Cruz	Nelson	TEX	**3.692**	1	**3.692**	6	11	19	60	0.317
195	2001	Vizquel	Omar	CLE	**3.678**	6	**0.613**	0	20	57	228	0.250
196	1988	Dempsey	Rick	LAN	**3.674**	3	**1.225**	1	7	20	66	0.303
197	1986	Garner	Phil	HOU	**3.667**	4	**0.917**	1	8	21	68	0.309
198	1972	Kaline	Al	DET	**3.663**	2	**1.832**	3	9	16	48	0.333
199	2001	Knoblauch	Chuck	NYA	**3.636**	5	**0.727**	2	20	63	244	0.258
200	2000	Canseco	Jose	NYA	**3.633**	5	**0.727**	7	18	19	103	0.184

Best Batting Postseason Careers

From 1884-2010

					Postseason							
				Final	**Career**	**Total**	**Career**					
Rank	**LPostYr**	**Name**	**First**	**PTeam**	**PEVA-B**	**Yrs**	**PerYr**	**HR**	**RBI**	**H**	**AB**	**AVG**
201	2010	Ross	Cody	SFN	**3.598**	1	**3.598**	5	10	15	51	0.294
202	1884	Hines	Paul	PRO	**3.597**	1	**3.597**	0	1	2	8	0.250
203	1996	Pendleton	Terry	ATL	**3.554**	6	**0.592**	3	23	58	230	0.252
204	1981	Smith	Reggie	LAN	**3.551**	4	**0.888**	6	17	25	107	0.234
205	1999	Strawberry	Darryl	NYA	**3.544**	5	**0.709**	9	22	32	126	0.254
206	2009	Damon	Johnny	NYA	**3.526**	7	**0.504**	9	30	68	244	0.279
207	1996	Lemke	Mark	ATL	**3.523**	5	**0.705**	1	25	63	232	0.272
208	1999	Alomar	Sandy	CLE	**3.523**	5	**0.705**	5	28	37	173	0.214
209	1981	Otis	Amos	KCA	**3.517**	5	**0.703**	3	11	23	78	0.295
210	2004	Borders	Pat	MIN	**3.516**	5	**0.703**	2	13	35	111	0.315
211	2001	Gant	Ron	OAK	**3.480**	6	**0.580**	8	28	43	189	0.228
212	2003	Alou	Moises	CHN	**3.446**	4	**0.861**	5	24	37	134	0.276
213	1945	Hack	Stan	CHN	**3.444**	4	**0.861**	0	5	24	69	0.348
214	1981	Aikens	Willie	KCA	**3.443**	2	**1.722**	4	10	15	40	0.375
215	1892	McCarthy	Tommy	BSN	**3.433**	2	**1.716**	1	11	18	62	0.290
216	1910	Sheckard	Jimmy	CHN	**3.411**	4	**0.853**	0	5	15	77	0.195
217	1982	Yount	Robin	ML4	**3.388**	2	**1.694**	1	7	22	64	0.344
218	1988	Hatcher	Mickey	LAN	**3.386**	1	**3.386**	2	8	12	40	0.300
219	1890	Collins	Hub	BRO	**3.379**	2	**1.690**	1	3	22	64	0.344
220	1974	Wynn	Jimmy	LAN	**3.375**	1	**3.375**	1	4	5	26	0.192
221	1905	Donlin	Mike	NY1	**3.357**	1	**3.357**	0	1	5	19	0.263
222	1941	Herman	Billy	BRO	**3.350**	4	**0.838**	1	7	16	66	0.242
223	1886	Pfeffer	Fred	CHN	**3.331**	2	**1.666**	2	4	17	48	0.354
224	1909	Wagner	Honus	PIT	**3.331**	2	**1.665**	0	9	14	51	0.275
225	1923	Ward	Aaron	NYA	**3.330**	3	**1.110**	3	9	18	63	0.286
226	1910	Tinker	Joe	CHN	**3.307**	4	**0.827**	1	6	16	68	0.235

227	2008	Garciaparra	Nomar	LAN	**3.302**	5	**0.660**	7	24	36	112	0.321
228	1940	Ripple	Jimmy	CIN	**3.271**	3	**1.090**	2	9	16	50	0.320
229	1985	Guerrero	Pedro	LAN	**3.269**	3	**1.090**	4	16	20	89	0.225
230	2009	Giambi	Jason	COL	**3.258**	8	**0.407**	7	19	40	138	0.290
231	1981	Watson	Bob	NYA	**3.240**	2	**1.620**	2	9	23	62	0.371
232	1985	Yeager	Steve	LAN	**3.234**	6	**0.539**	5	14	27	107	0.252
233	2003	Alfonzo	Edgardo	SFN	**3.231**	3	**1.077**	4	22	35	117	0.299
234	2010	Howard	Ryan	PHI	**3.225**	4	**0.806**	7	27	42	151	0.278
235	1990	Evans	Dwight	BOS	**3.224**	4	**0.806**	4	19	27	113	0.239
236	1966	Wills	Maury	LAN	**3.212**	4	**0.803**	0	4	19	78	0.244
237	1974	Davis	Tommy	BAL	**3.184**	5	**0.637**	0	5	21	67	0.313
238	1990	Barrett	Marty	BOS	**3.155**	3	**1.052**	0	9	25	75	0.333
239	1991	Gladden	Dan	MIN	**3.135**	2	**1.568**	1	15	29	104	0.279
240	1913	Murray	Red	NY1	**3.118**	3	**1.039**	0	5	14	68	0.206
241	1966	Johnson	Lou	LAN	**3.116**	2	**1.558**	2	4	12	42	0.286
242	1969	Agee	Tommie	NYN	**3.089**	1	**3.089**	3	5	8	32	0.250
243	1959	Furillo	Carl	LAN	**3.080**	7	**0.440**	2	13	34	128	0.266
244	2010	Victorino	Shane	PHI	**3.079**	4	**0.770**	6	28	41	156	0.263
245	1909	Cobb	Ty	DET	**3.076**	3	**1.025**	0	9	17	65	0.262
246	2010	Renteria	Edgar	SFN	**3.058**	7	**0.437**	3	23	61	242	0.252
247	1888	Robinson	Yank	SL4	**3.054**	4	**0.764**	0	14	34	124	0.274
248	2010	Molina	Bengie	TEX	**3.051**	4	**0.763**	5	20	36	132	0.273
249	1940	Werber	Billy	CIN	**3.035**	2	**1.517**	0	4	14	43	0.326
250	1978	White	Roy	NYA	**2.996**	3	**0.999**	2	8	22	79	0.278

Best Batting Postseason Careers

From 1884-2010

Rank	LPostYr	Name	First	Final PTeam	Postseason Career PEVA-B	Total Yrs	Career PerYr	HR	RBI	H	AB	AVG
251	2006	Delgado	Carlos	NYN	**2.979**	1	**2.979**	4	11	13	37	0.351
252	1954	Wertz	Vic	CLE	**2.972**	1	**2.972**	1	3	8	16	0.500
253	1916	Meyers	Chief	BRO	**2.970**	4	**0.742**	0	5	18	62	0.290
254	1923	Bancroft	Dave	NY1	**2.960**	4	**0.740**	0	7	16	93	0.172
255	1940	Fox	Pete	DET	**2.956**	3	**0.985**	0	6	18	55	0.327
256	1889	Ward	John	NY1	**2.947**	2	**1.473**	0	13	26	65	0.400
257	1909	Clarke	Fred	PIT	**2.934**	2	**1.467**	2	9	13	53	0.245
258	1995	Larkin	Barry	CIN	**2.933**	2	**1.467**	0	3	24	71	0.338
259	1983	Fisk	Carlton	CHA	**2.914**	2	**1.457**	2	7	14	54	0.259
260	2010	Upton	B.J.	TBA	**2.914**	2	**1.457**	7	18	23	87	0.264
261	1887	Welch	Curt	SL4	**2.906**	3	**0.969**	1	9	23	105	0.219
262	2010	Glaus	Troy	ATL	**2.901**	4	**0.725**	9	16	25	78	0.321
263	2008	Griffey	Ken	CHA	**2.893**	3	**0.964**	6	11	20	69	0.290
264	2005	Sanders	Reggie	SLN	**2.876**	6	**0.479**	7	25	43	221	0.195
265	1924	Kelly	George	NY1	**2.864**	4	**0.716**	1	11	25	101	0.248
266	2009	Lowell	Mike	BOS	**2.863**	4	**0.716**	4	21	29	115	0.252
267	1940	Goodman	Ival	CIN	**2.861**	2	**1.431**	0	6	13	44	0.295
268	1946	Walker	Harry	SLN	**2.837**	3	**0.946**	0	6	10	36	0.278
269	1890	Burns	Oyster	BRO	**2.826**	2	**1.413**	3	16	14	62	0.226
270	1976	Tolan	Bobby	PHI	**2.819**	5	**0.564**	2	13	21	83	0.253
271	1923	Stengel	Casey	NY1	**2.778**	3	**0.926**	2	4	11	28	0.393
272	1917	Robertson	Dave	NY1	**2.772**	1	**2.772**	0	1	11	22	0.500
273	1986	Boone	Bob	CAL	**2.766**	7	**0.395**	2	13	33	106	0.311
274	1908	Rossman	Claude	DET	**2.760**	2	**1.380**	0	5	13	38	0.342
275	1940	Bartell	Dick	DET	**2.742**	3	**0.914**	1	7	20	68	0.294
276	1925	Harris	Bucky	WS1	**2.727**	2	**1.363**	2	7	13	56	0.232
277	1979	Belanger	Mark	BAL	**2.710**	6	**0.452**	1	7	23	126	0.183
278	1931	Bottomley	Jim	SLN	**2.704**	4	**0.676**	1	10	18	90	0.200
279	1997	Pena	Tony	HOU	**2.680**	5	**0.536**	1	5	24	71	0.338

280	1905	McGann	Dan	NY1	**2.669**	1	**2.669**	0	4	4	17	0.235
281	2010	Granderson	Curtis	NYA	**2.662**	2	**1.331**	4	13	22	81	0.272
282	1964	Lopez	Hector	NYA	**2.644**	5	**0.529**	1	7	8	28	0.286
283	2006	Eckstein	David	SLN	**2.644**	4	**0.661**	2	18	49	176	0.278
284	1988	Hernandez	Keith	NYN	**2.634**	3	**0.878**	2	21	31	117	0.265
285	1925	Carey	Max	PIT	**2.624**	1	**2.624**	0	2	11	24	0.458
286	1927	Waner	Lloyd	PIT	**2.623**	1	**2.623**	0	0	6	15	0.400
287	1981	Gamble	Oscar	NYA	**2.619**	3	**0.873**	2	8	13	43	0.302
288	2003	Conine	Jeff	FLO	**2.617**	2	**1.309**	1	8	31	102	0.304
289	2005	Jordan	Brian	ATL	**2.606**	5	**0.521**	6	27	35	140	0.250
290	1998	Gaetti	Gary	CHN	**2.588**	3	**0.863**	5	16	22	93	0.237
291	1903	Parent	Freddy	BOS	**2.579**	1	**2.579**	0	4	9	31	0.290
292	1954	Irvin	Monte	NY1	**2.569**	2	**1.284**	0	4	13	33	0.394
293	1966	Parker	Wes	LAN	**2.561**	2	**1.280**	1	2	10	36	0.278
294	1928	Thevenow	Tommy	SLN	**2.557**	2	**1.278**	1	4	10	24	0.417
295	1998	Gwynn	Tony	SDN	**2.539**	3	**0.846**	1	11	33	108	0.306
296	1920	O'Neill	Steve	CLE	**2.537**	1	**2.537**	0	2	7	21	0.333
297	1933	Schulte	Fred	WS1	**2.536**	1	**2.536**	1	4	7	21	0.333
298	1890	Foutz	Dave	BRO	**2.534**	5	**0.507**	1	19	34	151	0.225
299	1921	Burns	George	NY1	**2.522**	3	**0.841**	0	6	19	74	0.257
300	1992	Wilson	Willie	OAK	**2.503**	6	**0.417**	1	10	40	150	0.267

Best Batting Postseason Careers

From 1884-2010

				Final	**Postseason Career**	**Total**	**Career**					
Rank	**LPostYr**	**Name**	**First**	**PTeam**	**PEVA-B**	**Yrs**	**PerYr**	**HR**	**RBI**	**H**	**AB**	**AVG**
301	1979	Concepcion	Dave	CIN	**2.499**	5	**0.500**	2	13	30	101	0.297
302	1964	Blanchard	Johnny	NYA	**2.484**	5	**0.497**	2	5	10	29	0.345
303	1887	Thompson	Sam	DTN	**2.484**	1	**2.484**	2	7	21	58	0.362
304	1918	Merkle	Fred	CHN	**2.483**	5	**0.497**	1	9	21	88	0.239
305	2008	Pierzynski	A.J.	CHA	**2.476**	4	**0.619**	5	17	30	100	0.300
306	2001	Martinez	Edgar	SEA	**2.465**	4	**0.616**	8	24	34	128	0.266
307	1911	Davis	Harry	PHA	**2.454**	3	**0.818**	0	7	15	61	0.246
308	1997	Fernandez	Tony	CLE	**2.433**	5	**0.487**	1	23	49	150	0.327
309	2005	Walker	Larry	SLN	**2.423**	3	**0.808**	7	15	23	100	0.230
310	1940	Campbell	Bruce	DET	**2.415**	1	**2.415**	1	5	9	25	0.360
311	1988	Sax	Steve	LAN	**2.405**	4	**0.601**	0	4	24	87	0.276
312	1996	Smith	Ozzie	SLN	**2.404**	4	**0.601**	1	10	34	144	0.236
313	2006	Spiezio	Scott	SLN	**2.404**	2	**1.202**	3	25	23	81	0.284
314	1936	Terry	Bill	NY1	**2.404**	3	**0.801**	2	7	18	61	0.295
315	1997	Daulton	Darren	FLO	**2.401**	2	**1.201**	3	10	18	64	0.281
316	1990	Brunansky	Tom	BOS	**2.401**	2	**1.200**	2	12	13	54	0.241
317	1946	Higgins	Pinky	BOS	**2.400**	2	**1.200**	1	8	13	48	0.271
318	2009	Pierre	Juan	LAN	**2.389**	3	**0.796**	0	7	24	79	0.304
319	1997	Boggs	Wade	NYA	**2.387**	6	**0.398**	2	16	42	154	0.273
320	1998	Blauser	Jeff	CHN	**2.385**	7	**0.341**	5	16	35	168	0.208
321	1919	Felsch	Happy	CHA	**2.377**	2	**1.188**	1	6	11	48	0.229
322	1999	Valentin	John	BOS	**2.372**	3	**0.791**	5	19	25	72	0.347
323	1981	Milbourne	Larry	NYA	**2.369**	1	**2.369**	0	4	17	52	0.327
324	2000	Baines	Harold	CHA	**2.367**	6	**0.394**	5	16	33	102	0.324
325	2006	Klesko	Ryan	SDN	**2.366**	7	**0.338**	10	22	39	165	0.236
326	2007	Nixon	Trot	CLE	**2.362**	6	**0.394**	6	25	39	138	0.283
327	1916	Hoblitzel	Dick	BOS	**2.357**	2	**1.178**	0	3	9	33	0.273
328	2003	Aurilia	Rich	SFN	**2.342**	3	**0.781**	6	18	22	98	0.224
329	1941	Medwick	Joe	BRO	**2.337**	2	**1.169**	1	5	15	46	0.326
330	1990	Hassey	Ron	OAK	**2.332**	3	**0.777**	1	6	10	31	0.323
331	1910	Kling	Johnny	CHN	**2.315**	4	**0.579**	0	4	12	65	0.185
332	1932	Cuyler	Kiki	CHN	**2.306**	3	**0.769**	2	12	18	64	0.281

333	1976	Hendricks	Ellie	NYA	**2.306**	5	**0.461**	2	10	18	66	0.273
334	2001	Buhner	Jay	SEA	**2.304**	4	**0.576**	8	12	26	85	0.306
335	2008	Dye	Jermaine	CHA	**2.292**	6	**0.382**	5	17	44	163	0.270
336	2006	Monroe	Craig	DET	**2.288**	1	**2.288**	5	9	12	50	0.240
337	1906	Donahue	Jiggs	CHA	**2.286**	1	**2.286**	0	4	5	18	0.278
338	1889	Connor	Roger	NY1	**2.276**	2	**1.138**	0	15	19	58	0.328
339	1913	Doyle	Larry	NY1	**2.266**	3	**0.755**	1	5	18	76	0.237
340	1928	Maranville	Rabbit	SLN	**2.245**	2	**1.122**	0	3	8	26	0.308
341	2001	Gonzalez	Luis	ARI	**2.243**	3	**0.748**	4	12	22	87	0.253
342	1931	Haas	Mule	PHA	**2.239**	3	**0.746**	2	9	10	62	0.161
343	2008	Konerko	Paul	CHA	**2.236**	3	**0.745**	7	17	18	74	0.243
344	1972	Javier	Julian	CIN	**2.235**	4	**0.559**	1	7	18	54	0.333
345	1985	McRae	Hal	KCA	**2.227**	9	**0.247**	1	15	42	143	0.294
346	1975	Fosse	Ray	OAK	**2.224**	3	**0.741**	2	7	10	58	0.172
347	1926	Southworth	Billy	SLN	**2.223**	2	**1.112**	1	4	10	30	0.333
348	1983	Roenicke	Gary	BAL	**2.219**	2	**1.110**	1	5	6	32	0.188
349	1978	Ferguson	Joe	LAN	**2.209**	2	**1.104**	1	4	7	35	0.200
350	1931	Bishop	Max	PHA	**2.207**	3	**0.736**	0	1	12	66	0.182
351	2001	Boone	Bret	SEA	**2.205**	3	**0.735**	3	12	34	118	0.288
352	2000	Zeile	Todd	NYN	**2.200**	4	**0.550**	4	14	33	113	0.292
353	1972	Horton	Willie	DET	**2.179**	2	**1.089**	1	3	8	33	0.242
354	1979	Davis	Willie	CAL	**2.178**	4	**0.544**	0	3	10	56	0.179

Abbreviation Code: LPostYr - Last Postseason Year; FinalPost Team - Final Postseason Team.

Per Postseason Rankings

Best Batting Postseason Careers

From 1884-2010

					Postseason				
				LPost	**Career**	**Total**	**Career**	**PRanks**	
Rank	**LPostYr**	**Name**	**First**	**Team**	**PEVA-B**	**Yrs**	**PerYr**	**Career**	**C+Per**
1	1968	Brock	Lou	SLN	10.684	3	**3.561**	22	23
2	1943	Keller	Charlie	NYA	13.928	4	**3.482**	11	13
3	1939	Simmons	Al	CIN	13.759	4	**3.440**	12	15
4	1938	Gehrig	Lou	NYA	22.452	7	**3.207**	4	8
5	1916	Lewis	Duffy	BOS	9.120	3	**3.040**	33	38
6	1934	Martin	Pepper	SLN	8.696	3	**2.899**	34	40
7	1918	Hooper	Harry	BOS	11.353	4	**2.838**	20	27
8	1932	Ruth	Babe	NYA	28.065	10	**2.806**	1	9
9	1931	Foxx	Jimmie	PHA	8.127	3	**2.709**	40	49
10	1984	Garvey	Steve	SDN	13.368	5	**2.674**	13	23
11	1964	Tresh	Tom	NYA	7.954	3	**2.651**	44	55
12	1945	Greenberg	Hank	DET	10.574	4	**2.644**	24	36
13	1969	Aaron	Hank	ATL	7.915	3	**2.638**	45	58
14	1971	Robinson	Frank	BAL	13.183	5	**2.637**	14	28
15	1993	Molitor	Paul	TOR	7.901	3	**2.634**	46	61
16	1922	Baker	Frank	NYA	15.186	6	**2.531**	9	25
17	1920	Speaker	Tris	CLE	7.464	3	**2.488**	52	69
18	1984	Matthews	Gary	CHN	7.439	3	**2.480**	53	71
19	1937	Ott	Mel	NY1	7.438	3	**2.479**	54	73
20	1978	Munson	Thurman	NYA	7.181	3	**2.394**	59	79

*Note: Per Year Rank with at least 3 postseasons

*Note: PRanks (Postseason Ranks) for Career Total and Career plus Per Postseason

Total Postseason Rankings

When you begin the conversation, or any list, about the Best Ever Postseason Careers by a pitcher, the first thing we'll note about this particular ranking is that the Top Ten is not predominantly filled by Yankees. Oh, they are there. In fact, they take the two at the top, one completely deserving of it and the other a bit high due to the amount of postseasons pitched for the Bronx Bombers. But what you'll also see in the Top Ten is a broad range of pitchers who helped the Cardinals and the Braves and the Giants and the Red Sox make it to the top of the World Series heap.

The major part of this list contains the pitchers who tallied the highest total of ratings points over their careers, but unlike with the regular season rankings, that tells a whole lot less of the story. Yes, we know it's important with the regular season, too, but you get our point. So at the end of the list, we've included a Top Top Twenty list ranking Postseason pitchers by Per Season average, too, like we did with the position players. Because just like them batters, to really figure out who the best of the best was in postseason play is a combination of both.

Best Postseason Careers by Pitchers

1. Mariano Rivera

Yes, it's finally happened. This year, Mariano Rivera passed Whitey Ford into the #1 spot on the best postseason career list. And who can argue. A 0.71 ERA in 139.7 innings pitched. That's ridiculous! Mariano Rivera is the best Yankee pitcher of All-Time and that says a lot. His rate stats during postseason play stand out as unparalleled in playoff history. And while we usually reserve our accolades for the starters and have a significant problem with the current notion of six inning starters being great and thus one inning pitchers being fantastic, that has to be put to rest when you're talking about Mariano. He doesn't have the the highest ranking on a per year basis, mostly due to the lack of innings that a relief pitcher hurls, and some years when the Yankees didn't get to the World Series bring that number down, but it's still pretty high. And let's just talk about those rate stats for a minute. 6.89 Hits and Walks per 9 innings, a Strikeout to Walk Ratio of 5.19, and that ERA, all well below his regular season averages. Just spectacular! When Mariano got the ball in a playoff game, it was game over, Rivera to the rescue, and Rivera to victory.

Mariano Rivera - Postseason (14 Years)			
Category			**Rank**
TOTAL POSTSEASON PEVA		**18.817**	**1**
POST PEVA PER YEAR*		**1.254**	**36**
POST WHIP9	6.89		
POST SO/W	5.19		

2. Whitey Ford

It's a matter of accumulation and skill for Ford as he tops the list of Best Postseason Careers in the counting stat PEVA rankings. We don't think Ford should be considered the best or even second best, but he was very good. In the eleven series Whitey pitched, the Yankees won six times, with Ford winning more games than any other pitcher in World Series history, 10. His ERA of 2.71 was consistent with his career average of the regular season at 2.75, something some other top pitchers had trouble doing. There were more dominant World Series pitchers, like Gibson coming soon, but Whitey Ford was very good.

Whitey Ford - Postseason (11 Years)			
Category			**Rank**
TOTAL POSTSEASON PEVA		**18.789**	**2**
POST PEVA PER YEAR*		**1.708**	**13**
POST WHIP9	10.23		
POST SO/W	2.76		

*Note: Per Year Rank with at least 3 postseasons

3. Bob Gibson

Here's the man we consider to be the best postseason pitcher in history, despite coming in third in the total postseason PEVA rankings; he does top out at #1 in per season numbers, however, more than 1.5 times higher than the man in second place, George Earnshaw. Gibson was as combative a pitcher on the mound as there has ever been in baseball. He pitched inside, tight, but hit few batters. And while he was very good during the regular season, he was great in the playoffs. A WHIP9 in the World Series at 8.00 compared to 10.71 for his regular season career, SO/W Ratio of 5.41 vs. 2.33 and an ERA of 1.89 vs. 2.91. Gibson only got three chances to show his stuff in the playoffs so he falls behind Ford in the counting categories, but if you had one game to win in the postseason, there would be a whole lot of managers who would choose this Bob.

Bob Gibson - Postseason (3 Years)			
Category			**Rank**
TOTAL POSTSEASON PEVA		**13.470**	**3**
POST PEVA PER YEAR*		**4.490**	**1**
POST WHIP9	8.00		
POST SO/W	5.41		

4. Curt Schilling

Schilling can get himself in trouble with his mouth, but get out of it on the mound. From bloody sock fame to four other postseason heroics, Schilling established himself as the #4 best Postseason Pitcher of All-Time, no matter whether you take a Total or Per Average take on the question. Pitching for Philadelphia, Arizona, and Boston, Schilling racked up strikeouts, wins, and World Series titles, one with Arizona and two with the Red Sox. There are some who don't think that Schilling will make the Hall of Fame. Too few regular season wins, they say. Too short a time span of dominance to others. We disagree on both accounts. And when you add in the heroics of Curt in the postseason, like Whitey Ford whose regular season stats look very similar, we think there will be a day very soon when Red Sox nation will be flocking to Cooperstown for one of their own World Series heros, and his name will be Curt Schilling.

Curt Schilling - Postseason (5 Years)			
Category			**Rank**
TOTAL POSTSEASON PEVA		**12.460**	**4**
POST PEVA PER YEAR*		**2.492**	**4**
POST WHIP9	8.71		
POST SO/W	4.80		

5. Orel Hershiser

The tale of Hershiser's postseason career really gets told in two postseasons; the first with the Dodgers in 1988 and the second with the Indians in 1995. Those were the years in which Orel dominated with both seasons ranked in the Top 11 of best postseason performances. And that was enough to put him into the Top Ten for his career as well. Overall, he was 8-3 with a 2.59 ERA, plus that one save, over 132 innings pitched.

Orel Hershiser - Postseason (6 Years)		
Category		**Rank**
TOTAL POSTSEASON PEVA	**12.424**	**5**
POST PEVA PER YEAR*	**2.071**	**9**

POST WHIP9	9.95
POST SO/W	2.26

6. John Smoltz

One of the few pitchers to garner postseason headlines as a starting and relief pitcher, Smoltz had a combination of talents that land him in the #6 spot on the All-Time list. 15 wins and 4 saves with an ERA of 2.79 over 14 seasons in the playoffs helped Smoltz climb the ladder, even though he was less than consistent in his approach. In fact, there were really only three postseasons where Smoltz stood out. Overall, Smoltz likely is much further down the line in the postseason rankings when you consider the average year and the higher WHIP9, raising folks like Beckett, Koufax, and Earnshaw above him. Atlanta made so many postseaons, but only five of them made it to the World Series, which causes a lower per postseason average than players of the past who played in one round, and that's not his fault. But then Smoltz had opportunities others would not have had as well. You can decide for yourself where you stand on that issue.

John Smoltz - Postseason (14 Years)			
Category			**Rank**
TOTAL POSTSEASON PEVA		**12.303**	**6**
POST PEVA PER YEAR*		**0.879**	**66**
POST WHIP9	10.30		
POST SO/W	2.97		

7. Andy Pettitte

Andy Pettitte wins baseball games, but does benefit from the prowess of the teams he plays on. With the highest postseason ERA of anyone in the Top 75, Pettitte is the consummate professional, taking the ball and giving his team innings while the offense attacks. We don't think of Pettite in the terms of best playoff pitchers, despite all the wins he accumulates, but more of a steady performer that any team in the World Series would want on their squad. And this year that moves Pettitte up one spot to #7 on the list.

Andy Pettitte - Postseason (12 Years)			
Category			**Rank**
TOTAL POSTSEASON PEVA		**11.474**	**7**
POST PEVA PER YEAR*		**0.883**	**66**
POST WHIP9	11.74		
POST SO/W	2.40		

8. Christy Mathewson

If you weren't going to choose Bob Gibson to pitch that one game, another man in the conversation would have to be Mathewson. He is one of only three pitchers with an ERA under 1.00 who have pitched 50 postseason innings; the others are Rivera and Koufax. His WHIP9 at 7.52 is the second lowest among the Top Ten here behind Rivera. Now Mathewson lost some games; he was only a 0.500 pitcher in the World Series, but with those gawdy stats, it's hardly his fault. Overall, Mathewson ranks as the #8 pitcher in Total Postseason Career PEVA and #3 per season. And although we're still going to choose Mr. Gibson as our Number One starter in the playoffs, we'll be just pleased as punch to name Mathewson as Number Two.

Christy Mathewson - Postseason (4 Years)			
Category			**Rank**
TOTAL POSTSEASON PEVA		**11.260**	**8**
POST PEVA PER YEAR*		**2.815**	**3**
POST WHIP9	7.52		

POST SO/W	4.80

9. Greg Maddux

Although Maddux makes the Top Ten here, his postseason performance is likely the main reason most don't consider him as the best pitcher in history. Not that Maddux was bad in the postseason; it's just that outside of the back to back seasons of 1995 and 1996, he was not great. He had a below 0.500 record at 11-14, an ERA of 3.27, higher than his regular season average of 3.16, an WHIP9 of 11.18 that was higher than the regular season average of 10.29, and a SO/W Ratio of 2.45, worse than the 3.37 ca[illegible]
this occurred, we don't know. But it did, so despite the high ranking in Total PEVA Postseason Points culled from all those seasons Maddux pitched in the playoffs, and the major reason Maddux being that they got there, it is beyond the true measure of his postseason prowess.

Greg Maddux - Postseason (13 Years)			
Category			**Rank**
TOTAL POSTSEASON PEVA		**10.679**	**9**
POST PEVA PER YEAR*		**0.821**	**71**
POST WHIP9	11.18		
POST SO/W	2.45		

10. Herb Pennock

Pennock is one of those pitchers we just don't recall, not on the tip of the postseason tongue of most baseball fans, even those who follow the history of the game fairly well. However, that shouldn't be the case. Pennock pitched so well during the five World Series in which he participated, that his WHIP9 was only 7.64 hits and walks over 9 inning pitched and his ERA only 1.95. Plus, Pennock never lost a game, going 5-0 and pitched in 3 saves, too, for the Philadelphia A's and New York Giants between 1914 and 1932.

Herb Pennock - Postseason (5 Years)			
Category			**Rank**
TOTAL POSTSEASON PEVA		**9.987**	**10**
POST PEVA PER YEAR*		**1.997**	**11**
POST WHIP9	7.64		
POST SO/W	3.00		

Best Pitching Postseason Careers
From 1884-2010

Rank	LPostYr	Name	First	FinalPo Team	Postseason Career PEVA-P	Total Yrs	Career PerYr	W	L	SV	IP	ERA
1	2010	Rivera	Mariano	NYA	**18.817**	15	**1.254**	8	1	42	139.7	0.71
2	1964	Ford	Whitey	NYA	**18.789**	11	**1.708**	10	8	0	146.0	2.71
3	1968	Gibson	Bob	SLN	**13.470**	3	**4.490**	7	2	0	81.0	1.89
4	2007	Schilling	Curt	BOS	**12.460**	5	**2.492**	11	2	0	133.3	2.23
5	1999	Hershiser	Orel	NYN	**12.424**	6	**2.071**	8	3	1	132.0	2.59
6	2009	Smoltz	John	SLN	**12.303**	14	**0.879**	15	4	4	209.0	2.67
7	2010	Pettitte	Andy	NYA	**11.474**	13	**0.883**	19	10	0	263.0	3.83
8	1913	Mathewson	Christy	NY1	**11.260**	4	**2.815**	5	5	0	101.7	0.97
9	2008	Maddux	Greg	LAN	**10.679**	13	**0.821**	11	14	1	198.0	3.27
10	1932	Pennock	Herb	NYA	**9.987**	5	**1.997**	5	0	3	55.3	1.95
11	2009	Beckett	Josh	BOS	**9.779**	4	**2.445**	7	3	0	93.7	3.07
12	1931	Hoyt	Waite	PHA	**9.692**	7	**1.385**	6	4	0	83.7	1.83

13	1993	Stewart	Dave	TOR	**9.627**	6	**1.604**	10	6	0	133.0	2.84
14	1953	Reynolds	Allie	NYA	**9.036**	6	**1.506**	7	2	4	77.3	2.79
15	1966	Koufax	Sandy	LAN	**9.019**	4	**2.255**	4	3	0	57.0	0.95
16	2006	Glavine	Tom	NYN	**8.977**	12	**0.748**	14	16	0	218.3	3.42
17	1931	Earnshaw	George	PHA	**8.877**	3	**2.959**	4	3	0	62.7	1.58
18	2007	Clemens	Roger	NYA	**8.507**	12	**0.709**	12	8	0	199.0	3.75
19	1953	Raschi	Vic	NYA	**8.348**	6	**1.391**	5	3	0	60.3	2.24
20	2010	Lee	Cliff	TEX	**8.250**	2	**4.125**	7	2	0	76.0	2.13
21	1942	Ruffing	Red	NYA	**7.632**	7	**1.090**	7	2	0	85.7	2.63
22	1953	Lopat	Ed	NYA	**7.252**	5	**1.450**	4	1	0	52.0	2.60
23	2006	Johnson	Randy	NYA	**7.168**	9	**0.796**	9	10	0	146.3	3.32
24	1978	Hunter	Catfish	NYA	**7.143**	7	**1.020**	9	6	1	132.3	3.26
25	1937	Hubbell	Carl	NY1	**7.005**	3	**2.335**	4	2	0	50.3	1.79
26	1928	Alexander	Pete	SLN	**6.974**	3	**2.325**	3	2	1	43.0	3.56
27	2010	Hamels	Cole	PHI	**6.877**	4	**1.719**	6	4	0	75.7	3.45
28	1939	Gomez	Lefty	NYA	**6.732**	5	**1.346**	6	0	0	50.3	2.86
29	1974	McNally	Dave	BAL	**6.716**	6	**1.119**	7	4	0	90.3	2.49
30	1929	Nehf	Art	CHN	**6.409**	5	**1.282**	4	4	0	79.0	2.16
31	1983	Palmer	Jim	BAL	**6.291**	8	**0.786**	8	3	0	124.3	2.61
32	1983	Carlton	Steve	PHI	**6.267**	8	**0.783**	6	6	0	99.3	3.26
33	1884	Radbourn	Charley	PRO	**6.157**	1	**6.157**	3	0	0	22.0	0.00
34	1986	Sutton	Don	CAL	**6.128**	5	**1.226**	6	4	0	100.3	3.68
35	1910	Overall	Orval	CHN	**5.995**	4	**1.499**	3	1	0	51.3	1.58
36	1931	Grove	Lefty	PHA	**5.945**	3	**1.982**	4	2	2	51.3	1.75
37	1914	Bender	Chief	PHA	**5.878**	5	**1.176**	6	4	0	85.0	2.44
38	1977	Gullett	Don	NYA	**5.817**	6	**0.969**	4	5	1	94.3	3.72
39	1992	Morris	Jack	TOR	**5.807**	4	**1.452**	7	4	0	92.3	3.80
40	1983	McGregor	Scott	BAL	**5.717**	2	**2.859**	3	3	0	49.7	1.63
41	1889	Caruthers	Bob	BR3	**5.655**	4	**1.414**	7	8	1	147.0	2.51
42	2005	Hernandez	Orlando	CHA	**5.639**	7	**0.806**	9	3	0	106.0	2.55
43	1947	Newsom	Bobo	NYA	**5.542**	2	**2.771**	2	2	0	28.3	2.86
44	2007	Mussina	Mike	NYA	**5.532**	9	**0.615**	7	8	0	139.7	3.42
45	1925	Coveleski	Stan	WS1	**5.366**	2	**2.683**	3	2	0	41.3	1.74
46	1947	Chandler	Spud	NYA	**5.323**	4	**1.331**	2	2	1	33.3	1.62
47	1982	John	Tommy	CAL	**5.281**	5	**1.056**	6	3	0	88.3	2.65
48	2006	Wells	David	SDN	**5.269**	11	**0.479**	10	5	0	125.0	3.17
49	2010	Lidge	Brad	PHI	**5.179**	5	**1.036**	2	4	18	43.3	2.48
50	1963	Podres	Johnny	LAN	**5.142**	4	**1.286**	4	1	0	38.3	2.11

Best Pitching Postseason Careers

From 1884-2010

Rank	LPostYr	Name	First	FinalPo Team	Postseason Career PEVA-P	Total Yrs	Career PerYr	W	L	SV	IP	ERA
51	1984	Gossage	Rich	SDN	**5.105**	4	**1.276**	2	1	8	31.3	2.87
52	1974	Cuellar	Mike	BAL	**5.041**	5	**1.008**	4	4	0	85.3	2.85
53	1892	Clarkson	John	CL4	**4.928**	3	**1.643**	2	5	0	64.0	2.67
54	2010	Contreras	Jose	PHI	**4.861**	4	**1.215**	4	3	0	49.0	3.49
55	1925	Johnson	Walter	WS1	**4.854**	2	**2.427**	3	3	0	50.0	2.16
56	1922	Mays	Carl	NYA	**4.808**	4	**1.202**	3	4	1	57.3	2.35
57	2010	Lowe	Derek	ATL	**4.653**	7	**0.665**	5	7	1	95.3	3.21
58	1892	Stivetts	Jack	BSN	**4.525**	1	**4.525**	2	0	0	29.0	0.93
59	2009	Martinez	Pedro	PHI	**4.457**	5	**0.891**	6	4	0	96.3	3.46
60	2002	Stanton	Mike	NYA	**4.452**	11	**0.405**	5	2	1	55.7	2.10
61	1998	Eckersley	Dennis	BOS	**4.450**	7	**0.636**	1	3	15	36.0	3.00
62	1889	Keefe	Tim	NY1	**4.408**	3	**1.469**	4	3	1	61.0	2.66
63	1939	Pearson	Monte	NYA	**4.383**	4	**1.096**	4	0	0	35.7	1.01
64	1910	Brown	Mordecai	CHN	**4.337**	4	**1.084**	5	4	0	57.7	2.97
65	1958	Burdette	Lew	ML1	**4.311**	2	**2.156**	4	2	0	49.3	2.92

66	1972	Lolich	Mickey	DET	**4.262**	2	**2.131**	3	1	0	46.0	1.57
67	1932	Moore	Wilcy	NYA	**4.254**	2	**2.127**	2	0	1	16.0	0.56
68	1914	Rudolph	Dick	BSN	**4.229**	1	**4.229**	2	0	0	18.0	0.50
69	1935	Warneke	Lon	CHN	**4.207**	2	**2.104**	2	1	0	27.3	2.63
70	1916	Shore	Ernie	BOS	**4.137**	2	**2.069**	3	1	0	34.7	1.82
71	1946	Brecheen	Harry	SLN	**4.136**	3	**1.379**	4	1	0	32.7	0.83
72	1981	Hooton	Burt	LAN	**4.098**	3	**1.366**	6	3	0	59.7	3.17
73	1993	Guzman	Juan	TOR	**4.087**	3	**1.362**	5	1	0	51.7	2.44
74	1964	Terry	Ralph	NYA	**4.051**	5	**0.810**	2	4	0	46.0	2.93
75	2001	Charlton	Norm	SEA	**4.043**	5	**0.809**	5	1	4	38.3	1.17
76	1975	Carroll	Clay	CIN	**4.001**	4	**1.000**	4	2	2	32.3	1.39
77	1996	Avery	Steve	ATL	**3.783**	5	**0.757**	5	3	0	77.7	2.90
78	2009	Carpenter	Chris	SLN	**3.751**	3	**1.250**	5	2	0	58.3	2.93
79	1962	Larsen	Don	SFN	**3.719**	5	**0.744**	4	2	0	36.0	2.75
80	1944	Cooper	Mort	SLN	**3.704**	3	**1.235**	2	3	0	45.0	3.00
81	1914	Plank	Eddie	PHA	**3.655**	4	**0.914**	2	5	0	54.7	1.32
82	1975	Holtzman	Ken	OAK	**3.644**	4	**0.911**	6	4	0	70.3	2.30
83	1958	Spahn	Warren	ML1	**3.636**	3	**1.212**	4	3	0	56.0	3.05
84	2006	Rogers	Kenny	DET	**3.626**	4	**0.907**	3	3	0	43.3	4.15
85	1938	Dean	Dizzy	CHN	**3.581**	2	**1.790**	2	2	0	34.3	2.88
86	1987	Blyleven	Bert	MIN	**3.560**	3	**1.187**	5	1	0	47.3	2.47
87	1974	Odom	Blue Moor	OAK	**3.504**	3	**1.168**	3	1	0	42.0	1.07
88	1981	Guidry	Ron	NYA	**3.408**	4	**0.852**	5	2	0	62.7	3.02
89	1981	Fingers	Rollie	ML4	**3.407**	6	**0.568**	5	3	9	54.7	2.63
90	1987	Dayley	Ken	SLN	**3.256**	2	**1.628**	1	0	5	20.7	0.44
91	1947	Casey	Hugh	BRO	**3.253**	2	**1.627**	2	2	1	15.7	1.72
92	2009	Papelbon	Jonathan	BOS	**3.235**	4	**0.809**	2	1	7	27.0	1.00
93	1909	Phillippe	Deacon	PIT	**3.206**	2	**1.603**	3	2	0	50.0	2.70
94	1990	Boddicker	Mike	BOS	**3.168**	3	**1.056**	2	2	0	28.7	2.51
95	1934	Haines	Jesse	SLN	**3.167**	4	**0.792**	3	1	0	32.3	1.67
96	2010	Sabathia	C.C.	NYA	**3.135**	5	**0.627**	7	4	0	77.3	4.66
97	1918	Vaughn	Hippo	CHN	**3.106**	1	**3.106**	1	2	0	27.0	1.00
98	1998	Myers	Randy	SDN	**3.096**	5	**0.619**	2	2	8	30.7	2.35
99	1923	Bush	Joe	NYA	**3.043**	5	**0.609**	2	5	1	60.7	2.67
100	1979	Seaver	Tom	CIN	**3.033**	3	**1.011**	3	3	0	61.7	2.77

Best Pitching Postseason Careers

From 1884-2010

Rank	LPostYr	Name	First	FinalPo Team	Postseason Career PEVA-P	Total Yrs	Career PerYr	W	L	SV	IP	ERA
101	1953	Sain	Johnny	NYA	**3.018**	4	**0.755**	2	2	0	30.7	2.64
102	1959	Sherry	Larry	LAN	**3.013**	1	**3.013**	2	0	2	12.7	0.71
103	1996	Jackson	Danny	SLN	**2.994**	5	**0.599**	4	3	0	57.3	3.30
104	1932	Pipgras	George	NYA	**2.978**	3	**0.993**	3	0	0	26.0	2.77
105	2008	Rodriguez	Francisco	LAA	**2.965**	5	**0.593**	5	4	3	31.7	3.13
106	1988	Tudor	John	LAN	**2.947**	3	**0.982**	5	4	0	63.3	3.41
107	2010	Oswalt	Roy	PHI	**2.941**	3	**0.980**	5	1	0	66.3	3.39
108	1909	Mullin	George	DET	**2.924**	3	**0.975**	3	3	0	58.0	1.86
109	1973	Matlack	Jon	NYN	**2.908**	1	**2.908**	2	2	0	25.7	1.40
110	1947	Shea	Spec	NYA	**2.879**	1	**2.879**	2	0	0	15.3	2.35
111	1975	Tiant	Luis	BOS	**2.836**	2	**1.418**	3	0	0	34.7	2.86
112	1960	Labine	Clem	PIT	**2.824**	5	**0.565**	2	2	2	31.3	3.16
113	1928	Zachary	Tom	NYA	**2.814**	3	**0.938**	3	0	0	28.3	2.86
114	2005	Garcia	Freddy	CHA	**2.812**	3	**0.937**	6	2	0	55.0	3.11
115	1997	Key	Jimmy	BAL	**2.792**	6	**0.465**	5	3	0	68.7	3.15
116	1992	Moore	Mike	OAK	**2.756**	3	**0.919**	4	3	0	38.3	3.29
117	2001	Wohlers	Mark	NYA	**2.756**	7	**0.394**	1	2	9	38.3	2.35
118	1945	Derringer	Paul	CHN	**2.730**	4	**0.683**	2	4	0	52.7	3.42

119	2000	Cone	David	NYA	**2.727**	8	**0.341**	8	3	0	111.3	3.80
120	1954	Lemon	Bob	CLE	**2.704**	2	**1.352**	2	2	0	29.7	3.94
121	2001	Rocker	John	CLE	**2.701**	4	**0.675**	2	0	3	20.7	0.00
122	1929	Ehmke	Howard	PHA	**2.675**	1	**2.675**	1	0	0	12.7	1.42
123	1940	Rowe	Schoolboy	DET	**2.656**	3	**0.885**	2	5	0	46.0	3.91
124	2003	Nelson	Jeff	NYA	**2.650**	8	**0.331**	1	4	0	55.3	1.95
125	2008	Suppan	Jeff	MIL	**2.649**	4	**0.662**	3	4	0	57.0	3.63
126	2004	Brown	Kevin	NYA	**2.618**	3	**0.873**	5	5	0	81.7	4.19
127	2007	Corpas	Manuel	COL	**2.617**	1	**2.617**	1	0	5	10.3	0.87
128	1977	Lonborg	Jim	PHI	**2.588**	3	**0.863**	2	3	0	33.3	3.51
130	2009	Lackey	John	LAA	**2.562**	5	**0.512**	3	4	0	78.0	3.12
131	2004	Hampton	Mike	ATL	**2.514**	6	**0.419**	2	4	0	65.0	3.74
132	1988	Hurst	Bruce	BOS	**2.491**	2	**1.245**	3	2	0	51.0	2.29
133	2009	Wainwright	Adam	SLN	**2.483**	2	**1.242**	1	0	4	17.7	0.51
134	1989	Lefferts	Craig	SFN	**2.470**	3	**0.823**	2	0	1	15.7	1.15
135	1976	Nolan	Gary	CIN	**2.448**	4	**0.612**	2	2	0	59.3	3.34
136	1934	Hallahan	Bill	SLN	**2.447**	4	**0.612**	3	1	1	39.7	1.36
137	2010	Lincecum	Tim	SFN	**2.410**	1	**2.410**	4	1	0	37.0	2.43
138	1981	Lyle	Sparky	PHI	**2.399**	4	**0.600**	3	0	1	21.3	1.69
139	1916	Coombs	Jack	BRO	**2.377**	3	**0.792**	5	0	0	53.3	2.70
140	1889	Crane	Ed	NY1	**2.369**	2	**1.184**	5	2	0	55.0	3.27
141	1986	Ryan	Nolan	HOU	**2.364**	5	**0.473**	2	2	1	58.7	3.07
142	1945	Bridges	Tommy	DET	**2.331**	4	**0.583**	4	1	0	46.0	3.52
143	1999	Saberhagen	Bret	BOS	**2.329**	5	**0.466**	2	4	0	54.0	4.67
144	1982	Kison	Bruce	CAL	**2.322**	6	**0.387**	5	1	0	36.3	1.98
145	1985	Andujar	Joaquin	SLN	**2.320**	3	**0.773**	3	2	1	35.0	4.11
146	1995	Aguilera	Rick	BOS	**2.311**	4	**0.578**	2	1	5	24.0	2.62
147	1925	Adams	Babe	PIT	**2.298**	2	**1.149**	3	0	0	28.0	1.29
148	1990	Rijo	Jose	CIN	**2.288**	1	**2.288**	3	0	0	27.7	2.28
149	1969	Taylor	Ron	NYN	**2.273**	2	**1.136**	1	0	3	10.3	0.00
150	2008	Timlin	Mike	BOS	**2.251**	11	**0.205**	0	3	1	50.7	4.26

Best Pitching Postseason Careers

					Postseason							
From 1884-2010				**FinalPo**	**Career**	**Total**	**Career**					
Rank	**LPostYr**	**Name**	**First**	**Team**	**PEVA-P**	**Yrs**	**PerYr**	**W**	**L**	**SV**	**IP**	**ERA**
151	1966	Drysdale	Don	LAN	**2.250**	5	**0.450**	3	3	0	39.7	2.95
152	1956	Maglie	Sal	BRO	**2.250**	3	**0.750**	1	2	0	29.0	3.41
153	1996	Valenzuela	Fernando	SDN	**2.242**	4	**0.561**	5	1	0	63.7	1.98
154	1960	Turley	Bob	NYA	**2.231**	5	**0.446**	4	3	1	53.7	3.19
155	1992	Drabek	Doug	PIT	**2.219**	3	**0.740**	2	5	0	48.3	2.05
156	1890	Lovett	Tom	BRO	**2.212**	2	**1.106**	2	3	0	38.0	4.50
157	1940	Walters	Bucky	CIN	**2.211**	2	**1.105**	2	2	0	29.0	2.79
158	1987	Dravecky	Dave	SFN	**2.210**	2	**1.105**	1	1	0	25.7	0.35
159	2009	Lester	Jon	BOS	**2.205**	3	**0.735**	2	3	0	42.0	2.57
160	2007	Hernandez	Livan	ARI	**2.188**	4	**0.547**	7	3	0	68.0	3.97
161	1981	Gura	Larry	KCA	**2.185**	5	**0.437**	2	3	0	41.7	4.10
162	1903	Dinneen	Bill	BOS	**2.173**	1	**2.173**	3	1	0	35.0	2.06
163	1960	Law	Vern	PIT	**2.154**	1	**2.154**	2	0	0	18.3	3.44
164	1981	McGraw	Tug	PHI	**2.141**	7	**0.306**	3	3	7	52.3	2.24
165	2002	Millwood	Kevin	ATL	**2.133**	4	**0.533**	3	3	1	41.3	3.92
166	2005	Leiter	Al	NYA	**2.125**	5	**0.425**	2	3	0	81.7	4.63
167	1921	Douglas	Phil	NY1	**2.122**	2	**1.061**	2	2	0	27.0	2.00
168	1999	Nagy	Charles	CLE	**2.115**	5	**0.423**	3	4	0	84.7	4.46
169	1918	Ruth	Babe	BOS	**2.114**	2	**1.057**	3	0	0	31.0	0.87
170	1917	Faber	Red	CHA	**2.114**	1	**2.114**	3	1	0	27.0	2.33
171	1920	Marquard	Rube	BRO	**2.105**	5	**0.421**	2	5	0	58.7	3.07
172	2010	Madson	Ryan	PHI	**2.101**	3	**0.700**	2	1	1	30.7	2.35

173	1886	McCormick	Jim	CHN	**2.097**	2	**1.048**	3	3	0	44.0	2.86
174	1946	Beazley	Johnny	SLN	**2.096**	2	**1.048**	2	0	0	19.0	2.37
175	2009	Weaver	Jeff	LAN	**2.074**	5	**0.415**	4	4	0	39.3	3.89
176	1914	James	Bill	BSN	**2.071**	1	**2.071**	2	0	0	11.0	0.00
177	2010	Wilson	Brian	SFN	**2.055**	1	**2.055**	1	0	6	11.7	0.00
178	1946	Dobson	Joe	BOS	**2.054**	1	**2.054**	1	0	0	12.7	0.00
179	1944	Donnelly	Blix	SLN	**2.040**	1	**2.040**	1	0	0	6.0	0.00
180	1903	Young	Cy	BOS	**2.039**	2	**1.020**	2	3	0	61.0	2.36
181	1983	Holland	Al	PHI	**2.031**	1	**2.031**	0	0	2	6.7	0.00
182	2000	Neagle	Denny	NYA	**2.019**	6	**0.336**	1	4	0	64.3	3.64
183	1936	Malone	Pat	NYA	**2.015**	3	**0.672**	0	3	1	20.7	3.05
184	1956	Erskine	Carl	BRO	**2.000**	5	**0.400**	2	2	0	41.7	5.83
185	1949	Page	Joe	NYA	**1.979**	2	**0.990**	2	1	2	21.3	3.37
186	2008	Buehrle	Mark	CHA	**1.978**	3	**0.659**	2	1	1	30.7	4.11
187	1995	Pena	Alejandro	ATL	**1.962**	5	**0.392**	4	3	4	31.3	2.01
188	1983	Koosman	Jerry	CHA	**1.954**	3	**0.651**	4	0	0	40.3	3.79
189	2006	Williams	Woody	SDN	**1.952**	5	**0.390**	3	3	0	41.3	5.66
190	1948	Bearden	Gene	CLE	**1.943**	1	**1.943**	1	0	1	10.7	0.00
191	1954	Newhouser	Hal	CLE	**1.941**	2	**0.971**	2	1	0	20.7	6.53
192	2005	Morris	Matt	SLN	**1.932**	5	**0.386**	2	6	0	73.3	4.05
193	1943	Murphy	Johnny	NYA	**1.928**	6	**0.321**	2	0	4	16.3	1.10
194	1919	Cicotte	Eddie	CHA	**1.915**	2	**0.958**	2	3	0	44.7	2.22
195	2005	Mulder	Mark	SLN	**1.912**	3	**0.637**	3	4	0	42.3	2.34
196	2005	Backe	Brandon	HOU	**1.905**	2	**0.952**	1	0	0	36.7	2.95
197	2004	Foulke	Keith	BOS	**1.889**	3	**0.630**	1	2	3	21.3	2.53
198	2010	Pavano	Carl	MIN	**1.889**	3	**0.630**	2	2	0	32.3	2.51
199	1993	Ward	Duane	TOR	**1.888**	4	**0.472**	4	1	4	24.7	4.74
200	1944	Galehouse	Denny	SLA	**1.860**	1	**1.860**	1	1	0	18.0	1.50

Best Pitching Postseason Careers

From 1884-2010

Rank	LPostYr	Name	First	FinalPo Team	Postseason Career PEVA-P	Total Yrs	Career PerYr	W	L	SV	IP	ERA
201	1975	Blue	Vida	OAK	**1.851**	5	**0.370**	1	5	2	61.7	4.52
202	1985	Reuss	Jerry	LAN	**1.851**	5	**0.370**	2	8	0	62.7	3.59
203	1962	Sanford	Jack	SFN	**1.849**	1	**1.849**	1	2	0	23.3	1.93
204	1993	Cox	Danny	TOR	**1.843**	4	**0.461**	3	3	0	58.3	3.24
205	1990	Dibble	Rob	CIN	**1.842**	1	**1.842**	1	0	1	9.7	0.00
206	1975	Giusti	Dave	PIT	**1.837**	5	**0.367**	0	2	2	20.3	4.87
207	1943	Russo	Marius	NYA	**1.823**	2	**0.911**	2	0	0	18.0	0.50
208	1996	Honeycutt	Rick	SLN	**1.807**	7	**0.258**	3	0	1	24.7	6.93
209	1979	Jackson	Grant	PIT	**1.799**	5	**0.360**	3	0	0	17.7	2.55
210	2008	Wakefield	Tim	BOS	**1.790**	9	**0.199**	5	7	0	72.0	6.75
211	1953	Roe	Preacher	BRO	**1.779**	3	**0.593**	2	1	0	28.3	2.54
212	2001	Hitchcock	Sterling	NYA	**1.770**	3	**0.590**	4	0	0	30.7	1.76
213	1957	Byrne	Tommy	NYA	**1.768**	4	**0.442**	1	1	0	21.3	2.53
214	1944	Lanier	Max	SLN	**1.745**	3	**0.582**	2	1	0	31.7	1.71
215	1944	Kramer	Jack	SLA	**1.739**	1	**1.739**	1	0	0	11.0	0.00
216	2005	Isringhausen	Jason	SLN	**1.730**	5	**0.346**	1	1	11	26.7	2.36
217	2000	Franco	John	NYN	**1.720**	2	**0.860**	2	0	1	14.3	1.88
218	1999	Stottlemyre	Todd	ARI	**1.718**	7	**0.245**	3	5	0	53.3	5.91
219	1919	Kerr	Dickey	CHA	**1.713**	1	**1.713**	2	0	0	19.0	1.42
220	1945	Borowy	Hank	CHN	**1.711**	3	**0.570**	3	2	0	29.0	4.97
221	1890	Ehret	Red	LS2	**1.679**	1	**1.679**	2	0	1	20.0	1.35
222	1977	Torrez	Mike	NYA	**1.678**	1	**1.678**	2	1	0	29.0	3.10
223	2006	Radke	Brad	MIN	**1.677**	4	**0.419**	2	3	0	35.0	3.60
224	1987	Viola	Frank	MIN	**1.676**	1	**1.676**	3	1	0	31.3	4.31
225	1981	Rogers	Steve	MON	**1.676**	1	**1.676**	3	1	0	27.7	0.98

226	1916	Foster	Rube	BOS	**1.667**	2	**0.834**	2	0	0	21.0	1.71
227	1932	Grimes	Burleigh	CHN	**1.663**	4	**0.416**	3	4	0	56.7	4.29
228	1996	Worrell	Todd	LAN	**1.662**	3	**0.554**	1	1	4	23.3	1.93
229	1979	Tekulve	Kent	PIT	**1.657**	2	**0.829**	0	1	3	13.7	3.29
230	1978	Eastwick	Rawly	PHI	**1.655**	3	**0.552**	4	0	1	15.7	4.02
231	1972	Briles	Nelson	PIT	**1.633**	4	**0.408**	2	1	0	37.3	2.65
232	1989	Flanagan	Mike	TOR	**1.625**	3	**0.542**	3	2	0	35.3	4.33
233	2008	Jenks	Bobby	CHA	**1.615**	2	**0.807**	0	0	5	9.0	2.00
234	1981	Leonard	Dennis	KCA	**1.605**	5	**0.321**	3	5	0	50.0	4.32
235	1937	Schumacher	Hal	NY1	**1.594**	3	**0.531**	2	2	0	32.7	4.13
236	2008	Moyer	Jamie	PHI	**1.593**	4	**0.398**	3	3	0	41.3	4.14
237	2003	Schmidt	Jason	SFN	**1.591**	2	**0.795**	3	1	0	32.3	3.06
238	1958	Kucks	Johnny	NYA	**1.578**	4	**0.394**	1	0	0	19.0	1.89
239	1984	Wilcox	Milt	DET	**1.568**	2	**0.784**	3	1	0	19.0	1.42
240	1976	Billingham	Jack	CIN	**1.563**	4	**0.391**	2	1	1	42.0	1.93
241	1964	Bouton	Jim	NYA	**1.562**	2	**0.781**	2	1	0	24.3	1.48
242	1970	Drabowsky	Moe	BAL	**1.549**	2	**0.775**	1	0	0	10.0	0.90
243	1910	Reulbach	Ed	CHN	**1.544**	4	**0.386**	2	0	0	32.7	3.03
244	1998	Ogea	Chad	CLE	**1.532**	4	**0.383**	2	3	0	39.0	3.23
245	1934	Dean	Paul	SLN	**1.527**	1	**1.527**	2	0	0	18.0	1.00
246	2005	Tavarez	Julian	SLN	**1.524**	5	**0.305**	2	4	0	30.7	3.52
247	1909	Donovan	Bill	DET	**1.520**	3	**0.507**	1	4	0	50.0	2.88
248	1975	Drago	Dick	BOS	**1.513**	1	**1.513**	0	1	2	8.7	1.04
249	1927	Kremer	Ray	PIT	**1.510**	2	**0.755**	2	2	0	26.0	3.12
250	1971	Hall	Dick	BAL	**1.506**	3	**0.502**	2	1	2	8.7	0.00
251	1971	Grant	Mudcat	OAK	**1.489**	2	**0.745**	2	1	0	25.0	2.52
252	1991	Acker	Jim	TOR	**1.487**	3	**0.496**	0	0	0	13.0	0.69
253	1992	Welch	Bob	OAK	**1.487**	8	**0.186**	3	3	2	50.0	4.50

Abbreviation Code: LPostYr - Last Postseason Year; FinalPost Team - Final Postseason Team.

Note: Postseason PEVA reflects 10% of regular PEVA values for the Regular Season.

Note: Statistics above reflect postseason stats only.

Per Postseason Rankings

Best Pitching Postseason Careers

From 1884-2010

				LPost	Postseason Career	Total	Career	PRanks	
Rank	**LPostYr**	**Name**	**First**	**Team**	**PEVA-B**	**Yrs**	**PerYr**	**Career**	**C+Per**
1	1968	Gibson	Bob	SLN	**13.470**	3	**4.490**	3	4
2	1931	Earnshaw	George	PHA	**8.877**	3	**2.959**	17	19
3	1913	Mathewson	Christy	NY1	**11.260**	4	**2.815**	7	10
4	2007	Schilling	Curt	BOS	**12.460**	5	**2.492**	4	8
5	2009	Beckett	Josh	BOS	**9.779**	4	**2.445**	11	16
6	1937	Hubbell	Carl	NY1	**7.005**	3	**2.335**	25	31
7	1928	Alexander	Pete	SLN	**6.974**	3	**2.325**	26	33
8	1966	Koufax	Sandy	LAN	**9.019**	4	**2.255**	15	23
9	1999	Hershiser	Orel	NYN	**12.424**	6	**2.071**	5	14
10	1932	Pennock	Herb	NYA	**9.987**	5	**1.997**	10	20
11	1931	Grove	Lefty	PHA	**5.945**	3	**1.982**	36	47
12	2009	Hamels	Cole	PHI	**6.877**	4	**1.719**	27	39
13	1964	Ford	Whitey	NYA	**18.789**	11	**1.708**	2	15
14	1892	Clarkson	John	CL4	**4.928**	3	**1.643**	53	67
15	1993	Stewart	Dave	TOR	**9.627**	6	**1.604**	13	28
16	1953	Reynolds	Allie	NYA	**9.036**	6	**1.506**	14	30
17	1910	Overall	Orval	CHN	**5.995**	4	**1.499**	35	52
18	1889	Keefe	Tim	NY1	**4.408**	3	**1.469**	62	80
19	1992	Morris	Jack	TOR	**5.807**	4	**1.452**	39	58

20	1953	Lopat	Ed	NYA	**7.252**	5	**1.450**	22	42

*Note: Per Year Rank with at least 3 postseasons

*Note: PRanks (Postseason Ranks) for Career Total and Career plus Per Postseason

Chapter 9 - HALL OF FAME RANKS

It's five years after their playing careers have been over and the baseball writers, plus veterans committees have taken stock of their entire playing careers. Yes, it's time for the Hall of Fame in Cooperstown to determine whether you're worthy of enshrinement in their hall or not. But it's not an easy decision and there is not just one set of criteria. We know, a lot of talk is given to counting stats, and now to whether or how much the integrity of the game comes into play when you're talking about Rose or Bonds or Clemens. And then there's the determination of how the position played comes into play.

Below we have each position, according to the Hall of Fame, a Hall of Famer played and how they ranked among those peers. Kinda helps one determine whether someone really belongs on the list or not, and you can see what positions just don't seem to get their due.

And listed below each position are the current players who meet the minimum criteria of that position, at least according to PEVA ratings. Pretty interesting to see where the current players sit and how far up those rankings they'll have to go to gain inclusion once their playing careers are completed. We'll also list those that have retired within the past five years, but not yet been listed on the Hall of Fame ballot, plus a couple other interesting characters who might have gotten on the ballot recently.

First Base

Hall of Fame - Position Rank				PEVA		PEVA	Batting			Pitching		
Rank	Name	LastName	HOF Cat.	TOTAL	YRS.	PER YR.	HR	RBI	Ave.	W	SV	ERA
1	Lou	Gehrig	HOFP-1B	**479.522**	**17**	**28.207**	493	1995	0.340			
2	Cap	Anson	HOFP-1B	**432.886**	**27**	**16.033**	97	2076	0.333	0	1	4.50
3	Jimmie	Foxx	HOFP-1B	**359.936**	**20**	**17.997**	534	1922	0.325	1	0	1.52
4	Dan	Brouthers	HOFP-1B	**348.085**	**19**	**18.320**	106	1296	0.342	0	0	7.83
5	Roger	Connor	HOFP-1B	**336.978**	**18**	**18.721**	138	1322	0.317			
6	Eddie	Murray	HOFP-1B	**324.361**	**21**	**15.446**	504	1917	0.287			
7	Harmon	Killebrew	HOFP-1B	**270.652**	**22**	**12.302**	573	1584	0.256			
8	Willie	McCovey	HOFP-1B	**261.038**	**22**	**11.865**	521	1555	0.270			
9	Willie	Stargell	HOFP-1B	**241.847**	**21**	**11.517**	475	1540	0.282			
10	Tony	Perez	HOFP-1B	**215.659**	**23**	**9.376**	379	1652	0.279			
11	Hank	Greenberg	HOFP-1B	**215.163**	**13**	**16.551**	331	1276	0.313			
12	Jake	Beckley	HOFP-1B	**213.234**	**20**	**10.662**	86	1575	0.308	0	0	6.75
13	Orlando	Cepeda	HOFP-1B	**209.487**	**17**	**12.323**	379	1365	0.297			
14	George	Sisler	HOFP-1B	**197.119**	**15**	**13.141**	102	1175	0.340	5	3	2.35
15	Johnny	Mize	HOFP-1B	**191.547**	**15**	**12.770**	359	1337	0.312			
16	Bill	Terry	HOFP-1B	**157.226**	**14**	**11.230**	154	1078	0.341			
17	Jim	Bottomley	HOFP-1B	**145.525**	**16**	**9.095**	219	1422	0.310			
18	Frank	Chance	HOFP-1B	**108.846**	**17**	**6.403**	20	596	0.296			
19	George	Kelly	HOFP-1B	**93.677**	**16**	**5.855**	148	1020	0.297	1	0	0.00

First Base Possibles of Today

Hall of Fame - Position Rank				PEVA		PEVA	Batting			Pitching		
Rank	Name	LastName	HOF Cat.	TOTAL	YRS.	PER YR.	HR	RBI	Ave.	W	SV	ERA
	Frank	Thomas		**333.924**	**19**	**17.575**	521	1704	0.301			
	Albert	Pujols		**308.345**	**10**	**30.835**	408	1230	0.331			
	Jeff	Bagwell		**293.606**	**15**	**19.574**	449	1529	0.297			
	Mark	McGwire		**261.187**	**16**	**16.324**	583	1414	0.263			
	Rafael	Palmeiro		**260.423**	**20**	**13.021**	569	1835	0.288			
	Jim	Thome		**252.394**	**20**	**12.620**	589	1624	0.278			
	Fred	McGriff		**224.159**	**19**	**11.798**	493	1550	0.284			
	Jason	Giambi		**210.366**	**16**	**13.148**	415	1365	0.281			
	Carlos	Delgado		**203.096**	**17**	**11.947**	473	1512	0.280			
	Todd	Helton		**183.922**	**14**	**13.137**	333	1239	0.324			

Second Base

Hall of Fame - Position Rank				PEVA		PEVA	Batting			Pitching		
Rank	**Name**	**LastName**	**HOF Cat.**	**TOTAL**	**YRS.**	**PER YR.**	**HR**	**RBI**	**Ave.**	**W**	**SV**	**ERA**
1	Rogers	Hornsby	HOFP-2B	**361.117**	**23**	**15.701**	301	1584	0.358			
2	Eddie	Collins	HOFP-2B	**349.132**	**25**	**13.965**	47	1300	0.333			
3	Nap	Lajoie	HOFP-2B	**337.934**	**21**	**16.092**	83	1599	0.338			
4	Joe	Morgan	HOFP-2B	**287.830**	**22**	**13.083**	268	1133	0.271			
5	Charlie	Gehringer	HOFP-2B	**233.202**	**19**	**12.274**	184	1427	0.320			
6	Rod	Carew	HOFP-2B	**208.083**	**19**	**10.952**	92	1015	0.328			
7	Ryne	Sandberg	HOFP-2B	**177.295**	**16**	**11.081**	282	1061	0.285			
8	Bid	McPhee	HOFP-2B	**166.066**	**18**	**9.226**	53	1067	0.271			
9	Bobby	Doerr	HOFP-2B	**157.073**	**14**	**11.219**	223	1247	0.288			
10	Nellie	Fox	HOFP-2B	**155.463**	**19**	**8.182**	35	790	0.288			
11	Jackie	Robinson	HOFP-2B	**146.209**	**10**	**14.621**	137	734	0.311			
12	Frankie	Frisch	HOFP-2B	**141.796**	**19**	**7.463**	105	1244	0.316			
13	Red	Schoendiens	HOFP-2B	**131.155**	**19**	**6.903**	84	773	0.289			
14	Billy	Herman	HOFP-2B	**131.084**	**15**	**8.739**	47	839	0.304			
15	Joe	Gordon	HOFP-2B	**127.893**	**11**	**11.627**	253	975	0.268			
16	Tony	Lazzeri	HOFP-2B	**126.796**	**14**	**9.057**	178	1191	0.292			
17	Johnny	Evers	HOFP-2B	**102.257**	**18**	**5.681**	12	538	0.270			
18	Bill	Mazeroski	HOFP-2B	**98.836**	**17**	**5.814**	138	853	0.260			

Second Base Possibles of Today

	Name	LastName	HOF Cat.	PEVA TOTAL	YRS.	PEVA PER YR.	HR	RBI	Ave.
	Roberto	Alomar		**184.493**	**17**	**10.853**	210	1134	0.300
	Jeff	Kent		**177.365**	**17**	**10.433**	377	1518	0.290

Third Base

Hall of Fame - Position Rank				PEVA		PEVA	Batting			Pitching		
Rank	**Name**	**LastName**	**HOF Cat.**	**TOTAL**	**YRS.**	**PER YR.**	**HR**	**RBI**	**Ave.**	**W**	**SV**	**ERA**
1	Mike	Schmidt	HOFP-3B	**383.574**	**18**	**21.310**	548	1595	0.267			
2	Eddie	Mathews	HOFP-3B	**360.187**	**17**	**21.187**	512	1453	0.271			
3	George	Brett	HOFP-3B	**301.897**	**21**	**14.376**	317	1595	0.305			
4	Wade	Boggs	HOFP-3B	**293.035**	**18**	**16.280**	118	1014	0.328	0	0	3.86
5	Brooks	Robinson	HOFP-3B	**232.881**	**23**	**10.125**	268	1357	0.267			
6	Frank	Baker	HOFP-3B	**204.500**	**13**	**15.731**	96	987	0.307			
7	Jimmy	Collins	HOFP-3B	**167.223**	**14**	**11.944**	65	983	0.294			
8	Pie	Traynor	HOFP-3B	**141.423**	**17**	**8.319**	58	1273	0.320			
9	George	Kell	HOFP-3B	**134.828**	**15**	**8.989**	78	870	0.306			
10	Freddie	Lindstrom	HOFP-3B	**95.076**	**13**	**7.314**	103	779	0.311			

Third Base Possibles of Today

	Name	LastName	HOF Cat.	PEVA TOTAL	YRS.	PEVA PER YR.	HR	RBI	Ave.
	Alex	Rodriguez		**346.418**	**17**	**20.378**	613	1831	0.303
	Chipper	Jones		**261.173**	**17**	**15.363**	436	1491	0.306
	Scott	Rolen		**176.523**	**15**	**11.768**	304	1212	0.284

Shortstop

Hall of Fame - Position Rank				PEVA		PEVA	Batting			Pitching		
Rank	**Name**	**LastName**	**HOF Cat.**	**TOTAL**	**YRS.**	**PER YR.**	**HR**	**RBI**	**Ave.**	**W**	**SV**	**ERA**
1	Honus	Wagner	HOFP-SS	**451.472**	**21**	**21.499**	101	1732	0.327	0	0	0.00
2	Cal	Ripken	HOFP-SS	**320.188**	**21**	**15.247**	431	1695	0.276			
3	Robin	Yount	HOFP-SS	**291.976**	**20**	**14.599**	251	1406	0.285			
4	Ernie	Banks	HOFP-SS	**264.788**	**19**	**13.936**	512	1636	0.274			
5	George	Davis	HOFP-SS	**245.441**	**20**	**12.272**	73	1437	0.295	0	1	15.75
6	John	Ward	HOFP-SS	**212.855**	**17**	**12.521**	26	867	0.275	164	3	2.10
7	Luke	Appling	HOFP-SS	**210.671**	**20**	**10.534**	45	1116	0.310			
8	Joe	Cronin	HOFP-SS	**196.868**	**20**	**9.843**	170	1424	0.301			

Rank	Name	LastName	HOF Cat.	PEVA TOTAL	YRS.	PEVA PER YR.	HR	RBI	Ave.	W	SV	ERA
9	Arky	Vaughan	HOFP-SS	**191.281**	**14**	**13.663**	96	926	0.318			
10	Bobby	Wallace	HOFP-SS	**186.535**	**25**	**7.461**	34	1121	0.268	24	1	3.87
11	Lou	Boudreau	HOFP-SS	**168.772**	**15**	**11.251**	68	789	0.295			
12	Pee Wee	Reese	HOFP-SS	**166.690**	**16**	**10.418**	126	885	0.269			
13	Ozzie	Smith	HOFP-SS	**162.052**	**19**	**8.529**	28	793	0.262			
14	Joe	Sewell	HOFP-SS	**159.048**	**14**	**11.361**	49	1055	0.312			
15	Luis	Aparicio	HOFP-SS	**156.787**	**18**	**8.710**	83	791	0.262			
16	Hughie	Jennings	HOFP-SS	**152.306**	**17**	**8.959**	18	840	0.311			
17	Rabbit	Maranville	HOFP-SS	**148.551**	**23**	**6.459**	28	884	0.258			
18	Dave	Bancroft	HOFP-SS	**122.761**	**16**	**7.673**	32	591	0.279			
19	Joe	Tinker	HOFP-SS	**119.627**	**15**	**7.975**	31	782	0.262			
20	Phil	Rizzuto	HOFP-SS	**111.989**	**13**	**8.615**	38	563	0.273			
21	Travis	Jackson	HOFP-SS	**98.655**	**15**	**6.577**	135	929	0.291			

Shortstop Possibles of Today

Name	LastName	PEVA TOTAL	YRS.	PEVA PER YR.	HR	RBI	Ave.
Derek	Jeter	**233.157**	**16**	**14.572**	234	1135	0.314
Miguel	Tejada	**173.278**	**14**	**12.377**	300	1256	0.287
Barry	Larkin	**165.994**	**19**	**8.737**	198	960	0.295
Omar	Vizquel	**145.546**	**22**	**6.616**	80	936	0.273

Catcher

Hall of Fame - Position Rank

				PEVA		PEVA	Batting			Pitching		
Rank	**Name**	**LastName**	**HOF Cat.**	**TOTAL**	**YRS.**	**PER YR.**	**HR**	**RBI**	**Ave.**	**W**	**SV**	**ERA**
1	Johnny	Bench	HOFP-C	**256.662**	**17**	**15.098**	389	1376	0.267			
2	Yogi	Berra	HOFP-C	**251.624**	**19**	**13.243**	358	1430	0.285			
3	Gary	Carter	HOFP-C	**233.612**	**19**	**12.295**	324	1225	0.262			
4	Carlton	Fisk	HOFP-C	**214.953**	**24**	**8.956**	376	1330	0.269			
5	Bill	Dickey	HOFP-C	**202.680**	**17**	**11.922**	202	1209	0.313			
6	Mickey	Cochrane	HOFP-C	**173.444**	**13**	**13.342**	119	832	0.320			
7	Roy	Campanella	HOFP-C	**162.658**	**10**	**16.266**	242	856	0.276			
8	Gabby	Hartnett	HOFP-C	**155.361**	**20**	**7.768**	236	1179	0.297			
9	Buck	Ewing	HOFP-C	**146.265**	**18**	**8.126**	71	883	0.303	2	0	3.45
10	Ernie	Lombardi	HOFP-C	**126.894**	**17**	**7.464**	190	990	0.306			
11	Roger	Bresnahan	HOFP-C	**120.459**	**17**	**7.086**	26	530	0.279	4	0	3.93
12	Rick	Ferrell	HOFP-C	**109.810**	**18**	**6.101**	28	734	0.281			
13	Ray	Schalk	HOFP-C	**102.242**	**18**	**5.680**	11	594	0.253			

Catcher Possibles of Today

Name	LastName	PEVA TOTAL	YRS.	PEVA PER YR.	HR	RBI	Ave.
Mike	Piazza	**255.854**	**16**	**15.991**	427	1335	0.308
Ivan	Rodriguez	**191.791**	**20**	**9.590**	309	1313	0.298
Jorge	Posada	**156.632**	**16**	**9.790**	261	1021	0.275

Outfielders - Right Field

Hall of Fame - Position Rank

				PEVA		PEVA	Batting			Pitching		
Rank	**Name**	**LastName**	**HOF Cat.**	**TOTAL**	**YRS.**	**PER YR.**	**HR**	**RBI**	**Ave.**	**W**	**SV**	**ERA**
1	Babe	Ruth	HOFP-RF	**690.379**	**22**	**31.381**	714	2217	0.342	94	4	2.28
2	Hank	Aaron	HOFP-RF	**535.808**	**23**	**23.296**	755	2297	0.305			
3	Frank	Robinson	HOFP-RF	**411.403**	**21**	**19.591**	586	1812	0.294			
4	Mel	Ott	HOFP-RF	**369.518**	**22**	**16.796**	511	1860	0.304			
5	Reggie	Jackson	HOFP-RF	**301.318**	**21**	**14.348**	563	1702	0.262			
6	Al	Kaline	HOFP-RF	**297.259**	**22**	**13.512**	399	1583	0.297			
7	Sam	Crawford	HOFP-RF	**295.985**	**19**	**15.578**	97	1525	0.309			
8	Roberto	Clemente	HOFP-RF	**264.635**	**18**	**14.702**	240	1305	0.317			
9	Harry	Heilmann	HOFP-RF	**249.516**	**17**	**14.677**	183	1539	0.342			
10	Paul	Waner	HOFP-RF	**242.638**	**20**	**12.132**	113	1309	0.333			
11	Tony	Gwynn	HOFP-RF	**238.944**	**20**	**11.947**	135	1138	0.338			

12	King	Kelly	HOFP-RF	**235.100**	**16**	**14.694**	69	950	0.308	2	0	4.14
13	Andre	Dawson	HOFP-RF	**230.234**	**21**	**10.964**	438	1591	0.279			
14	Willie	Keeler	HOFP-RF	**220.111**	**19**	**11.585**	33	810	0.341			
15	Sam	Thompson	HOFP-RF	**206.207**	**15**	**13.747**	127	1299	0.331			
16	Harry	Hooper	HOFP-RF	**191.097**	**17**	**11.241**	75	817	0.281	0	0	0.00
17	Sam	Rice	HOFP-RF	**182.580**	**20**	**9.129**	34	1078	0.322	1	0	2.52
18	Enos	Slaughter	HOFP-RF	**180.421**	**19**	**9.496**	169	1304	0.300			
19	Chuck	Klein	HOFP-RF	**169.818**	**17**	**9.989**	300	1201	0.320			
20	Ross	Youngs	HOFP-RF	**110.998**	**10**	**11.100**	42	592	0.322			
21	Tommy	McCarthy	HOFP-RF	**100.436**	**13**	**7.726**	44	735	0.292	0	0	4.93

Outfielders - Centerfield

1	Ty	Cobb	HOFP-CF	**569.252**	**24**	**23.719**	117	1937	0.366	0	1	3.60
2	Willie	Mays	HOFP-CF	**520.998**	**22**	**23.682**	660	1903	0.302			
3	Tris	Speaker	HOFP-CF	**479.576**	**22**	**21.799**	117	1529	0.345	0	0	9.00
4	Mickey	Mantle	HOFP-CF	**455.611**	**18**	**25.312**	536	1509	0.298			
5	Joe	DiMaggio	HOFP-CF	**297.630**	**13**	**22.895**	361	1537	0.325			
6	Duke	Snider	HOFP-CF	**253.177**	**18**	**14.065**	407	1333	0.295			
7	Billy	Hamilton	HOFP-CF	**231.802**	**14**	**16.557**	40	736	0.344			
8	Hugh	Duffy	HOFP-CF	**215.369**	**17**	**12.669**	106	1302	0.324			
9	Zack	Wheat	HOFP-CF	**213.678**	**19**	**11.246**	132	1248	0.317			
10	Elmer	Flick	HOFP-CF	**207.666**	**13**	**15.974**	48	756	0.313			
11	Richie	Ashburn	HOFP-CF	**207.541**	**15**	**13.836**	29	586	0.308			
12	Kirby	Puckett	HOFP-CF	**204.074**	**12**	**17.006**	207	1085	0.318			
13	Max	Carey	HOFP-CF	**194.330**	**20**	**9.717**	70	800	0.285			
14	Earl	Averill	HOFP-CF	**187.958**	**13**	**14.458**	238	1164	0.318			
15	Larry	Doby	HOFP-CF	**183.611**	**13**	**14.124**	253	970	0.283			
16	Edd	Roush	HOFP-CF	**173.402**	**18**	**9.633**	68	981	0.323			
17	Hack	Wilson	HOFP-CF	**149.587**	**12**	**12.466**	244	1063	0.307			
18	Earle	Combs	HOFP-CF	**131.405**	**12**	**10.950**	58	632	0.325			
19	Lloyd	Waner	HOFP-CF	**107.291**	**18**	**5.961**	27	598	0.316			

Outfielders - Left Field

1	Ted	Williams	HOFP-LF	**493.074**	**19**	**25.951**	521	1839	0.344	0	0	4.50
2	Stan	Musial	HOFP-LF	**481.184**	**22**	**21.872**	475	1951	0.331	0	0 -	
3	Ed	Delahanty	HOFP-LF	**351.660**	**16**	**21.979**	101	1464	0.346			
4	Carl	Yastrzemski	HOFP-LF	**339.140**	**23**	**14.745**	452	1844	0.285			
5	Rickey	Henderson	HOFP-LF	**331.490**	**25**	**13.260**	297	1115	0.279			
6	Jim	O'Rourke	HOFP-LF	**329.815**	**23**	**14.340**	62	1203	0.311	0	2	4.12
7	Dave	Winfield	HOFP-LF	**327.768**	**22**	**14.899**	465	1833	0.283			
8	Jesse	Burkett	HOFP-LF	**276.693**	**16**	**17.293**	75	952	0.338	3	0	5.56
9	Billy	Williams	HOFP-LF	**275.510**	**18**	**15.306**	426	1475	0.290			
10	Fred	Clarke	HOFP-LF	**255.011**	**21**	**12.143**	67	1015	0.312			
11	Al	Simmons	HOFP-LF	**237.604**	**20**	**11.880**	307	1827	0.334			
12	Joe	Kelley	HOFP-LF	**226.052**	**17**	**13.297**	65	1194	0.317			
13	Jim	Rice	HOFP-LF	**224.084**	**16**	**14.005**	382	1451	0.298			
14	Goose	Goslin	HOFP-LF	**220.227**	**18**	**12.235**	248	1609	0.316			
15	Ralph	Kiner	HOFP-LF	**208.610**	**10**	**20.861**	369	1015	0.279			
16	Joe	Medwick	HOFP-LF	**186.522**	**17**	**10.972**	205	1383	0.324			
17	Lou	Brock	HOFP-LF	**179.216**	**19**	**9.432**	149	900	0.293			
18	Kiki	Cuyler	HOFP-LF	**161.663**	**18**	**8.981**	128	1065	0.321			
19	Heinie	Manush	HOFP-LF	**156.227**	**17**	**9.190**	110	1183	0.330			
20	Chick	Hafey	HOFP-LF	**92.982**	**13**	**7.152**	164	833	0.317			
21	Monte	Irvin	HOFP-LF	**60.284**	**8**	**7.535**	99	443	0.293			

Outfield Possibles of Today

	Barry	Bonds		**606.700**	22	**27.577**	762	1996	0.298

Manny	Ramirez		**288.532**	**18**	**16.030**	555	1830	0.313
Ken	Griffey		**286.056**	**22**	**13.003**	630	1836	0.284
Sammy	Sosa		**232.726**	**18**	**12.929**	609	1667	0.273
Bobby	Abreu		**218.502**	**15**	**14.567**	276	1265	0.296
Vladimir	Guerrero		**208.999**	**15**	**13.933**	436	1433	0.320
Bernie	Williams		**201.884**	**16**	**12.618**	287	1257	0.297
Luis	Gonzalez		**196.581**	**19**	**10.346**	354	1439	0.283
Jim	Edmonds		**188.605**	**16**	**11.788**	382	1176	0.284
Brian	Giles		**185.683**	**15**	**12.379**	287	1078	0.291

Designated Hitter

Paul	Molitor	HOFP-DH	**242.403**	**21**	**11.543**	234	1307	0.306

Designated Hitter Possibles of Today

Edgar	Martinez		**225.436**	**18**	**12.524**	309	1261	0.312
David	Ortiz		**150.947**	**14**	**10.782**	349	1170	0.281

Pitcher

Hall of Fame - Position Rank				**PEVA**		**PEVA**	**Batting**			**Pitching**		
Rank	**Name**	**LastName**	**HOF Cat.**	**TOTAL**	**YRS.**	**PER YR.**	**HR**	**RBI**	**Ave.**	**W**	**SV**	**ERA**
1	Cy	Young	HOFP-P	**505.634**	**22**	**22.983**	18	290	0.210	511	17	2.63
2	Walter	Johnson	HOFP-P	**488.778**	**21**	**23.275**	24	255	0.235	417	34	2.17
3	Pete	Alexander	HOFP-P	**387.680**	**20**	**19.384**	11	163	0.209	373	32	2.56
4	Warren	Spahn	HOFP-P	**372.497**	**21**	**17.738**	35	189	0.194	363	29	3.09
5	Lefty	Grove	HOFP-P	**342.629**	**17**	**20.155**	15	121	0.148	300	55	3.06
6	Tom	Seaver	HOFP-P	**336.144**	**20**	**16.807**	12	86	0.154	311	1	2.86
7	Steve	Carlton	HOFP-P	**328.713**	**24**	**13.696**	13	140	0.201	329	2	3.22
8	Kid	Nichols	HOFP-P	**316.988**	**15**	**21.133**	16	278	0.226	361	17	2.95
9	Robin	Roberts	HOFP-P	**309.046**	**19**	**16.266**	5	103	0.167	286	25	3.41
10	Carl	Hubbell	HOFP-P	**302.043**	**16**	**18.878**	4	101	0.191	253	33	2.98
11	Christy	Mathewson	HOFP-P	**293.500**	**17**	**17.265**	7	165	0.215	373	28	2.13
12	Gaylord	Perry	HOFP-P	**286.329**	**22**	**13.015**	6	47	0.131	314	11	3.11
13	Nolan	Ryan	HOFP-P	**270.539**	**27**	**10.020**	2	36	0.110	324	3	3.19
14	Phil	Niekro	HOFP-P	**269.835**	**24**	**11.243**	7	109	0.169	318	29	3.35
15	Don	Sutton	HOFP-P	**266.168**	**23**	**11.573**	0	64	0.144	324	5	3.26
16	Fergie	Jenkins	HOFP-P	**250.825**	**19**	**13.201**	13	85	0.165	284	7	3.34
17	Bob	Feller	HOFP-P	**246.598**	**18**	**13.700**	8	99	0.151	266	21	3.25
18	Jim	Palmer	HOFP-P	**240.182**	**19**	**12.641**	3	41	0.174	268	4	2.86
19	Jim	Bunning	HOFP-P	**232.753**	**17**	**13.691**	7	75	0.167	224	16	3.27
20	Juan	Marichal	HOFP-P	**231.781**	**16**	**14.486**	4	75	0.165	243	2	2.89
21	Don	Drysdale	HOFP-P	**231.166**	**14**	**16.512**	29	113	0.186	209	6	2.95
22	Bob	Gibson	HOFP-P	**227.176**	**17**	**13.363**	24	144	0.206	251	6	2.91
23	Early	Wynn	HOFP-P	**223.795**	**23**	**9.730**	17	173	0.214	300	15	3.54
24	Hal	Newhouser	HOFP-P	**219.148**	**17**	**12.891**	2	81	0.201	207	26	3.06
25	Dennis	Eckersley	HOFP-P	**217.107**	**24**	**9.046**	3	12	0.133	197	390	3.50
26	Ed	Walsh	HOFP-P	**214.819**	**14**	**15.344**	3	68	0.193	195	34	1.82
27	Dazzy	Vance	HOFP-P	**212.748**	**16**	**13.297**	7	75	0.150	197	11	3.24
28	Amos	Rusie	HOFP-P	**210.560**	**10**	**21.056**	8	176	0.247	245	5	3.07
29	Sandy	Koufax	HOFP-P	**206.787**	**12**	**17.232**	2	28	0.097	165	9	2.76
30	Eppa	Rixey	HOFP-P	**202.106**	**21**	**9.624**	3	111	0.191	266	14	3.15
31	Whitey	Ford	HOFP-P	**201.182**	**16**	**12.574**	3	69	0.173	236	10	2.75
32	John	Clarkson	HOFP-P	**197.394**	**12**	**16.449**	24	232	0.219	328	5	2.81
33	Burleigh	Grimes	HOFP-P	**187.970**	**19**	**9.893**	2	168	0.248	270	18	3.53
34	Ted	Lyons	HOFP-P	**180.468**	**21**	**8.594**	5	149	0.233	260	23	3.67
35	Tim	Keefe	HOFP-P	**178.455**	**14**	**12.747**	12	134	0.187	342	2	2.62
36	Red	Faber	HOFP-P	**177.411**	**20**	**8.871**	3	70	0.134	254	28	3.15
37	Stan	Coveleski	HOFP-P	**176.253**	**14**	**12.589**	1	81	0.159	215	21	2.89

38	Joe	McGinnity	HOFP-P	**176.188**	**10**	**17.619**	0	90	0.194	246	24	2.66
39	Eddie	Plank	HOFP-P	**174.524**	**17**	**10.266**	3	122	0.206	326	23	2.35
40	Catfish	Hunter	HOFP-P	**172.508**	**15**	**11.501**	6	51	0.226	224	1	3.26
41	Bob	Lemon	HOFP-P	**171.875**	**15**	**11.458**	37	147	0.232	207	22	3.23
42	Mordecai	Brown	HOFP-P	**168.239**	**14**	**12.017**	2	74	0.206	239	49	2.06
43	Hoyt	Wilhelm	HOFP-P	**167.483**	**21**	**7.975**	1	21	0.088	143	227	2.52
44	Red	Ruffing	HOFP-P	**166.258**	**22**	**7.557**	36	273	0.269	273	16	3.80
45	Dizzy	Dean	HOFP-P	**163.005**	**12**	**13.584**	8	76	0.225	150	30	3.02
46	Vic	Willis	HOFP-P	**159.269**	**13**	**12.251**	1	84	0.166	249	11	2.63
47	Pud	Galvin	HOFP-P	**154.695**	**15**	**10.313**	5	220	0.201	364	2	2.86
48	Charley	Radbourn	HOFP-P	**153.761**	**12**	**12.813**	9	259	0.235	309	2	2.67
49	Waite	Hoyt	HOFP-P	**150.713**	**21**	**7.177**	0	100	0.198	237	52	3.59
50	Rich	Gossage	HOFP-P	**138.013**	**22**	**6.273**	0	2	0.106	124	310	3.01
51	Lefty	Gomez	HOFP-P	**135.569**	**14**	**9.683**	0	58	0.147	189	9	3.34
52	Rollie	Fingers	HOFP-P	**130.458**	**17**	**7.674**	2	9	0.172	114	341	2.90
53	Rube	Waddell	HOFP-P	**129.158**	**13**	**9.935**	4	83	0.161	193	5	2.16
54	Herb	Pennock	HOFP-P	**124.406**	**22**	**5.655**	4	103	0.191	240	32	3.60
55	Addie	Joss	HOFP-P	**121.462**	**9**	**13.496**	1	51	0.144	160	5	1.89
56	Jack	Chesbro	HOFP-P	**120.772**	**11**	**10.979**	5	82	0.197	198	5	2.68
57	Jesse	Haines	HOFP-P	**116.440**	**19**	**6.128**	3	79	0.186	210	10	3.64
58	Mickey	Welch	HOFP-P	**115.650**	**13**	**8.896**	12	202	0.224	307	4	2.71
59	Chief	Bender	HOFP-P	**109.459**	**16**	**6.841**	6	116	0.212	212	34	2.46
60	Bruce	Sutter	HOFP-P	**108.819**	**12**	**9.068**	0	6	0.088	68	300	2.83
61	Rube	Marquard	HOFP-P	**104.505**	**18**	**5.806**	1	64	0.179	201	19	3.08

Pitcher Possibles of Today

	Greg	Maddux		**594.209**	**23**	**25.835**				355	0	3.16
	Roger	Clemens		**487.448**	**24**	**20.310**				354	0	3.12
	Randy	Johnson		**408.708**	**22**	**18.578**				303	2	3.29
	Pedro	Martinez		**328.493**	**18**	**18.250**				219	3	2.93
	Tom	Glavine		**325.047**	**22**	**14.775**				305	0	3.54
	John	Smoltz		**301.456**	**21**	**14.355**				213	154	3.33
	Curt	Schilling		**282.056**	**20**	**14.103**				216	22	3.46
	Mike	Mussina		**276.995**	**18**	**15.389**				270	0	3.68
	Kevin	Brown		**263.325**	**19**	**13.859**				211	0	3.28
	Roy	Halladay		**257.438**	**13**	**19.803**				169	1	3.32
	Mariano	Rivera		**215.110**	**16**	**13.444**				74	559	2.23
	Johan	Santana		**205.233**	**11**	**18.658**				133	1	3.10
	Andy	Pettitte		**198.187**	**16**	**12.387**				240	0	3.88
	Jamie	Moyer		**191.352**	**24**	**7.973**				267	0	4.24
	C.C.	Sabathia		**190.081**	**10**	**19.008**				157	0	3.57

Executive

1	Al	Spalding	HOFEP	**233.591**	**8**	**29.199**	2	327	0.313	253	11	2.14
2	George	Wright	HOFEP	**155.730**	**12**	**12.978**	11	330	0.302	0	0	1.80
3	Clark	Griffith	HOFEP	**135.858**	**21**	**6.469**	8	166	0.233	237	6	3.31
4	Charlie	Comiskey	HOFEP	**86.636**	**13**	**6.664**	29	883	0.264	0	0	0.73
5	Candy	Cummings	HOFEP	**78.342**	**6**	**13.057**	0	107	0.212	145	0	2.49

Manager

1	Miller	Huggins	HOFM	**110.533**	**13**	**8.503**	9	318	0.265
2	John	McGraw	HOFM	**106.609**	**16**	**6.663**	13	462	0.334
3	Al	Lopez	HOFM	**75.328**	**19**	**3.965**	51	652	0.261
4	Casey	Stengel	HOFM	**74.105**	**14**	**5.293**	60	535	0.284
5	Bucky	Harris	HOFM	**64.359**	**12**	**5.363**	9	506	0.274
6	Leo	Durocher	HOFM	**53.611**	**17**	**3.154**	24	567	0.247
7	Wilbert	Robinson	HOFM	**52.028**	**17**	**3.060**	18	722	0.273

Historic baseball statistics were provided courtesy of the Sean Lahman Database and Baseballdatabank.org. Other statistic sources used include Baseball-reference.com & the Chuck Roscium Catcher Database.

Chapter 10 - FANTASY BASEBALL CHEATSHEETS

From Stat Geek Baseball and baseballevaluation.com

PEVA Player Rating Boxscore		
64.000	Maximum	Maximum Player Rating
32.000	Fantastic	MVP/CY Young Candidate
20.000	Great	All-League
15.000	Very Good	All-Star Caliber
10.000	Good	Plus Starter
3.500	Average	Bench Player
0.200	Minimum	Minimum Player Rating

Below is our initial fantasy ranking cheatsheet for 2011, based on the PEVA rating system and ranks for 2010. Hopefully this can help you get a leg up on your strategy for 2011. Use the data in any way you wish in your fantasy calculations, but remember, there is a defensive component included in a batter's PEVA number and if your fantasy league does not take that into account, it may skew the results for you. Take that into account and adjust according to your specific league rules.

TOP 400 Pitchers and Position Players
Sorted by 2010 PEVA Player Grades

Rank	Year	Name	Team	Lg	Age	2010 PEVA
1	2010	Roy Halladay	PHI	NL	34	**43.340**
2	2010	Albert Pujols	SLN	NL	31	**36.842**
3	2010	Adam Wainwright	SLN	NL	29	**35.709**
4	2010	Miguel Cabrera	DET	AL	28	**34.146**
5	2010	Joey Votto	CIN	NL	27	**32.034**
6	2010	Ubaldo Jimenez	COL	NL	27	**31.355**
7	2010	Jose Bautista	TOR	AL	30	**29.447**
8	2010	Felix Hernandez	SEA	AL	25	**28.666**
9	2010	CC Sabathia	NYA	AL	30	**24.810**
10	2010	Adrian Gonzalez	SDN	NL	29	**24.739**
11	2010	Matt Holliday	SLN	NL	31	**24.006**
12	2010	Tim Hudson	ATL	NL	35	**22.997**
13	2010	Evan Longoria	TBA	AL	25	**22.846**
14	2010	Josh Hamilton	TEX	AL	30	**22.732**
15	2010	Robinson Cano	NYA	AL	28	**22.110**
16	2010	Roy Oswalt	TOT	NL	33	**21.861**
17	2010	Josh Johnson	FLO	NL	27	**21.832**
18	2010	Cliff Lee	TOT	AL	32	**21.660**
19	2010	Jered Weaver	LAA	AL	28	**21.199**
20	2010	Jayson Werth	PHI	NL	32	**20.813**
21	2010	Shin-Soo Choo	CLE	AL	28	**20.642**
22	2010	Paul Konerko	CHA	AL	35	**20.625**
23	2010	Justin Verlander	DET	AL	28	**20.326**
24	2010	Matt Cain	SFN	NL	26	**19.857**
25	2010	Trevor Cahill	OAK	AL	23	**19.816**
26	2010	David Wright	NYN	NL	28	**19.724**
27	2010	Chris Carpenter	SLN	NL	36	**19.484**
28	2010	Jon Lester	BOS	AL	27	**19.275**
29	2010	Carlos Gonzalez	COL	NL	25	**19.178**
30	2010	David Price	TBA	AL	25	**19.172**
31	2010	Clay Buchholz	BOS	AL	26	**19.094**
32	2010	Ryan Braun	MIL	NL	27	**19.028**

33	2010	Adrian Beltre	BOS	AL	32	**18.902**
34	2010	Carl Crawford	TBA	AL	29	**18.841**
35	2010	Prince Fielder	MIL	NL	27	**18.812**
36	2010	Joe Mauer	MIN	AL	28	**18.296**
37	2010	Aubrey Huff	SFN	NL	34	**17.612**
38	2010	Rickie Weeks	MIL	NL	28	**16.876**
39	2010	Ryan Zimmerman	WAS	NL	26	**16.813**
40	2010	Adam Dunn	WAS	NL	31	**16.634**
41	2010	Dan Uggla	FLO	NL	31	**16.179**
42	2010	Vernon Wells	TOR	AL	32	**16.100**
43	2010	Mark Teixeira	NYA	AL	31	**16.017**
44	2010	Corey Hart	MIL	NL	29	**15.769**
45	2010	Cole Hamels	PHI	NL	27	**15.694**
46	2010	Mat Latos	SDN	NL	23	**15.624**
47	2010	Brett Myers	HOU	NL	30	**15.454**
48	2010	Brian McCann	ATL	NL	27	**15.186**
49	2010	C.J. Wilson	TEX	AL	30	**15.168**
50	2010	Clayton Kershaw	LAN	NL	23	**15.059**

TOP 400 Pitchers and Position Players
Sorted by 2010 PEVA Player Grades

Rank	Year	Name	Team	Lg	Age	2010 PEVA
51	2010	Tim Lincecum	SFN	NL	27	**14.927**
52	2010	Gio Gonzalez	OAK	AL	25	**14.861**
53	2010	Bronson Arroyo	CIN	NL	34	**14.638**
54	2010	Troy Tulowitzki	COL	NL	26	**14.536**
55	2010	Billy Wagner	ATL	NL	39	**14.512**
56	2010	Rafael Soriano	TBA	AL	31	**14.495**
57	2010	Joakim Soria	KCA	AL	27	**14.368**
58	2010	Jonathan Sanchez	SFN	NL	28	**14.284**
59	2010	Torii Hunter	LAA	AL	35	**14.188**
60	2010	Carl Pavano	MIN	AL	35	**14.077**
61	2010	Andre Ethier	LAN	NL	29	**13.962**
62	2010	Brian Wilson	SFN	NL	29	**13.696**
63	2010	Ryan Dempster	CHN	NL	34	**13.669**
64	2010	David Ortiz	BOS	AL	35	**13.659**
65	2010	Nick Swisher	NYA	AL	30	**13.648**
66	2010	John Danks	CHA	AL	26	**13.629**
67	2010	Hunter Pence	HOU	NL	28	**13.609**
68	2010	Andrew McCutcher	PIT	NL	24	**13.539**
69	2010	Mariano Rivera	NYA	AL	41	**13.402**
70	2010	Ryan Howard	PHI	NL	31	**13.258**
71	2010	Jay Bruce	CIN	NL	24	**13.149**
72	2010	Nick Markakis	BAL	AL	27	**13.133**
73	2010	Billy Butler	KCA	AL	25	**13.109**
74	2010	Chris Young	ARI	NL	32	**13.085**
75	2010	Alex Rodriguez	NYA	AL	35	**13.058**
76	2010	Casey McGehee	MIL	NL	28	**12.973**
77	2010	Vladimir Guerrero	TEX	AL	36	**12.936**
78	2010	Scott Rolen	CIN	NL	36	**12.920**
79	2010	Delmon Young	MIN	AL	25	**12.794**
80	2010	Tommy Hanson	ATL	NL	24	**12.780**
81	2010	Mike Young	TEX	AL	34	**12.723**
82	2010	Francisco Liriano	MIN	AL	27	**12.599**
83	2010	Ervin Santana	LAA	AL	28	**12.576**
84	2010	Dan Haren	TOT	MLB	30	**12.562**
85	2010	Kelly Johnson	ARI	NL	29	**12.520**